Current Research in Holography: Imaging and Metrology

Current Research in Holography: Imaging and Metrology

Edited by **Wynne Davis**

NY RESEARCH
PRESS

New York

Published by NY Research Press,
23 West, 55th Street, Suite 816,
New York, NY 10019, USA
www.nyresearchpress.com

Current Research in Holography: Imaging and Metrology
Edited by Wynne Davis

International Standard Book Number: 978-1-63238-111-8 (Hardback)

Printed in China.

Contents

	Preface	**VII**
Part 1	**Digital Holographic Interferometry**	**1**
Chapter 1	**Three-Dimensional Displacement and Strain Measurements by Windowed Phase-Shifting Digital Holographic Interferometry** Yoshiharu Morimoto and Motoharu Fujigaki	**3**
Chapter 2	**Real-Time Colour Holographic Interferometry (from Holographic Plate to Digital Hologram)** Jean-Michel Desse	**27**
Chapter 3	**Digital Holographic Interferometric Characterization of Optical Waveguides** Hamdy Wahba and Mamdouh Shams El-Din	**53**
Chapter 4	**Multiple-Wavelength Holographic Interferometry with Tunable Laser Diodes** Atsushi Wada	**75**
Chapter 5	**Single-Shot Phase-Shifting Digital Holography Based on the Spatial Carrier Interferometry and Its Tolerance Analysis** Yasuhiro Harada, Aizuddin Wan and Hiroyasu Sone	**91**
Chapter 6	**Multi-Channel Adaptive Interferometers Based on Dynamic Hologram Multiplexing** Roman Romashko and Yuri Kulchin	**103**
Chapter 7	**Infrared Holography for Wavefront Reconstruction and Interferometric Metrology** Sergio De Nicola, Andrea Geltrude, Massimiliano Locatelli, Kais Al-Naimee, Riccardo Meucci and F.Tito Arecchi	**137**

Chapter 8	**Incoherent Holographic Interferometry** Kaige Wang	**161**
Part 2	**Digital Holographic Microscopy**	**181**
Chapter 9	**Alternative Reconstruction Method and Object Analysis in Digital Holographic Microscopy** Francisco Palacios, Oneida Font, Jorge Ricardo, Guillermo Palacios, Mikiya Muramatsu, Diogo Soga, Daniel Palacios, José Valin and Freddy Monroy	**183**
Part 3	**Imaging**	**207**
Chapter 10	**Study of Liquid Crystal on Silicon Displays for Their Application in Digital Holography** Angel Lizana, Laura Lobato, Andrés Márquez, Claudio Iemmi, Ignacio Moreno, Juan Campos and María J. Yzuel	**209**
Chapter 11	**Synthetic Image Holograms** Jakub Svoboda, Marek Škereň and Pavel Fiala	**233**
Chapter 12	**Holoimages on Diffraction Screens** José J. Lunazzi	**257**
Chapter 13	**Two and Three Dimensional Extreme Ultraviolet Holographic Imaging with a Nanometer Spatial Resolution** P. W. Wachulak and M. C. Marconi	**277**
Chapter 14	**Computer-Generated Phase-Only Holograms for Real-Time Image Display** Edward Buckley	**299**
Part 4	**Seeing**	**327**
Chapter 15	**A Contribution to Virtual Experimentation in Optics** Javier Gamo	**329**
Chapter 16	**The Visual Language of Holograms** Paula Dawson	**347**
	Permissions	
	List of Contributors	

Preface

An elucidative and advanced study regarding holography imaging and metrology has been provided in this book. It covers digital holographic microscopy and interferometry, inclusive of interferometry in the infra-red. Other subjects comprise of synthetic imaging, the application of reflective spatial light modulators for writing dynamic holograms and image display using holographic screens. Holography is talked about as a means of artistic display and the book offers applications of software for the acquisition of skills in optics and holography. It presents an all-inclusive introduction to their specific topics, with a survey of developments to date.

After months of intensive research and writing, this book is the end result of all who devoted their time and efforts in the initiation and progress of this book. It will surely be a source of reference in enhancing the required knowledge of the new developments in the area. During the course of developing this book, certain measures such as accuracy, authenticity and research focused analytical studies were given preference in order to produce a comprehensive book in the area of study.

This book would not have been possible without the efforts of the authors and the publisher. I extend my sincere thanks to them. Secondly, I express my gratitude to my family and well-wishers. And most importantly, I thank my students for constantly expressing their willingness and curiosity in enhancing their knowledge in the field, which encourages me to take up further research projects for the advancement of the area.

Editor

Part 1

Digital Holographic Interferometry

1

Three-Dimensional Displacement and Strain Measurements by Windowed Phase-Shifting Digital Holographic Interferometry

Yoshiharu Morimoto and Motoharu Fujigaki
Moire Institute Inc. & Wakayama University
Japan

1. Introduction

It is important to measure displacement and strain distributions to prevent failures of structures. However, one point measurement method such as strain gage is widely used in industrial practice. It is time–consuming and difficult for one-point method to find the point with the maximum strain on the structures. Whole-field optical methods such as moire method, digital image correlation, speckle method and holography are effective to find the point with the maximum strain (Sharpe, 2008). Especially, holography provides three dimensional displacement information and high resolution.

In conventional holography, an object beam and a reference beam interfere each other on a high-resolution photographic plate. It is necessary to develop the photographic plate (hologram) and it is time-consuming and bothersome work. In conventional holographic interferometry, holograms recorded before and after deformation are superposed on a photographic plate and the displacement is obtained from the interference fringe pattern (Valery et al., 1996; Ranson et al., 1993; Hayashi et al. 1986). In digital holography, a hologram is usually recorded on a CCD or C-MOS sensor instead of a photographic plate (Yaroslavskii et al., 1980; Pedriniet al., 1998; Schnars, et al. 2005). In phase-shifting digital holographic interferometry (PSDHI), the complex amplitude of the object is analyzed from the phase-shifted holograms obtained by shifting the phase of the reference beam. The reconstructed image can be calculated from the complex amplitudes of the hologram using the Fresnel diffraction integral with a computer (Yamaguchi et al., 1997; Zhang et al., 1998). The displacement at each point of an object is obtained from the phase-difference between the reconstructed images analyzed from the digital holograms recorded before and after deformation. The equipment is simple and the analysis is fast. It is useful for practical field measurement of displacement and strain distributions.

Holograms and reconstructed images have speckle noise and they provide large error in the calculation of displacement and strain analysis. In order to reduce the effect of speckle noise, the authors developed a novel method, i.e., the windowed phase-shifting digital holographic interferometry (Windowed PSDHI). In holography, any part of a hologram has the optical information about the whole reconstructed image. By using this feature of holograms, a hologram is divided into several parts by superposing many different windows. The phase-difference values at the same reconstructed point obtained from any other different part of the hologram should be the same. If there is speckle noise, the phase-difference with higher intensity at a reconstructed point is more reliable than that with lower intensity. Therefore, the phase-difference at each point is selected when the intensity is the largest at the same point (Morimoto et al. 2004, 2005a), or the phase-difference is calculated by averaging the phase-differences obtained from all the windowed holograms by considering the weight of the intensity (Morimoto et al. 2005b, 2005c, 2007).

In this study, in order to check the effect of the number n of the windowed holograms on accuracy, the number n is changed. The weight for averaging is also changed. When the weight is the m-th power of the absolute amplitude of the complex amplitude of the reconstructed object, the accuracy, that is, the standard deviation of the analyzed displacement error changes according to the power m. The optimal weight is studied experimentally for out-of-plane movement of a flat plate. The resolution becomes better when the number of windows becomes larger. However, the spatial resolution may be lower when the number of windows becomes larger. Therefore, the effects of the number of windows, the window size, the displacement resolution and the spatial resolution on the accuracy are also studied.

Holography has basically three-dimensional information. It is useful to analyze three-dimensional displacement and strain distributions of an object. (Zhang et al., 1998; Kolenovic et. al., 2003) In this study, the Windowed PSDHI is extended to analyze three-dimensional displacement components. The authors developed three systems for digital holographic interferometry using three-directional illuminations. The first one was for a laboratory bench system using a microscope (Morimoto et al., 2008a). The second one used collimated light beams from one laser source and 3 shutters (Fujigaki et al., 2005). It provided a stable system for static measurement (Shiotani et al., 2008). The third one used three spherical waves from three laser sources. It provided a compact system for static and dynamic measurement. In this study, after discussing the relationship between phase differences and displacement components, examples of three measurements using microscope, collimated light beams, and three spherical waves from three laser sources, are introduced.

2. Principle of phase-shifting digital holographic interferometry

2.1 Principle of phase-shifting digital holography (Yamaguchi et al., 1997; Zhang et al., 1998)

As an example of phase-shifting digital holography, a Twyman-Green type interferometer shown in Figure 1 is used. A collimated light from a laser is divided into an object wave and a reference wave by a beam splitter. The phase of the reference wave is shifted by α with a PZT stage. The value α is set as 0, $\pi/2$, π and $3\pi/2$ in this phase-shifting digital holography. The four phase-shifted digital holograms are recorded on the CCD plane in a CCD camera without any focusing lens. The intensity of the hologram with a phase-shift value α at the pixel coordinates (X, Y) on the CCD plane is expressed as $I(X, Y, \alpha)$. The amplitude $a_0(X, Y)$, the phase $\Phi_0(X, Y)$ and the complex amplitude $g(X, Y)$ of the object wave are expressed at the pixel coordinates (X, Y) on the CCD plane as follows, respectively.

$$a_o(X,Y)=\frac{1}{4}\sqrt{\left\{I(X,Y,\frac{3\pi}{2})-I(X,Y,\frac{\pi}{2})\right\}^2+\{I(X,Y,0)-I(X,Y,\pi)\}^2} \tag{1}$$

$$\tan\Phi_o(X,Y)=\frac{I(X,Y,\frac{3\pi}{2})-I(X,Y,\frac{\pi}{2})}{I(X,Y,0)-I(X,Y,\pi)} \tag{2}$$

$$g(X,Y)=a_o(X,Y)\exp\{i\Phi_o(X,Y)\} \tag{3}$$

By calculating the Fresnel diffraction integral from the complex amplitudes $g(X, Y)$ on the CCD plane, the complex amplitude $u(x, y)$ of the reconstructed image at the position (x, y) on the reconstructed object surface being at the distance R from the CCD plane is expressed as follows.

$$u(x,y)=\exp\left[\frac{ik(x^2+y^2)}{2R}\right]F\left\{\exp\left[\frac{ik(X^2+Y^2)}{2R}\right]g(X,Y)\right\} \tag{4}$$

where k and F denote the wave number of the light and the operator of Fourier transform, respectively. The optical axis is normal to the CCD plane and the origin is at the center of the CCD plane. By calculating the intensities of these complex amplitudes on the reconstructed object surface, a holographically reconstructed image is obtained.

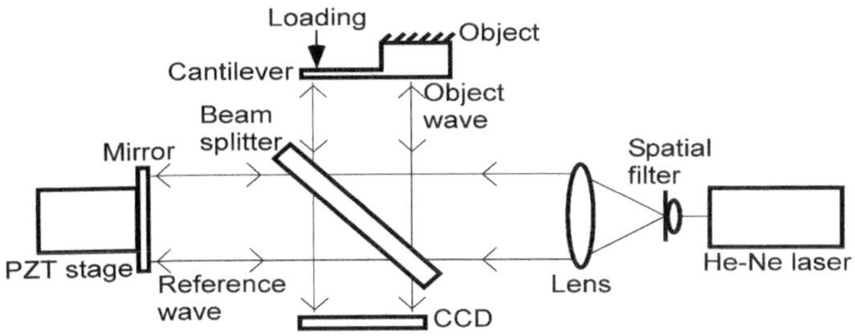

Fig. 1. Twyman-Green type interferometer for phase-shifting digital holography

2.2 Principle of holographic interferometry

Let us consider the deformation of an object (cantilever) shown in Figure 1. If the out-of-plane displacement $w(x, y)$ of the object is small, the amplitudes of the reconstructed object before and after deformation are almost the same and only the phase changes by $\Delta\phi(x, y)$ i.e. the phase-difference before and after the deformation. The relationship between the out-of-plane displacement $w(x, y)$ and the phase-difference $\Delta\phi(x, y)$ is expressed as follows.

$$w(x,y)=\frac{\lambda}{4\pi}\Delta\phi(x,y) \tag{5}$$

where λ is the wavelength of the light source.

2.3 Experiment of displacement measurement (Morimoto et al., 2007)

As an experiment, the deflection of a cantilever is measured using the optical system shown in Figure 1. The light source is a He-Ne laser. The power is 8 mW and the wavelength λ is 632.8

nm. The pixel size of the CCD sensor is 4.65 μm by 4.65 μm. A captured image is sampled by 1280(V) by 960(H) pixels and the image is digitized with 8 bits. The image of 960(V) by 960(H) pixels near the center of the recorded image is used for the analysis in this study.

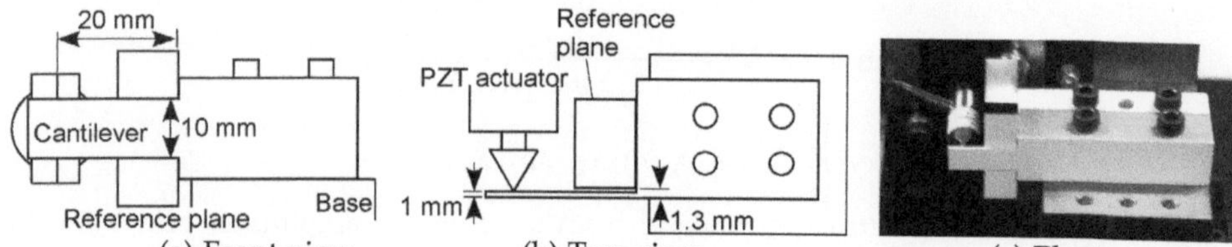

(a) Front view (b) Top view (c) Photograph

Fig. 2. Specimen (Cantilever)

As a specimen, a cantilever shown in Figure 2 is analyzed. (In this experiment, there is not a reference plane shown in the figure. The reference plane is used in Section 4). The cantilever is cut out from a thick stainless steel plate. The cantilever size is 10 mm wide, 25 mm long and 1mm thick. The loading point is 20 mm from the fixed end. The displacement at the loading point is given by a micrometer with a wedge. To improve the reflection from the specimen, lusterless white lacquer is sprayed on the surface of the cantilever. The distance R from the CCD to the cantilever is 280 mm. The phase of the reference wave is shifted by every $\pi/2$ using a mirror controlled with a PZT stage. Then four phase-shifted digital holograms for one cycle of the phase-shifting are recorded in the memory of a personal computer. The complex amplitude of the brightness at each pixel on the hologram is calculated using the phase-shifting method expressed in Equations (1) to (3). The reconstructed complex amplitude of the object wave at a point of the reconstructed object surface is calculated from the complex amplitudes of the holograms using the Fresnel diffraction integral expressed in Eq. (4).

After the cantilever is deformed, the reconstructed complex amplitude of the object waves at the same point of the reconstructed object is obtained similarly. The reconstructed images and the phase distributions are obtained from the holograms with 960 x 960 pixels before and after deformation. The phase-difference distribution before and after deformation, that is, the out-of-plane displacement distribution is shown in Figure 3(a).

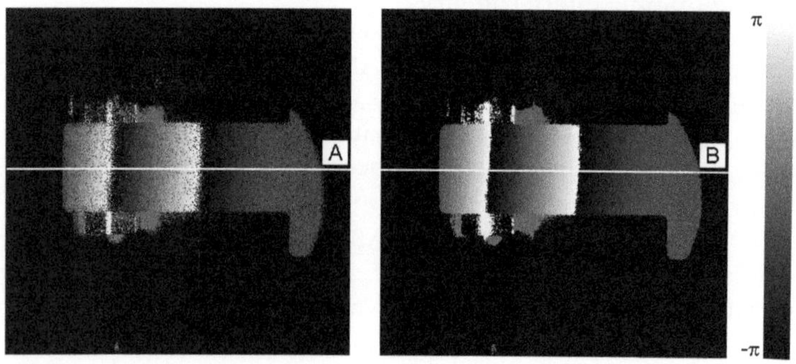

(a) From whole hologram (n=1) (b) From divided holograms (n=16)

Fig. 3. Phase-difference distributions obtained by digital holographic interferometry

3. Reduction method of speckle noise error by averaging phase-differences obtained by different windowed holograms (Morimoto et al., 2007)

3.1 Windowed hologram

As mentioned above, the phase-difference is obtained from the complex amplitudes of the reconstructed holographic object before and after deformation. However, because coherent light is reflected from a rough surface, it provides random speckle patterns in the reconstructed image. And also, any measurement system has measurement error basically. Therefore, the obtained displacement distribution has also noise and the results are not so accurate. At the point where the intensity of the speckle is weak, the accuracy of the phase value of the light at the point becomes low. In holography, any part of a hologram has the optical information of the whole reconstructed image. By using this feature of holograms, the hologram is divided into many parts. The phase-difference at the same reconstructed point obtained from any part of the hologram should be the same if there is no error. The phase-difference obtained from the complex amplitude with high intensity is more reliable than the phase-difference obtained from the complex amplitude with low intensity. If there is speckle noise, among the phase-differences obtained from each divided hologram, the phase-difference with higher intensity at a reconstructed point is more reliable. Therefore, in our previous papers, the phase-difference was obtained by selecting the phase-difference with the maximum intensity at the same point (Morimoto et al. 2004, 2005a), or the phase-difference at the same point was obtained by averaging the phase-differences obtained from all the divided holograms by considering the weight of the intensity (Morimoto et al. 2005b, 2005c, 2007). It provided the displacement distribution with high-resolution.

In this section, the divided holograms are considered as windowed holograms using some window functions. A window function with value 1 in a small part of the whole hologram area and value 0 in the remaining area is superposed on an original hologram. By multiplying the window function with values 1 and 0 by the complex amplitude of the original hologram, the windowed hologram is obtained. By changing the position of the area with value 1 in the window function, many windowed holograms are formed. The reconstructed object image is calculated from each windowed hologram using Eq. (4). A point of the reconstructed object image has a speckle pattern. The speckle patterns obtained by different windowed holograms are all different from each other. However, the speckle pattern does not move as a result of small deformation but the phase is changed by the deformation. The intensity distributions of the reconstructed object and the phase-difference distribution before and after deformation are obtained from the windowed holograms with the same window function before and after deformation. After calculating the average intensity before and after deformation from n sets of the windowed holograms obtained from the different window functions, the average value of the n phase-difference values weighted proportional to the average intensity before and after deformation is calculated at each reconstructed point. The resultant average phase-difference value is highly reliable.

In this study, especially, the effect of the size or the number n of the windows is examined. At first, for an example, let us consider the case of n=16. The hologram is divided into 16 square areas. That is, a window with value 1 in a square of 240 x 240 pixels and value 0 in the other area in the 960 x 960 pixels is superposed on the original hologram with 960 x 960 pixels. By moving the area with value 1 in the window, 16 windowed holograms are obtained and they are numbered as shown in Figure 4(a). The reconstructed object image is calculated from each windowed hologram using Eq. (4). The 16 intensity distributions of the reconstructed object images before deformation are shown in Figure 4(b). At each point on the reconstructed object, 16 complex amplitudes are obtained. Sixteen phase-difference values before and after deformation at each point are obtained. The average phase-difference is calculated by considering the weight of the average intensity before and after deformation. The result is shown in Figure 3(b) though the details of the displacement are described later. It provides more accurate phase-difference distribution than that shown in Figure 3(a).

In order to check the effect of the number of the windows on accuracy, the hologram is divided into n square areas by changing n (n=1, 4, 16, 64, 256 and 1024). That is, the original hologram with 960 x 960 pixels is windowed with n window functions whose small square area has value 1 and the remaining area has value 0 in the each window as in Figure 4(a). The average values of the n phase-difference values obtained from the windowed holograms are calculated. The results are shown later.

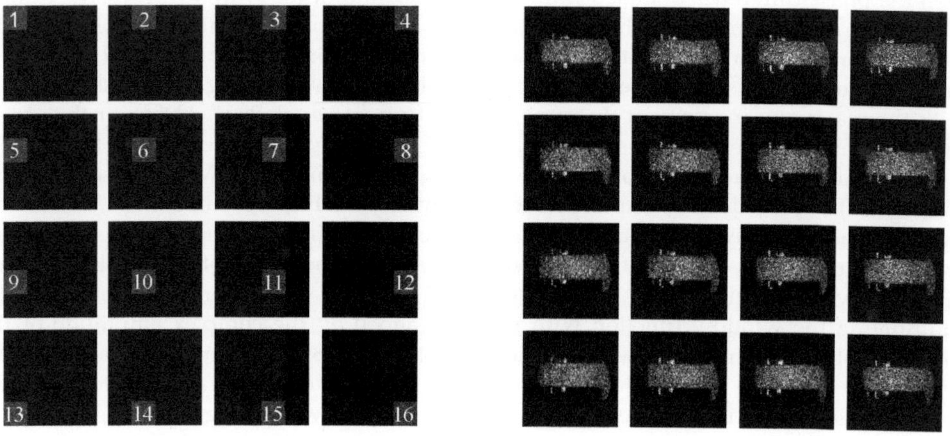

(a) Divided holograms (b) Reconstructed object images

Fig. 4. Divided holograms and reconstructed object images

3.2 Effect of number of windowed holograms on accuracy

The effect on accuracy of changing the number of windowed holograms is examined. By changing the number n into 1, 4, 16, 64, 256 and 1024, the displacement distributions are obtained. The displacement distributions along the centerline of the beam, shown as lines A and B in Figure 3 are shown in Figures 5(a) to 5(f). The theoretical displacement distribution for a cantilever is obtained by fitting a cubic curve with the minimum error by the least square method to each obtained distribution from the fixed point to the loading point. The standard deviations of the errors from the theoretical cubic curves are shown in Table 1 and Figure 6. The standard deviation decreases according to the number n, and the value becomes 670 pm when n=1024. It shows the proposed method provides very high-resolution measurement.

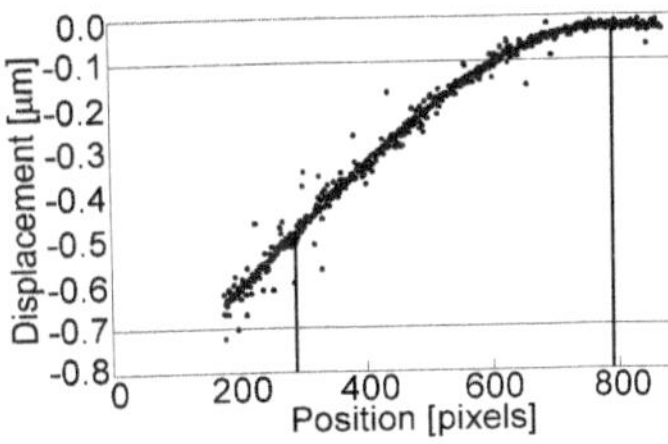

(a) n=1

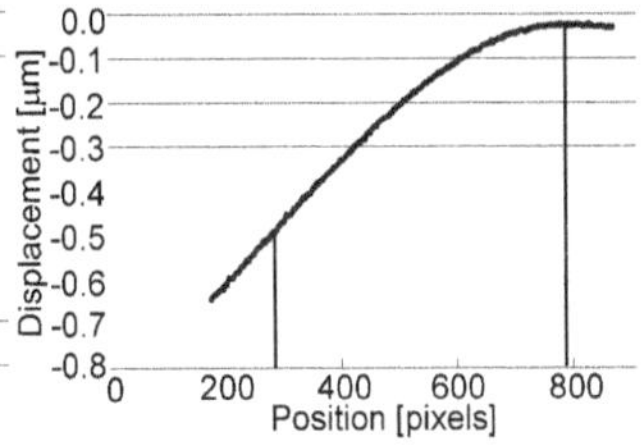

(c) n=16

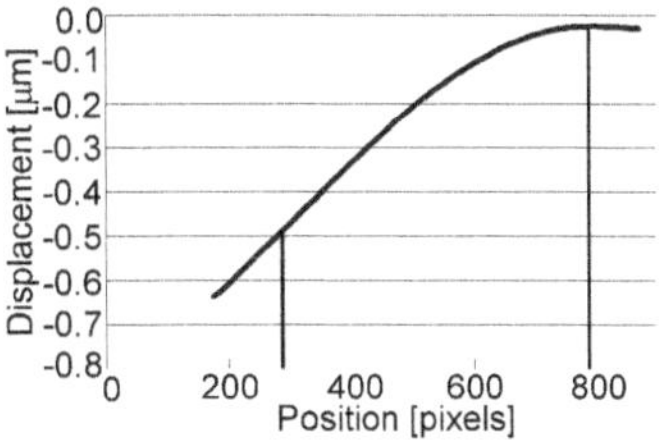

(f) n=1024

Fig. 5. Displacement distributions along lines A and B shown in Figure3

3.3 Effect of weight of averaging on accuracy

The effect on accuracy of changing the number of windowed holograms and the weight of the averaging is examined. When the weight is proportional to the m-th power of the amplitude of the complex amplitude, the standard deviations of errors are compared by changing the number m. In Section 3.2, the case of m=2 as the weight was examined by considering that reliability is larger according to the intensity, i.e., the 2nd power of the absolute amplitude of the complex amplitude. In this section, though the theoretical curve of the cantilever mentioned in the previous chapter is a cubic function, the experimental curve is not cubic in a precise sense because of the anisotropy of the materials, the actual boundary conditions, etc. In this section, parallel movement of a flat plate is adopted because the anisotropy of the materials, the actual boundary conditions, etc. are almost cut off. The parallel movement is a half wavelength, that is, about 316 nano- meters along the direction of the normal to the flat surface. The displacement distributions are shown in Figure 7 when n= 1 and 1024 and m=2.

By changing 1, 4, 16, 64, 256 and 1024 as the number n, and 1/4, 1/2, 1, 2, 4 and 8 as m for the m-th power of the amplitude of the complex amplitude for the weight of the averaging of the phase-difference, i.e., the displacement distributions, the errors are examined. The displacement distributions along the centerline of the flat plate are analyzed by phase-shifting digital holographic interferometry. The theoretical displacement distribution for the centerline is obtained by fitting a linear expression with the minimum error by the least square method to each obtained distribution. The standard deviations of the errors from the theoretical linear expressions are shown in Figure 8. The standard deviation decreases according to the number n, as same as in the case of Section 3.2. By changing n and m, the standard deviation is examined. In the same n, the standard deviation is the minimum when m is 2, that is, when the weight is proportional to the intensity of the complex amplitude. It is appropriate to adopt m=2 as the weight of averaging. The minimum standard deviation of errors is 88 pico-meters when n=1024 and m=2.

Number of windows n	1	4	16	64	256	1024
Standard deviation [nm]	16.39	4.02	1.95	1.09	0.78	0.67

Table 1. Relationship between number of windows and standard deviation of errors

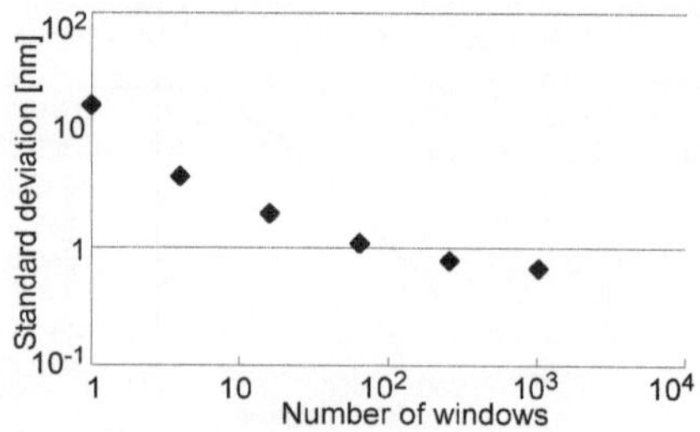

Fig. 6. Relationship between number of windows and standard deviation of errors

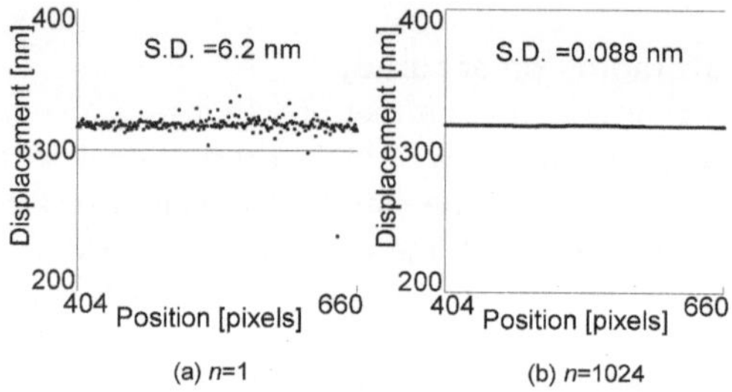

Fig. 7. Displacement distributions when flat plate is moved out of plane

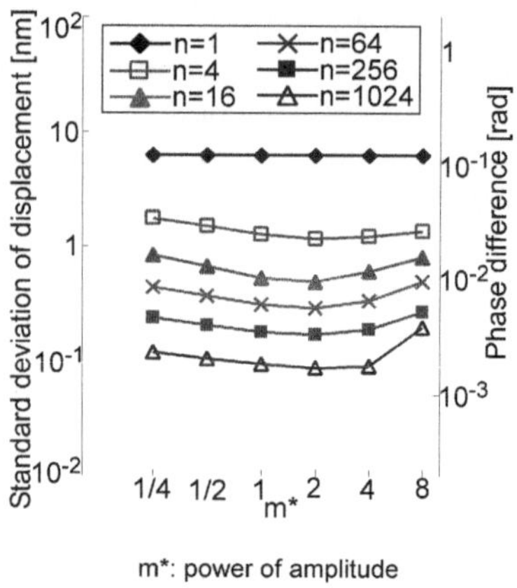

Fig. 8. Standard deviations of errors when number of windows and power of the amplitude as weight of averaging are changed

4. Effect of window size on accuracy and spatial resolution in windowed phase-shifting digital holographic interferometry (Morimoto, 2008b)

As shown in Figure 6, the accuracy becomes better according to the number of windows. However, the spatial resolution may be lower when the number of windows becomes larger. In this section, therefore, the effects of the number of windows, the window size, the displacement resolution and the spatial resolution on the accuracy are analyzed.

4.1 Displacement measurement of cantilever by phase-shifting digital holographic interferometry

As an experiment, the deflection of a cantilever is also measured using the optical system shown in Figure 1. As a specimen, a same size cantilever shown in Figure 2 is also analyzed. In this experiment, a fixed reference plate is set at 1.3 mm behind the cantilever to check spatial resolution, as shown in Figure 1.

The reconstructed images (the intensity distribution, i.e., the amplitude squared) and the phase distributions are obtained from the holograms with 960 x 960 pixels before and after deformation. The reconstructed image before deformation is shown in Figure 9(a). The phase-difference distribution before and after deformation, that is, the out-of-plane displacement distribution is shown in Figure 10(a).

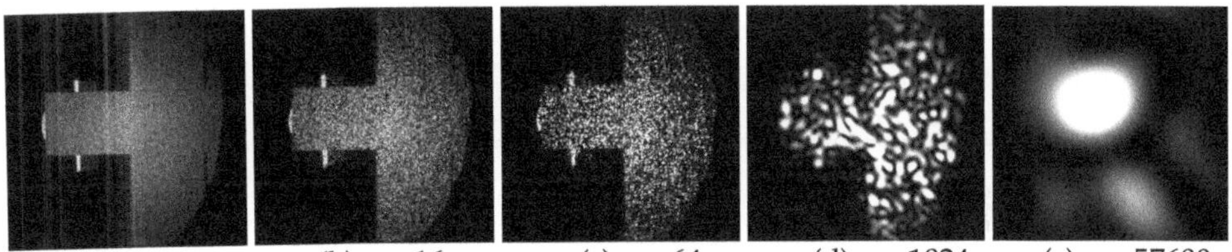

(a) $n = 1$ (b) $n = 16$ (c) $n = 64$ (d) $n = 1024$ (e) $n = 57600$

Fig. 9. Reconstructed image obtained by Windowed PSDH

(a) $n = 1$ (b) $n = 16$ (c) $n = 64$ (d) $n = 1024$ (e) $n = 57600$

Fig. 10. Phase difference distribution obtained by Windowed PSDH

4.2 Effect of number of windowed holograms on accuracy and computation processing time

The effect on accuracy by changing the number n of windowed holograms into 1, 4, 16, 64, 256, 1024, 4096, 14400, 25600 and 57600 is examined. Some reconstructed images obtained from each one of the windowed holograms are shown in Figure 9. When the number of windows becomes larger, that is, the window size becomes smaller, the speckle size becomes larger. It is considered that the spatial resolution would be worse when the speckle size becomes larger. Figure 10 shows some of the results. The displacement distributions along the centerline of the beam are shown in Figure 11. The theoretical displacement distribution for a cantilever is obtained by fitting a cubic curve with the minimum error by means of the least square method to each obtained distribution from the fixed point to the loading point. The standard deviations of the errors from the theoretical cubic curves are shown in Table 2 and Figure 11. The standard deviation changes according to the number n. Though the standard deviation when n =1 is 22 nm, it is 2.4 nm when n =16. The value becomes the minimum value of 680 pm when n =1024. It shows the proposed method provides very high-resolution measurement by selecting appropriate number n. The calculation time is, however, almost proportional to the number n as shown in Table 2 and Figure 13. In practice, 16, 64 or 256 windowed holograms are recommended by considering the balance of the computation time and accuracy.

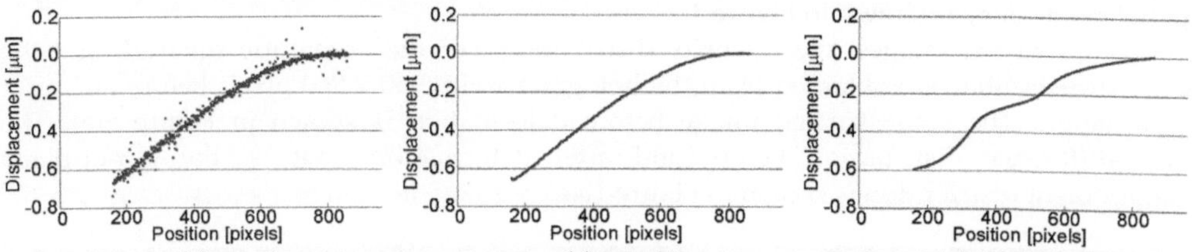

(a) $n = 1$ (c) $n = 64$ (e) $n = 57600$

Fig. 11. Displacement distributions along centerline of cantilever of Figure 4

Number of window n	Window size [pixels]	Standard deviation [nm]	Standard deviation [rad]	Processing time[s]	Spatial resolution [pixel]
1	960	21.93	0.435	3	0.36
4	480	4.87	0.097	9	0.44
16	240	2.35	0.047	29	1.5
64	120	1.61	0.032	110	2.8
256	60	0.89	0.018	430	6.66
1024	30	0.68	0.013	1704	16.24
4096	15	0.94	0.019	6791	27.68
14400	8	2.29	0.045	23751	-
25600	6	2.8	0.056	42237	-
57600	4	15.7	0.312	95430	-
Smoothing 11x11	--	1.69	0.034	4	6.77

Table 2. Relationship among number of windows, standard deviation of error and processing time

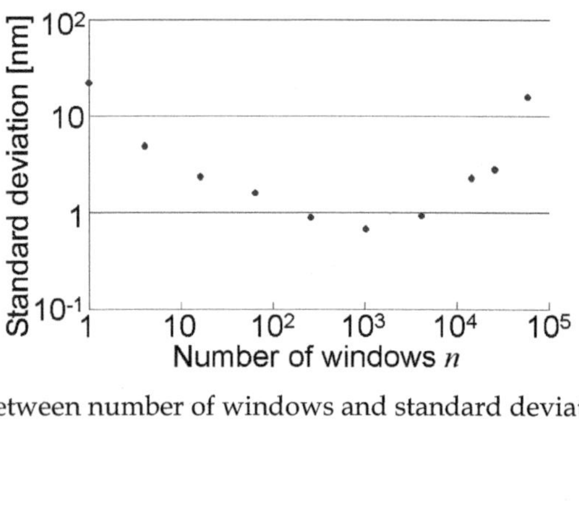

Fig. 12. Relationship between number of windows and standard deviation

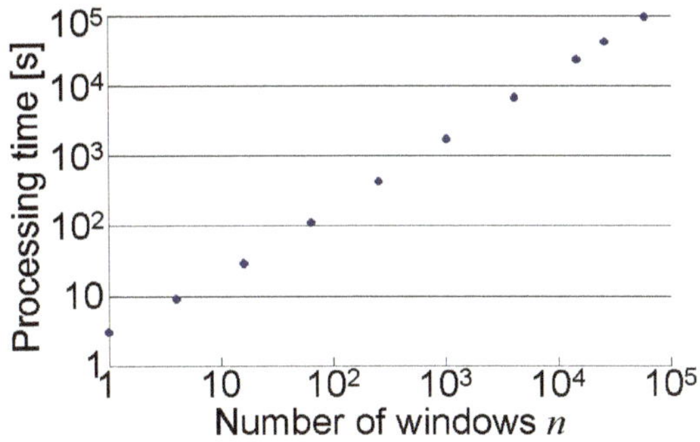

Fig. 13. Relationship between number of windows and processing time

4.3 Effect of window size on standard deviation of errors and spatial resolution

As shown in Figures 9(e) and 10(e), large speckles cannot provide accurate results. In this section, the relationship between window size and spatial resolution is examined. The phase-difference distributions along the vertical white line A (x = 632 pixels, y = 260–380 pixels) in Figure 10(a) are shown in Figure 14. The theoretical phase difference along the line A on the cantilever is $\pi/2$. The theoretical phase difference along the line A on the fixed back plate is 0. Although the theoretical phase-difference distribution has a discontinuity at the edge of the cantilever (y = 318), the analysed phase-difference distributions are continuous slopes. Figure 14 also shows the result obtained by a conventional averaging method using values at the peripheral 11 × 11 pixels. However, the spatial resolution obtained from a small window may be bad. That is, bad spatial resolution changes a step function along line A into a continuous function. In order to define the spatial resolution quantitatively in this study, it is defined as the distance (in pixels) between the positions at the 25% and 75% values of the step of the phase difference as shown in Figure 15.

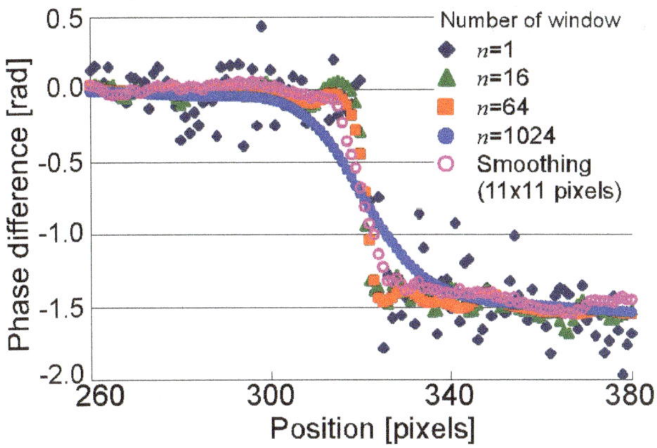

Fig. 14. Phase difference distribution along line A in Figure 10

The results of the relationships between the number of windows, the standard deviation of error and the spatial resolution are shown in Table 2 and Figure 16. When the window size becomes smaller, the standard deviation of errors becomes better and the spatial resolution becomes worse. The computation time becomes worse. By comparing the spatial resolution (6.77 pixels) of the conventional averaging method using 11 × 11 pixels and the spatial resolution (2.80 pixels) when the number of windows is 64, which have almost the same standard deviation of error, the windowed PSDHI is better than the conventional averaging method.

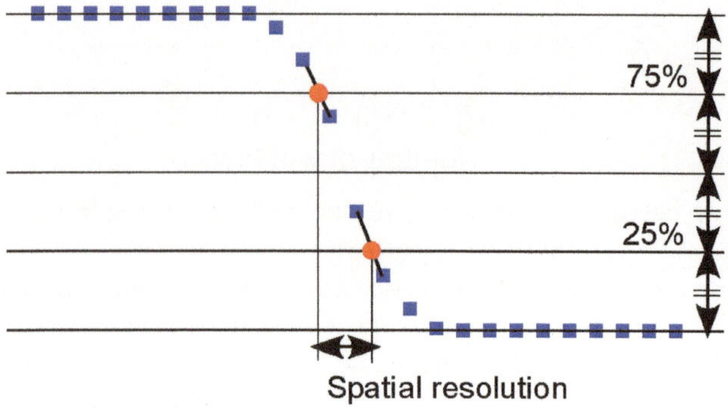

Fig. 15. Definition of special resolution

Although the standard deviation of error when n = 1024 is the best in Table 2, the data when n = 1024 in Figure 14 show a bad slope function far from the step function. Although the result from the hologram with a small window has good accuracy in an area with almost constant values of intensity near the centerline, it is not very accurate in the area with non-constant values near the edge because of large speckle size.

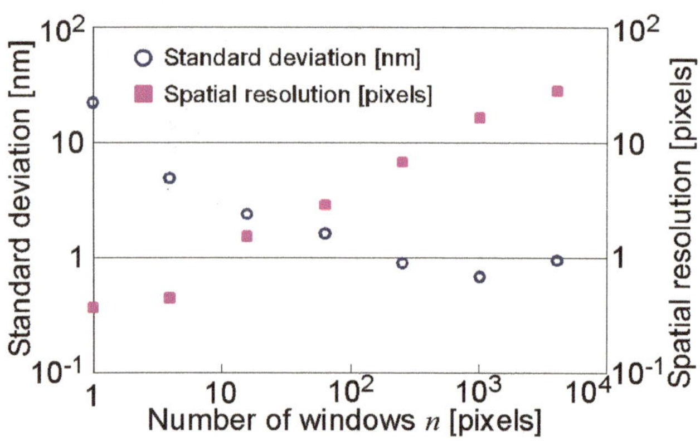

Fig. 16. Standard deviation and spatial resolution against number of windows in Table 2

4.4 Effect of larger window size on standard deviation of error and spatial resolution

In the previous section, the effect of window size on the standard deviation of error and the spatial resolution were examined. However, the numbers of windows are different. In this section, the effect of window size on the standard deviation of error and the spatial resolution are examined when the numbers of the windows are the same (n = 256). The standard deviation and the spatial resolution when the window sizes are 480, 240, 120, 60 and 30 are shown in Table 3 and Figure 17. The spatial resolution becomes worse when the window size becomes smaller. The standard deviation becomes also better when the window size becomes smaller. However, the standard deviation is the best when the window size is 60. The standard deviation when the window size is 30 is worse than that when the window size is 60. Although the case w = 60 and n = 256 uses all the data of the hologram, the case w = 30 and n = 256 does not use all the data of the hologram. Large speckles cannot provide accurate results.

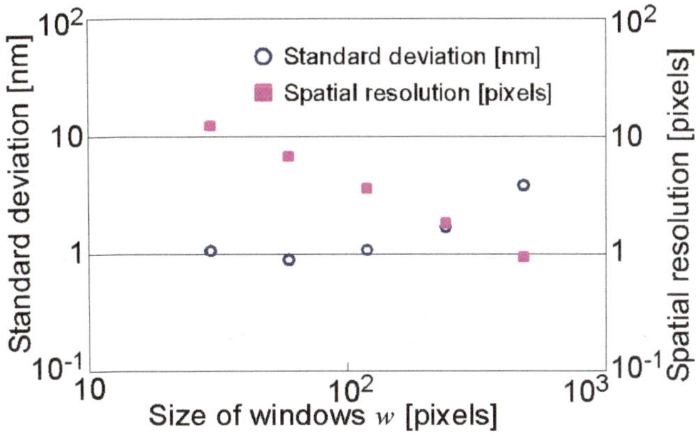

Fig. 17. Relationship between standard deviation and spatial resolution in Table 3

Window size w	Number of windows n	Standard deviation [nm]	Spatial resolution [pixels]	Processing time [s]
480	256	3.81	0.93	493
240	256	1.68	1.81	457
120	256	1.07	3.55	439
60	256	0.89	6.66	430
30	256	1.06	12.16	428

Table 3. Relationship among Window size, standard deviation of error and spatial resolution

In this windowed PSDHI study, the effect of the number of windows or window size on accuracy, spatial resolution and computation time were examined. When the number of windows increases, accuracy becomes better at first because of speckle noise reduction and then it becomes worse because of larger speckle size. The best accuracy is of the subnanometer scale when the number of windows is 1024 in our experiment. However, the accuracy is better than that when using the conventional averaging method with 11 × 11 pixels. The spatial resolution becomes worse when the number of windows becomes larger, and correspondingly the window size becomes smaller. If the number of windows is constant, the spatial resolution is better when the window size becomes larger and the accuracy is better when all the data are used and the window size is smaller. The computation time increases according to the number of windows. In practice, it is useful when the number of windows is 16, 64 or 256, as in our experiment by considering the balance of the computation time and accuracy.

5. Three-dimensional displacement analysis by digital holographic interferometry

In order to measure three-dimensional displacement components, a three-directional-illumination method or three-directional-observation method is usually employed (Zhang et al., 1998; Kolenovic et al. 2003). However, in order to analyze strain distributions, the displacement is spatially differentiated which requires accurate measurement of displacement distributions. Windowed PSDHI is useful for measuring quantitatively nano-meter displacement of objects (Morimoto et al., 2007; 2008b). The authors previously proposed a method of simultaneous measurement of in-plane and out-of-plane displacements using two beam illuminations. (Okazawa et al. 2005) A three-directional illumination method was also proposed (Fujigaki et al. 2005; Morimoto et al. 2008a, 2008c). To miniaturize the equipment for practical use, laser beams with spherical wavefronts were used. However, if spherical waves are used, the incident angles are different for each point and then each point has different sensitivity vectors. The incident angle for each point on an object is determined by the three-dimensional position of the point and the point source of the laser beam. It is, however, difficult to measure the incident angle accurately. The authors proposed a calibration method with a reference flat plane. The reference plane was installed on an XYZ three-axis piezo stage which was movable in the XYZ directions by a very small amount (Morimoto et al., 2008a).

By calculating the each phase-difference between before and after deformation using digital holography, the parameter for the relationship between the displacement and the phase-difference can be obtained. Tabulation of parameters for each point helps to measure the displacement in high speed from the phase-difference of a specimen. Displacement measurement using spherical waves can be realized with this calibration method. The theoretical treatment and experimental results of some three-dimensional displacement measurements using this method are shown in this study.

5.1 Relationship between phase-differences and displacement components

The schematic positions of an object and an observation direction when the object is illuminated at a particular incident angle are shown in Figure 18. The phase-difference for a unit displacement depends on the position of the object (Morimoto et. al., 2008; Fujigaki et al., 2006). When the positional relationship is as expressed in Figure 18, the equation at a point P before deformation on an object is expressed as shown in Equation (6).

$$\Delta\phi = e \cdot d \tag{6}$$

where, e is the sensitivity vector which depends on the half of the angle between the incident angle θ_i and the observation angle θ_0 of the point P, and d is the displacement vector for the point P before deformation which moves to a point P′ after deformation. As the displacement is very small compared with the distance between the light source and the object, it is assumed that the incident angle θ_i and the observation angle θ_0 do not change. And $\Delta\phi$ is the phase-difference resulted from the displacement at the point P.

The displacement vector d and the sensitivity vector e each have components in the x, y, and z directions. Then Eq. 6 is written as

$$\Delta\phi = \begin{bmatrix} e_x & e_y & e_z \end{bmatrix} \begin{bmatrix} d_x \\ d_y \\ d_z \end{bmatrix} = e_x d_x + e_y d_y + e_z d_z \tag{7}$$

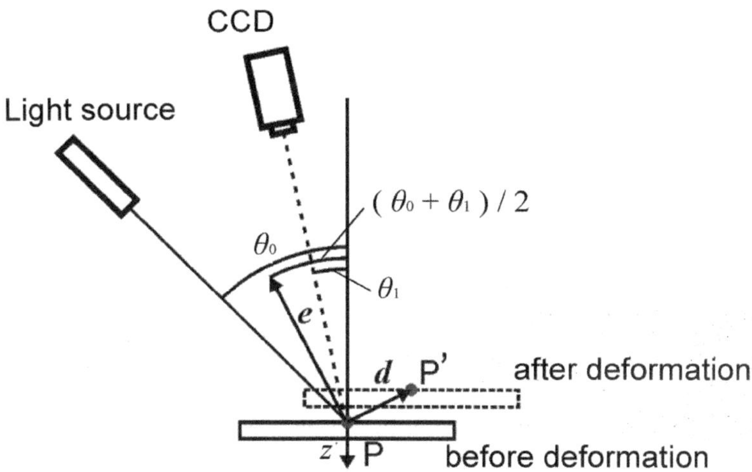

Fig. 18. Relationship between displacement of object, observation direction and incident angle at Point P

When an object is illuminated from three different directions, the number of parameters of the sensitivity vector components increases, and Equation (7) can be extended as Equation (8);

$$\begin{bmatrix} \Delta\phi_1 \\ \Delta\phi_2 \\ \Delta\phi_3 \end{bmatrix} = \begin{bmatrix} e_{1x} & e_{1y} & e_{1z} \\ e_{2x} & e_{2y} & e_{2z} \\ e_{3x} & e_{3y} & e_{3z} \end{bmatrix} \begin{bmatrix} d_x \\ d_y \\ d_z \end{bmatrix} \tag{8}$$

$$S = \begin{bmatrix} e_{1x} & e_{1y} & e_{1z} \\ e_{2x} & e_{2y} & e_{2z} \\ e_{3x} & e_{3y} & e_{3z} \end{bmatrix} \tag{9}$$

where the suffixes 1, 2 and 3 show the corresponding illumination directions.
Each component of the sensitivity vector matrix S is obtained by the geometric parameters of the optical system. When each component of the matrix S of Equation (7) is specified, the displacement components d_x d_y and d_z can be obtained from the phase-difference $\Delta\phi_1$, $\Delta\phi_2$ and $\Delta\phi_3$ for each incident light, respectively, using the inverse matrix S^{-1} of the sensitivity vector matrix S as follows;

$$\begin{bmatrix} d_x \\ d_y \\ d_z \end{bmatrix} = \begin{bmatrix} f_{x1} & f_{x2} & f_{x3} \\ f_{y1} & f_{y2} & f_{y3} \\ f_{z1} & f_{z2} & f_{z3} \end{bmatrix} \begin{bmatrix} \Delta\phi_1 \\ \Delta\phi_2 \\ \Delta\phi_3 \end{bmatrix} \tag{10}$$

$$S^{-1} = \begin{bmatrix} f_{x1} & f_{x2} & f_{x3} \\ f_{y1} & f_{y2} & f_{y3} \\ f_{z1} & f_{z2} & f_{z3} \end{bmatrix} \tag{11}$$

where f_{ij} is the (i, j) component of S^{-1}.

5.2 Displacement and strain measurement of L-shaped cantilever

5.2.1 Laboratory system for microscope

Figure 19 shows a schematic view of an optical system for digital holographic interferometry with a microscope. A parallel collimated laser beam illuminates an object. The light source is a He-Ne laser (Output: 8mW, Wavelength: 632.8mm). The laser beam from the light source is separated into three object beams ch.1, ch.2 and ch.3 and one reference beam by using three beam splitters, one half mirror and several mirrors. The reflected object beam arrives in a CCD plane through the microscope. A parallel reference beam is incident from a beam splitter between the object and the microscope. The reference beam also arrives in the same CCD plane. Each object beam is interfered with the reference beam by cutting off the other object beams. The phase of the reference beam is changed by using the PZT stage.

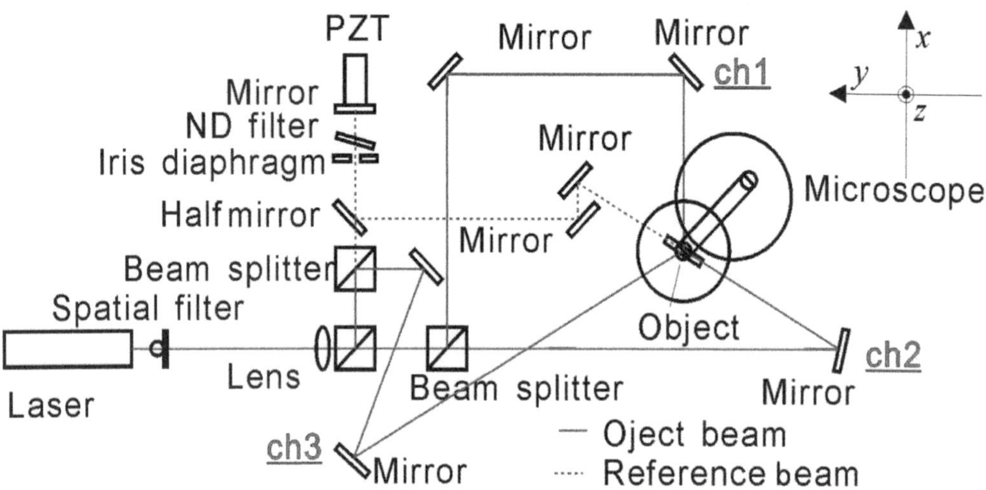

Fig. 19. Optical setup for experiment

The pixel size of the used CCD (SONY XCD-900) is 4.65 μm×4.65 μm. The analyzed image size is 960×960 pixels. A stereoscopic microscope (NIKON) using a CCD camera adapter is used. In the analysis, windowed PSDHI with 64 windows is used in order to decrease the effect of speckle noise.

Displacement measurement of an L-shaped cantilever is performed by using the experimental setup shown in Figure 19.

An L-shaped cantilever with a fixed end, 2 mm width, 8 mm length, and 1 mm thickness shown in Figure 20(a) is measured. Figure 20(b) shows the reconstructed image. Figures 20(c), (d) and (e) shows the phase-difference distribution respectively obtained by a laser beams ch1, ch2 and ch3 in Figure 19. Figures 20 (f) and (g) show the resultant three-dimensional displacement component distributions along the lines x_0x_1 and x_1x_2 shown in Figs. 20 (c), (d) and (e).

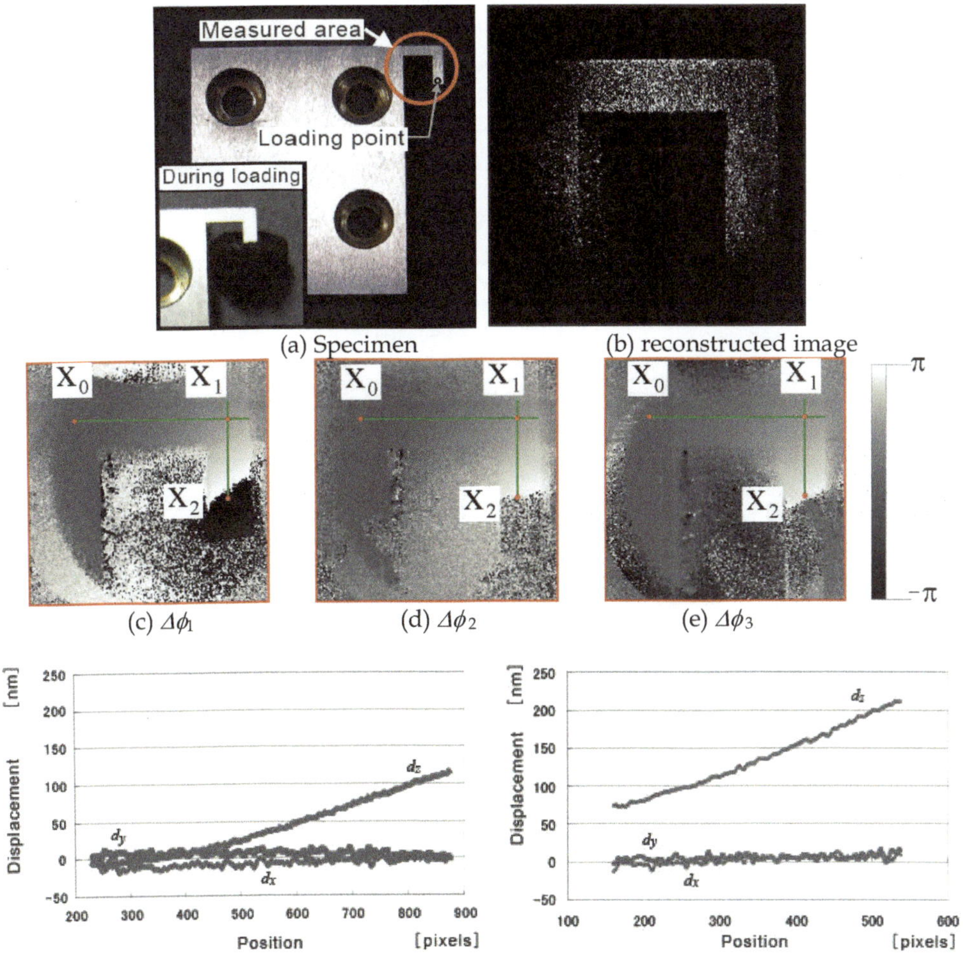

(a) Specimen (b) reconstructed image

(c) $\Delta\phi_1$ (d) $\Delta\phi_2$ (e) $\Delta\phi_3$

(f) Displacement components along x_0x_1 (g) Displacement components along x_1x_2

Fig. 20. Displacement measurement of L-shaped cantilever

5.2.2 Collimated wave system

This system has two sets of two-beam illuminations. One set, that is, an optical setup for displacement measurement using two-beam illuminations is shown in Figure 21. A collimated light beam from a laser is divided into an object wave and a reference wave by beam splitter 1. Additionally, the object wave is divided into object wave 1 and object wave 2 by beam splitter 2. Each object wave has a shutter on the light path. From the two holograms, the x-directional in-plane and the z-directional out-of-plane displacements are analyzed. Another set of two-beam illuminations provides the y-directional in-plane displacement.

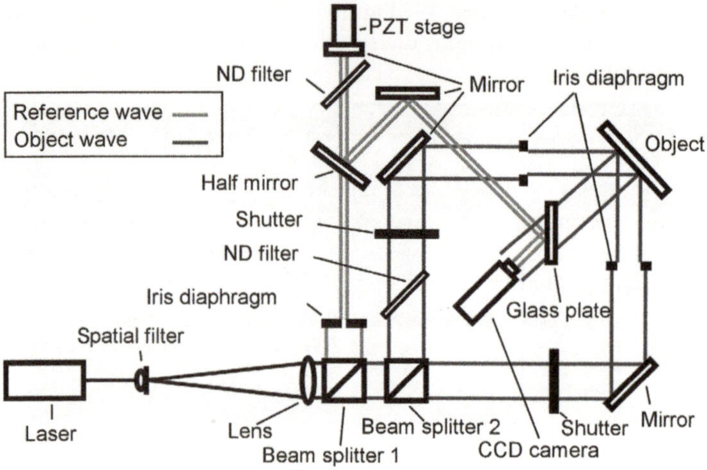

Fig. 21. Optical using two beam illuminations

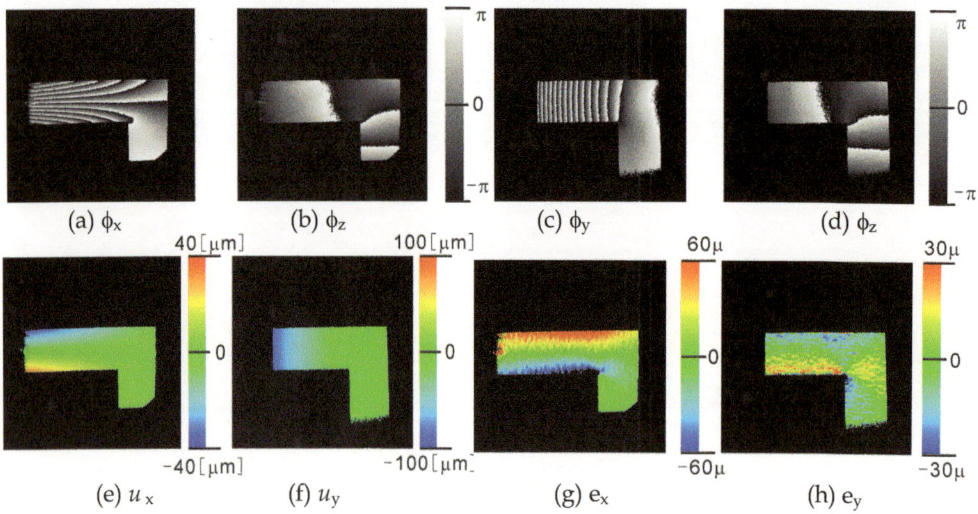

Fig. 22. Phase-difference displacement and strain distributions measured by collimated beam system

An L-shaped cantilever with a fixed line is measured by the system shown in Figure 21. Figures 22(a) and (b) show the phase-difference distributions along the x- and z-directions, respectively, obtained by the x-directional set. Figures 22(c) and (d) show the phase-difference distributions along the y- and z-directions, respectively, obtained by the y-directional set. Figures 22(e) and (f) show the x- and y-directional displacement distributions, respectively. Figures 22(g) and (h) show the x- and y- directional strain distributions, respectively.

5.2.3 Spherical wave system

In this experiment, three holograms using light sources 1, 2 and 3 are recorded on a CCD simultaneously as shown in Figure 23. Although the phases of the three reference waves are shifted simultaneously seven times, the phase-shift amounts during the seven times are different for each light source. The total phase-shifts are 2π, 4π and 6π for the light sources 1, 2 and 3, respectively. The each fringe pattern by the light sources 1, 2 and 3 is extracted from the continuous seven holograms using the Fourier transformation of the brightness change at each pixel of the holograms.

Let us explain the procedures. Figure 24 illustrates the captured brightness changes at a pixel and the brightness change corresponding to each of the three light sources. The discrete Fourier transformation of the captured brightness along the time axis provides the Fourier spectrum shown in Figure 25. It has seven frequency components from -3 to 3. Here, the components of the frequencies 1, 2 and 3 arise from the light sources 1, 2 and 3, respectively. By extracting these components and calculating the amplitudes and the phases of the components, the complex amplitudes of the three holographic fringe patterns are obtained separately.

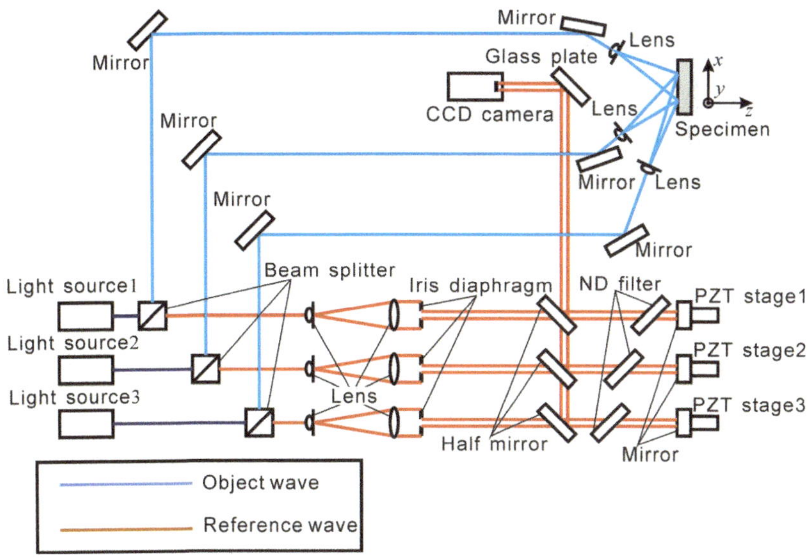

Fig. 23. Optical setup using three spherical waves

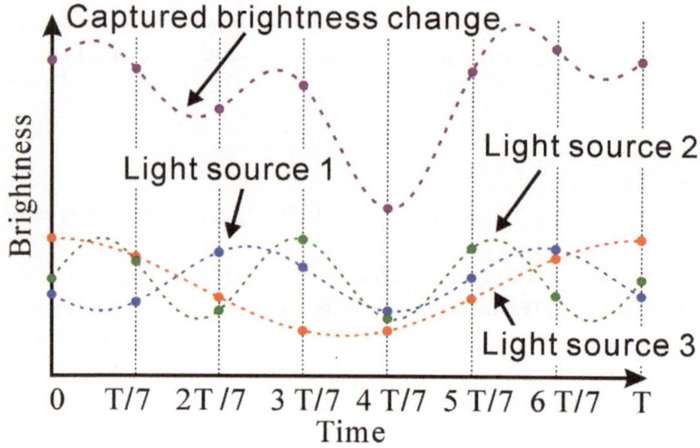

Fig. 24. Brightness change of composite wave

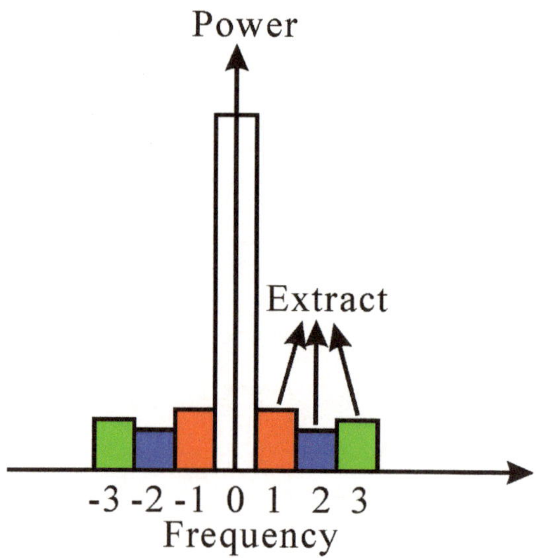

Fig. 25. Fourier spectrum of composite wave

A strain distribution measurement of a T-shaped cantilever with a fixed line is performed by the system shown in Figure 23. The size of the cantilever is 5 mm width, 15 mm length, and 5 mm thickness. A load is applied on the cantilever at the point 12.5 mm from the base of the cantilever.
Seven phase-shifted holograms are captured before and after deformation. The Fourier components corresponding to each light source are separated by the phase-shifting method using Fourier transformation and each phase value is obtained. The phase-difference distributions of the light sources 1, 2 and 3 are shown in Figs. 28 (a), (b) and (c), respectively. The strain distribution in the x-direction is shown in Figure 27.

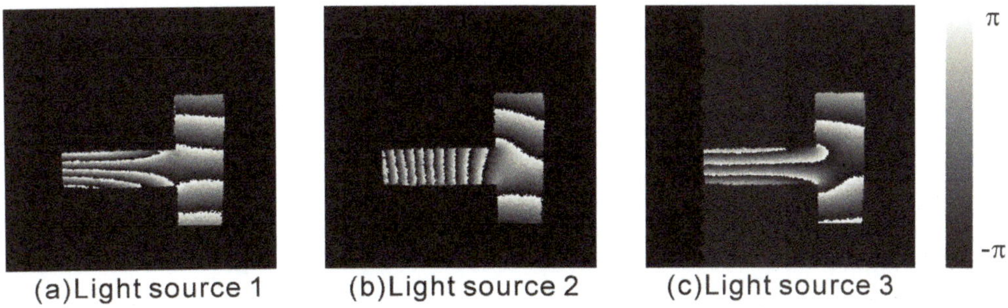

Fig. 26. Phase-difference distributions by each light sources

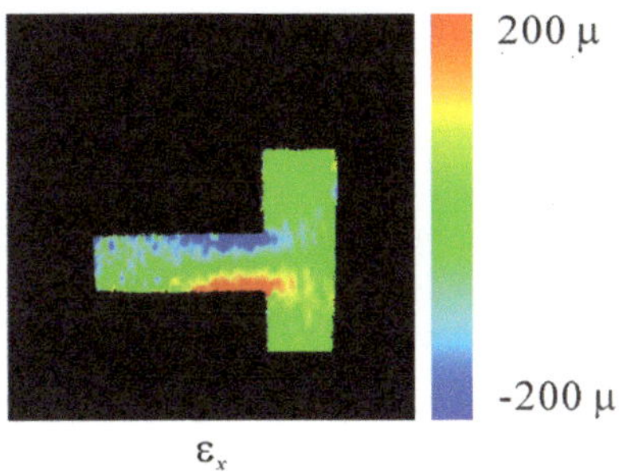

Fig. 27. Strain distribution in x-direction

6. Conclusions

In order to reduce the effect of speckle noise, the windowed digital holographic interferometry was proposed. In the method, the effect of number of windows, or window size on accuracy, spatial resolution and calculation time were examined. The results are that the number of windows increases, the accuracy becomes better at first because of speckle noise reduction and after that it becomes worse because of larger speckle size. The best accuracy is sub-nanometer when the number of windows is 1024 in our experiment.

However, the accuracy is better than the conventional averaging method with 11 by 11 pixels. The spatial resolution becomes worse when the number of windows becomes larger, that is, the window size becomes smaller. If the number of windows is constant, the accuracy and spatial resolution is better when the window size becomes larger. The calculation time increases according to the number of windows. In practice, it is useful when the number of windows is 16, 64 or 256 in our experiment. Three holographic systems were developed and applied to measure 3D displacement and strain distributions.

7. Acknowledgment

A part of this study was supported by a grant aided project for creation of new regional industry of Kansai Bureau of Economy Trade and Industry. The authors also appreciate the support by Mr. Kita, Mr. Okazawa, Mr. Kawagishi, Mr. Kido and Mr. Shiotani, Graduate School of Systems Engineering, Wakayama University, and Mr. Kitagawa and Mr. Nakatani of Hitachi Zosen Corporation.

8. References

Fujigaki, M., Matui, T., Morimoto, Y., Kita, T., Nakatani, M. & Kitagawa, A., (2005) Development of Real-Time Displacement Measurement System Using Phase-Shifting Digital Holography, *Proceedings of the 5th International Conference on Mechanics of Time Dependent Materials (MTDM05)*, pp. 160-163. Sep. 2005

Fujigaki, M., Kita, T., Okazawa, S., Matui, T. & Morimoto, Y. (2006) Calibration method with reference plane for phase-shifting digital holographic interferometry using spherical wave. *Int. Symp. Adv. Fluid/Sol. Sci. Technol. Exp. Mech., Sapporo, Japan*, pp. 11–14.

Hayashi, T., Ugo, R. & Morimoto, Y. (1986). Experimental observation of stress waves propagating in laminated composites, *Experimental Mechanics*, Vol. 26, No. 2, pp. 169-174, ISSN 1741-2765

Kolenovic, E., Osten, W., Klattenhoff, R., Lai, S., Von Kopylow, C. and Ju¨ptner, W. (2003) Miniaturized digital holography sensor for distal three-dimensional endoscopy. Appl. Opt. 42, pp. 5167–5172 ,ISSN 0003-6935, doi:10.1364/AO.42.005167

Morimoto, Y., Nomura, T., Fujigaki, M & Takahashi, I., (2004). "Reduction of speckle noise effect by divided holograms in phase-shifting digital holography", *Proc. of 12th Inter. Conf. on Experimental Mechanics* (CD-ROM), Politecnico di Bari, Italy, 29 August-2 Sept..

Morimoto, Y., Nomura, T., Fujigaki, M., Yoneyama, S. and Takahashi, I. (2005a) Deformation measurement by phase-shifting digital holography. *Exp. Mech.* Vol. 45, No. 1, pp. 65–70, ISSN: 0014-4851, DOI: 10.1007/BF02428991

Morimoto, Y., Matui, T.,. Fujigaki, M & Kawagishi, N., (2005b) Accurate displacement measurement by windowed holograms in digital holographic interferometry, *Proc. of 2005 SEM Annual Conference & Exposition on Experimental Mechanics* (CD-ROM), Portland, Oregon, USA, June 2005, pp. 7-9,.

Morimoto, Y., Matui, T., Fujigaki, M. & Kawagishi, N. (2005c), Effect of weight in averaging of phases on accuracy in windowed digital holographic interferometry for picometer displacement measurement, *Proc. of SPIE*, Vol. 6049: Optomechatronic Sensors and Instrumentation, Yasuhiro Takaya, Ed, Dec. 2005

Morimoto, Y., Matui, T., Fujigaki, M. and Kawagishi, N. (2007) Subnano-meter displacement measurement by averaging of phase-difference in windowed digital holographic interferometry,. *Opt. Eng.* Vol. 46, No. 2,, 025603. ISSN 0091-3286

Morimoto, Y., Matui, T., Fujigaki, M. and Matsui, A., (2008a). Three-dimensional Displacement Analysis by Windowed Phase-shifting Digital Holographic Interferometry, *Strain*, Vol. 44, No. 1, 49-56

Morimoto, Y. Matui, T, Fujigaki, M. & Kawagishi, N., (2008b) Effects of Window Size on Accuracy and Spatial Resolution in Windowed Phase-Shifting Digital Holographic Interferometry, *Strain* (2008) 44, 366–373.

Morimoto, Y. Matui, T, & Fujigaki, M. (2008c), Application of Three-Dimensional Displacement and Strain Distribution Measurement by Windowed Phase-Shifting Digital Holographic Interferometry, Alan K.T. Lau, J. Lu, Vijay K. Varadan, F.K. Chang, J.P. Tu and P.M. Lam (Eds), *Multi-functional Materials and Structures, Advanced Materials Research*, Vol.47-50, pp1262-1265, (2008) Trans Tech Publications, Switzerland, doi: 10. 4028, http://www.scientific.net/AMR.47-50.1262

Okazawa, S., Fujigaki, M., Morimoto, Y. & Matui, T., (2006) Simultaneous Measurement of Out-of-Plane and In-Plane Displacements by Phase-Shifting Digital Holographic Interferometry, *Applied Mechanics and Materials (Volumes 3 - 4), Advances in Experimental Mechanics IV*, J.M. Dulieu-Barton and S. Quinn (Eds.), pp. 223-228, 10.4028/www.scientific.net/AMM.3-4.223, August, 2006

Pedrini G., Fröning P., Fessler H., & Tiziani H.J. (1998). In-Line Digital Holographic Interferometry. *Appl. Opt.*, Vol. 37, No. 26, pp. 6262-6269, ISSN 1559-128X, 2155-3165. doi:10.1364/AO.37.006262

Pryputniewicz, R. J., (2008), Chap. 24 Holography, In: *Springer Handbook of Experimental Solid Mechanics*, Sharpe, W. N. Jr., (Ed.), pp. 675-700, Springer, ISBN: 978-0-387-34362-4, Verlag: Springer, US

Ranson, W. F., Sutton, M. A. & Peters, W. H. (1993). Chap. 8 Holographic and laser speckle interferometry," In: *Handbook on Experimental Mechanics*, 2nd ed., A.S. Kobayashi, (Ed.), pp. 365-405, VCH Publishers, ISBN: 978-0-471-18864-3.

Schnars, U. & Ju¨ptner, W. (2005) *Digital Holography*. Springer, ISBN: 978-3-540-21934-7, Berlin.

Sharpe, W. N. Jr., (Ed.), (2008), *Springer Handbook of Experimental Solid Mechanics*, Verlag: Springer, ISBN: 978-0-387-34362-4, US

Shiotani, K., Fujigaki, M., Matui, T. and Morimoto, Y., (2008) Strain Measurement Method for Tilted Planar Object by Phase-Shifting Digital Holography Combined with Grating Projection Method, *Journal of JSEM*, Vol. 8, Special Issue, pp. 108-113.

Valery, P. S., Vladimir, S. P., Sergey, A. N., Vitaly, V.B., Igor, N. O. & Mikhail, M. B. (1996). *Strain and Stress Analysis by Holographic and Speckle Interferometry*, John Wiley & Sons, ISBN 0471960772

Yamaguchi, I. & Zhang, T. (1997), Phase-shifting digital holography, *Optics Letters*, Vol. 22, No.16, pp. 1268-1270, ISSN 0146-9592

Zhang, T. and Yamaguchi, I. (1998) Three-dimensional microscopy with phase-shifting digital holography, *Optics Letters*, Vol. 23, Issue 15, pp. 1221-1223, ISSN 0146-9592, doi:10.1364/OL.23.001221

Yaroslavskii, L. P. & Merzlyakov, N. S. (1980) *Methods of Digital Holography*, Consultants Bureau, ISBN 0306109638 9780306109638, New York

2

Real-Time Colour Holographic Interferometry (from Holographic Plate to Digital Hologram)

Jean-Michel Desse
Office National d'Etudes et Recherches Aérospatiales (ONERA) Lille France

1. Introduction

In the area of Fluids Mechanics, detailed analysis and characterization of complex, unsteady flows require non-invasive optical methods to measure smaller and smaller quantities over space or time, or even both at once. Therefore, many researchers have spent considerable time over the last fifty years to develop metrology tools adapted to quantitative flow visualization. Some of these methods such as shadowgraph or schlieren method are based on measuring the light deviation through the test section (Merzkirch, 1974), other methods such as interferometry or holography are based on optical interferences and on measurement of the optical path difference or the signal phase (Vest, 1979). When qualitative measurements of the flow are sought, the former techniques can be used. The concepts and the many applications of shadowgraph or schlieren techniques can be found in (Settles, 2001). If quantitative data are required, Mach-Zehnder or Michelson interferometers have been developed, but these instruments are very sensitive to external vibrations, especially when the two arms of the interferometer have unequal length (Merzkirch, 1974). To avoid this problem, differential interferometry or Wollaston-prism shearing interferometry using a polarized white light source can be implemented (Philbert, 1958; Merzkirch, 1965; Smeets, 1975), but these techniques visualize the first derivative of the refractive index in the test section. The same optical technique equipped with high speed camera can be also used to analyze high speed flows (Desse, 1990, 2006). In this case, a sequence of colour interferograms is recorded at a high framing rate from which the derivative of the gas density can be extracted. The interferograms are analyzed nearly automatically by an image processing software specially designed for modelling the light intensity of the interference fringes as the path difference varies (Desse, 1997a). As the method gives a differential measurement, integration is necessary to get the full gas density field, whence a certain imprecision arises in the measurements. To avoid such imprecision related to integration and to maintain the advantage of colour interferograms[1], real-time colour holographic interferometry has been developed. One of the two variants of colour holographic interferometry is perfectly suitable for analyzing unsteady aerodynamic

[1] Colour interferometry has the property to exhibit a unique white fringe visualizing the zero order of interferences

phenomena (i.e., of transparent objects) and for real-time analysis of mechanical deformations (diffusive objects). The usual double-exposure method consists in recording the holograms of a transparent or diffusive object in two different states in succession, on the same photographic plate. This proven technique has yielded good results for several years, but it does have the disadvantage of not allowing the interferogram of the phenomenon being studied to be observed immediately and without interruption. Real-time colour holographic interferometry, on the other hand, allows direct observation through a reference hologram and makes it possible to take an ultra-high speed movie of the interferogram of a changing phenomenon (Surget, 1973). This was the method used.
After presenting the advantages associated with the use of a polychromatic light source rather than monochromatic, the principles of three-wavelength holographic interferometry in real-time are detailed. The feasibility of the method is shown when silver-halide panchromatic holographic plates are used either in transmission or reflection. The advantages and disadvantages of these techniques for recording and reconstruction, though substantially different, are presented through an application that examines the unsteady wake flow downstream of a cylinder at a subsonic Mach number. To conclude this chapter, colour digital holographic interferometry is presented as a method preferable to holographic techniques using holographic plates even if the new generation of CMOS or CCD sensors are far from having the spatial resolution of holographic plates.

2. Advantage of using one source to multiple wavelengths

In monochromatic interferometry (for instance, λ=647nm), it is well known that the classical interference pattern is represented by a succession of dark and bright red fringes. For two successive fringes, the optical path difference is equal to the wavelength of the laser source (Fig.1a). Unfortunately, the zero order of interferences fringes can never be identified and it

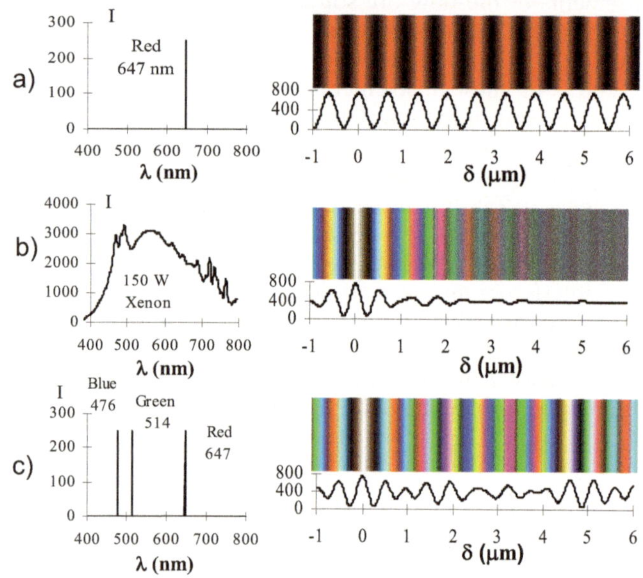

Fig. 1. Spectra and interference fringes given by three different light sources

is one of the major difficulties with interferences fringes in monochromatic light. Sometimes, it is not possible to follow the displacement of the fringes through a shock wave, for example, or to count the fringe number in a complex flow. When the light source is a continuous source (500 Watt xenon, see Fig. 1b), the interference pattern is a coloured fringe pattern in a sequence approximately matching Newton's colour scale. This fringes diagram exhibits a unique white fringe, visualizing the zero order of interference and it allows one to measure very small path differences, because six or seven different colours define the interval 0-0.8 microns. But, when the path difference is greater than three or four microns, the colours can no longer be separated and the larger path differences cannot be correctly measured (Desse, 1997b). Fig. 1c shows the fringes obtained with a laser that emits three different wavelengths (one blue line, one green line and one red line). One can see that the disadvantages of the two others sources have disappeared. The zero order is always identifiable and the colours always remain distinguishable for the small and the large path differences. The interference pattern also presents the following peculiarity: while the white fringe is not visible on the interferogram, the sequence of three successive colours in the diagram is unique.

3. Principle of real-time colour holographic interferometry

The various holographic interferometry methods - double exposure, time-averaged, or real-time holography - are the main scientific applications of holography. Until recent years, experiments in holographic interferometry were performed with a single laser, i.e., they were monochromatic. Most experiments found in the literature relate to transmission holograms (Rastogi, 1994) and few experiments have been performed to date using holographic interferometry with reflected white light (Smigielski et al., 1976; Vikram, 1992). It should be said that, in monochromatic mode, experiments in reflected white-light holography have little advantage over holographic interferometry in transmitted light. Some publications mention the use of three-wavelength differential interferometry (Desse, 1997b) and holographic interferometry by reflection (Harthong, 1997; Jeong, 1997) and all show that the essential advantage of colour is that the achromatic fringe can be located in the observed field.

Real-time true colour holographic interferometry uses three primary wavelengths (red, green, blue) to record the interference between the three object beams and the three reference beams simultaneously on a single reference hologram. Under no-flow conditions, the undisturbed object waves ΣRO, ΣGO and ΣBO are recorded in the hologram by virtue of their interference with the three reference waves ΣRR, ΣGR and ΣBR. As can be seen in Fig. 2, step 1, at recording and for a plate recorded in transmission, the three reference waves and the three object waves arrive on the same side of the plate while in reflection, they come from opposite sides of the holographic plate. After treatment of the plate and resetting in the optical bench, the three reference waves ΣRR, ΣGR and ΣBR are diffracted by transmission or by reflection according to the recording mode used to form the three diffracted object waves ΣROD, ΣGOD and ΣBOD (Step 3, Fig. 2).

Then the hologram is illuminated simultaneously by the three reference beams and three object beams, from which we get the three object beams ΣROD, ΣGOD and ΣBOD reconstructed by the holographic plate simultaneously with the three live object waves

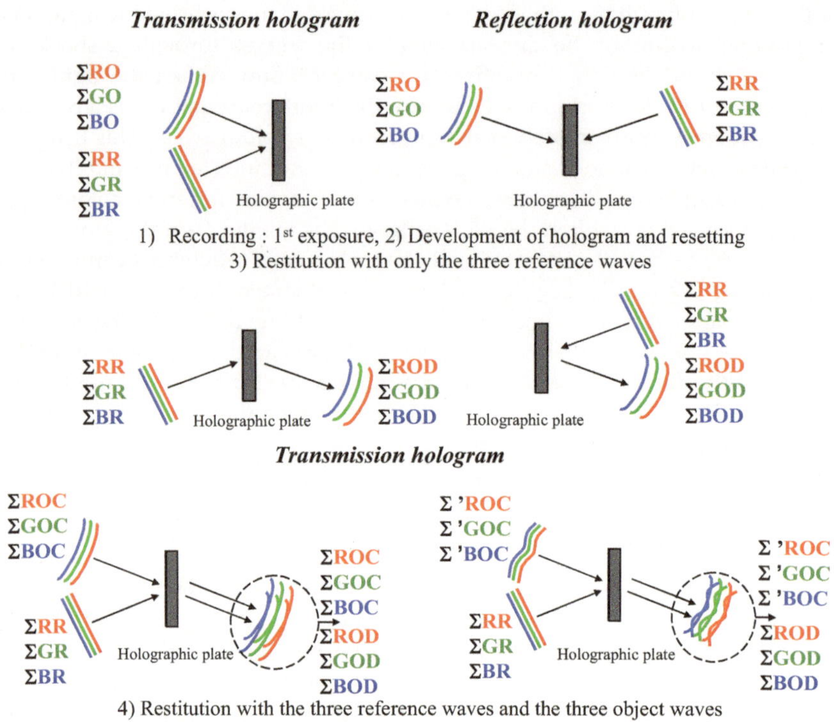

Fig. 2. Formation of colour interference fringes

transmitted ΣROC, ΣGOC and ΣBOC. The profiles of the ΣROD and ΣROC, ΣGOD and ΣGOC waves, and the ΣBOD and ΣBOC waves are strictly identical to each other if no change has occurred between the two exposures and if the hologram gelatine has not contracted during development. So there will be three simultaneous interferences among the object waves constructed by the hologram and the live object waves. In this case, a flat uniform colour can then be observed behind the hologram. If a change in optical path is created in the test section of wind tunnel, the three live waves will deform and adopt the profiles Σ'ROC, Σ'GOC and Σ'BOC while the waves reconstructed in the hologram, ΣROD, ΣGOD and ΣBOD, remain unchanged. Any colour variations representing optical path variations will thus be visualized in real time behind the hologram (Step 4, Fig. 2).

4. Real-time colour transmission holographic interferometry

4.1 Laboratory study of feasibility

4.1.1 Choice of laser and colour base

The following points have to be addressed to show the feasibility of real-time colour holographic interferometry. Firstly, a laser has to be found that will supply the three primary wavelengths forming as extensive as possible a base triangle[2]. It is an Innova

---

[2] The three wavelengths chosen define the three vertices of a triangle chromaticity diagram (MacAdam, 1985).

Spectrum 70 ionized gas laser (mixed argon and krypton) that produces approximately 10 visible lines with a total power of 4.7 W. The three wavelengths retained are 647 nm for the red line of krypton and 514 nm and 476 nm for the green and blue lines of argon. All three exhibit a TEM_{00} mode because thee laser is equipped with a Fabry-Perot etalon to increase the coherence length of the three lines selected. The etalon is treated on both faces to get 66% transmittance for the blue line, 63% for the green line and 61% for the red line. This treatment increases the blue line's coherence length 2 or 3 centimetres at a range of several tens of centimetres. This is sufficient because, in our study, the reference and measurement paths can be roughly equalized and the optical path variations to be measured are no greater than a few microns. In the other hand, a large light energy is needed to record the studied phenomena at ultra-high speed (35,000 f/s with an exposure time of 750 10^{-9} s).

4.1.2 Panchromatic holographic plates

Since Russian panchromatic plates came on the market twenty years ago, progress has been made in true colour holograms (Hubel, 1991; Bjelkhagen & Vukicevic, 1991; Bjelkhagen et al., 1996). The various chemical treatments applied to these plates are explained in detail by several authors (Bjelkhagen, 1993; Sasomov, 1999). The plates used are silver-coated single-film PFG 03C plates from the Slavich company in Moscow. Their chemical treatment first includes a hardening of the gelatine, development, bleaching, a series of rinses in alcohol and slow drying. The hologram's spectral characteristics were analyzed by taking a double exposure holographic interferogram and placing a small mirror near the object to be analyzed, in order to make a spectral analysis of the reconstructed light waves. The spectrograms of the reconstructed waves in our very first validation tests showed that the three peaks corresponding to the reconstructed colours are slightly shifted by a few nanometres, which corresponds to the contraction of the gelatine thickness:

- 471 nm for the blue, or $\Delta\lambda = -5$ nm,
- 511 nm for the green, or $\Delta\lambda = -3$ nm,
- 640 nm for the red, or $\Delta\lambda = -7$ nm.

These differences are reduced practically to zero in wind tunnel tests when the hologram is placed normal to the bisector of the angle formed by the object and reference beams.

4.1.3 Laboratory results

Fig. 3 shows the feasibility setup implemented in the laboratory. The Innova Spectrum laser emits eleven lines in the visible simultaneously. The red, green, and blue lines we want are diffracted by an acousto-optic cell in which are generated three frequencies f_1, f_2 and f_3 appropriate to the three wavelengths λ_1, λ_2 and λ_3. A beamsplitter cube splits the reference beams and three object beams. The three reference beams are collimated onto the holographic plate by an achromatic lens located a focal length from the pinhole (diameter 25 μm) of a spatial filter having a microscope objective lens (x60). The three object beams are collimated the same way to form three parallel light beams between two large achromatic lenses and illuminate the test section.

The hologram is thus illuminated on the same side by the three parallel reference beams and the three convergent measurement waves. A diaphragm is placed in the focal plane just in front of the camera in order to be able to filter out any parasitic interference. The hologram is first illuminated in the absence of flow and is then developed and placed back in exactly its original position. When the hologram is illuminated with the reference beam, nine diffraction images are seen. They are clearly visible in Fig. 4. Of the nine, three coincide and

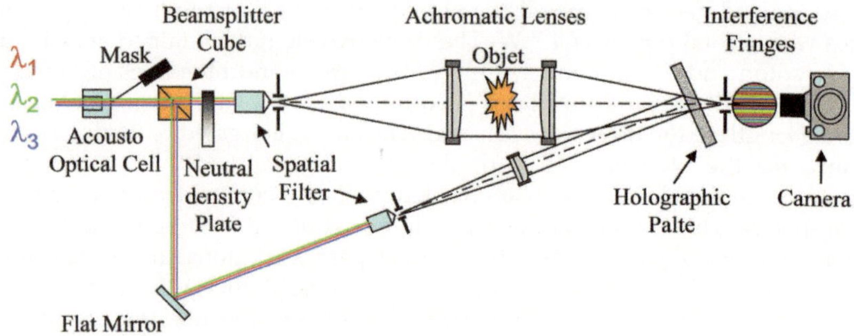

Fig. 3. Real-time colour holographic interferometry setup (Transmission mode)

focus exactly at a single point if the setup is perfectly achromatic. These three images correspond to the diffraction of the blue image by the blue beam, that of the green image by the green beam, and that of the red image by the red beam. They are exactly superimposed. The other six images are parasitic and have to be filtered out. They correspond to the diffraction of the blue image by the green and red lines, that of the green image by the blue and red lines, and that of the red image by the blue and green lines. Fig. 4 shows how all the images are diffracted behind the hologram, and the spatial filter system that is used to select only the focal point of interest to us.

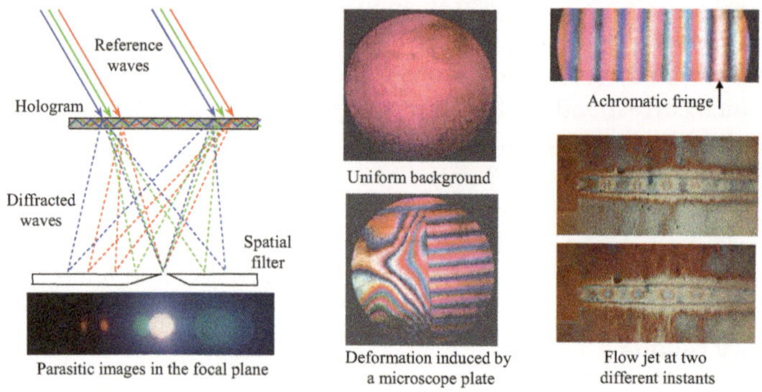

Fig. 4. Real-time colour holographic interferometry setup (Transmission mode)

Moreover, an adjustable neutral density filter made it possible to balance the power of the measurement beams with that of the reference beams. For reference, the power of the three lines at the acousto-optic output is 50 mW for the red line and 90 mW for the green and blue. In the hologram plane, the reference beam powers are 240 μW/cm² for the red line, 270 μW/cm² for the green, and 290 μW/cm² for the blue. For the object beams, one measured 100 μW/cm² for the red and green lines and 330 μW/cm² for the blue. The hologram exposure time for the first exposure was 4 s. Fig. 4 shows some images obtained. First, the uniform background colour was obtained over the entire surface of the test section, which shows that it is possible to re-position the three live waves simultaneously with the three waves contained in the hologram. The tint is the purple of the first order of interference, which is one of the most sensitive. The interference pattern is obtained by slightly displacing one of the two large achromatic lenses. So a horizontal, vertical, or even circular fringe pattern can be formed. The achromatic central white fringe can be made out very clearly. One also can see the deformation due to an microscope plate and a small jet at two different instants. Lastly, the diffraction efficiency was measured behind the hologram. The hologram exhibits a diffraction efficiency of about 0.8% for the red and green lines and 0.5% for the blue line. Although these values are very low, they do allow a good visualization.

4.2 Wind tunnel adaptation of the method

The optical setup for testing the feasibility of the technique had to be modified and adapted around ONERA's wind tunnel at Lille centre. For reference, this wind tunnel is equipped with a 2D test section 200 mm high and 42 mm wide. The Mach number can be varied from 0.3 to 1.1. The flow studied was the unsteady flow downstream of a cylinder 20 mm in diameter D placed crosswise in the test section.

4.2.1 Optical setup around the wind tunnel

The optical setup is shown in Fig. 5. The beam power as it leaves the laser is 1.80 W when the etalon is set perpendicular to the laser beam axis and 1.20 W when it is tilted. The polarization of the three beams rotates 90° at the exit from the acousto-optic modulator, so that the polarization vectors lie parallel to the reflecting surfaces of the mirrors. This arrangement makes it possible to have beams of the same polarization interfere on the hologram. The three wavelengths downstream of the acousto-optic cell are split into a reference beam and object beam by a beam splitter cube. A right angle prism is used to adjust the reference and object path lengths on the hologram. A spatial filter is used to expand the beam for its passage through the test section. A pair of achromatic lenses converts the beam into parallel light in the test section and then focuses it on the hologram. The reference beam passes over the test section, and then another achromatic lens is used to illuminate the hologram with a parallel light beam. For reference, the object beam diameter is 40 mm at the hologram and that of the reference beam is 60 mm. At the acousto-optic cell, the power of the three light waves is practically the same (of the order of 70 mW per channel). The beam splitter cube distributes 85% of this power to the reference path and 15% to the measurement path.

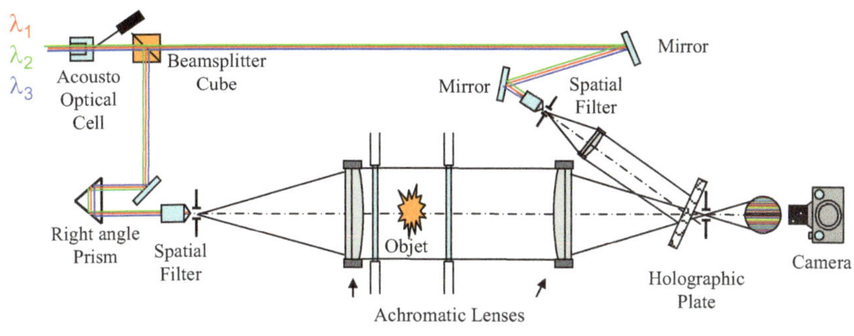

Fig. 5. Optical setup implemented around the wind tunnel

At the hologram, one measures 250 μW/cm² in the red and blue lines and 280 μW/cm² in the green line for the reference beam, while the object beam powers are 30 μW/cm² in the red line and 40 μW/cm² in the green and blue lines. These proportions can be used to obtain a perfect balance among the powers of the three waves diffracted by the hologram when repositioning it, and the three live waves. For reference, the hologram diffracts 70 μW/cm² in the red line, 65 μW/cm² in the green line, and 90 μW/cm² in the blue line. The first exposure lasts 2 s. The holograms are then subjected to treatments to harden the gel, develop it, and bleach it. When the hologram is put back in place, the light power at the camera entrance is 1.5 10^{-3} W at the focal point, which is sufficient to record interferograms at an ultra-high speed of 35,000 frames per second with an exposure time of 750 ns per shot.

4.2.2 Results and analysis of interferograms at Mach 0.37

Figure 6a gives a sequence of six interferograms of the flow around the cylinder at Mach 0.37. The time interval between each picture is 100 μs. One can see that each vortex is represented by concentric rings of different colours where each colour represents an isochoric line. The vortex formation and dissipation phases can be seen very clearly while the fringes oscillate between the upper and lower surfaces of the cylinder. Several types of measurements were made by analyzing a sequence of some 100 interferograms. First, the vortex centre defined by the centre of the concentric rings was located in space for each interferogram, which made it possible to determine the mean paths for the vortices issuing from the upper and lower surfaces. The results of this are shown in Fig. 6b. The "o" symbols represent the positions of the vortex centres from the upper surface, and the "•" symbols those of the lower surface. Remarkably, the two paths exhibit a horizontal symmetry about the x = 0 axis passing through the cylinder centre. One may also point out that even at x/D = 4 downstream of the cylinder, the upper and lower vortex paths do not come together and line up.

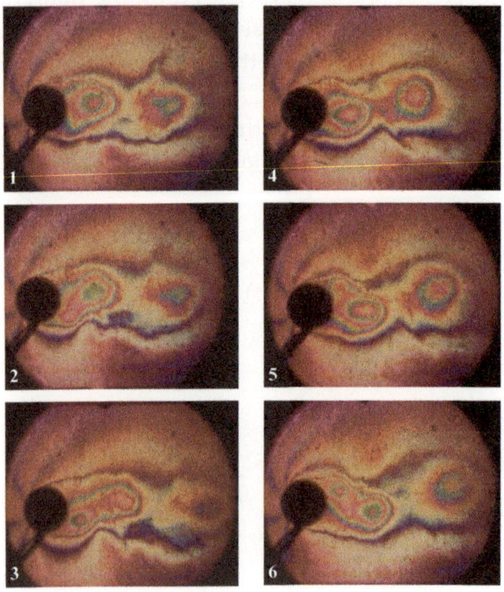

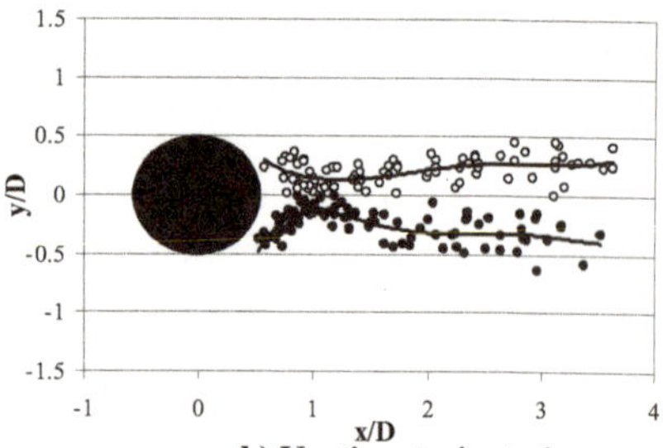

b) Vortices trajectories

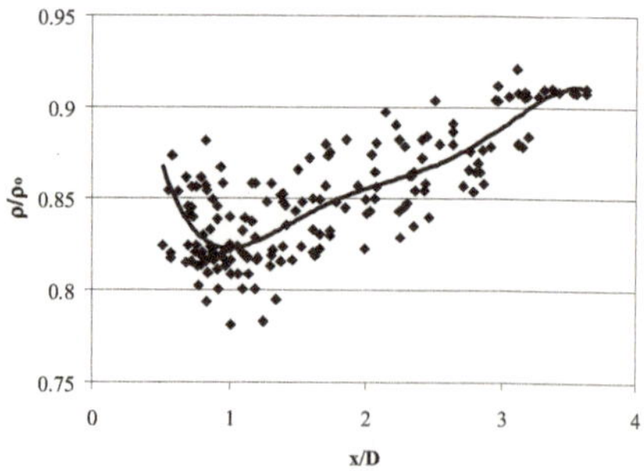

a) High speed holographic interferograms recorded by transmission – Δt = 100μs

c) Vortices gas density

Fig. 6. Unsteady wake flow around the cylinder -Results and analysis - Mach 0.37

Lastly, the colours of each interferogram were analyzed using the "MIDI" software[3], which models the light intensity and experimental interference fringe colours as a function of the path difference (Desse, 1997a). The gas density measured under free stream conditions is the same as that measured at the outer flow of the wake (measured in the vicinity of the wind tunnel's upper and lower walls).
If e is the test section width and n the refractive index, the optical thickness E can be expressed by :

$$E = (n-1).e \tag{1}$$

Knowing the Gladstone-Dale relationship :

$$K = \frac{(n-1)}{\rho / \rho_s} \tag{2}$$

where ρ_s is the gas density under standard conditions (1.29 kg/m³) and K the Gladstone-Dale constant (296.10⁻⁶), the relative gas density variation is written :

$$\frac{\Delta\rho}{\rho_0} = \frac{\Delta E}{e.K}\frac{\rho_s}{\rho_0} \tag{3}$$

with $\Delta\rho = \rho - \rho_\infty$ and $\Delta E = E - E_\infty$, ρ_∞ being the gas density measured under free stream conditions and ρ_0 the stagnation gas density. Starting from a point external to the wake, one can work back to the gas density at the stagnation point (at the nose of cylinder) and measure the gas densities at the centre of the vortices. The graph of Fig. 6c shows how ρ/ρ_0 varies for the vortices emanating from the upper and lower surfaces. The trend curves plotted show the same variations. For $0.5 < x/D < 1$, the vortices are in a formation or agglomeration phase because the gas density decreases at their centre. Then, when $x/D > 1$, the vortices enter a dissipation phase because the gas density increases again at their centres. The drop in gas density is large, reaching about 20% ρ_0. A rather large dispersion may nonetheless be noticed in the data. This is due mainly to our determination of the vortex centre locations, which are not very easy to determine when the vortices are in the dissipation phase. Moreover, taking the hologram transfer function into account in the exact modelling of the interference fringes should greatly improve the modelled colours between the orders of interference, and thereby the measurement precision.

5. Real-time colour reflection holographic interferometry

Since transmission holograms are used, the diffraction efficiency of holograms just reaches between 10% and 20%, which limits the quality and the contrast of interferences fringes. On the other hand, if reflection holograms are used, the theoretical diffraction efficiency can reach 100% with a monochromatic light. The development of real-time true colour reflection holographic interferometry also offers two important advantages. The first one concerns the analysis of the three-dimensional (3D) flows and the second one lies in the comparison with digital colour holographic interferometry. In fact, ONERA is looking towards analyzing unsteady 3D flows, and the optical setup to be designed must be based on several crossings

[3] Modelling of Luminous Interferences and Analysis of Interferograms

of the flow along different view angles. It is very evident that classical optical setup based on monochromatic holographic interferometry defined in section 4, for instance, for analyzing two-dimensional (2D) flows, cannot be reproduced three or four times. Moreover, as the optical path differences to be measured are smaller in 3D flows than in the 2D case, it is preferable that each optical ray crosses the phenomena twice in order to increase the sensitivity. Also, to simplify the setup, all the optical pieces have to be located on the same side of the wind tunnel, except the flat mirror which reflects the light rays back into the test section.

In literature, several authors have analyzed 3D flows using multidirectional tomography (Cha & Cha, 1996; Yan & Cha, 1998). They present holographic interferometric tomography for limited data reconstruction to measure an asymmetric temperature field. Other researchers designed an optical scheme for obtaining specklegrams simultaneously in four directions (Fomin, 1998; Fomin et al., 2002) or built an interferometric tomography apparatus with six viewing directions from which multidirectional data sets were analyzed following a method of examining spatial coherence (Pellicia-Kraft & Watt, 2000, 2001). The same approach is used by researchers developing digital holographic interferometric techniques. Timmerman & Watt (1995) developed a dual-reference beam holographic interferometer providing six simultaneous views of a compressible flow. One can also note the optical tomograh using six views interferometers for the measurement of 3-dimensional distribution of temperature in an evaporating liquid (Joannes et al, 2000). All the measurement techniques yield either the derivative of the refractive index (speckle holography, differential interferometry or back oriented schlieren) or the refractive index itself (holographic interferometry) and, very often, the spatial resolution of recording camera is very low compared to that of a holographic plate. As the reconstructed field depends strongly on the measured quantity, on the number of the viewing directions and on the spatial resolution, ONERA wanted to develop a metrological tool having limited viewing directions (three or four), high spatial resolution and yielding absolute value of the gas density in the field.

In colour holographic image and panchromatic holographic materials, one can note the recent work of Bjelkhagen & Mirlis, 2008 who produce highly realistic three-dimensional images. They show that the quality of a colour hologram depends on the properties of the recording material and the demand on a panchromatic material for colour holography is described.

5.1 Description of optical bench

The optical setup could be named "Denisyuk" because it uses a holographic plate in a classical Lippmann-Denisyuk in-line experiment. To obtain a very simple setup, all the optical pieces are located on the same side of the wind tunnel, except the flat mirror which reflects the light rays back into the test section. Due to these considerations, the optical setup based on real-time colour reflection holographic interferometry has been designed. It is presented in Fig. 7.

Here, the light source used behind the interferometer is constituted with three different lasers. An argon-krypton laser delivers the red line (λ_1 = 647 nm), and a green line (λ_2 = 532 nm) and a blue line (λ_3 = 457 nm) are given by two diode pumped solid state lasers. A spatial filter SP is just located at the focal point of the large achromatic lens AL3 (see photograph, Fig. 7) which is set in the front of the test section TS so that the object under

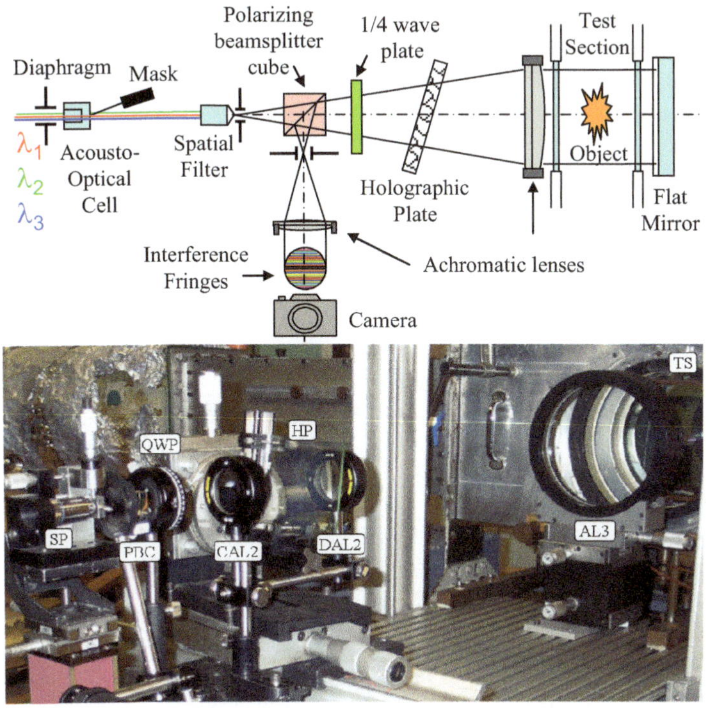

Fig. 7. Real-time three-color reflection holographic interferometer

analysis is crossed by a parallel light beam of 200 mm in diameter. A flat mirror located just behind the test section returns the three beams on the hologram HP inserted between the quarter wave plate QWP and the large achromatic lens. The hologram is illuminated on the two sides by the three collimated reference and measurement waves which are formed by the convergent and divergent achromatic CAL2 and DAL2 lenses (not shown in scheme of Fig.7). This arrangement allows one to easily obtain before the test a uniform background colour (infinite fringes) or narrowed fringes (finite fringes). In this setup, a polarizing beam splitter cube PBC is inserted between the spatial filter and the quarter wave plate which transforms the waves polarization twice (from P parallel to circular and from circular to S parallel) so that, when the rays are returning, the beam splitter cube returns the rays towards the camera. A diaphragm is placed in the focal plane just in front of the camera in order to filter out any parasitic interference. The interferences fringes produced by the phenomenon under analysis can be directly recorded using high speed camera. Here, the camera used is a CORDIN 350 Dynafax. The size of each recording is 10x8 mm^2 and the pictures are taken in a staggered pattern on a 35mm film. High sensitivity (800/1600 ASA) daylight reversible colour films are suitable.

5.2 Principle of real-time three-color in-line holographic interferometry

Fig. 8 details how the interferences fringes are generated in the real-time three-color reflection holographic interferometer.

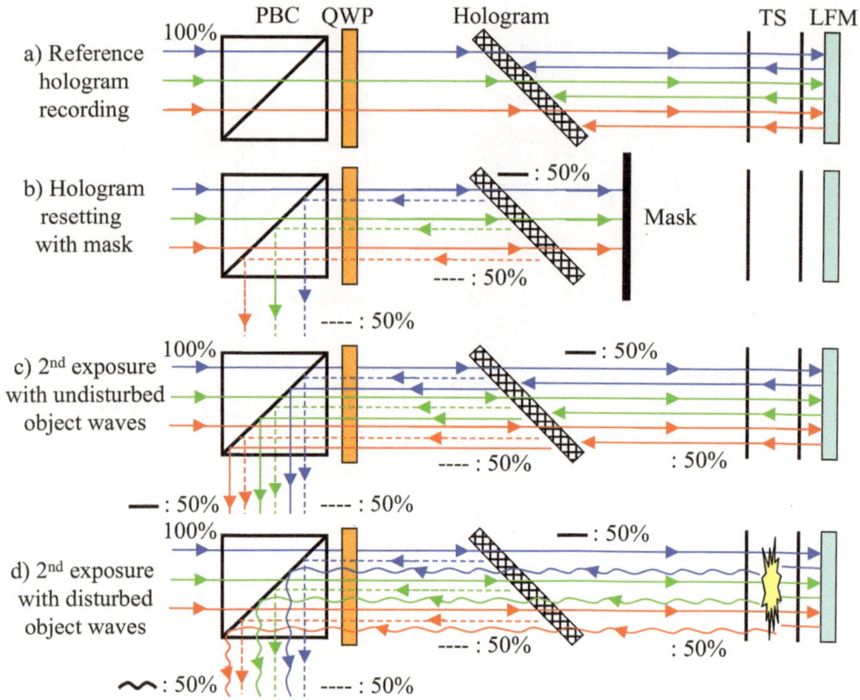

Fig. 8. Formation of colour interference fringes in optical setup

First, the holographic plate HP is simultaneously illuminated with the three wavelengths (Fig. 8a). The panchromatic hologram records simultaneously the three sets of interference fringes produced by the three incident waves and the three waves reflected by the flat mirror LFM (first exposure). Then the hologram is developed and it is reset in the optical bench at the same location. At the second exposure, if the diffraction efficiency of the holographic plate is near to 50% for the three lines, 50% of the light is reflected by the hologram (dashed lines) and 50% crosses the holographic plate (solid lines). If a mask is inserted in the front of the test section, one can observe on the screen the three images diffracted by the plate (Fig. 8b). This operation allows for verifying the quality of the holograms diffracted. When the mask is moved, 50% of the light crosses the test section twice and interferes in real-time with the three references waves (solid lines). Interference fringes are not localized because they can be observed from the holographic plate to the camera. If no disturbances exist in the test section, a uniform background colour is obtained in the camera (Fig. 8c). If variation in refractive index exists in the test section, colour fringes will be seen on the screen. As the luminous intensities of reference and measurement waves are basically equal, the contrast of colour interferences fringes will be maximum (Fig. 8d).

This optical setup is very simple but it presents some advantage and some inconvenience. The advantage lies in the small number of optical pieces which are used. The reference beams and the measurement beams are co-linear and there is just a flat mirror behind the test section. The contrast of colour interferences fringes depends on the diffraction efficiency of the holographic plate and the colours saturation depends on the luminous intensity of the three wavelengths which can be adjusted with the acousto-optic cell. A main inconvenience resides in the fact that it is not possible to adjust the diffraction efficiency of the holographic plate. It is only fixed by the chemical treatment and it is a function of gelatine thickness. The unique solution to solve this problem will consist in a specific treatment of the surface of the flat mirror and this operation implies prior knowledge of the diffraction efficiency of the hologram.

Finally, the three interference fringe patterns will exist and can be recorded if the coherence length of the three wavelengths is more than twice the distance between the holographic plate and the flat mirror located just behind the test section. Compared to the setup of transmission holographic interferometry, here it is not possible to adjust the length between the reference and measurement rays. In this experiment, two types of holographic plates have been tested: Russian plates (PFG03C) from Slavich and French plates (Ultimate 08) from Gentet, typically 10 μm thick. For information, energies at each wavelength applied at the first exposure are given in Table 1 for Slavich and Gentet plates.

PFG03c (Slavich)	Ultimate 08 (Gentet)
1.0 10^{-3} J/cm^2 @ 457nm	0.8 10^{-3} J/cm^2 @ 457nm
1.3 10^{-3} J/cm^2 @ 532nm	0.8 10^{-3} J/cm^2 @ 532nm
1.0 10^{-3} J/cm^2 @ 647nm	0.8 10^{-3} J/cm^2 @ 647nm

These values can be compared with tests and results found in Petrova et al., 2000 who determine the holographic characteristics of panchromatic light sensitive material for reflective 3D display.

5.3 Problem of gelatine shrinkage

The problem of gelatine shrinkage is described in detail in Desse, 2006. In Fig. 9, one can see how the interference fringes are inscribed into the gelatine when the holographic image is recorded by transmission or reflection. In transmission, the interference fringes are perpendicular to the plate and a small variation in the gelatine thickness caused by the chemical treatment of the hologram does not modify the three inter fringe distances. On the other hand, in reflection, interference fringes are recorded parallel to the plate surface and the inter fringe distance is very sensitive to a small variation of the gelatine thickness. Fig. 9 presents the effects of the gelatine contraction when a reflection hologram is recorded with a green wavelength (514 nm). At reconstruction, a white light source (xenon source) illuminates three different holograms at the incidence angle that the reference wave had at recording. One can see that if the gelatine thickness is kept constant (Δe=0), the hologram only diffracts the recording wavelength, i.e. for the green hologram, the green wavelength contained in the xenon spectrum. If the gelatine thickness has decreased by 5%, (Δe = - 0.5 μm), the fringe spacing will be proportionally reduced and the diffracted wavelength will be shifted by a quantity equal to

$$\Delta\lambda = \frac{\lambda}{e}\Delta e \tag{4}$$

where e is the gelatine layer thickness (about 10 μm). The hologram will diffract a wavelength equal to 488.3 nm corresponding to a blue line and, if the gelatine thickness increases by 10% (Δe>0), the hologram illuminated in white light will diffract a wavelength close to yellow (565.4 nm).

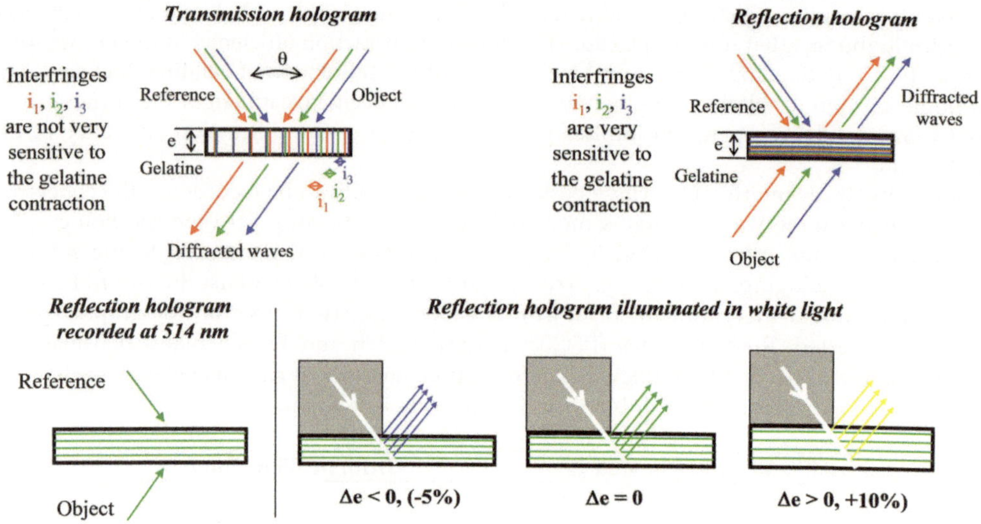

Fig. 9. Effect of the gelatine contraction on the different waves

On the other hand, it is well known that the chromatic perceptibility of eye δl varies with the wavelength. It is defined as being the variation δl between two different wavelengths perceived by the eye at constant luminosity. It is about 1 nm in the green and yellow colours and 6 nm in the blue and red colours, which corresponds to relative variations of 0.2% and 1.5% respectively. For the diffracted colour change not to be detected by a human eye, it is mandatory that $\delta l/\lambda$ has to be less than $\delta\lambda/\lambda$ that which implies that the variation in gelatine thickness should be less than 0.2%. This means changes in thickness of more than 20 nm are not acceptable. As the optical technique is based on the knowledge of the true colours diffracted by the hologram, variations of the gelatine thickness are a cause of large errors in the data analysis. It is for this reason that the gelatine shrinkage problem has to be perfectly mastered.

5.4 Gelatine contraction control

In this experiment, two types of holograms have been tested: Russian plates from Slavich and French plates from Gentet. Concerning the first ones, a specific treatment proposed by Kim, 2002, has allowed obtaining a gelatine contraction smaller than 20 nm by mixing 2 ml of glycerol in the last bath of ethanol (100% ethanol drying). One can mention that the treatment applied to the Russian plates includes about ten steps and it is very sensitive to the temperature and the PH of the solutions. About the French plates, following Gentet's recommendations, we have also obtained basically no variation in the gelatine thickness diffracted by Gentet and Slavich plates when they are illuminated in white light.

In these graphs of Fig. 10, the dashed lines represent the spectrum of white light diffracted by the holograms with no treatment of the plates. We can observe a shifting of about 1.8 % with Gentet holograms and about 1.1% with Slavich holograms which is not acceptable. It can be also seen that the spectrum diffracted by Gentet holograms is wider than the spectrum diffracted by Slavich holograms. For example, if one looks at the red line at a normalized intensity of 0.4, the width of the curve is near to 16 nm and 27 nm for the Slavich and Gentet holograms respectively. It is of 10 nm and 17 nm for the blue line. If a small variation of the gelatine thickness is allowed then the wider spectral response of Gentet holograms is advantageous. A bad adjustment of the Fabry-Perot etalon in the argon-krypton laser cavity, meaning that the coherence length was reduced, explains the weakness of the green response in the Slavich graph.

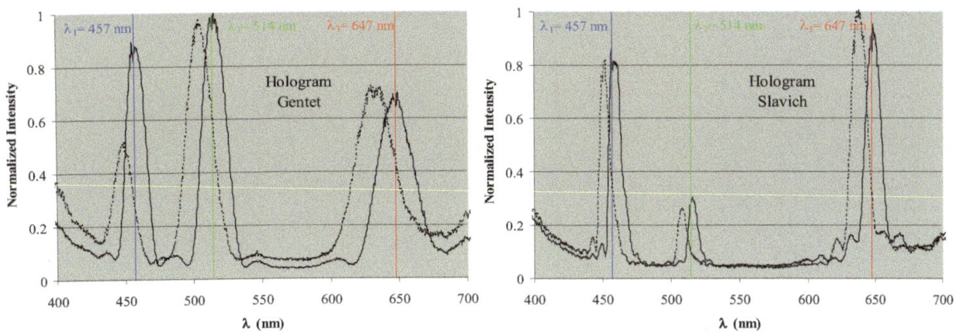

Fig. 10. Measurement of the gelatine shrinkage - Gentet and Slavich holograms

When no shrinkage exists, the curves are basically centred on the three different lines of the lasers and the best diffraction efficiency of hologram is reached. The three different gratings inscribed in the hologram gelatine diffract very well the three different wavelengths (blue, green and red) of the laser sources. In fact, the response of the hologram is the best one because the bell curves are centred on the three laser wavelengths. In these conditions, there is basically no difference in the gelatine thickness before and after the chemical treatment of the plates. Moreover, if the power of the three reference and measurement wavelengths are the same, it is possible to obtain very bright, high contrast fringes.

5.5 Diffraction efficiency measurement

The diffraction efficiency DE of the holographic plates can be evaluated when Russian or French plates are illuminated in white light. For example, Fig. 11 shows how the diffraction efficiency of Gentet holographic plate is determined from the spectrum of the xenon light source transmitted by the plates. It very easy to see the three hollows corresponding to the part of the white light diffracted by the Gentet holographic plate. For each line, the bandwidth can be determined when the diffraction efficiency is more than 35%. In fact, as regards luminous intensity, when the diffraction efficiency of the holographic plate is equal to or more than 35%, the visibility coefficient of interference fringes $((I_{max}-I_{min})/(I_{max}+I_{min}))$ is near to 0.21, that means the colour interference pattern will be of very high? contrast. For the blue line, the acceptable wavelength shift is 8 nm (1.75%), 9 nm (1.75%) for the green line and 13 nm (2%) for the red line. This particularity is very interesting because the theoretical constraints on the variation of the gelatine thickness (near 0.2%) become larger (about 1.75%) due to the spectral broadening of the transmission curve.

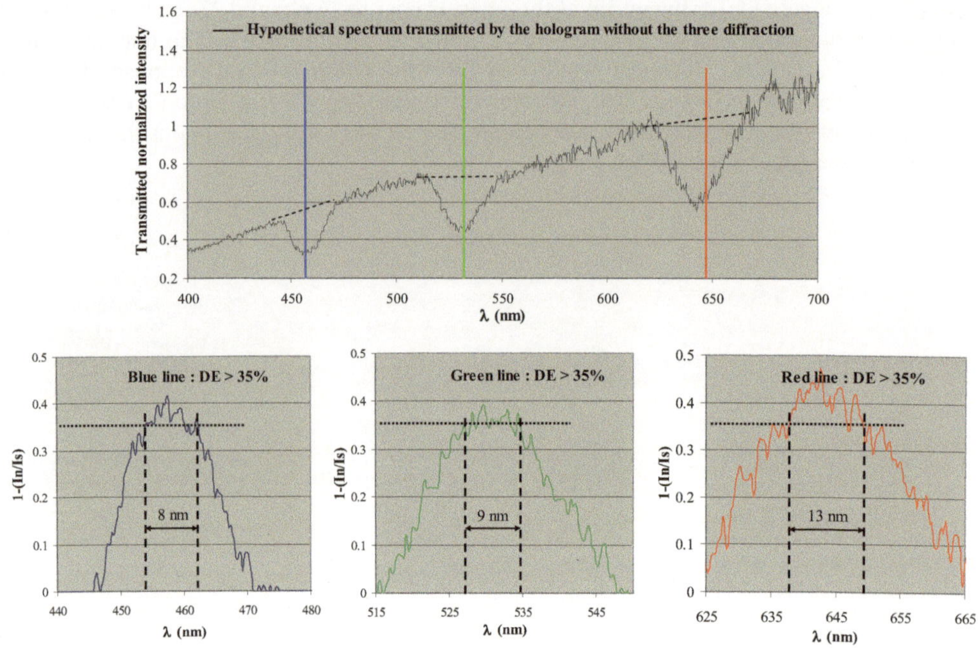

Fig. 11. Evaluation of diffraction efficiency of Gentet plate

5.6 Wind tunnel implementation of optical technique

A real-time colour reflection holographic interferometer has been implemented in the ONERA transonic wind tunnel for analyzing the same unsteady wake flow around the cylinder. In this experiment, the infinite Mach number was fixed at 0.45 and the high speed interferograms were recorded with the rotating drum camera which is equipped with a 400 ASA colour film. The time interval between two successive frames is 117 μs. The time exposure (750 ns) of each interferogram is given by a small window size inside the camera and the number of recorded interferograms is about 220. The images are $8x10mm^2$ in size and they are digitalized with a SONY 325P video camera through a Matrox image processing board. Several movies have been recorded with uniform background colour (infinite fringes), circular and narrowed fringes (finite fringes). As the optical setup is very sensitive to external vibrations, the uniform background colour is difficult to adjust when the wind tunnel is running, but the fringe formation can be observed on the hologram surface so that it is possible to adjust the uniform background colour with the wind tunnel operating. Fig. 12 shows three of twelve interferograms covering about a period of the vortex street. They are recorded in infinite fringes. The interferogram colours are well saturated and of higher contrast than those obtained in previous experiments performed with transmission holograms (see interferograms of Fig. 6).

When the background colour is uniform, it is very easy to follow the vortices emitted from the upper and lower side. For instance, if one looks at the colours coming out in the vortex cores, one can easily see that the first vortex emitted from the upper side enters a formation phase where the gas density decreases in the vortex centre. A second phase of dissipation is

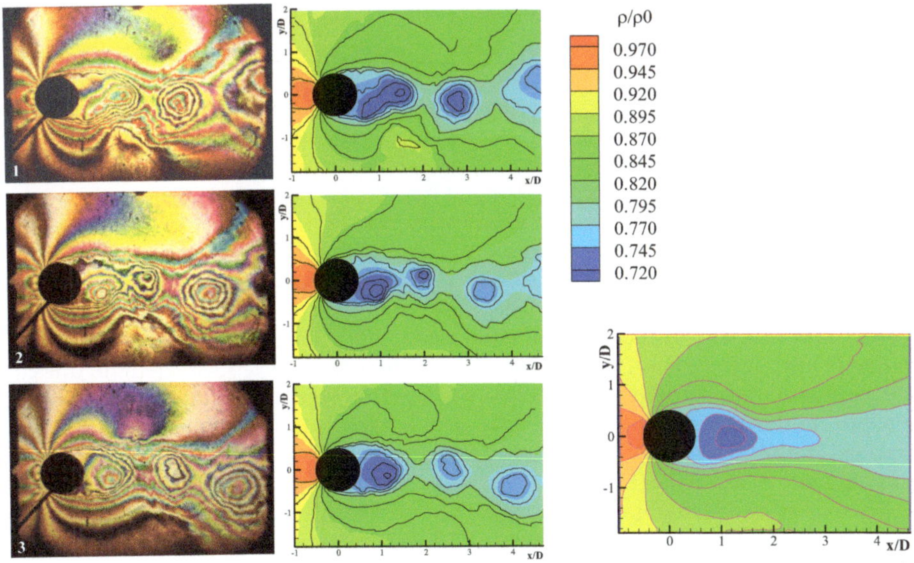

Fig. 12. Interferogram analysis : instantaneous and average gas density field – $\Delta t = 117\mu s$

observed looking at the last vortex leaving the observed field. Finally, as with transmission holographic interferometry, each colour represents a value of the gas density. In analysis, the gas density field was referred to ρ_0, the stagnation gas density. One can see that the instantaneous gas density varies from 0.70 to 0.98. The average gas density in the field has been calculated from twelve successive interferograms. The interferogram number is not very significant, but the obtained field is already symmetrical enough and the gas density varies from 0.72 to 0.97. Finally, if the colour scale of interference pattern is very well known to the user, the image of interferograms is sufficient to correctly evaluate the evolution of the gas density field.

6. Digital colour holographic interferometry

The fast development of technology, such as high resolution sensors, various DPSS lasers with large coherence, data post processing and computation power provide now opportunities to conceive new optical methods capable of simultaneous full field measurements with high spatial and temporal resolutions and giving absolute data. Digital holography with matrix sensors appeared in the last decade with cheap high resolution CCD cameras and the increasing power of computers. Image sensors have now size and spatial resolutions compatible with the needs for digital holographic recording. For example, matrices with 1636×1238 pixels of size $3.9\times3.9\mu m^2$ are now available (Yamaguchi & Zhang, 1997). In the literature, only few papers concern works in digital colour holographic interferometry. In 2002, Yamaguchi et al. & Kato et al. demonstrated phase shifting digital colour holography using a multi-wavelength HeCd continuous wave laser (636nm, 537.8nm, 441.6nm) and a colour CCD equipped with a Bayer mosaic[4]. The authors demonstrated the

[4] Matrix structure in front of a sensor to create a colour information from a panchromatic monochrome

possibility for the reconstruction of colour images but experimental results have a relatively low spatial resolution since the effective pixel number at each wavelength was 818×619, leading to an effective pixel pitch of 7.8μm. Demoli et al., 2003, presented the first study on fluids using digital colour Fourier holography. They used a monochrome CCD sensor and three wavelengths from three continuous wave lasers (647nm, 532nm, 476nm). Their results show the evolution of thermal dissipation in an oil tanker with an excellent resolution, by using the amplitude image of the reconstructed holograms. Note that historically the use of several wavelengths for holographic interferometry was first described by Jeong et al., 1997. Furthermore Ecole Popytechnique Fédérale de Lausanne's researchers (Kuhn et al., 2007) and INOA (Ferraro et al., 2007) proposed the use of several wavelengths in digital holographic microscopy. For quantitative phase microscopy, Ferraro et al., 2004, showed that severe chromatic aberration can be eliminated. Since the main effect of the chromatic aberration is to shift the correct focal image plane differently at each wavelength, this can be readily compensated by adjusting the corresponding reconstruction distance for each wavelength. In these works the recording at each wavelength is sequential. Such approach was applied in three dimensional image fusion using sequential colour recording at several distances (Javidi et al., 2005). Such strategy was also considered by J. Zhao et al., 2008, by using three laser wavelengths as an imaging approach. All these methods use sequential recording at each wavelength and generally off axis reference waves are incident on the recording area at a constant angle. At the same time, two wavelength profilometry was proposed in a digital holographic microscope using two laser diodes (680nm, 760nm) and a monochrome sensor, giving a synthetic wavelength of 6.428μm (Kuhn et al., 2007). The method is based on a spatial multiplexing by incoherent addition of single-wavelength interferograms, in the same way as was developed by Picart et al., 2003, 2005, with a single wavelength, each having different propagation directions for the reference waves, and recording with a monochrome CCD.
As regards these works, ONERA and LAUM[5] decided to join theirs respective competences acquired in the past in order to develop adaptable and new optical imaging methods, firstly having properties such as full field imaging with high spatial and temporal resolutions, secondly giving absolute data after post processing and finally giving dynamic three dimensional measurements. These non-invasive optical methods are based on digital colour holography.

6.1 Theoretical basics

In the case of Fluids Mechanics, hologram analysis performed by direct and inverse two-dimensional FFT algorithms is very well adapted for the reconstruction of transparent phase objects. For any wavelength λ, the recorded image plane hologram can be expressed as:

$$H_\lambda(x,y) = O_0(x,y) + R(x,y)O^*(x,y) + R^*(x,y)O(x,y) \tag{5}$$

where $O_0(x,y)$ and $R(x,y)$ are the zero order and the reference wave respectively and $O = b_\lambda exp(i\varphi_\lambda)$ is the object wave. For convenience, $R(x,y)$ can be represented with unit amplitude and zero phase. Subscript λ refers to one of the three colours, that is $\lambda = R, G$ or B. In the case of in-line holography, computation of the Fourier transform gives a broad spectrum centred at the zero spatial frequency. So, no relevant information can be extracted from such a spectrum. Consider now off-axis holography in which a spatial carrier is introduced along x

[5] **L**aboratoire d'**A**coustique de l'**U**niversité du **M**aine, Prof. P. Picart, Le Mans, France

or y or along the two directions. In a general case, $2\pi(u_\lambda x+v_\lambda y)$ is the spatial carrier modulation along xy, the hologram becomes:

$$H_\lambda(x,y)=O_0(x,y)+O(x,y)\exp\left[2i\pi(u_\lambda x+v_\lambda y)\right]+O^*(x,y)\exp\left[-2i\pi(u_\lambda x+v_\lambda y)\right] \quad (6)$$

By developing complex exponentials one obtains:

$$\begin{aligned}H_\lambda(x,y)=O_0(x,y)+b_\lambda(x,y)\exp\left[i\varphi_\lambda(x,y)\right]\exp\left[2i\pi(u_\lambda x+v_\lambda y)\right]\\ +b_\lambda(x,y)\exp\left[-i\varphi_\lambda(x,y)\right]\exp\left[-2i\pi(u_\lambda x+v_\lambda y)\right]\end{aligned} \quad (7)$$

Fourier transform of Eq. (7) gives:

$$\tilde{H}_\lambda(u,v)=A_\lambda(u,v)+C_\lambda(u-u_\lambda,v-v_\lambda)+C_\lambda^*(u+u_\lambda,v+v_\lambda) \quad (8)$$

where $C_\lambda(u,v)$ and $A_\lambda(u,v)$ are respectively the Fourier transform of $b_\lambda(x,y)exp[i\varphi_\lambda(x,y)]$ and $O_0(x,y)$. If the spatial frequencies u_λ and v_λ are well chosen, the three orders are well separated in the Fourier plane. Applying a binary mask around the frequency $+u_\lambda$, $+v_\lambda$, respectively of width Δu and Δv, allows the extraction of the object optical phase $\varphi_\lambda(x,y)$. Inverse Fourier transform applied to the selected region gives an estimation of the object complex amplitude and of the impulse response corresponding to the filtering applied in the Fourier domain (Desse et al., 2008).

This is quite easy to do if the hologram is not too noisy or if the optical phase $\varphi\lambda$ has a narrow spectrum. Practically, the spectral lobe of the first order diffraction can be manually selected in order to make it more reliable. When the test object is modified, for instance by a flow, this induces a modification in the refractive index along the probe beam and thus this modifies the optical path and then the optical phase. At any wavelength λ, recording a new hologram $H_\lambda(x,y)$ leads to the computation of phase. The phase change between the two states of the object is simply obtained by computing. In a setup where the test section is crossed twice, the optical path difference due to the phenomena is given by:

$$\delta=\frac{\lambda}{4\pi}\Delta\varphi_\lambda \quad (9)$$

6.2 Digital Michelson holographic interferometer

The assembly shown in Fig. 13a is very simple. It is like a conventional Michelson interferometer in which a beamsplitter cube is inserted between the spatial filter and the test section. The spatial filter is placed at the focal point of the achromatic lens so that the test section is illuminated with a parallel light beam as in previous optical setups. 50% of the light is reflected from the concave spherical mirror to form the three reference beams and 50% of light passes through the test section to form the three measurement waves. The flat mirror, placed just behind the test section, returns the beams towards the beam splitter. 25% of light is focused on the diaphragm which is placed in front of the camera lens. So, 25% of the reference beam intensities are focused in the same diaphragm by the concave mirror.

The generation of micro-fringes used as spatial carrier frequencies is shown in Fig.14. When the focal points of the reference and object waves are superimposed in the diaphragm which is placed in front of the camera, a uniform background colour is observed on the screen. If the focusing point of reference waves is moved in the plane of the diaphragm, straight

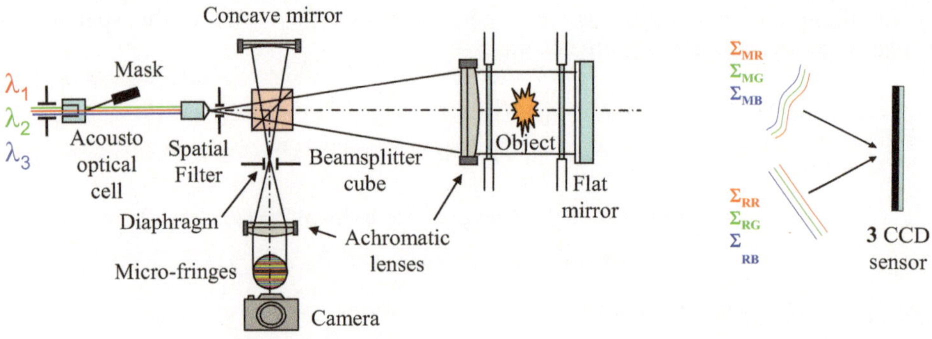

a) Digital optical bench *b) Micro-fringes recording*

Fig. 13. Digital colour holographic interferometer – Formation of spatial carrier frequencies

interference fringes are introduced into the field of visualization. These micro fringes are recorded on the CCD in order to calculate the three reference phase maps. Then the wind tunnel is started and the three object waves are distorted by the aerodynamic phenomenon. Micro-fringes interference is again recorded by the 3CCD sensor to enable calculation of the phase maps related to the object. For maps of phase difference, the reference phase is subtracted from the phase object.

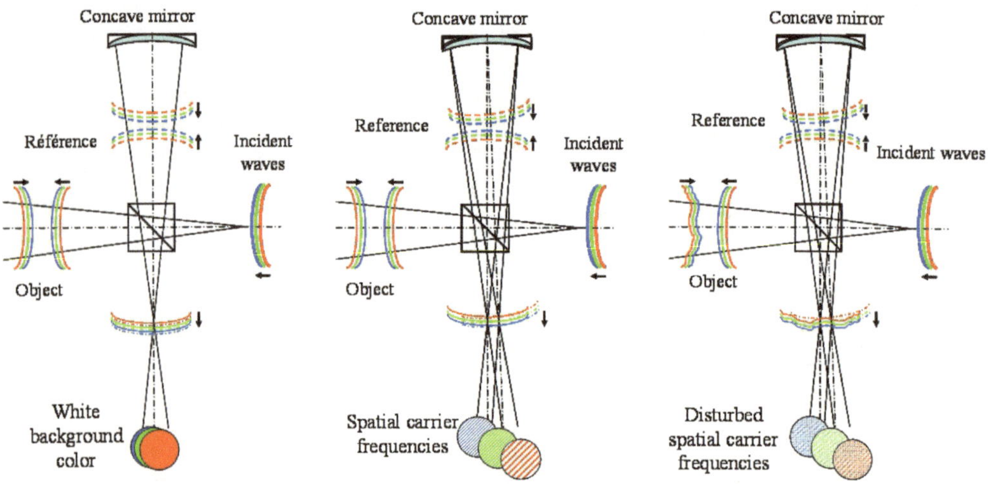

Fig. 14. Generation and micro-fringes formation by the phenomenon studied

6.3 Results obtained in subsonic wake flows

Again, near wake flow downstream from a circular cylinder has been studied at Mach 0.45. Here, one used a ORCA-3CCD Hamamatsu camera with 3 chips of 1344x1024 pixels, 6.45 µm x 6.45 µm in size. The framing rate is 9 f/s and the filters of RGB camera are very narrowband and centred on the three laser wavelengths. As the framing rate is very slow compared to the frequency of the vortex street, a transducer has been implemented in the cylinder at an azimuth of 90° (perpendicular to the flow axis) in order to synchronize the interferogram recording with the signal of the unsteady pressure measurement. The cycle of the vortex street was decomposed in eight different instants shifted by 76 μs and at each instant, five interferograms were recorded from several cycles to average the unsteady maps. First, Fig. 15 shows two micro fringes images recorded with and without the flow in order to constitute reference and object interferograms. It can be seen in the zoomed image that micro-fringes are deformed by the shear layer of the upper side.

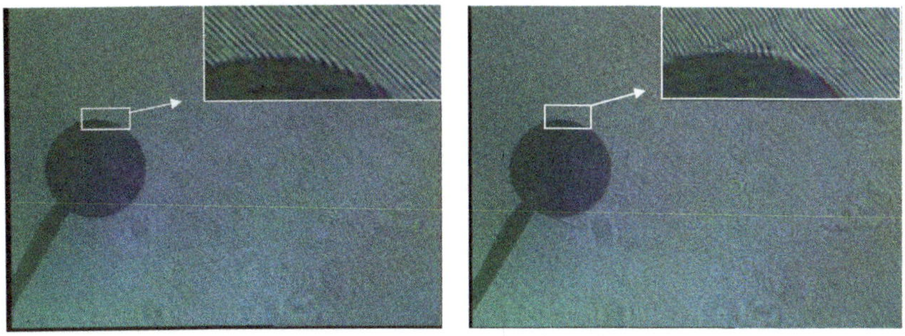

Fig. 15. Micro-fringes recording for the reference and object images

The three Fourier transforms are calculated from each image in order to reconstruct the phase maps with the +1 order (the zero order and the -1 order are filtered). An example of reference and measurement spectra is given in Fig. 16 for the green line. One can see that the spectrum only exhibits a spot corresponding to the green spatial carrier frequency. No parasitic frequencies due to the blue and red lines are found. By subtracting the reference phase maps from the measurement phase maps, one obtains the modulo2π phase difference maps. After unwrapping, it possible to compute the refractive index maps and the gas density field assuming the Gladstone-Dale relation.

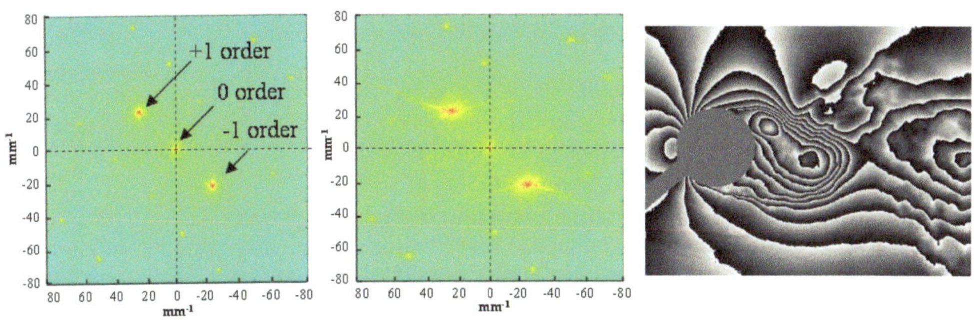

Reference and object spectra Modulo 2π phase difference

Fig. 16. FFT reference and object spectra and difference phase map for the green line

Colour interferences fringes and gas density field are shown in Fig. 17 for the first three images of one cycle of the vortex street.

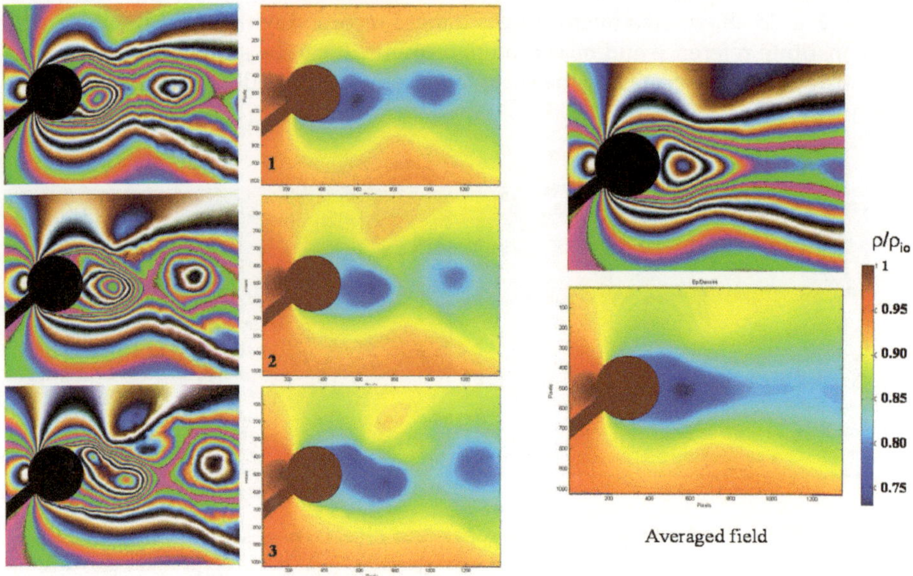

Instantaneous interferograms – Δt = 76 µs

Fig. 17. Evolution of colour interference fringes and gas density field - Mach 0.45

The intensity of the interference fringes is computed on the three channels R, G, B from the phase maps with following the following relationship :

$$I_{\lambda} = A_{\lambda}(1 + \cos(\Delta\phi_{\lambda}) \quad (10)$$

The gas density measured to the cylinder nose is particular as the gas density is equal to the stagnation gas density though the position of the vortex street, that means the colour found at this point has to be the same on each interferogram. Here, the intensity of colour interferences fringes is computed by imposing the white colour (δ = 0 µm) on each interferogram. Note this shifting is only made possible by the use of colour in the experiments. The time evolution of the gas density fields shows that the gas density decreases to 73% of ρ_{io} in the vortex core. Then, the averaged field of one cycle is calculated by averaging the 8 maps of instantaneous gas density field.

6.4 Comparison between holographic plate and digital holograms

As regards previous results obtained, silver-halide plate and digital holographic interferometry can be compared. The only possibility to compare plate and digital interferograms is to compare the interferograms displaying the interference fringes. Indeed, the technique of holographic interferometry in real time using panchromatic plates directly displays the colour density variations of the flow. It's a light intensity information that is obtained. With digital holography, the three monochromatic intensity maps are superimposed to obtain a colour map of the intensity of the interference fringes. This map can then be compared to that obtained using the reflection holographic plates. After locating an interferogram recorded at a phase very similar to that of digital interferogram, Fig. 18 shows that the correspondence is very good.

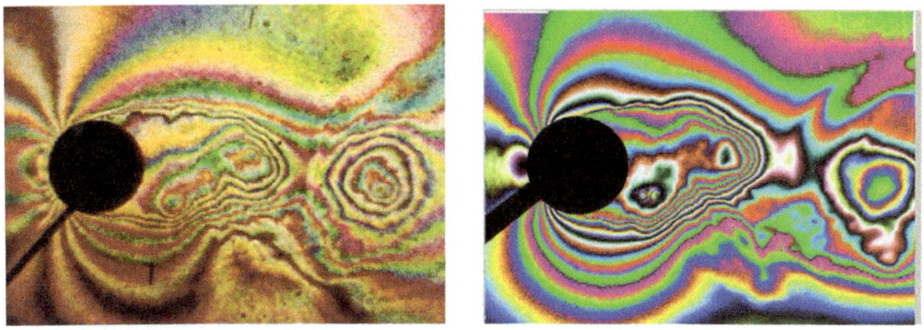

Fig. 18. Comparison between image and digital interferogram

So, in image colour holographic interferometry, a reflection panchromatic holographic plate (7,000 to 10,000 lines per mm in spatial resolution) has to be illuminated with a total energy of 600 μJ and the resetting of holographic plate is very sensitive and delicate. In digital technique, energy of 1μJ is sufficient to illuminate the sensor (155 lines per mm in resolution). The implementation is easy enough and the phase difference is entirely estimated with a computer. The coherence lengths of the three lasers must be more 2 meters in the two optical setups. In image holographic interferometry, about 220 successive frames of the phenomenon can be recorded at high framing rate (35,000 images per second with an exposure time of 750 ns for each). Each image has to be digitalized and processed. Also, it is important to obtain a reference hologram of about 50% diffraction efficiency for the three lines. In digital holographic interferometry, the framing rate is limited to 9 frames per second, full size and a synchronized triggering of interferograms recording has to be used to analyze unsteady phenomena.

7. Conclusion

The possibilities of image and digital colour holographic interferometry have been demonstrated. Colour holographic interferometry using panchromatic plates will continue to be used due to the high resolution of holographic plates. In near future, digital three-wavelength holographic interferometry seems the best candidate to characterize the future complex flows. Although CCD resolution and size are not as good as that of holographic plates, the digital approach is more accessible and versatile since the time for the hologram processing is greatly reduced and the processing is purely numerical. On the other hand, the value of using colour has been demonstrated as the zero order fringe can be easily determined and the variation in the background colour due to disturbances can be quantified. The limitations of the digital method seem to lie in the wide spectral sensitivity of the sensor which produces light diffusion in each monochromatic hologram. Work is currently in progress for removing the colour diffusion using a segmentation approach. Success in this strategy will allow increasing the spatial resolution in the reconstructed object. Future work will focus on the extension of the proposed technique for analyzing 3D unsteady wake flows. At present, a specific setup of digital holographic interferometry has been defined in a single sight direction, and the aim will be to reproduce the same optical setup along several sight directions, each shifted by a given angle. It is obvious that the optical setup can be reproduced no more than three or four times. But the lack of sight directions should be compensated by high tomographic interferogram resolution for the reconstruction of the 3D gas density field.

8. Acknowledgments

The author thanks Jean-Louis Tribillon, retired, from Délégation Générale à l'Armement and Félix Albe, retired, from Institut Franco-Allemand de Recherches de Saint-Louis for their collaboration for developing the transmission and reflection holographic interferometers. A great thank also to Professor Pascal Picart from Laboratoire d'Acoustique du Maine for the implementation of digital holographic interferometry in ONERA. The section 6 of this research has been funded from the French National Agency for Research (ANR) under grant agreement n° ANR 2010 BLAN 0302.

9. References

Bjelkhagen, H.I. & Vukicevic, D. (1992). Lippmann color holography in single-layer silver-halide emulsion, *5th International Symposium On Display Holography*, T. H. Jeong ed. Proc. SPIE Vol.2333, pp. 34-48

Bjelkhagen, H.I. (1993). Silver-halide recording-materials for holography and their processing, *Springer Series in Optical Sciences*, Vol.66, Springer Verlag New-York, ISBN 3-540-58619-9.

Bjelkhagen H.I., Jeong, T.H. & Vukicevic, D. (1996). Color reflection holograms recorded in a panchromatic ultra high-resolution single-layer silver-halide emulsion", *Journal of Imaging Sciences and Technology*, Vol.40, pp. 134-146

Bjelkhagen, H.I. & Mirlis, E. (2008). Color holography to produce highly realistic three-dimensional images, *Applied Optics*, Vol.47, pp. A123-A133

Cha, D.J. & Cha, S.S. (1996). Holographic interferometric tomography for limited data reconstruction, *AIAA Journal*, Vol.34, pp. 1019-1026

Demoli, N., Vukicevic, D. & Torzynski, M. (2003). Dynamic digital holographic interferometry with three wavelengths, *Optics Express*, Vol.11, pp. 767-774.

Desse, J.M. (1990). Instantaneous density measurement in two-dimensional gas flow by high speed differential interferometry, *Experiments in Fluids*, Vol.9, pp. 85-91

Desse, J.M. (1997a). Recording and processing of interferograms by spectral characterization of the interferometric setup, *Experiments in Fluids*, Vol.23, pp. 265-271

Desse, J.M. (1997b). Three-color differential interferometry, *Applied Optics*, Vol.36, pp. 7150-7156

Desse, J.M. (2006). Recent contribution in color interferometry and applications to high-speed flows, *Optics and Lasers in Engineering*, Vol.44, pp. 304-320

Desse, J.M., Picart, P. & Tankam, P. (2008). Digital three-color holographic interferometry for flow analysis, *Optics Express*, Vol.16, pp. 5471-5480

Desse, J.M., Picart, P. & Tankam, P. (2010). Digital Three-Color Holographic Interferometry Devoted to Fluid Mechanics, *Optical metrology*, Speckle 2010 7387, 73870S1-73870S6

Ferraro, P., De Nicola, S., Coppola, G., Finizio, A., Alfieri, D. & Pierattini, G. (2004). Controlling image size as a function of distance and wavelength in Fresnel-transform reconstruction of digital holograms, *Optics Letters*, Vol.29, pp. 884-886

Ferraro, P., Miccio, L., Grilli, S., Paturzo, M., De Nicola, S., Finizio, A., Osellame, R. & Laporta, P. (2007). Quantitative phase microscopy of microstructures with extended measurement range and correction of chromatic aberrations by multi wavelength digital holography, *Optics Express*, Vol.15, pp. 14591-14600

Fomin, N. (1998). Speckle photography for fluid mechanics measurements, *Springer*, Berlin Heidelberg New York, ISBN: 978-3-540-63767-7

Fomin, N., Lavinskaya, E. & Vitkin, D. (2002). Speckle tomography of turbulent flows with density fluctuations, *Experiments in Fluids*, Vol.33, pp. 160-169

Harthong, J., Sadi, J., Torzynski, M. & Vukicevic, D. (1997). Speckle phase averaging in high-resolution color hologhraphy, *Journal of the Optical Society of America A*, Vol.14, pp. 405-410.

Hubel, P.M. (1991). Recent advances in color reflection holography, *Practical Holography V*, S.A. Benton ed. Proceeding SPIE Vol.1461, pp. 167-174

Javidi, B., Ferraro, P., Hong, S., De Nicola, S., Finizio, A., Alfieri, D. & Pierattini, G. (2005). Three-dimensional image fusion by use of multiwavelength digital holography, Optics Letters, Vol.30, pp. 144-146

Jeong, T.H., Bjelkhagen, H.I. & Spoto, L.M. (1997). Holographic interferometry with multiple wavelengths, *Applied Optics*, Vol. 36, pp. 3686-3688

Joannes, L., Dupont, O., Dubois, F., Colinet, P. & Legros, J.C. (2000). Interferometric optical tomography for 3-dimensional investigation of liquids, In: Carlomagno GM, Grant I (eds) CD ROM *Proceedings of the 8th International Symposium on Flow Visualization*, Edinburgh, Heriot Watt, Scotland, pp. 428.1-428.9

Kato, J., Yamaguchi, I. & Matsumura, T. (2002). Multicolor digital holography with an achromatic phase shifter, *Optics Letters*, Vol.27, pp. 1403-1405.

Kim, J.M., Choi, B.S., Choi, Y.S., Kim, J.M., Bjelkhagen, H.I. & Phillips, N.J. (2002). Holographic Optical Elements Recorded in Silver Halide Sensitized Gelatin Emulsions. Part 2. Reflection Holographic Optical Elements, *Applied Optics*, Vol.41, pp. 1522-1533

Kuhn, J., Colomb, T., Montfort, F., Charrière, F., Emery, Y., Cuche, E., Marquet, P. & Depeursinge, C. (2007). Real-time dual-wavelength digital holographic microscopy with a single hologram acquisition, *Optics Express*, Vol.15, pp. 7231-7242

MacAdam, D.L. (1985). Color Measurement, *Springer Series in Optical Sciences*, Vol.27, Springer Verlag Berlin Heidelberg, ISBN 0-387-15573-2

Merzkirch, W. (1965). A simple schlieren interferometer system, *AIAA Journal*, Vol.3, pp. 1974-1976

Merzkirch, W. (1974). Flow visualization, *Academic Press New York*, ISBN 0-12-491350-4

Picart, P., Moisson, E., Mounier, D. (2003). Twin sensitivity measurement by spatial multiplexing of digitally recorded holograms, *Applied Optics*, Vol.42, pp. 1947-1957.

Picart, P., Leval, J., Grill, M., Boileau, J.P., Pascal, J.C., Breteau, J.M., Gautier, B. & Gillet, S. (2005). 2D full field vibration analysis with multiplexed digital holograms, *Optics Express*, Vol.13, pp. 8882-8892

Pellicia-Kraft, B.J. & Watt, D.W. (2000). Three-dimensional imaging of a turbulent jet using shearing interferometry and optical tomography, *Experiments in Fluids*, Vol.29, pp. 573-581

Pellicia-Kraft, B.J. & Watt, D.W. (2001). Visualization of coherent structure in scalar fields of unsteady jet flows with interferometric tomography and proper orthogonal decomposition, *Experiments in Fluids*, Vol.30, pp. 633-644

Petrova, T.S., Ivanov, B., Zdravkov, Z., Nazarova, D., Stoykova, E., Minchev, G. & Sainov, V. (2009). Basic Holographic Characteristics of a Panchromatic Light Sensitive Material for Reflective Autostereoscopic 3D Display, *EURASIP Journal on Advances in Signal Processing*, Vol. 2009, Article ID 217341

Philbert, M. (1958). Emploi de la strioscopie interférentielle en aérodynamique, *La Recherche Aérospatiale*, Vol.65, pp. 19-27

Rastogi, P.K. (1994). (Ed.) Holographic interferometry, principles and methods, *Springer Series in Optical Sciences*, Vol.68, Springer Verlag Berlin Heidelberg, ISBN 0-387-57354-2

Sasomov, Y., Kumonko, P., Ratcliffe, D., Skokov, G. & Grichine, M. (1999). Recent advances in holographic materials from Slavich, *Holographic materials V*, T.J. Trout ed. Proceeding SPIE Vol. 3638, pp. 42-53

Settles, G.S. (2001). Schlieren and shadowgraph techniques, *Springer-Verlag*, Berlin Heidelberg New-York, ISBN 3-540-66155-7

Smeets, G. ((1975). Observational techniques related to differential interferometry, In: *Proceeding of 11th International Congress on High Speed Photography*, Chapman & Hall, UK, pp. 283-288

Smigielski, P., Fagot, H. & Albe, F. (1976). Application de l'holographie ultra rapide à référence arrière à l'étude de déformations dynamiques, *Proceeding of the 12th International Congress of High Speed Photography*, Toronto

Surget, J. (1973). Etude quantitative d'un écoulement aérodynamique, *La Recherche Aérospatiale*, Vol.3, pp. 167-171

Timmerman, B. & Watt, D.W. (1995). Tomographic high-speed digital holographic interferometry, *Measurement Science and Technology*, Vol.6, pp. 1270-1277

Vest, C.M. (1979). Holographic interferometry, *Wiley-Interscience*, ISBN 0-471-90683-2, New-York

Vikram, C.S. & Witherow, W.K. (1992). Critical needs of fringe order accuracy in two-colour holographic interferometry", *Experiments in Mechanics*, pp. 74-77

Yamaguchi, I. & Zhang, T. (1997). Phase-shifting digital holography, *Optics Letters*, Vol.22 pp. 1268-1270

Yamaguchi, I., Matsumura, T. & Kato, J. (2002). Phase-shifting color digital holography, *Optics Letters*, Vol.27, pp. 1108-1110.

Yan, D. & Cha, S.S. (1998). Computational and interferometric system for real-time limited-view tomography of flow fields, *Applied Optics*, Vol.37, pp. 1159-1164

Zhao, J., Jiang, H. & Di, J. (2008). Recording and reconstruction of a color holographic image by using digital lensless Fourier transform holography, Optics Express, Vol.16, pp. 2514-2519.

3

Digital Holographic Interferometric Characterization of Optical Waveguides

Hamdy Wahba and Mamdouh Shams El-Din
Physics Department, Faculty of Science, 34517 Damietta,
University of Mansoura
Egypt

1. Introduction

Holography was developed by Dennis Gabor 1947. Gabor presented holography as a lensless process for image formation by reconstructed wavefronts (Gabor, 1948, 1949, 1951). Holography can be defined as a method for recording and reconstructing whole optical wavefields, which means intensity and phase (Gabor, 1948, 1949), thus it exhibits 3D characteristics like depth of field or parallax. Holographic interferometry (Powell & Stetson, 1965) is a very effective non-destructive, contactless tool to measure shape, deformation or refractive index distributions (Kreis, 1996).

In 1994 the modern digital holography was introduced (Schnars, 1994; Schnars, & Jüptner, 1994a; Schnars, & Jüptner, 1994b; Schnars et al., 1995; Schnars, & Jüptner, 2005). The digital holography can be defined as digital recording of the holograms and the numerical reconstruction of the wave fields in a computer, where, the charge coupled devices (CCDs) are the most frequently used devices to record the holograms. The digital holography in the last ten years was involved in a lot of applications due to the development of powerful computers, ultra large memories and smaller pixel size CCD targets (Kreis, 2005).

The phase shifting interferometric (PSI) technique was introduced by Hariharan et al. into the field of holography as an accurate method for real time fringe measurement (Hariharan et al., 1982). Furthermore, PSI combined with digital holography (Skarman et al., 1996). Yamaguchi and Zhang 1997 improved phase shifting digital holography (PSDH) (Yamaguchi & Zhang, 1997). The phase difference using PSDH is measured with an accuracy of $2\pi/200$ (Hariharan, 2002).

For a long time, the determination of refractive index distributions in fibres, optical waveguides or other transparent solids was performed by interferometric methods. Two-beam and multiple beam based interferometers were used as a non-destructive tool to determine the optical parameters of fibres (Faust, 1952, 1954; Marhic et al., 1975). While, the mathematics used in (Marhic et al., 1975; Saunder & Gardner, 1977; Barakat et al., 1985) neglect the non-straightforward refraction of the light beam inside the fibres. Hamza et al. (Hamza et al., 1994, 1995) constructed an accurate mathematical model (multilayer model) which considered the exact local refraction of the incident beam on its way through the graded index optical fibre, which is divided into a large number of thin concentric layers of constant refractive index. This model was verified with two-beam and multiple-beam interferometers. The consideration of incident beam refraction gave a better accuracy in the determination of the optical parameters of graded index optical fibres (Hamza et al., 1995, 2001), thick optical fibres (Hamza et al., 2004) and mechanically stressed optical fibres (Sokkar et al., 2008) than without this consideration. Automated Fizeau interferometric techniques were applied in studies of the optical and optothermal properties of fibres (El-Morsy et al., 2002a, 2002b; Hamza et al., 2007). These cited developments substantially increased the accuracy of the measured optical fibre parameters.

In addition, digital holographic interferometry (DHI) is used for determining refractive index profiles. Digital holographic microscopy (DHM) was used to measure the mean integral refractive index and thickness of living cells (Rappaz et al., 2005). The absolute accuracy of the mean refractive index measurement was about 0.0003. Also, Kemper et al. (Kemper & Carl, 2006; Kemper et al., 2007) used DHI to measure the refractive index and thickness of living cells. Kebbel et al. applied digital holography to refractive index variations within transparent media in microgravity experiments (Kebbel et al., 1999). Two-dimensional refractive index profiles of phase gratings have been investigated using DHI (De Angelis et al., 2006) as well as the refractive indices of liquids using lensless Fourier DHI (Hossain et al., 2006). The high performance of DHI was tested by measurements of low variation refractive indices of fluids in a comparative study with other techniques (Dubois et al., 1999; Owen & Zozulya, 2002) like traditional Mach-Zehnder interferometry. The coupling of digital holographic microscopy and polarization imaging digital holography was demonstrated in an investigation of induced birefringence in non striped bent optical fibres and the birefringence of stressed PMMA (Cuche et al., 1999; Colomb et al., 2002, 2005). The mathematics describing the refractive index of transparent materials used in (Dubois et al., 1999; Owen & Zozulya, 2002; De Angelis et al., 2006; Hossain et al., 2006) cannot be used directly to measure and configure the refractive index profile of GRIN optical waveguides, since it assumes a constant refractive index along the light path in a material. So it can be used only to determine the mean refractive index of the GRIN optical waveguides but it does not consider the varying refraction of the beam along its path inside the fibre. Large scale strongly refracting fields produce ray bending. This effect was recognized in holographic interferometric investigations combined with iterative calculations (Sweeney & Vest, 1973) as well as tomographic methods (Cha & Vest, 1981).

Recently, digital holographic phase shifting interferometry with the aid of mathematical models, which consider the refraction of the incident rays, were used to investigate some optical parameters of fibrous materials (Wahba & Kreis, 2009a, 2009b, 2009c; Yassien et al., 2010). The refractive index profile and optical parameters of graded index fibres as well as the refractive index profile of bent optical fibres were determined with accuracy 2.3×10^{-4}.

Organic or inorganic optical waveguides are made of transparent dielectric materials, e.g. graded index (GRIN) planar optical waveguides and GRIN optical fibres. The fabrication of integrated optics is a very crucial technology for optical communications and sensing devices in the near future. However, it is absolutely necessary to have accurate knowledge about the optical waveguide parameters to use it in technological applications.

In recent times, polymeric integrated optics has shifted to the focus of interest due to the low material cost and simple handing and processing of polymers during the manufacturing process. Polymethyl methacrylate (PMMA) is a very promising material to be used as a basic material to produce optical waveguides and gratings (Baker & Dyer, 1993; Eldada & Shacklette, 2000; Vollertsen & Wochnowski, 2004).

The UV-laser lithographic fabrication of integrated optical waveguides has been well-recognized in the last years. The UV excimer laser has been used to irradiate homogeneous slabs of PMMA in order to modify the refractive index locally in a controllable way at the polymeric substrate surface (Wochnowski et al., 2000; Shams-Eldin et al., 2004, 2005). Thus the integrated optical waveguides can be directly laser-written into the surface of polymeric substrates by employing lithographic masks. It is very important that the refractive index is modified in a controllable way by using specific conditions during the irradiation process in order to obtain the desired waveguide structures (Wochnowski et al., 2000). The UV photons of the incident laser radiation do interact with the molecules of the polymeric material due to a UV-photon-induced reaction (Shams-Eldin et al., 2005), so the refractive index at the surface of the PMMA slabs was changed in the irradiated area. Consequently, the refractive index modification in the irradiated area appears as a smooth variation, which depends on the waveguide depth.

The mode field distribution across the GRIN optical waveguides is very important in applications. In the case of GRIN optical waveguides, the variation in refractive index can be presented as a function of the form of a complementary error function, exponential, Gaussian, parabolic, or shape not described by any simple mathematical function (Chiang, 1985; Mathey & Jullien, 1996). To use such optical waveguides in telecommunication, it is useful to know the number of propagated modes and their propagation constants for any given profile. The number of modes depends on two factors; the first one is the difference between maximum and minimum refractive indices of the guiding region, the second one is the radius of the guiding region. Eguchi and Horinouchi (Eguchi & Horinouchi, 2004) used the finite-element method to determine the number of modes of optical fibres. Shemirani et al. (Shemirani et al., 2009) developed a field-coupling model for propagation in a graded index optical fibre, analogous to the principal states model for polarization mode dispersion in the single mode fibre. This model was based on the concept of the first order principal modes, which have well defined group delays that depend on the strength of the mode coupling. This first-order model predicts a linear relationship between the intensity distributions at the input and output. This model is extended to account for higher order modal dispersion (Shemirani & Kahn, 2009).

In this Chapter, digital holographic phase shifting interferometry is used to investigate the optical properties of graded index optical waveguides, e.g. graded index planar optical waveguides and optical fibres. A planar optical waveguide sample is prepared by the UV laser lithographic method. The approach of digital holography with the aid of phase shifting interferometry is applied. The reconstructed optical phase differences along the GRIN optical waveguides are extracted and then a simple algorithm is used to avoid the problem of tilted GRIN optical waveguides inside the optical field. The extracted optical phase differences due to the optical waveguides are aligned to be perpendicular to the x-axis. After this process, the mean optical phase differences across the samples are calculated and thus the errors in the calculated mean values of optical phase differences across the optical waveguides are reduced. The proposed method is used to construct the refractive index profiles across GRIN optical waveguides with the aid of the multilayer model. Also, a simple algorithm is used to reconstruct the 3D refractive index of the GRIN optical fibre considering the symmetrical distribution of the GRIN optical fibre layers. In addition, an analytical method is presented to calculate the effective indices and the mode field distribution across the symmetric and asymmetric GRIN optical waveguides. The digital holographic phase shifting interferometric approach affects the accuracy of the calculated parameters.

2. Digital holographic phase shifting interferometry

In digital holography the recorded wavefield is reconstructed by multiplying the stored hologram values by the complex conjugate of a numerical model of the reference wave $r^*(\zeta, \eta)$ and then calculating the resulting diffraction field $b'(x', y')$ in the image plane, see Fig. 1 (Kreis, 1996, 2005; Schnars & Jüptner, 2005). The numerical reconstruction of the digitally recorded hologram is defined as a numerical calculation of the phase and the intensity of the recorded wavefield. This is theoretically calculated by the diffraction integral;

$$b'(x',y') = \frac{1}{i\lambda} \int h \int (\zeta,\eta) r^*(\zeta,\eta) \frac{\exp\{ik\rho\}}{\rho} \, d\zeta \, d\eta, \tag{1}$$

with $\rho = \sqrt{d'^2 + (\zeta \ x')^2 + (\eta \ y')^2}$ and k is the wave number ($k=2\pi/\lambda$).

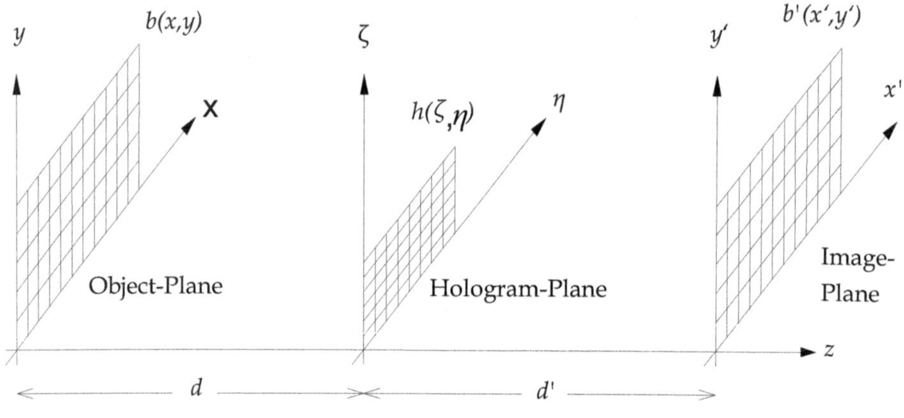

Fig. 1. Geometry of digital Fresnel holography.

The finite discrete form of the Fresnel approximation to the diffraction integral is

$$\begin{aligned} b'(n\Delta x', m\Delta y') &= A \sum_{j=0}^{N\ 1} \sum_{l=}^{M\ 1} h(j\Delta\zeta, l\Delta\eta) r^*(j\Delta\zeta, l\Delta\eta) \\ &\times \exp\left\{ \frac{i\pi}{d'\lambda} \left(j^2 \Delta\zeta^2 + l^2 \Delta\eta^2 \right) \right\} \exp\left\{ 2i\pi \left(\frac{jn}{N} + \frac{lm}{M} \right) \right\}. \end{aligned} \tag{2}$$

The parameters used in this formula for calculating the complex field in the image plane are given by the CCD array used, having $N \times M$ pixels of pixel pitches $\Delta\zeta$ and $\Delta\eta$ in the two orthogonal directions. The stored hologram is $h(j\Delta\zeta, l\Delta\eta)$. The distance between the object and the CCD is denoted by d, and normally $d'=d$. Complex factors not depending on the hologram under consideration are contained in a given specific CCD.

The pixel spacing in the reconstructed field is

$$\Delta x' = \frac{d'\lambda}{N\Delta\zeta} \qquad and \qquad \Delta y' = \frac{d'\lambda}{M\Delta\eta} \tag{3}$$

An alternative to the Fresnel approximation uses the fact that Eq. (1) describes a convolution of $h(\zeta, \eta)r^*(\zeta, \eta)$, with the impulse response $g(x', y', \zeta, \eta)=(exp\{ik\rho\})/i\lambda\rho$. The convolution theorem states that b' is given by

$$b' = A'F^{-1}\{F\{h.r^*\}.F\{g\}\}\ , \tag{4}$$

where F denotes the Fourier transform and F^{-1} is its inverse. In practice, both F and F^{-1} are calculated by the fast Fourier transform algorithm. The resulting pixel spacing (Kreis, 2005) for this convolution approach is

$$\Delta x' = \Delta\zeta \qquad and \qquad \Delta y' = \Delta\eta \tag{5}$$

The use of a real hologram in the Fresnel reconstruction or the convolution reconstruction, leads to a strong d.c. term, a focused real image, and a virtual image that is not sharp. The complex field can be recorded and calculated by phase-shifting digital holography. The calculated complex wavefield is used instead of a real hologram in the convolution approach to overcome the problems of the d.c. term and twin image. For this purpose several holograms, at least three, with known mutual phase shifts are recorded. These holograms are given by

$$I_n = a(\zeta,\eta) + b(\zeta,\eta)\cos(\varphi(\zeta,\eta) + \varphi_{Rn}), \qquad n = 1,2,3,..... \tag{6}$$

where $a(\zeta, \eta)$ and $b(\zeta, \eta)$ are the additive and the multiplicative distortions and φ_{Rn} is the phase shift performed in the reference wave during recording of the holograms. In our case the phase shift is 90°, and it starts with $\varphi_{Rn} = 0^o$. In this case we get a set of four linear equations that are point wise solved by a Gaussian least squares method(Kreis, 1996). The complex wavefield in the hologram plane can be calculated from

$$H(\zeta,\eta) = [I_1(\zeta,\eta) - I_3(\zeta,\eta)] + i[I_4(\zeta,\eta) - I_2(\zeta,\eta)] \tag{7}$$

Finally, the reconstruction process is based on the use of the complex wavefield in the convolution algorithm. The intensity distribution in the reconstruction plane is given by

$$I(x',y') = |b'(x',y')|^2\ , \tag{8}$$

and the phase distribution is given by

$$\varphi(x',y') = \arctan\left\{\frac{\mathrm{Im}|b'(x',y')|}{\mathrm{Re}|b'(x',y')|}\right\} \tag{9}$$

Then, the optical phase difference due to the used phase object such as GRIN optical waveguides can be extracted.

3. Refractive index profile of GRIN optical fibre using the multilayer model

The GRIN optical fibre sample can be considered as a phase object. When it is implemented in the path of one of the interfering beams of a Mach-Zehnder interferometer, the produced fringes suffer shifts in the GRIN optical fibre region. These shifts represent the optical path difference between the light beam traversing the fibre and that traversing the surrounding medium.

In GRIN optical fibre, the core is surrounded by a homogeneous cladding of refractive index n_{cl}. In the multilayer model, the core and cladding of the graded index optical fibre are assumed to consist of a large number of thin layers. Each layer has an annular cross section and is considered to have a thickness a and refractive index n_j where the layers are numbered by $j = 1, \ldots, N$ with $N = R/a$. R is the core radius. If this optical fibre is immersed in a liquid of refractive index n_L, there will be a refraction of the incident beam at the boundary between the liquid and the cladding. Hamza et al. obtained a recurrence relation which predicts the optical path of the refracted beam through Q layers of the fibre. The GRIN fibre is illuminated by a collimated beam, with a ray crossing the centre of the core defining the optical axis (Hamza et al., 1995). An arbitrary beam transverses the fibre at a distance d_Q from the optical axis and leaves the fibre at a distance x_Q. Assume a coordinate system whose origin is at the fibre centre, then the corresponding fringe shift is defined as Z_Q and the optical path difference δ_Q is given by the recurrence formula

$$\frac{\lambda Z_Q}{h} = \frac{\lambda}{2\pi}\Delta\phi_Q = \sum_{j=1}^{Q-1} 2n_j\left(\sqrt{(R-(j-1)a)^2 - d_Q^2 n_L^2 / n_j^2} - \sqrt{(R-ja)^2 - d_Q^2 n_L^2 / n_j^2}\right) \\ + 2n_Q\left(\sqrt{(R-(Q-1)a)^2 - d_Q^2 n_L^2 / n_Q^2}\right) - n_L\left(\sqrt{R^2 - d_Q^2} + \sqrt{R^2 - x_Q^2}\right), \qquad (10)$$

with Q running from 1 to N, h is the interfringe spacing and $\Delta\varphi_Q$ is the interference phase difference.

This recurrence relation also describes the shape of the fringes which is produced as a result of the refraction of the incident beam through the Q layers of the fibre. Without loss of generality, we can assume that the incident light beam passes through the middle of the Qth layer. Thus the value of d_Q in Eq. (10) can be obtained from the following relation

$$d_Q = \frac{n[R-(Q-0.5)a]}{n_L} \qquad (11)$$

Eq. (10) can now be used to get the refractive index profile for GRIN optical fibres ($N \rightarrow \infty$). Hamza et al. have shown that inhomogeneous refraction must be considered in the measurement and calculation of optical fibre parameters, so that the accuracy of the measured parameters of GRIN optical fibre is increased (Hamza et al., 1994, 1995).

4. Characterization of TE modes in the GRIN optical waveguides

The study of the GRIN optical waveguides using the wave equation to achieve the allowed propagating modes is familiar from basic quantum mechanics. The analogy between the particle in box and optical waveguide problem is very strong: both situations describe waves, which are confined between two reflecting boundaries. In both cases, the waves partially tunnel into the surrounding potential barrier before turning around. Only certain allowed energies (in the case of the particle), or propagation coefficients β (in the case of the optical waveguide) create a standing wave in one-dimensional system.

4.1 Symmetric optical waveguide

The squared value of the refractive index distribution at the core of the GRIN optical waveguide is shown in Fig. 2. At the centre of the core the refractive index has a maximum value n_o and then decreases to reach n_{clad} at $x=\pm R$ (R is the radius of the core). The squared value of the refractive index distribution is given by,

$$n^2(x) = \begin{Bmatrix} n_{clad}^2\,, & -R > x,\ R < x \\ \left(n_o - \Delta n\left(\dfrac{x}{R}\right)^2\right)^2, & -R < x > R \end{Bmatrix} \quad , \tag{12}$$

where $\Delta n = n_o - n_{clad}$.

The guided wave problem requires only the knowledge of the square of the refractive index distribution.

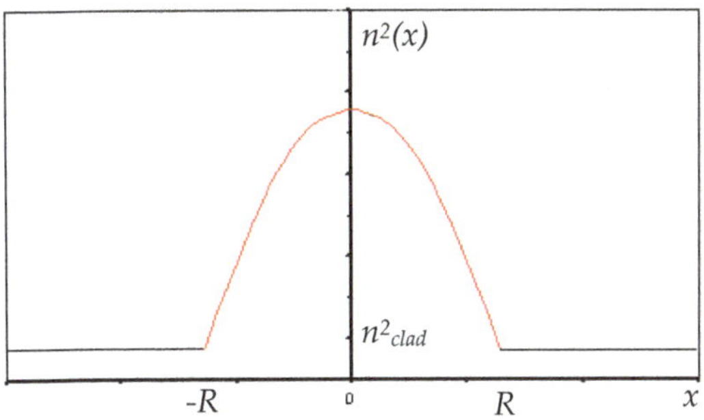

Fig. 2. The core of the GRIN optical fibre is confined between cladding media, arranged such that the squared refractive index $n^2(x)$ is larger than that of the surrounding media $\left(n_{clad}^2\right)$.

In the TE case, the E field is polarized along the y- axis and propagating along the z- axis. The TE modes of the GRIN optical waveguide are obtained as the solutions of the following equations (Conwell, 1973; Marcuse, 1973),

$$\frac{d^2E_y}{dx^2} + \left(k_o^2 n^2(x) - \beta^2\right)E_y = 0 \quad , \tag{13}$$

$$-\frac{i}{w\mu_o}\frac{\partial E_y}{\partial z} = H_x \quad , \tag{14}$$

$$\frac{i}{w\mu_o}\frac{\partial E_y}{\partial x} = H_z \quad , \tag{15}$$

where

$$k_o^2 = \omega^2 \varepsilon_o \mu_o \tag{16}$$

k_o, μ_o, ω and β are the wave number, the permeability constant, the angular frequency and the allowed propagation coefficient along the z-axis, respectively.
Applying the continuity boundary conditions that connect the solutions at the interfaces $x = R$ and $x = -R$ and for $\Delta n \ll n_o$, the solutions of the Eqs. (13)-(15) in the three regions (symmetric waveguide) are given in terms of Hermite polynomials (H_q) of degree q, as

$$E_y(x) = \begin{cases} A e^{-\frac{k_o (\Delta\varepsilon)^{\frac{1}{2}} R}{2}} H_q\left(R\sqrt{\frac{k_o (\Delta\varepsilon)^{\frac{1}{2}}}{2R}}\right) e^{-\gamma(x-R)} & , \ x > R \\ A e^{-\frac{k_o (\Delta\varepsilon)^{\frac{1}{2}} x^2}{2R}} H_q\left(\sqrt{\frac{k_o (\Delta\varepsilon)^{\frac{1}{2}}}{2R}}\, x\right) & , \ -R < x < R \\ A e^{-\frac{k_o (\Delta\varepsilon)^{\frac{1}{2}} R}{2}} H_q\left(-R\sqrt{\frac{k_o (\Delta\varepsilon)^{\frac{1}{2}}}{2R}}\right) e^{\gamma(x+R)} & , \ x < -R \end{cases} \tag{17}$$

where q is an integer that identifies the mode. H_q is the appropriate Hermite polynomial defined by

$$H_q(x) = (-1)^q e^{x^2} \frac{d^q}{dx^q} e^{-x^2} \tag{18}$$

The first three Hermite polynomials in x are

$$\begin{aligned} H_0(x) &= 1, \\ H_1(x) &= 2x, \\ H_2(x) &= 4x^2 - 2 \end{aligned} \tag{19}$$

Also γ in Eq. (17) is the attenuation coefficient in the clad region and $\Delta\varepsilon = 2\, n_o \Delta n$. The allowed propagation coefficient and the effective index for every mode are given respectively by

$$\beta = \sqrt{k_o^2 \varepsilon_o - \frac{k_o \sqrt{\Delta\varepsilon}}{R} (1 + 2q)} \tag{20}$$

and

$$n_{eff} = \frac{\beta}{k_o} . \tag{21}$$

where ε_o is the dielectric constant at the center of the symmetric optical waveguide, and it is related to the refractive index by $\varepsilon_o = n_o^2$.

4.2 Asymmetric optical waveguide

On the other hand, to find the allowed TE modes and the propagation coefficients of the asymmetric GRIN optical waveguide, we assume that, the GRIN optical waveguide, as shown in Fig. 3, consists of two graded index zones; both have a Gaussian refractive index profile. The first zone is in the range $0 < x < b_1$, and the second one is in the range $b_2 < x < b_4$. There is a very thin layer of constant refractive index n_{s1} confined between the two zones in the range $b_1 < x < b_2$. The plane at $x=0$ represents the interface between the surface of the first zone of the optical waveguide and the immersion liquid (of refractive index n_L) and the plane at $x = b_4$ represents the interface between the second zone of the waveguide and the substrate, which has refractive index n_s.

4.2.1 TE modes in the first zone of waveguide

The refractive index profile for each zone is fitted using a Gaussian function which could be used to determine the allowed TE modes for the optical waveguide sample. The first zone of the waveguide in the range $0 < x < b_1$ is represented by the Gaussian refractive index profile given by

$$n(x) = n_{s_1} + \Delta n_1 \, e^{-\left(\frac{x}{b}\right)^2} \quad , \qquad 0 < x < b_1 \tag{22}$$

where b is the depth of the first zone (between 0 and b_1), and Δn_1 is the difference between the refractive indices at $x = 0$ and $x = b_1$. The conditions for the wave propagation along the first zone of the waveguide are: (1) a zigzag of the beam path and (2) the total phase change must be a multiple of 2π. Using WKB approximation (Marcuse, 1973; Mathey, 1996), the lateral resonance condition for the first zone is given by

$$2\int_0^{b_1} \sqrt{k_o^2\, n^2(x) - \beta^2}\, dx - 2\Phi_{1/2} - 2\Phi_{1/3} = 2m\pi \, , \quad 1 \le m \le M \tag{23}$$

where $\beta = k_o n_{eff}$ is the propagation constant of the mth mode; m is an integer. M is the total number of modes, $k_o = 2\pi/\lambda$ is the wave-vector, λ is the wavelength, and n_{eff} is the effective

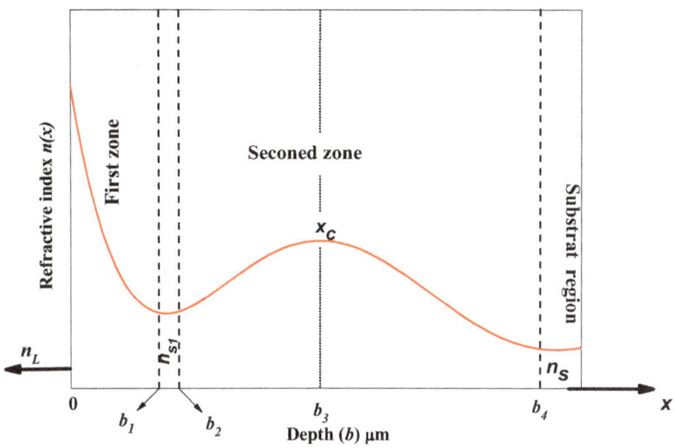

Fig. 3. The waveguide confined between liquid and substrate media, arranged such that the refractive index $n(x)$ is larger than that of the surrounding media (n_L) and (n_s).

mode index for light that propagates along the optical waveguide. $\Phi_{1/2}$ and $\Phi_{1/3}$ are the phase changes at the film-liquid interface and at the first and second zones interface, respectively. Generally these last quantities are approximated to $\Phi_{1/2} \approx \pi/2$ and $\Phi_{1/3} \approx \pi/4$ (Mathey, 1996). The effective indices (n_{eff}) for the first zone are obtained numerically by substituting Eq. (22) into Eq. (23).

4.2.2 TE modes of second zone waveguide

Referring to Fig. 3, we suppose that, without loss of generality, the second region of the waveguide in the range $b_2 < x < b_4$ is divided into two asymmetric parts which have an individually Gaussian refractive index profile. The first and the second parts within the ranges are $b_2 < x < b_3$ and $b_3 < x < b_4$, respectively. The refractive index profile in these two parts of the second zone of the waveguide is given by

$$n(x) = \begin{cases} n_{s_1} + \Delta n_2 \, e^{-\left(\frac{x - x_c}{b}\right)^2} & , b_2 < x < b_3 \\ n_{s_2} + \Delta n_3 \, e^{-\left(\frac{x - x_c}{b}\right)^2} & , b_3 < x < b_4 \end{cases} , \quad (24)$$

where b is the irradiation depth for both, the first part between b_2 and b_3, and the second part between b_3 and b_4. Δn_2 is the difference between the refractive indices at the interface between the first and the second parts at $x = b_3$ and that at $x = b_2$, whereas Δn_3 is the difference between the refractive indices at the interface between the first and the second parts at $x = b3$ and that of the substrate at $x = b4$.
Similarly, by analogy with Eq. (23) we have for the first and second parts

$$2 \int_{b_2}^{b_3} \sqrt{k_o^2 \, n^2(x) - \beta^2} \, dx - 2\Phi_{1/3} = 2m\pi , \quad 1 \le m \le M \quad (25)$$

$$2 \int_{b_3}^{b_4} \sqrt{k_o^2 \, n^2(x) - \beta^2} \, dx - 2\Phi_{1/3} = 2m\pi , \quad 1 \le m \le M \quad (26)$$

since $\Phi_{1/2} = 0$ at x_c.
The effective indices (n_{eff}) for the two parts of the second zone are obtained numerically by using Eq. (24) in Eqs. (25) and (26).

5. Digital holographic set-up

The digital holographic setup is a Mach-Zehnder system, see Fig. 4. The optical waveguide sample is immersed in a liquid of refractive index n_l near the cladding refractive index n_{clad} of the GRIN optical fibre or the refractive index of the substrate of the waveguide sample. An Abbe refractometer with an accuracy of ±0.0001 is used to measure the refractive indices of the immersion liquids. The collimated laser beam crosses the sample and passes through the microscope objective MO1 with magnification 10x and N.A. 0.25. An identical microscope objective MO2 is installed in the reference arm to eliminate the curvature of the optical field. The position of MO2 is precisely adjusted. These two beams are recombined at the beam splitter *BS2*, which is identical to *BS1*.

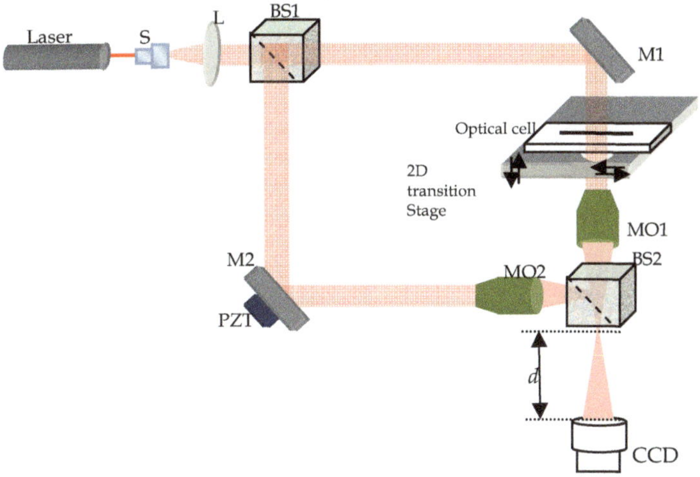

Fig. 4. Digital holographic interferometric set-up.

The mirror M2 in the digital holographic set-up reflects the reference beam. This mirror can be tilted to control the position of the reference beam on the CCD camera. The reference beam reaches the CCD camera at a small incident angle with respect to the propagation direction of the object wave. Therefore, the off axis holographic configuration is used (Wahba & Kreis, 2009a, 2009b, 2009c; Yassien et al., 2010). In addition, the mirror M2 in the digital holographic setup is mounted on a piezoelectric transducer (PZT), which acts as the phase shifting tool. It is positioned in the reference arm and enables us to obtain four phase shifted holograms. The phase shifting steps start with 0° and the holograms are shifted mutually by $\pi/2$. These phase shifted holograms are recorded by an Allied Vision Marlin F145B2 CCD camera with pixel pitch 4.65 µm × 4.65 µm and pixel numbers 1392, and 1040 in the horizontal and vertical directions, respectively.

6. Results and discussion

6.1 Refractive index profile of GRIN optical waveguide

Phase shifting digital holographic interferometry is used to investigate the optical parameters of GRIN optical waveguide samples. Fresnel off-axis holograms have been produced by using a Mach-Zehnder holographic arrangement; see Fig. 4. A piezoelectric transducer (PZT), acting as a phase-shifting tool, is applied in the reference arm and helps us to obtain four phase-shifted holograms. Fig. 5 represents the phase shifted digital holograms for a GRIN optical fibre sample. The GRIN optical fibre sample is immersed in a liquid of refractive index 1.46, which is greater than that of its cladding. The phase of the hologram (Fig. 5(a)) is assumed to be zero, whereas the phase shifts of the holograms (Fig. 5(b) to 5(d)) are $\pi/2$, π, and $3\pi/2$; respectively.

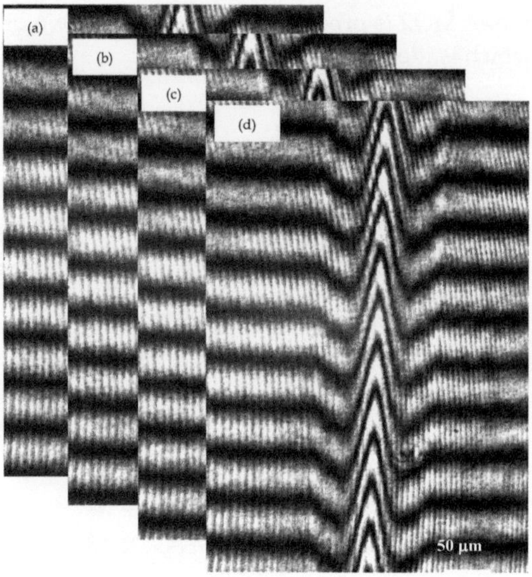

Fig. 5. Phase shifted digital holograms of the GRIN optical fibre sample at wavelength λ = 632.8 nm, (a) 0, (b) $\pi/2$, (c) π, and (d) $3\pi/2$.

According to the phase shift algorithm, these four phase shifted holograms are used to produce the complex field in the hologram plane. From this field, the interference phase distribution is reconstructed by the convolution algorithm, and the result is shown in Fig. 6. In this figure we have the wrapped interference phase. The phase has been unwrapped and then the linear increase of the background phase from left to right has been approximated by a linear regression and subtracted from the unwrapped interference phase. The result with normalized background is shown in Fig. 7(a). This image represents the interference phase map of the GRIN optical fibre sample, whose interference phase varies across the sample but remains nearly constant along the optical fibre sample. This is due to the fact that the refractive index of each layer of the optical fibre sample remains constant along the optical fibre sample. One notes that the GRIN optical fibre sample is tilted inside the optical phase map, see Fig. 7(a). The tilt angle of the sample is small but it increases the error in the calculated mean values of optical phase differences along the fibre. Also, it might cause a broadening of the calculated mean value of the fibre radius. A simple algorithm is used to avoid this tilting and its corresponding errors. The optical phase differences across the fibre are shifted in each row of the phase map matrix. The optical interference phase differences along each layer of the optical fibre are shifted to be in the same column of the phase map matrix. After the shifting of all rows, the optical interference phase differences along the fibre appear to be perpendicular to x-axis. Fig. 7(b) represents the perpendicular optical phase differences along the GRIN optical fibre sample.

The modified interference optical phase map (Fig. 7(b)) is used to calculate the mean values of the interference optical phase differences across the GRIN optical fibre sample. Then the mean values of interference phase differences across the GRIN optical fibre in combination with the multilayer model are used to estimate the refractive index profile of the GRIN optical fibre. The refractive index profile of the GRIN optical fibre cladding is determined using the multilayer model.

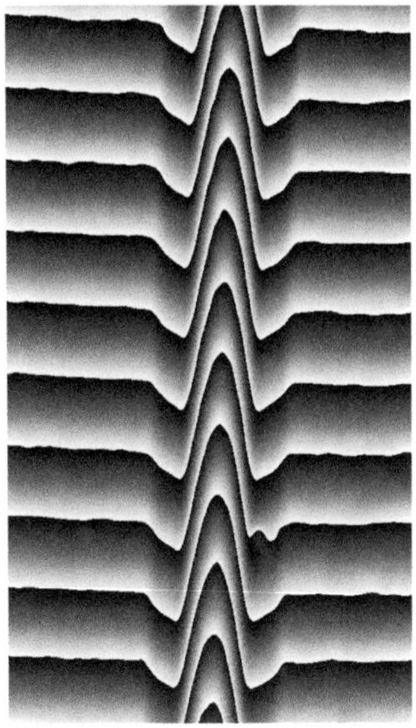

Fig. 6. Reconstructed interference phase modulo 2π from the phase shifted digital holograms of Fig. 5.

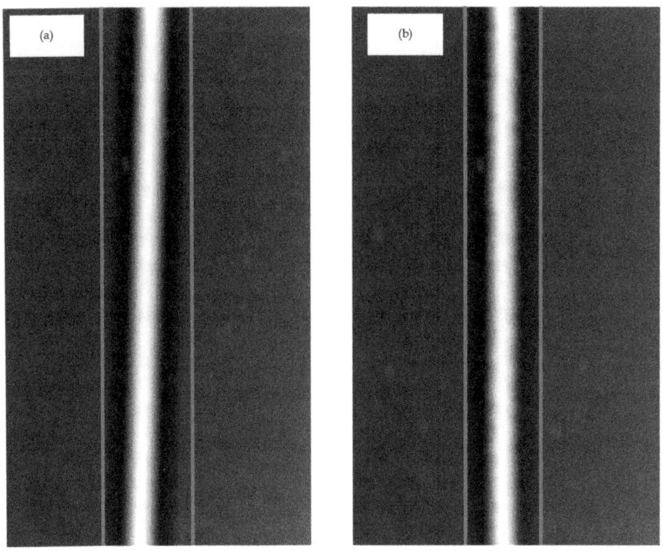

Fig. 7. Unwrapped interference phase distribution with normalized background of the GRIN optical fibre, (a) tilted and (b) perpendicular to x-axis.

Fig. 8 displays the calculated core refractive index profile of the GRIN optical fibre sample. These refractive indices are fitted using a parabolic function which could be used to determine the TE modes for the GRIN optical fibre.

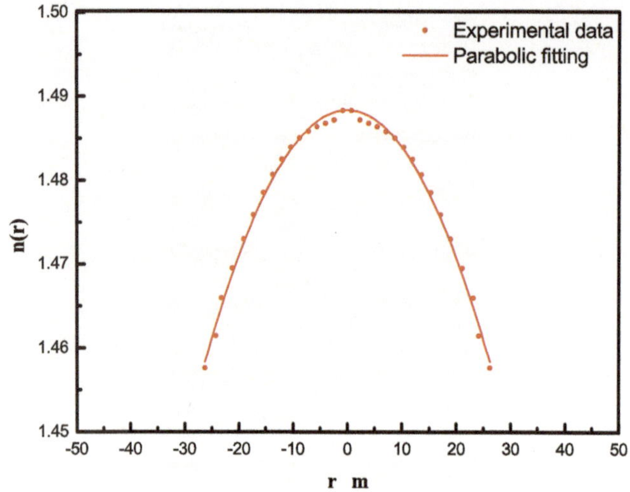

Fig. 8. The calculated refractive index profile of GRIN optical fibre.

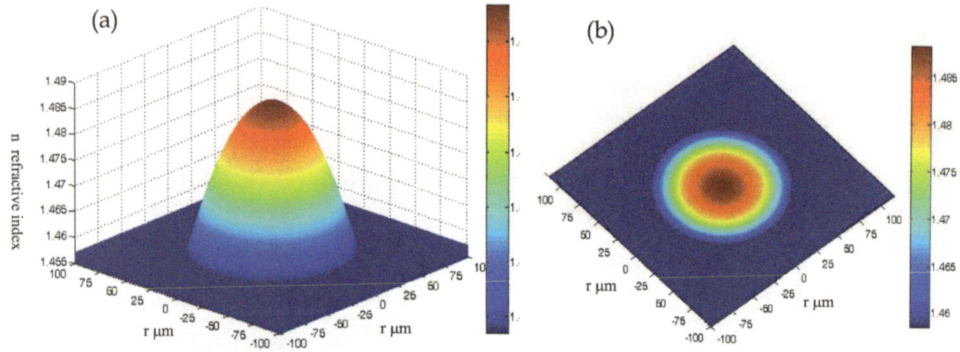

Fig. 9. (a) The reconstructed 3D refractive index profile, (b) the projection of the 3D refractive index profile of GRIN optical fibre.

According to the symmetry of GRIN optical fibre, where the GRIN optical fibre consists of a large number of coaxial layers, all points of each layer have the same refractive index. Whereas, the 3D refractive index can be presented in spherical polar coordinates as $n(r,\theta,\rho)=n_0+\Delta n(r/R)^2$ with $r=0{:}\ R$, $\theta=0{:}\pi$ and $\rho=0{:}2\pi$, then a simple algorithm based on the assumptions of the multilayer model is used to reconstruct the refractive index profile across the GRIN optical fibre. Hence, Fig. 9(a) shows the three dimensional view of the refractive index across the GRIN optical fibre. The projection of the three dimensional refractive index profile is presented in Fig. 9(b).

In addition the same procedures are applied to a fabricated GRIN optical waveguide. Hence the UV irradiation lithographic method is used to inscribe an optical waveguide in a slab of PMMA. The waveguide sample is prepared at Λ = 248 nm, with laser pulse energy density 32 mJ/cm^2, R (repetition rate) = 5 Hz, N (number of pulses) = 1250 pulse, and temperature of 22°C. This sample is immersed in a fluid of n_L=1.4917 perfectly matching the refractive index of the substrate. The photo-induced refractive index variations in the waveguide area can be divided into two zones. The refractive index in the first zone is high at the top surface and decreases gradually until reaching a minimum value, so that the refractive index distribution in the first zone can be classified as a graded index profile. The refractive index distribution in the second zone looks like Gaussian shape. The original homogeneous material surrounds the modified area of the PMMA slab, where the refractive index of the PMMA slab is n_S. The cross sectional area of the waveguide is divided into Q layers. The refractive index of each layer is constant. The refractive index of the ith layer is n_i, where $i=1,2,\ldots,Q$. In addition, the thickness of each layer is constant. The refractive index variations extend in the polymer slab up to depth b. The layer thickness is equal to b/Q.

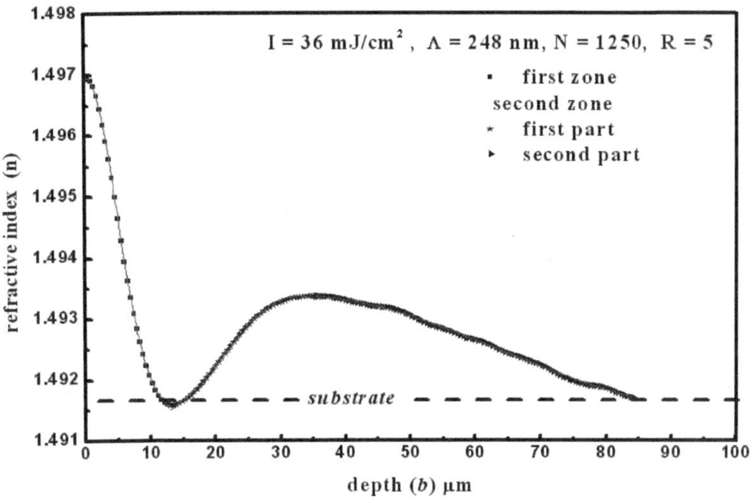

Fig. 10. The measured refractive index profile of fabricated GRIN optical waveguide.

The irradiated PMMA slab is immersed in a fluid and then it is used in the holographic set-up. The object beam crosses the irradiated PMMA slab and then is superposed on reference beam. The produced interference pattern contains a field distribution related to the refractive index variations inside the modified area. The optical interference phase difference due to each layer depends on the refractive index n_i. The optical interference phase difference $\Delta\varphi_i$ across each layer is given by

$$\frac{\Delta\varphi_i}{2\pi}\lambda = (n_i - n_s)\,a_i \quad , i = 1, 2, 3, \ldots\ldots.Q \tag{27}$$

where a_i is the width of the ith layer of the waveguide, and λ is the wavelength of the used light beam. The refractive index of PMMA slab (non-irradiated area) n_s is equal to the refractive index of the immersion liquid n_L. This equation is valid in this case because the refractive index difference between the maximum and minimum values is very small in comparison with the refractive index difference of GRIN optical fibre. The calculated refractive index profile of the fabricted optical waveguide is shown in Fig.10.

6.2 Mode field distribution of GRIN optical waveguides

The parabolic fitting of the refractive index profile for the investigated symmetric GRIN optical fibre core is shown in Fig. 8. The calculated optical parameters n_o, Δn with their standard error (*Sd*) and $\Delta\varepsilon$ are presented in table 1. The accuracy of the measurements of these optical parameters using the digital holographic method is increased in comparison to the accuracy of the multiple-beam Fizeau fringes and the two-beam interference Pluta polarizing microscope (Hamza et al., 2001).

The corresponding parameters ε_0 and ε_c for the centre of the core and the cladding of the optical waveguide respectively are determined. The solutions must be oscillatory in the core radius of the optical fibre in order to represent guided modes.

The modes are guided as long as the condition

$$\sqrt{\varepsilon_o} > \frac{\beta}{k_o} > \sqrt{\varepsilon_o - \Delta\varepsilon} \ , \tag{28}$$

is satisfied.

The values of β can be calculated directly by using Eq. (20) for every mode number q. The coefficient A in Eq. (17) is related to the power carried in the core of the optical waveguide. The power is calculated by integrating the z component of the time averaged Poynting vector over the cross-sectional area of the waveguide:

$$S_z = -\frac{1}{2}\,\mathrm{Re}\left(E \times H.\vec{z}\right) \tag{29}$$

The average power in TE modes is

$$P_z = -\frac{1}{2}\int_{-\infty}^{\infty} E_y H_x^* \, dx = \frac{\beta}{2\omega\mu_o}\int_{-\infty}^{\infty} \left|E_y\right|^2 dx \ . \tag{30}$$

The parameter P does not indicate the total power but the power per unit length in the y direction. The numbers of TE modes is related to the number of nodes, since the GRIN optical waveguide supports thirty nine modes, starting from q = 0 to 38. Each propagated mode has an effective index given by Eq. (21).

	n_o	Δn	$\Delta\varepsilon$
	1.48824	0.02974	0.088526
Sd	$\pm 1.42441\times10^{-4}$	$\pm 3.04453\times10^{-4}$	

Table 1. The calculated optical parameters of GRIN optical fibre

Fig. 11 represents the effective indices for all modes of the GRIN optical waveguide sample. The values of the effective indices are confined between the core centre and clad refractive indices according to Eq. (21).

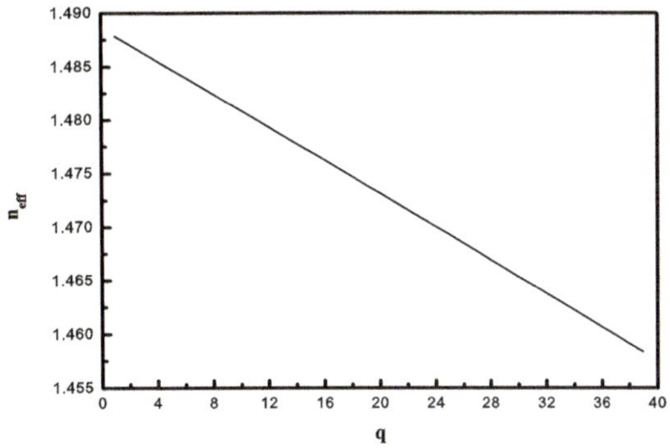

Fig. 11. Effective indices against the mode number q for GRIN optical fibre sample.

Fig. 12 represents the first mode for the sample. Fig. 13 represents the second and the sixth modes. The oscillations are featured by an amplitude increasing with x approaching x_t^i , where i is the interface order between the modes. Landau and Lifshitz proved that the increase in wave amplitude, when the reflection point is reached, is a typical property in the presence of a varying index of refraction (Landau & Lifshitz, 1960).

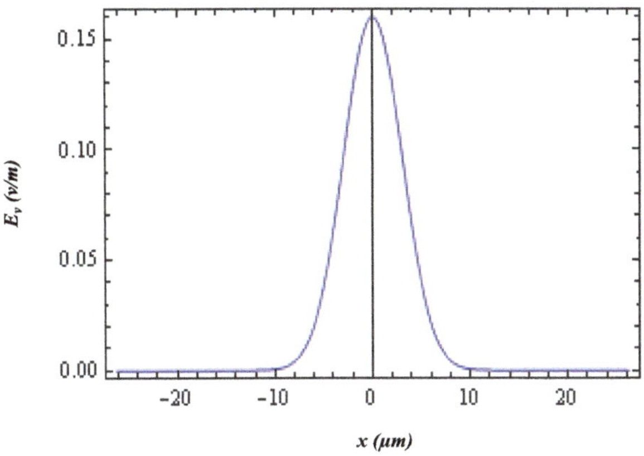

Fig. 12. Plot of the first mode *(q=0)* calculated for GRIN optical waveguide sample.

Now we can calculate the effective indices n_{eff} for each TE mode for an asymmetric (fabricated) optical waveguide. The guiding condition for the first zone is represented by

$$n_{s1} < \frac{\beta}{k_o} < n_{s1} + \Delta n_1 \ , \tag{31}$$

Also the guiding condition for the second zone, for the first and second parts are given respectively by

$$n_{s1} < \frac{\beta}{k_o} < n_{s1} + \Delta n_2 \tag{32}$$

and

$$n_s < \frac{\beta}{k_o} < n_s + \Delta n_3 \tag{33}$$

Eqs. (24), (25) and (26) can be used to calculate the effective indices n_{eff} of each mode for every part of the second zone. The first zone of the sample supports three modes. The first zone of the effective index n_{eff} for each mode is calculated. The numbers of modes in the first and second parts are four and eight modes respectively.

Generally, the number of the modes and the values of the effective indices n_{eff} depend on two factors; the first one is the depth and the second one is Δn.

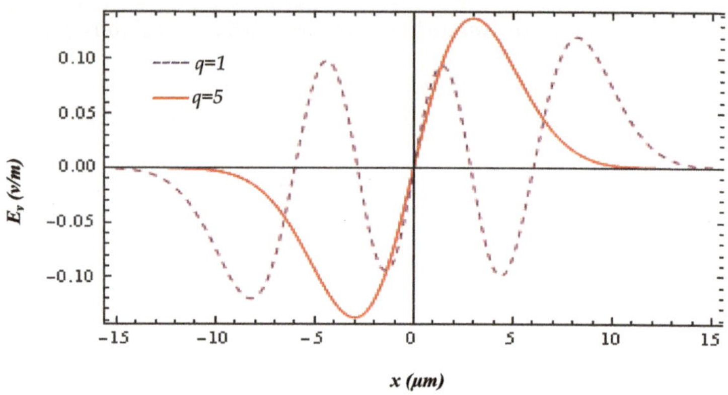

Fig. 13. Plots of the second mode *(q=1)* and sixth mode *(q=5)* calculated for GRIN optical waveguide sample.

7. Conclusions

In this chapter the digital holographic phase shifting interferometric method is used to investigate and characterize GRIN optical waveguide samples. Phase shifting digital holography has the advantages of no dc term and no twin image in the reconstructed field as well as high accuracy due to the redundant data in the multiple holograms. The optical interference phase differences due to the variation of refractive indices of GRIN optical waveguides are extracted. A simple algorithm is used to recover the tilting of the extracted optical interference phase differences along the GRIN optical wavegide, to avoid the errors of the calculated mean values of optical phase differences due to the GRIN optical waveguides. The broadening of optical waveguide radius or width is also reduced.

The advances of accurate reconstructed interference phase differences in combination of multilayer model are used to calculate the refractive index profiles of GRIN optical waveguide. Also, a simple algorithm based on the assumptions of the multilayer model is used to reconstruct the 3D refractive index profile of GRIN optical fibre.

The refractive index profiles of symmetric and asymmetric GRIN optical waveguides and their associated modes are calculated. The propagation coefficient β and hence the effective indices n_{eff} associated with every mode are determined.

8. Acknowledgment

The authors would like to express their gratitude to Professor A A Hamza for his fruitful discussions and comments. In addition, the authors are greatly indebted to the Unit of Research Management of Mansoura University for financial support. We are also acknowledging the encouragement and support of the Vice president of Mansoura University for Higher studies & Research."

9. References

Baker, A.K. & Dyer, P.E. (1993). Refractive-index modification of polymethylmethacrylate (PMMA) thin films by KrF-laser irradiation, *Appl. Phys. A*, Vol. 57, pp. 543

Barakat, N., Hamza, A. A. & Ganied, A. S. (1985). Multiple-beam interference fringes applied to GRIN optical waveguides to determine fiber characteristics, *Appl. Opt.*, Vol. 24, pp. 4383-4386

Cha, S. S. & Vest, C. M. (1981). Tomographic reconstruction of strongly refracting fields and its applications to interferometric measurements of boundary layers, *Appl. Opt.*, Vol. 20, pp. 2787–2794

Chiang, K. S. (1985). Construction of refractive-index profiles of planar dielectric waveguides from the distribution of effective indexes, *Journal of Lightwave Tech.*, Vol. 3, pp. 385

Colomb, T., Dahlgren, P., Beghuin, D., Cuche, E., Marquet, P. & Depeursinge, C. (2002). Polarization Imaging by Use of Digital Holography, *Appl. Opt.*, Vol. 41, pp. 27–37

Colomb, T., Dürr, F., Cuche, E., Marquet, P., Limberger, H. G., Salath, R. P. & Depeursinge, C. (2005). Polarization microscopy by use of digital holography: application to optical-fiber birefringence measurements, *Appl. Opt.*, Vol. 44, pp. 4461–4469

Conwell, e. m. (1973). Modes in optical waveguides formed by diffusion, *Appl. Phys. Lett.*, Vol. 26, pp. 328

Cuche, E., Marquet, P. & Depeursinge, C. (1999). Simultaneous amplitude-contrast and quantitative phase-contrast microscopy by numerical reconstruction of Fresnel off-axis holograms, *Appl. Opt.*, Vol. 38, pp. 6994–7001

De Angelis, M., De Nicola, S., Finizio, A., Pierattini, G., Ferraro, P., Pelli, S., Righini, G. & Sebastiani, S. (2006). Digital-holography refractive-index-profile measurement of phase gratings, *Appl. Phys. Lett.*, Vol. 88, pp. 111–114

Dubois, F., Joannes, L., Dupont, O., Dewandel, J. L. & Legros, J. C. (1999). An integrated optical set-up for fluid-physics experiments under microgravity conditions, *Meas. Sci. Technol.*, Vol. 10, pp. 934–945

Eguchi, M. & Horinouchi, S. (2004). Finite-Element Modal Analysis of Large-Core Multimode Optical Fibers, *Appl. Opt.*, Vol. 43, pp. 2163-2167

Eldada, L. & Shacklette, L. W. (2000). Advances in polymer integrated optics, *IEEE J. Sel. Top. Quantum Electron.*, Vol. 6 (1), pp. 54-68

El-Morsy, M. A., Yatagai, T., Hamza, A., Mabrouk, M. A. & Sokkar, T. Z. N. (2002a). Multiple-beam Fizeau fringe-pattern analysis using Fourier transform method for accurate measurement of fiber refractive index profile of polymer fiber, *J. Appl. Polymer Sci.*, Vol. 85, pp. 475–484

El-Morsy, M. A., Yatagai, T., Hamza, A., Mabrouk, M. A. & Sokkar, T. Z. N. (2002b). Automatic refractive index profiling of fibers by phase analysis method using Fourier transform, *Opt. and Lasers in Eng.*, Vol. 38, pp. 509–525

Faust, R. C. (1952).An Interferometric Method of Studying Local Variations in the Refractive Indices of a Solid, *Proc. Phys. Soc. B*, Vol. 65, pp. 48–62

Faust, R. C. (1954).The Determination of the Refractive Indices of inhomogeneous Solids by Interference Microscopy, *Proc. Phys. Soc. B*, Vol. 67, pp. 138–148

Gabor, D. (1948). A new microscopic principle, *Nature*, Vol. 161, pp. 777–778

Gabor, D. (1949). Microscopy by reconstructed wavefronts, *Proc. Royal Society*, A 197, pp. 454–487

Gabor, D. (1951). Microscopy by reconstructed wavefronts: II, *Proc. Royal Society*, B 64, pp. 449–469

Hamza, A. A. , Sokkar, T.Z.N., Mabrouk, M. A., Ghandar, A. M. & Ramadan, W. A. (1994). On the determination of the refractive index of a fibre: I. Skin-core fibre, *Pure Appl. Opt.*, Vol. 3, pp. 943-961

Hamza, A. A., Ghander, A. M., Sokkar, T. Z. N., Mabrouk, M. A. & Ramadan, W. A. (1995). On the determination of the refractive index of a fibre II. Graded index fibre ,*Pure Appl. Opt.*, Vol. 4, pp. 161-177

Hamza, A. A., Mabrouk, M. A., Ramadan, W. A. & Shams-Eldin, M. A. (2001). Determination of GRIN optical fibre parameters from transverse interferograms considering the refraction of the incident ray by the fibre, *Opt. Commun.*, Vol. 200, pp. 131–138

Hamza, A.A., Belal, A.E., Sokkar, T.Z.N., EL-Dessouky, H.M. & Agour, M.A. (2007). Interferometric studies on the influence of temperature on the optical and dispersion parameters of GRIN optical fibre, *Opt. and Lasers in Eng.*, Vol. 45, pp. 145-152

Hamza, A.A., Mabrouk, M.A., Ramadan, W.A. & Wahba, H.H. (2004). Core-index determination of a thick fibre using lens-fibre interference (LFI) technique, *Opt. and Lasers in Eng.*, Vol. 42, pp. 121–130

Hariharan, P. (2002). *Basics Of Holography*, Cambridge University Press, ISBN 0521807417, Cambridge, United Kingdom

Hariharan, P., Oreb, B. F. & Brown, N. (1982). A digital phase-measurement system for real-time holographic interferometry, *Opt. Commun.*, Vol. 41, pp. 393–396

Hossain, M. M., Mehta, D. S. & Shakher, C. (2006). Refractive index determination: an application of lensless Fourier digital holography, *Opt. Eng.*, Vol. 45, pp. 106203

Kebbel, V., Adams, M., Hartmann, H. J. & Jüptner, W. (1999). Digital holography as a versatile optical diagnostic method for microgravity experiments, *Meas. Sci. Technol.*, Vol. 10, pp. 893–899

Kemper, B. & Carl, D. (2006). Investigation of living pancreas tumor cells by digital holographic microscopy , *J. Biomed. Opt.*, Vol. 11, pp. 034005

Kemper, B., Kosmeier, S., Langehanenberg, P. & Bally, G. (2007). Integral refractive index determination of living suspension cells by multifocus digital holographic phase contrast microscopy, *J. Biomed. Opt.*, Vol. 12, pp. 054009

Kreis, Th. (1996). *Holographic Interferometry: Principles and Methods*, Akademie-Verlag, ISBN 30-05-501644-0, Berlin, Germany

Kreis, Th. (2005). *Handbook of Holographic Interferometry, Optical and Digital Methods*, Wiley-VCH, ISBN: 3-527-40546-1, Weinheim, Germany

Landau, L. D. and Lifshitz, E. M. (1960). *Electrodynamics of Continuous Media*, Pergamon, New York, USA

Marcuse, D. (1973). TE modes of graded-index slab waveguides, *IEEE J. Quantum Electron*, Vol. 9, pp. 1000

Marhic, M. E., Ho, P. S. & Epstein, M. (1975). Nondestructive refractive-index profile measurements of clad optical fibers, *Appl. Phys. Lett.*, Vol. 26, pp. 574–575

Mathey, P. & Jullien, P. (1996). Numerical analysis of a WKB inverse method in view of index profile reconstruction in diffused waveguides, *Opt. Commun.*, Vol. 122, pp. 127

Mathey, P., and Jullien, P. (1996). Numerical analysis of a WKB inverse method in view of index profile reconstruction in diffused waveguides, *Opt. Commun.*, Vol. 122, pp. 127

Owen, R. B. & Zozulya, A. A. (2002). Comparative Study with Double-Exposure Digital Holographic Interferometry and a Shack-Hartmann Sensor to Characterize Transparent Material, *Appl. Opt.*, Vol. 41, pp. 5891–5895

Powell, R. L. & Stetson, K. A. (1965).Interferometric vibration analysis by wavefront reconstruction, *J. Opt. Soc. Am.*, Vol. 55, pp. 1593–1608

Rappaz, B., Marquet, P., Cuche, E., Emery, Y., Depeursinge, C. & Magistretti, P. (2005). Measurement of the integral refractive index and dynamic cell morphometry of living cells with digital holographic microscopy, *Opt. Exp.*, Vol. 13, pp. 9361–9373

Saunder, M. J. & Gardner, W. B. (1977). Nondestructive interferometric measurement of the delta and alpha of clad optical fibers, *Appl. Opt.*, Vol. 16, pp. 2368- 2371

Schnars, U. & Jüptner, W. (1994a). Direct recording of holograms by a CCD target and numerical reconstruction, *Appl. Opt.*, Vol. 33, pp. 179–181

Schnars, U. & Jüptner, W. (1994b). Digital recording and reconstruction of holograms in hologram interferometry and shearography, *Appl. Opt.*, Vol. 33, pp.4373–4377

Schnars, U. & Jüptner, W. (2005).*Digital holography: digital hologram recording, numerical reconstruction, and related techniques*, Published by Springer, ISBN 3-540-21934-X , Germany

Schnars, U. (1994). Direct phase determination in hologram interferometry with use of digitally recorded holograms, *J. Opt. Soc. Amer. A* , Vol. 11, pp. 2011–2015

Schnars, U., Kreis, Th. & Jüptner, W. (1995).Digital recording and numerical reconstruction of holograms: Reduction of the spatial frequency spectrum, Opt. Eng., Vol. 35, pp. 977–982

Shams-Eldin, M.A. , Wochnowski, C., Metev, S., Hamza, A.A. and Jüptner, W. (2004). Determination of the refractive index depth profile of an UV-laser generated waveguide in a planar polymer chip, *Appl. Surf. Sci.*, Vol. 236, pp. 31

Shams-Eldin, M.A., Wochnowski, C., Koerdt, M., Metev, S., Hamza, A.A. and Jüptner, W. (2005). Characterisation of the optical-functional properties of a waveguide written by an UV-laser into a planar polymer chip, *Opt. Mater.*, Vol. 27, pp.1138

Shemirani, M.B. & Kahn, J.M. (2009). Higher-Order Modal Dispersion in Graded-Index Multimode Fiber, *J. Lightwave Tech.*, Vol. 27, 5461

Shemirani, M.B., Wei Mao Panicker, R.A. & Kahn, J.M. (2009). Principal Modes in Graded-Index Multimode Fiber in Presence of Spatial- and Polarization-Mode Coupling, *J. Lightwave Tech.*, Vol. 27, pp. 1248

Skarman, B.,Becker, J. &Wozniak, K. (1996). Simultaneous 3D-PIV and temperature measurements using a new CCD-based holographic interferometer, *Flow Meas. Instrum.*, Vol. 7, pp. 1-6

Sokkar, T. Z. N., El-Morsy, M. A. & Wahba, H. H. (2008). Automatic fringe analysis of the induced anisotropy of bent optical fibres, *Opt. Commun.*, Vol. 281, pp. 1915–1923

Sweeney, D. W. & Vest, C. M. (1973). Reconstruction of three dimensional refractive index fields from multidirectional interferometric data, *Appl. Opt.*, Vol. 12, pp. 2649–2663

Vollertsen, F. & Wochnowski, C. (2004). UV-laser assisted Fabrication of integrated-optical Waveguides, *CIRP Annals - Manufacturing Technology*, Vol. 53, pp. 199-202

Wahba, H. H. & Kreis, Th. (2009c). Digital holographic interferometric characterization of bent optical fibers, *J. Opt. A: Pure Appl. Opt.*, Vol. 11, pp.105407

Wahba, H. H. & Kreis, Th. (2009a). Characterization of graded index optical fibers by digital holographic interferometry, *Appl. Opt.*, Vol. 48, pp. 1573-1582

Wahba, H. H. & Kreis, Th. (2009b). Characterization of optical fibers by digital holographic interferometry'', *Proc. SPIE 7389*, 73890K, Munich, Germany, June, 2009

Wochnowski, C. , Metev, S. and Sepold, G.(2000). UV-laser-assisted Modification of the optical Properties of Polymethylmethacrylate,"*Appl. Surf. Sci.*, Vol.154–155 , pp. 706-711

Yamaguchi, I. & Zhang, T. (1997). Phase-shifting digital holography, *Opt. Lett.*, Vol. 22, pp. 1268-1270

Yassien, K., Agour, M., V. Koyplow, C. & Dessouky, H. (2010). On the Digital Holographic Interferometry of Fibrous Materials, I: Optical Properties of Polymer and Optical Fibers, *Opt. and Lasers Eng.*, Vol. 48, pp.555

4

Multiple-Wavelength Holographic Interferometry with Tunable Laser Diodes

Atsushi Wada
National Defense Academy
Japan

1. Introduction

Two-wavelength holographic interferometry is an effective technique to generate a contour map of a diffusely reflecting surface (Friesem & Levy (1976); Heflinger & Wuerker (1969); Hildebrand & Haines (1967); Yonemura (1985)). In this technique, two holograms are recorded with two wavelengths. An interference fringe pattern is generated by superposing two object images reconstructed from the holograms.

In digital holography, a hologram is recorded by an image sensor and saved into a computer. An object image can be reconstructed by numerical calculation. Several reconstruction methods were reported. Some of these have adjustablity of position and scale of a reconstruction image (Yu & Kim (2006); Zhang et al. (2004)). An object phase distribution can be obtained by the numerical reconstruction of digital holograms. Therefore, two-wavelength digital holographic interferometry makes it possible to generate a contour map by numerical extraction of a phase difference between two reconstructed images (Gass et al. (2003); Parshall & Kim (2006); Wagner et al. (2000; 1999); Yamaguchi (2001); Yamaguchi et al. (2006)).

A phase difference extracted from reconstructed object images is wrapped into a half-open interval $(-\pi, \pi]$. If a measured object height was large with respect to a synthetic wavelength, 2π ambiguities of the phase difference should be eliminated for retrieving the object profile. Common phase unwrapping algorithms (Asundi & Wensen (1998); Servin et al. (1998)) which use phase information of neighbor pixels can be applied when an object structure has no discontinuity. However the algorithms can not work correctly for an object profile having isolated region surrounded by discontinuous step.

An object profile with discontinuous structure can be measured by two-wavelength interferometry with a sufficiently large synthetic wavelength. For example, two-wavelength holographic interferometry with a ruby laser and a synthetic wavelength of ~ 2 cm was reported (Heflinger & Wuerker (1969); Pedrini et al. (1999)). Nevertheless the measurement error tends to be amplified linearly with an increase in the synthetic wavelength since most of the error sources are the product of the synthetic wavelength and an error of the extracted phase difference (Cheng & Wyant (1984)).

A technique which eliminates 2π ambiguities by using a phase difference with a large synthetic wavelength was reported (Cheng & Wyant (1985); de Groot (1991); Wagner et al. (2000)). This technique makes it possible to measure a large step-height with high depth resolution. Wagner *et. al.* reported multiple-wavelength holographic interferometry using a dye laser. They combined three phase differences with synthetic wavelengths of 3.04 mm,

1.53 mm and 0.76 mm, and realized the measurement with a measurable step-height of 0.6 mm and an error of 36 μm (Wagner et al. (2000)).
In this paper, multiple-wavelength digital holographic interferometry using tunablity of laser diodes for measuremnt of a large step-height with high accuracy is presented. Using the high-resolution wavelength tunability of laser diodes, a pair of holograms with a small wavelength difference less than 0.01 nm can be recorded, and used to extract a phase difference with a large synthetic wavelength of more than 120 mm. Several holograms are recorded through the change in the wavelength of a laser diode. Phase differences with synthetic wavelengths from 0.4637 mm to 129.1 mm are extracted from the holograms. The synthetic wavelength of 129.1 mm allows us to measure a step-height of 32 mm. The synthetic wavelength of 0.463 mm presents us with a measurement with an rms error of 0.01 mm. 2π ambiguities of the phase difference with a small synthetic wavelength are eliminated by a recursive calculations using the phase difference with a larger synthetic wavelength. The elimination of 2π ambiguities realizes the measurement with the measurable step-height of 32 mm and the rms error of 0.01. The requirements for performing the phase unwrapping are discussed. We found that precise knowledge of the recording wavelengths is required for correctly performing the phase unwrapping. The required precision of the knowledge is derived.
The Fresnel-transform based hologram reconstruction is fast and commonly used. However, the pixel size of reconstructed image increases with the recording wavelength. If the pixel size variation becomes large, the extraction of a phase difference cannot be performed correctly. In order to obtain a correct phase difference, the pixel size should be adjusted. A simple and fast algorithm for pixel size adjustment is described.

2. Hologram recording and reconstruction

Figure 1 shows a geometry of holographic interferometry. A setup for off-axis lensless Fourier transform holography is used for hologram recording. A spherical wave from a point light source is used as a reference wave. An object is located in the same plane as the point light source. A distance between the object and a hologram satisfies the Fresnel approximation (Goodman (1968)). Let the point light source be at $x = 0, y = 0, z = 0$ and the reference wave $u_r(x, y, z)$ is given by

$$u_r(x,y,z) = \frac{1}{z} \exp\left(i\pi \frac{x^2 + y^2}{\lambda z}\right), \tag{1}$$

where λ is wavelength.
Let $u_o(x, y, z)$ is a complex amplitude of the object light at the hologram. By using a complex amplitude distribution $u_o(x', y', 0)$ of the object light in the plane of $z = 0$, $u_o(x, y, z)$ is given by

$$u_o(x,y,z) = \frac{1}{z} \exp\left(i\pi \frac{x^2 + y^2}{\lambda z}\right) \iint dx'dy' u_o(x',y',0) \exp\left\{i2\pi \left(\frac{x'^2 + y'^2}{2\lambda z} - \frac{xx' + yy'}{\lambda z}\right)\right\}. \tag{2}$$

We define u'_o as the product of $u_o(x', y', 0)$ and a spherical wave phase factor, and U'_o as the Fourier spectrum of u'_o:

$$u'_o(x',y') = u_o(x',y',0) \exp\left(i\pi \frac{x'^2 + y'^2}{\lambda z}\right), \tag{3}$$

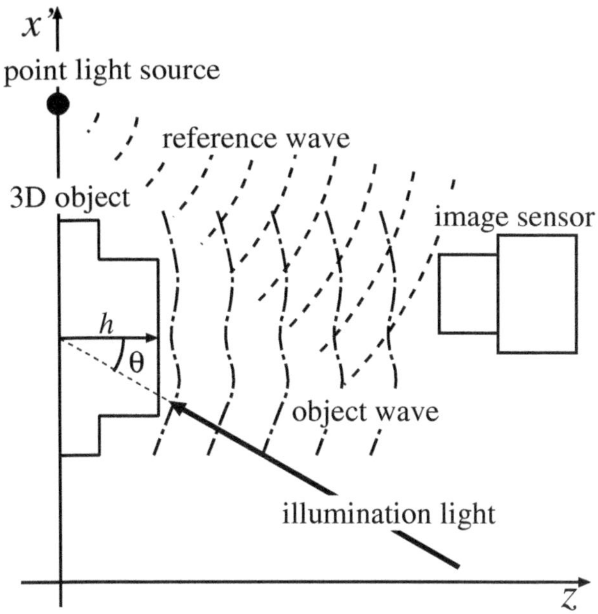

Fig. 1. Geometry of holographic interferometry.

and

$$U_o'(f_x', f_y') = \mathcal{F}(u_o') = \iint dx'dy'u_o'(x', y') \exp[-i2\pi(f_x'x' + f_y'y')]. \tag{4}$$

Equation (2) can be written in terms of U_o' by substituting the spatial frequencies $f_x' = x/\lambda z$ and $f_y' = y/\lambda z$:

$$u_o(x, y, z) = \frac{1}{z} \exp\left(i\pi \frac{x^2 + y^2}{\lambda z}\right) U_o'\left(\frac{x}{\lambda z}, \frac{y}{\lambda z}\right). \tag{5}$$

An intensity distribution $I(x, y, z)$ of an interference fringe pattern formed by the object wave and the reference wave is given by

$$\begin{aligned} I(x, y, z) &= |u_r(x, y, z) + u_o(x, y, z)|^2 \\ &= |u_r|^2 + |u_o|^2 + u_r^* u_o + u_r u_o^* \\ &= \frac{1}{z^2}\left\{1 + \left|U_o'\left(\frac{x}{\lambda z}, \frac{y}{\lambda z}\right)\right|^2 + U_o'\left(\frac{x}{\lambda z}, \frac{y}{\lambda z}\right) + U_o'^*\left(\frac{x}{\lambda z}, \frac{y}{\lambda z}\right)\right\} \\ &= \frac{1}{z^2}\left\{1 + \left|U_o'(f_x', f_y')\right|^2 + U_o'(f_x', f_y') + U_o'^*(f_x', f_y')\right\}. \end{aligned} \tag{6}$$

In Eq. (6), the third term includes the Fourier spectrum of the object light. The object image can be obtained by applying an inverse Fourier transform to $I(x, y, z)$. The reconstructed wave field u_{rec} obtained by the transformation is given by

$$
\begin{aligned}
u_{\text{rec}}(x',y') =& \mathcal{F}^{-1}\{I(x,y,z)\} \\
=& \mathcal{F}^{-1}\{I(\lambda z f'_x, \lambda z f'_y, z)\} \\
=& \iint df'_x df'_y \, I(\lambda z f'_x, \lambda z f'_y, z) \exp[i2\pi(f'_x x' + f'_y y')] \\
=& \lambda^2 \left\{ \frac{\delta(x',y')}{\lambda^2 z^2} + u'_{\text{o}}(x',y') \otimes u'^*_{\text{o}}(-x',-y') + u'_{\text{o}}(x',y') + u'^*_{\text{o}}(-x',-y') \right\},
\end{aligned} \tag{7}
$$

where the symbol $\otimes$ stands for the convolution operator. In Eq. (7), the first term is the reference point light source. The second term is an autocorrelation of the object light. The third term expresses the object light produced by a spherical wave phase factor. The fourth term is the complex conjugate of the third term. Since a phase factor does not affect intensity, the intensity distribution of the object light can be reconstructed by the third term of Eq. (7). Let consider that $I(x,y,z)$ is recorded by an image sensor having $N_x \times N_y$ pixels with pixel size of $\Delta_x \times \Delta_y$. Numerical reconstruction of the digital hologram is realized through the inverse discrete Fourier transform. Let the pixel size of the numerically reconstructed object image be $\Delta_x{}' \times \Delta_y{}'$. The discrete formulation of Eq. (7) is then

$$
u_{\text{rec}}(s\Delta_x{}', t\Delta_y{}') = \sum_{p=-N_x/2}^{N_x/2-1} \sum_{q=-N_y/2}^{N_y/2-1} I(p\Delta_x, q\Delta_y, z) \exp\left\{ i2\pi \left(\frac{p\Delta_x}{\lambda z} s\Delta'_x + \frac{q\Delta_y}{\lambda z} t\Delta'_y \right) \right\}, \tag{8}
$$

where p, q are integers, and s, t are integers and fulfill $x' = s\Delta_x{}'$, $y' = t\Delta_y{}'$. The right side of Eq. (8) is arranged in the form, in which fast Fourier transform can be applied:

$$
u_{\text{rec}}(s\Delta_x{}', t\Delta_y{}') = \sum_{p=-N_x/2}^{N_x/2-1} \sum_{q=-N_y/2}^{N_y/2-1} I(p\Delta_x, q\Delta_y, z) \exp\left\{ i2\pi \left(\frac{ps}{N_x} + \frac{qt}{N_y} \right) \right\}, \tag{9}
$$

where s, t, p and q are integers, and Δ'_x, Δ'_y satisfy the following condition

$$
\Delta'_x = \frac{\lambda z}{\Delta_x N_x}, \; \Delta'_y = \frac{\lambda z}{\Delta_y N_y}, \tag{10}
$$

and

$$
-N_x/2 \le s < N_x/2, \; -N_y/2 \le t < N_y/2. \tag{11}
$$

The field of view of the reconstructed image is $\lambda z/\Delta_x \times \lambda z/\Delta_y$.

3. Multiple-wavelength interferometry

The reference plane is assumed to be placed at $z = 0$. Let the illumination light irradiate to the object at an angle θ to the reference plane, and the height of the object surface respect to the reference plane be $h(x',y')$. The phase distribution $\phi(x',y')$ of the reconstructed image is given by

$$
\phi(x',y') = \frac{2\pi}{\lambda} \left\{ -L + \frac{x'^2 + y'^2}{2z} \right\}, \tag{12}
$$

where L is an optical path difference caused by the object surface structure, giving

$$L = (1 + \cos\theta)h(x', y') - x' \sin\theta. \tag{13}$$

The second term of Eq. (12) is the spherical wavefront introduced in Eq. (3).
Holograms are recorded with wavelengths of λ_n satisfying $\lambda_n < \lambda_{n+1}$ and the phase distribution of the object wave at the plane of $z = 0$ is ϕ_n. The phase difference $\Delta\phi_n$ between ϕ_n and ϕ_0 is given by

$$\Delta\phi_n(x', y') = \phi_n - \phi_0 = \frac{2\pi}{\Lambda_n}\left(L + \frac{x'^2 + y'^2}{2z}\right) \simeq \frac{2\pi}{\Lambda_n}L, \tag{14}$$

where Λ_n are synthetic wavelengths and

$$\Lambda_n = \frac{\lambda_0\lambda_n}{\lambda_n - \lambda_0} \simeq \frac{\bar{\lambda}_n{}^2}{\Delta\lambda_n}, \tag{15}$$

where $\bar{\lambda}_n$ and $\Delta\lambda_n$ are the average and difference of wavelengths given by

$$\bar{\lambda}_n = \frac{\lambda_0 + \lambda_n}{2}, \; \Delta\lambda_n = \lambda_n - \lambda_0. \tag{16}$$

In Eq. (8), the condition $z \gg (x'^2 + y'^2)/\Lambda_n$ is assumed and a quadratic phase term is neglected.
Let Ψ_n be the phase difference extracted from the image reconstructed from the hologram with wavelength of λ_n. The phase difference Ψ_n are given by

$$\Psi_n = \tan^{-1}\frac{\mathrm{Im}(u'_{on}u'^{*}_{o0})}{\mathrm{Re}(u'_{on}u'^{*}_{o0})}. \tag{17}$$

Ψ_n is wrapped into $(-\pi, \pi]$ and the relation between Ψ_n and $\Delta\phi_n$ is given by

$$\Delta\phi_n = \Psi_n + 2\pi m_n, \tag{18}$$

where m_n is an integer. If the conditions of $\Delta\phi_1 = \Psi_1$ and $\Lambda_n < \Lambda_{n-1}$ are satisfied, $\Delta\phi_n$ can be retrieved (Nadeborn et al. (1996); Paulsson et al. (2000); Wagner et al. (2000)) through recursive calculation of

$$\Delta\phi_n = \Psi_n + 2\pi\mathrm{NINT}\left(\frac{\alpha_n\Delta\phi_{n-1} - \Psi_n}{2\pi}\right), \tag{19}$$

where $\mathrm{NINT}(a)$ denotes the function returning the nearest neighbor integer of argument a, and α_n is the sensitivity ratio between each phase difference $\Delta\phi_n$ and $\Delta\phi_{n-1}$ and given by

$$\alpha_n = \frac{\Lambda_{n-1}}{\Lambda_n}. \tag{20}$$

4. Requirements for the phase unwrapping

In this section, we discuss the requirements for correctly performing the phase unwrapping. First, the requirements of the measurement error and the sensitivity ratio between the phase differences are discussed based on the assumption that the error of the sensitivity ratio is negligible small. Next, it is pointed out that the sensitivity ratio should be estimated for the phase unwrapping. Then, the required accuracy of the sensitivity ratio is derived. Lastly, it is explained how precise measurement of the wavelengths is required for high accuracy estimation of the sensitivity ratio.

Let the true phase differences and the sensitivity ratio are defined as $\Delta\phi'_n$, Ψ'_n and α'_n, and the measurement errors of the phase differences and the sensitivity ratio are defined as $\Psi_{\epsilon,n}$ and $\alpha_{\epsilon,n}$. The relations between these parameters and $\Psi, \Delta\phi$ and α are given by

$$\Psi_n = \Psi'_n + \Psi_{\epsilon,n},\ \Delta\phi'_n = \Psi'_n + 2\pi m_n = \Delta\phi_n - \Psi_{\epsilon,n},\ \alpha_n = \alpha'_n + \alpha_{\epsilon,n}. \tag{21}$$

By substituting Eqs. (21) into Eq. (19), we get

$$\Delta\phi_n = \Psi'_n + \Psi_{\epsilon,n} + 2\pi \text{NINT}\left(\frac{(\alpha'_n + \alpha_{\epsilon,n})(\Delta\phi'_{n-1} + \Psi_{\epsilon,n-1}) - \Psi'_n - \Psi_{\epsilon,n}}{2\pi}\right). \tag{22}$$

For the true phase difference and sensitivity ratio, the condition

$$\frac{\alpha'_n \Delta\phi'_{n-1} - \Psi'_n}{2\pi} = \frac{\Delta\phi'_n - \Psi'_n}{2\pi} = m_n, \tag{23}$$

is satisfied. Substituting Eqs. (23) into Eq. (22) gives

$$\Delta\phi_n = \Delta\phi'_n + \Psi_{\epsilon,n} + 2\pi \text{NINT}\left(\frac{\alpha_{\epsilon,n}(\Delta\phi'_{n-1} + \Psi_{\epsilon,n-1}) + \alpha_n \Psi_{\epsilon,n-1} - \Psi_{\epsilon,n}}{2\pi}\right). \tag{24}$$

In order to correctly perform the phase unwrapping, the third term in the right side of Eq. (24) should be zero, and the condition

$$\left|\frac{\alpha_{\epsilon,n}\Delta\phi'_{n-1} + \alpha_n \Psi_{\epsilon,n-1} - \Psi_{\epsilon,n}}{2\pi}\right| < \frac{1}{2}, \tag{25}$$

should be satisfied. In Eq. (25), the condition $\Psi_{\epsilon,n-1} \ll \Delta\phi'_{n-1}$ is assumed. Let the maximum of $|\Psi_{\epsilon,n}|$ for all n be $2\pi\epsilon$:

$$|\Psi_{\epsilon,n}| \leq 2\pi\epsilon. \tag{26}$$

Let the true synthetic wavelength be defined as Λ'_n. Substituting $\Delta\phi'_{n-1}$, α'_n and Λ'_n into Eq. (8) gives

$$\Delta\phi'_{n-1} = \frac{2\pi L}{\alpha'_n \Lambda'_n}. \tag{27}$$

Substitution of Eqs. (26) and (27) into Eq. (25) gives

$$\epsilon < \frac{1}{\alpha'_n + 1}\left(\frac{1}{2} - \frac{|\alpha_{\epsilon,n}|}{\alpha'_n}\frac{L}{\Lambda'_n}\right). \tag{28}$$

If the assumption $|\alpha_{\epsilon,n}| \ll \alpha'_n$ is introduced, we have the requirements

$$\epsilon < \frac{1}{2(\alpha'_n + 1)}, \tag{29}$$

and

$$\alpha'_n < \frac{1-2\epsilon}{2\epsilon}. \tag{30}$$

The requirements of ϵ and α'_n seem to be relaxed as compared with the requirements derived in previous works (Nadeborn et al. (1996); Wagner et al. (2000)). For example, if $\alpha'_n = 2$ then the upper limit of ϵ is $1/6$. In other words, if $\epsilon = 1/12$ then the upper limit of α'_n is 5. It should be noted that the requirements denoted in Eqs. (29) and (30) are based on the assumption of $|\alpha_{\epsilon,n}| \ll \alpha'_n$. If there has been a small error of the phase difference sensitivity ratio such as $|\alpha_{\epsilon,n}|/\alpha'_n = 1/40$, the upper limit of ϵ and α'_n become small to $1/12$ and 2 respectively with $\Lambda'_n = 2.5$ mm and $L = 25$ mm.

Let us consider how accurate an estimation of the phase difference sensitivity ratio is required for the phase unwrapping. By solving Eq. (28) for $|\alpha_{\epsilon,n}|$, we obtain the condition

$$|\alpha_{\epsilon,n}| < \alpha'_n \frac{\Lambda'_n}{L}\left(\frac{1}{2} - \epsilon(\alpha'_n + 1)\right). \tag{31}$$

This shows that the upper limit of the measurement error of the sensitivity ratio for correctly performing the phase unwrapping become small in proportion to $\alpha'_n\Lambda'_n/L$. This means that α_n should be precisely estimated for measuring a large object height by using a small synthetic wavelength. For example, if $\alpha'_n = 2$, $\Lambda'_n = 2.5$ mm, $L = 25$ mm and $\epsilon = 1/12$, then $|\alpha_{\epsilon,n}|$ should be less than $1/20$.

Let us consider how precise the measurement of the wavelengths is required to be for the phase unwrapping. Let the true value of wavelength be λ'_n, and the true and the measurement error of the wavelength difference are defined as $\Delta\lambda'_n$ and $\Delta\lambda_{\epsilon,n}$. By calculating the error propagation of $\Delta\lambda_n$ in Eq. (15), we have

$$\frac{\alpha_{\epsilon,n}}{\alpha'_n}\frac{L}{\Lambda'_n} = \left(\frac{\Delta\lambda_{\epsilon,n-1}}{\Delta\lambda'_{n-1}} - \frac{\Delta\lambda_{\epsilon,n}}{\Delta\lambda'_n}\right)\frac{L\Delta\lambda'_n}{\bar{\lambda}_n^{2}}. \tag{32}$$

Let the maximum of $|\Delta\lambda_{\epsilon,n}|$ for all n be $\Delta\lambda_\epsilon$:

$$|\Delta\lambda_{\epsilon,n}| \le \Delta\lambda_\epsilon. \tag{33}$$

By substituting Eqs. (32) and (33) into Eq. (28) and solving for $\Delta\lambda_\epsilon$, we obtain the condition

$$\Delta\lambda_\epsilon < \frac{\bar{\lambda}_n^{2}}{L}\left(\frac{1}{2(\alpha'_n+1)} - \epsilon\right). \tag{34}$$

For example, $\alpha'_n = 2$, $\bar{\lambda}_n = 800$ nm, $L = 25$ mm, and $\epsilon = 1/12$, $\Delta\lambda_\epsilon$ should be less than 0.002 nm.

5. Scale adjustment for phase calculation

In the Fresnel-transform based method (FTM), the pixel size of the reconstructed image increases with the reconstruction distance and the recording wavenumber. The variation of the pixel size poses problems in holographic interferometry (Yamaguchi et al. (2006)). In contrast, the convolution method (CM) (Yamaguchi et al. (2002)) keeps the pixel size of the reconstructed image the same as the pixel size of an image sensor used for hologram recording. However, the CM requires huge computational time with respect to the FTM. A zero padding method was proposed to control the pixel size of the image reconstructed by the FTM (ALFIERI et al. (2006); Ferraro et al. (2004)). In this method, an increase in the total pixel number of the hologram decreases the pixel size of the reconstructed image and can cause an explosive increase of the computational time as a result of the impossibility of the fast Fourier transform (FFT) application in the reconstruction. Algorithms based on an approach of splitting the reconstruction process into two diffraction processes were reported (Yu & Kim (2006); Zhang et al. (2004)). The methods require almost the twice computational time with respect to the FTM.

We present a simple and fast adjustment of pixel size in the reconstructed image. We adjust the pixel size by magnifying a recorded hologram before reconstruction of the hologram. The total pixel number of the hologram can be maintained for the possibility of the FFT application. Only one time calculation of the diffraction is required. Let consider that a hologram is recorded with the wavelength of λ_1, and an another hologram is recorded with the wavelength of λ_2. By application of a image magnification into the second hologram plane, the pixel sizes of the images reconstructed from the holograms are adjusted to be the same. Let the rescaled pixel size of the second hologram in a hologram plane be $\Delta_{x2} \times \Delta_{y2}$. For the adjustment, Δ_{x2} and Δ_{y2} should satisfy

$$\Delta_{x2} = \Delta_x \frac{\lambda_2}{\lambda_1}, \ \Delta_{y2} = \Delta_y \frac{\lambda_2}{\lambda_1}. \tag{35}$$

Experiments are performed to demonstrate the effectiveness of our pixel adjustment method. A test object was a Japanese one yen coin. Intensity and phase distributions of an image reconstructed from a hologram are shown in Fig. 2. As shown in Fig. 2 (b), the phase distribution has a random structure due to the light scattering from the diffusely surface of the object.

(a) Intensity distribution. (b) Phase distribution.

Fig. 2. Reconstructed image of hologram.

Holograms were recorded through changing the wavelength by injection current control of the laser diode. Three holograms with the recording wavelengths of 783.39 nm , 783.59 nm and 785.14 nm were used for two-wavelength holographic interferometry. Figure 3 shows the phase differences extracted from the images reconstructed from the holograms using no scale adjustment. As shown in Fig. 3 (a), it is clear that the phase subtraction was correctly performed on the entire object with the wavelength difference of 0.20 nm. In contrast, the phase difference with the wavelength difference of 1.75 nm shown in Fig. 3 (b) has random structures in the right and the lower side. The random structures were caused by incorrect phase subtraction due to the changes in the pixel size of the reconstructed image.

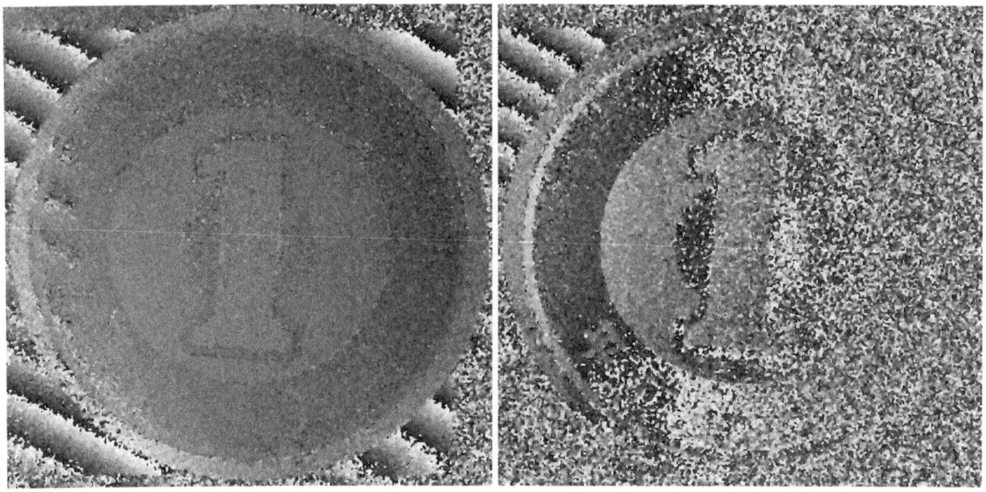

(a) $\Delta\lambda = 0,20$ nm (b) $\Delta\lambda = 1.75$ nm

Fig. 3. Phase difference distributions calculated using no scale adjustment.

The present method was applied to solve the problem in the phase subtraction for two-wavelength holographic interferometry. The bilinear interpolation (Kreis (2005)) was used for image magnification. Figure 4 shows phase difference distributions with a wavelength difference of 1.75 nm calculated using scale adjustment by application of the bilinear interpolation after and before reconstruction of the hologram. It is seen that a random structure appears on a part of the phase difference shown in Fig. 4 (a). As shown in Fig. 4 (b), it is clear that the phase subtraction was correctly performed on the entire object. This means that the present method completely solved the problem in the phase subtraction due to the changes in the pixel size.

6. Experiment

Figure 5 shows an optical setup for hologram recording. The light source was a laser diode (Hitachi HL7851G) with a wavelength of 785 nm and an output power of 50 mW. The image sensor was a CCD camera (Prosilica EC1350) having pixels of 1360 × 1024 with a pitch of 4.65 μm × 4.65 μm.

A beam emitted from the laser diode was collimated by a objective lens OL1 and then cleaned by a objective lens OL2 and a pinhole. The beam was again collimated by a lens L1 and separated into two beams by a half mirror. One was expanded by lenses L2 and L3 and reflected by a half mirror to illuminate the object. The angle of incidence θ was equal to zero.

(a) (b)

Fig. 4. Phase difference distributions with a wavelength difference of 1.75 nm calculated using scale adjustment by application of the bilinear interpolation (a) after- and (b) before-reconstruction of the hologram.

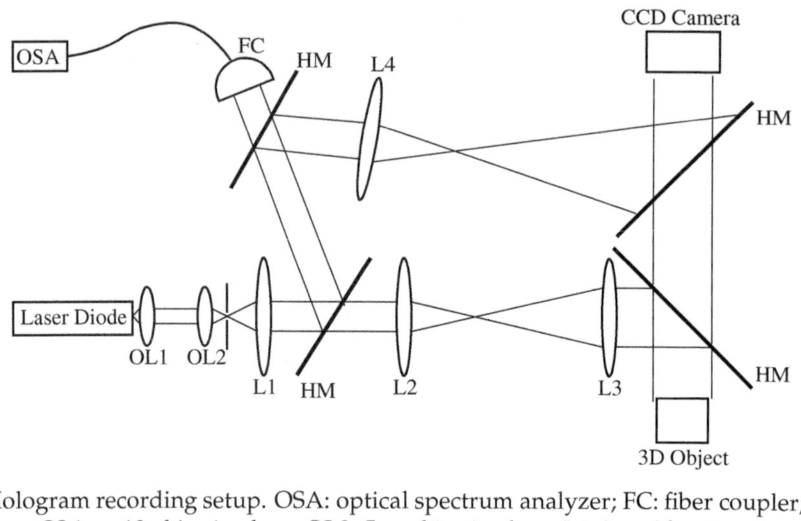

Fig. 5. Hologram recording setup. OSA: optical spectrum analyzer; FC: fiber coupler; HM: half mirror; OL1: × 10 objective lens; OL2: 5 × objective lens; L1: lens (f =60 mm); L2: lens (f=70 mm); L3: lens (f=170 mm); L4: lens (f=60 mm).

The other beam was again splited by a half mirror. The light transmitted through the half mirror was coupled into a fiber coupler and detected by an optical spectrum analyzer. The light reflected by the half mirror was focused by a lens L4 and became a reference point light source.

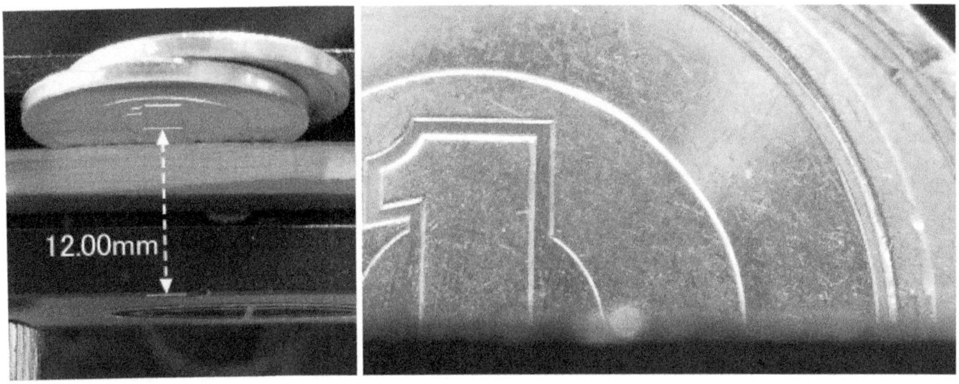

(a) Top side of view. (b) Front side of view.

Fig. 6. Photograph of a test object.

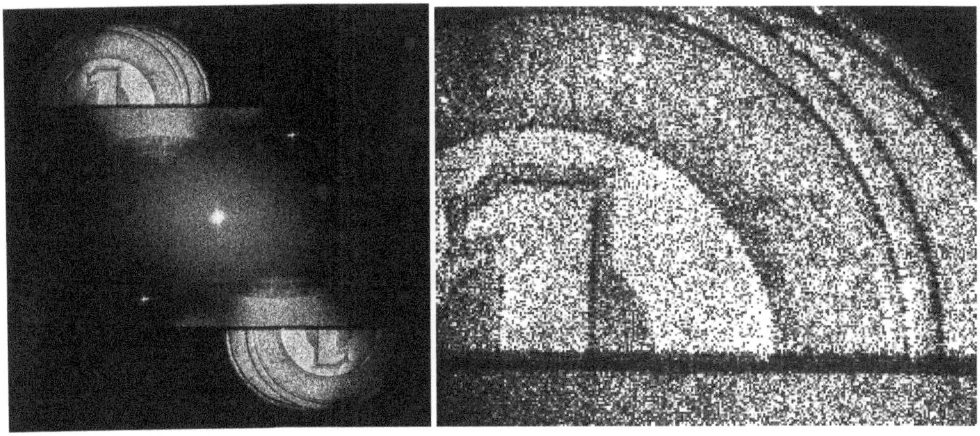

(a) The entire image. (b) Closeup in a 1st-order image.

Fig. 7. Image reconstructed from a hologram.

The distance between the object and the camera was 300 mm. The field of view of a reconstructed object image was 50 mm × 50 mm. As shown in Fig.6, a test object was a Japanese one yen coin.
For keeping an aspect ratio of a reconstructed object image equal to 1.0, the central region with 1024 × 1024 pixels of a hologram was used for the object image reconstruction.
The reconstructed image is shown in Fig. 7. The object light and its conjugate were reconstructed at the upper left and the lower right. Two bright points in the upper right and the lower left are produced by reflection at the back surface of a half mirror.
Holograms were recorded through changing the wavelength of a laser diode by injection current and operation temperature controls. The power spectral distribution of the laser light was measured using an optical spectrum analyzer(Ando Electric AQ-63515A) during hologram recording. The specification of the analyzer was as follows: the wavelength accuracy was 0.05 nm, the resolution was 0.05 nm, and the repeatability was 0.005 nm.

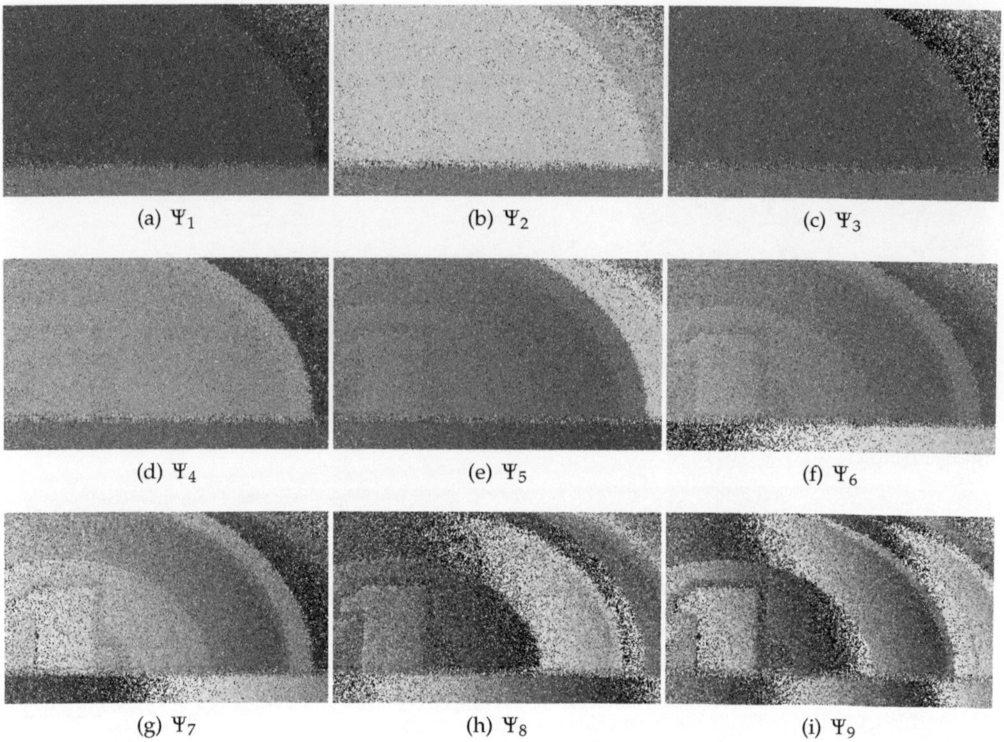

Fig. 8. Phase differences.

The central wavelength was obtained through the calculation of the center of gravity in the spectrum distribution measured using the optical spectrum analyzer.
Ten holograms for phase difference extraction were chosen from the holograms. The phase differences Ψ_n were extracted from the object images reconstructed from the holograms with wavelengths λ_n. Figure 8 shows Ψ_n. It can be seen in Fig. 8 that the phase differences have a salt-and-pepper noize produced by speckle noize. As shown in later, the noize can be suppressed by image processing method such as smoothing and median filtering (Yamaguchi (2001); Yamaguchi et al. (2006)).
The unwrapped phase differences $\Delta\phi_n$ were retrieved by the recursive calculations of Eq. (19). The synthetic wavelengths were calibrated (Wada et al. (2008)) by the comparison between the phase differences and the object height measured by using a slide caliper. The calibrated synthetic wavelengths were $\Lambda_1 = 129.1$mm, $\Lambda_2 = 34.02$mm, $\Lambda_3 = 11.47$mm, $\Lambda_4 = 8.593$mm, $\Lambda_5 = 4.914$mm, $\Lambda_6 = 2.849$mm, $\Lambda_7 = 1.380$mm, $\Lambda_8 = 0.9061$mm, $\Lambda_9 = 0.4637$mm.
The object profile was obtained using $\Delta\phi_n$. Since the incident angle θ for the object illumination was equal to zero, $L = 2h$ and

$$h = \frac{\Lambda_n \Delta\phi_n}{4\pi}. \tag{36}$$

Because $\Delta\phi_1 = \Psi_1$ and a phase difference within $(-\pi, \pi]$ corresponds to a object height within $(-\Lambda_n/4, \Lambda_n/4]$, a measurable step height is $\Lambda_1/4$ = 32 mm. The object height distributions calculated by $\Delta\phi_n$ with Λ_n are shown in Fig. 9. Figure 10 shows the plot of the object heights along a line denoted in white and black in Fig. 9 (i) as a function of the lateral positions. Figure 10 shows that the step-heights of 0.1 mm and 12 mm were correctly detected.

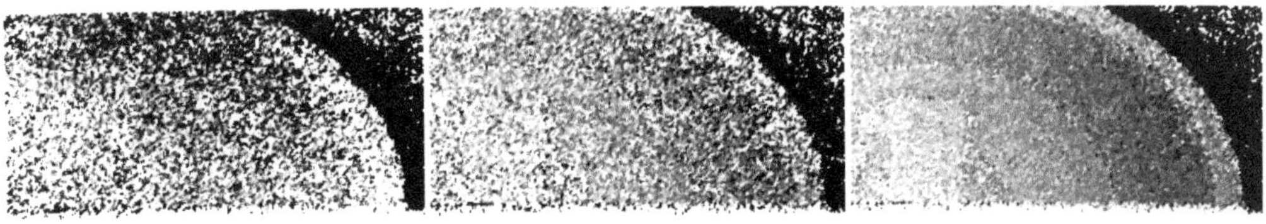

(a) $\Delta\phi_1$ (b) $\Delta\phi_2$ (c) $\Delta\phi_3$

(d) $\Delta\phi_4$ (e) $\Delta\phi_5$ (f) $\Delta\phi_6$

(g) $\Delta\phi_7$ (h) $\Delta\phi_8$ (i) $\Delta\phi_9$

Fig. 9. An object height calculated from $\Delta\phi_n$

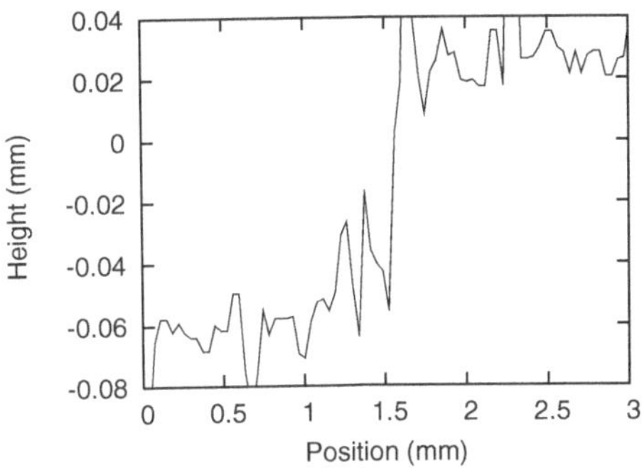

(a)

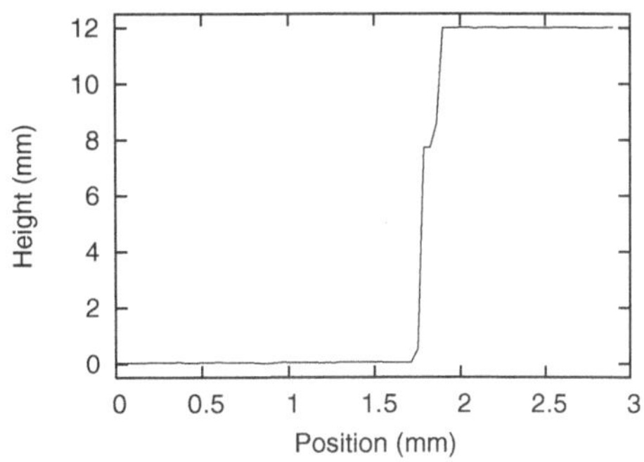

(b)

Fig. 10. Plot of the object heights along a line denoted in (a) white and (b) black in Fig.9 (i) as a function of lateral positions.

7. Conclusion

Multiple-wavelength digital holographic interferometry using tunablity of laser diodes for measuring a large step-height with high accuracy was presented. The requirements for performing the phase unwrapping were discussed. We have found that precise knowledge of the recording wavelengths is required for correctly performing the phase unwrapping. The required precision of the knowledge was derived. A simple and fast algorithm for pixel size adjustment was presented. It has been demonstrated that the problem of the phase subtraction in two-wavelength holographic interferometry can be solved by the present method.
Several holograms were recorded through the changes in the wavelength of a laser diode by injection current and operation temperature controls. A pair of holograms with a small wavelength difference less than 0.01 nm was recorded and used for realizing holographic interferometry with a large synthetic wavelength more than 120 mm. Phase differences with synthetic wavelengths from 0.4637 mm to 129.1 mm were extracted by using the holograms. The synthetic wavelengths were calibrated by the comparison between the phase differences and the object heights measured by using a slide caliper. The step-heights of 0.1 mm and 12 mm were correctly detected.

8. References

ALFIERI, D., COPPOLA, G., DENICOLA, S., FERRARO, P., FINIZIO, A., PIERATTINI, G. & JAVIDI, B. (2006). Method for superposing reconstructed images from digital holograms of the same object recorded at different distance and wavelength, *Optics Communications* **260**(1): 113–116.
URL: *http://dx.doi.org/10.1016/j.optcom.2005.10.055*

Asundi, A. & Wensen, Z. (1998). Fast Phase-Unwrapping Algorithm Based on a Gray-Scale Mask and Flood Fill, *Applied Optics* **37**(23): 5416.
URL: *http://ao.osa.org/abstract.cfm?URI=ao-37-23-5416*

Cheng, Y.-Y. & Wyant, J. C. (1984). Two-wavelength phase shifting interferometry, *Applied Optics* **23**(24): 4539.
URL: *http://ao.osa.org/abstract.cfm?URI=ao-23-24-4539*

Cheng, Y.-Y. & Wyant, J. C. (1985). Multiple-wavelength phase-shifting interferometry, *Applied Optics* **24**(6): 804.
URL: *http://ao.osa.org/abstract.cfm?URI=ao-24-6-804*

de Groot, P. (1991). Three-color laser-diode interferometer, *Applied Optics* **30**(25): 3612.
URL: *http://ao.osa.org/abstract.cfm?URI=ao-30-25-3612*

Ferraro, P., De Nicola, S., Coppola, G., Finizio, A., Alfieri, D. & Pierattini, G. (2004). Controlling image size as a function of distance and wavelength in Fresnel-transform reconstruction of digital holograms, *Optics Letters* **29**(8): 854.
URL: *http://ol.osa.org/abstract.cfm?URI=ol-29-8-854*

Friesem, A. A. & Levy, U. (1976). Fringe formation in two-wavelength contour holography, *Applied Optics* **15**: 3009–3020.
URL: *http://ao.osa.org/abstract.cfm?id=20668*

Gass, J., Dakoff, A. & Kim, M. K. (2003). Phase imaging without 2π ambiguity by multiwavelength digital holography, *Optics Letters* **28**(13): 1141–1143.
URL: *http://www.opticsinfobase.org/abstract.cfm?URI=ol-28-13-1141*

Goodman, J. W. (1968). *Introduction to Fourier Optics*, McGraw-Hill, New York.

Heflinger, L. O. & Wuerker, R. F. (1969). HOLOGRAPHIC CONTOURING VIA MULTIFREQUENCY LASERS, *Applied Physics Letters* **15**: 28.
URL: *http://scitation.aip.org/getabs/servlet/GetabsServlet?prog=normal&id=APPLAB000015000001000028000001&idtype=cvips&gifs=yes*

Hildebrand, B. P. & Haines, K. A. (1967). Multiple-wavelength and multiple-source holography applied to contour generation, *Journal of the Optical Society of America* **57**: 155–162.
URL: *http://www.opticsinfobase.org/abstract.cfm?id=75509*

Kreis, T. (2005). *Handbook of holographic interferometry: optical and digital methods*, Wiley-VCH.
URL: *http://books.google.com/books?id=H5nO1i7CR8cC&pgis=1*

Nadeborn, W., Andrä, P. & Osten, W. (1996). A Robust Procedure for Absolute Phase Measurement, *Optics and Lasers in Engineering* **24**(2): 245–260.
URL: *http://www.sciencedirect.com/science?_ob=GatewayURL&_origin=ScienceSearch&_method=citationSearch&_piikey=0143816695000178&_version=1&_returnURL=&md5=91f687b89ebc9a690103a69e85082d62*

Parshall, D. & Kim, M. K. (2006). Digital holographic microscopy with dual-wavelength phase unwrapping, *Applied Optics* **45**(3): 451–459.
URL: *http://www.ncbi.nlm.nih.gov/entrez/query.fcgi?cmd=Retrieve&db=pubmed&dopt=Abstract&list_uids=16463728*

Paulsson, L., Sjdahl, M., Kato, J. & Yamaguchi, I. (2000). Temporal Phase Unwrapping Applied to Wavelength-Scanning Interferometry, *Applied Optics* **39**(19): 3285–3288.
URL: *http://ao.osa.org/abstract.cfm?customerkey=422ab3c3-851b-405f-9df2-adcd7de575fb&id=62589*

Pedrini, G., Fröning, P., Tiziani, H. J. & Gusev, M. E. (1999). Pulsed digital holography for high-speed contouring that uses a two-wavelength method, *Applied Optics* **38**: 3460–3467.
URL: *http://www.opticsinfobase.org/abstract.cfm?id=44295*

Servin, M., Marroquin, J. L., Malacara, D. & Cuevas, F. J. (1998). Phase Unwrapping with a Regularized Phase-Tracking System, *Applied Optics* **37**(10): 1917.
URL: *http://ao.osa.org/abstract.cfm?URI=ao-37-10-1917*

Wada, A., Kato, M. & Ishii, Y. (2008). Multiple-wavelength digital holographic interferometry using tunable laser diodes, *Applied Optics* **47**(12): 2053.
URL: *http://ao.osa.org/abstract.cfm?URI=ao-47-12-2053*

Wagner, C., Osten, W. & Seebacher, S. (2000). Direct shape measurement by digital wavefront reconstruction and multiwavelength contouring, *Optical Engineering* **39**: 79–85.
URL: *http://spiedl.aip.org/getabs/servlet/GetabsServlet?prog=normal&id=OPEGAR000039000001000079000001&idtype=cvips&gifs=yes*

Wagner, C., Seebacher, S., Osten, W. & Jüptner, W. (1999). Digital Recording and Numerical Reconstruction of Lensless Fourier Holograms in Optical Metrology, *Applied Optics* **38**: 4812–4820.
URL: *http://ao.osa.org/abstract.cfm?id=44347*

Yamaguchi, I. (2001). Surface contouring by phase-shifting digital holography, *Optics and Lasers in Engineering* **36**(5): 417–428.
URL: *http://dx.doi.org/10.1016/S0143-8166(01)00069-0*

Yamaguchi, I., Ida, T., Yokota, M. & Yamashita, K. (2006). Surface shape measurement by phase-shifting digital holography with a wavelength shift, *Applied Optics* **45**(29): 7610–7616.

URL: *http://www.ncbi.nlm.nih.gov/entrez/query.fcgi?dispmax=50&DB=pubmed&term=Surface+shape+measurement+by+phase-shifting+digital+holography*

Yamaguchi, I., Matsumura, T. & Kato, J.-i. (2002). Phase-shifting color digital holography, *Optics Letters* **27**(13): 1108.
URL: *http://ol.osa.org/abstract.cfm?URI=ol-27-13-1108*

Yonemura, M. (1985). Wavelength-change characteristics of semiconductor lasers and their application to holographic contouring, *Optics L etters* **10**: 1–3.
URL: *http://ol.osa.org/abstract.cfm?id=8352*

Yu, L. & Kim, M. K. (2006). Pixel resolution control in numerical reconstruction of digital holography, *Optics Letters* **31**(7): 897.
URL: *http://ol.osa.org/abstract.cfm?URI=ol-31-7-897*

Zhang, F., Yamaguchi, I. & Yaroslavsky, L. P. (2004). Algorithm for reconstruction of digital holograms with adjustable magnification, *Optics Letters* **29**(14): 1668.
URL: *http://ol.osa.org/abstract.cfm?URI=ol-29-14-1668*

Single-Shot Phase-Shifting Digital Holography Based on the Spatial Carrier Interferometry and Its Tolerance Analysis

Yasuhiro Harada, Aizuddin Wan and Hiroyasu Sone
Kitami Institute of Technology
Japan

1. Introduction

In digital holography (Kreis, 2005; Schnars & Jueptner, 2005), holograms are recorded using an image sensor such as CCD cameras, and reconstruction of images of the object is performed by a computer. From this, digital holography has following attractive features: A wet process for developing recording media is not required; qualitative evaluation is easy for three-dimensional images of objects; and focused images of three-dimensional objects at a desired depth can be instantaneously acquired without a mechanical focusing process. However, this technology suffers from lower resolution of image sensors than photographic materials, and therefore, object size to be recorded is restricted to be large and far from the image sensor in off-axis recording geometry. These limitations can be moderated by introducing the phase-shifting method (Schreiber et al., 2007) that enables us to derive complex amplitude of the object wave on the image sensor directly from three or more phase-shifted holograms recorded in the in-line geometry (Yamaguchi & Zhang, 1997). However, since this method requires to record at least three or four phase-shifted hologram, it is useless for instantaneous measurement of the dynamic object.

In order to resolve this problem, various single-shot technologies with phase-shifting features in subsequent holograms are proposed(Awatsuji et al., 2004; Nomura et al., 2006; Toge et al., 2008; Wyant, 2003). All of these methods are based on space-division multiplexing of the phase-shifted holograms in the single recording but device and technique used are different each other. In the former three methods, special micro-element arrays of phase retarders (Awatsuji et al., 2004) or polarizers (Nomura et al., 2006; Wyant, 2003) suitable to a CCD array for recording are strictly required, and therefore, any satisfactory result for practical instrumentations has not been reported until now. On the other hand, the last method only require an introduction of off-axis plane wave illumination to an image sensor for recording single holograms (Toge et al., 2008). From its simplicity, this method seems to be one of most practical candidate for single-shot digital holography. However, it should be point out that we must align the incident angle of the reference wave plane wave so that the relative phase difference between adjacent pixels of the image sensor becomes $2\pi/3$ or $\pi/2$ radian. This means that very rigorous alignment in reference wave illumination is required. Furthermore, this requirement may be more serious problem in the case of color digital holography where multicolor laser light is utilized. In this situation, optical alignment should be independently performed for each wavelength but there are no description about care of the angle of incidence of each reference wave (Toge et al., 2008).

In the present chapter, we concentrate our attention on the problem of the tolerance of the incidence angle of the plane reference wave in the method proposed by (Toge et al., 2008). We theoretically analyze the problem in the Fourier domain using the principle of single-shot phase-shifting digital holography based on the spatial carrier interferometry (Harada et al., 2008; 2009). From our theoretical analysis, importance in the alignment of the plane reference wave is brought to light and special procedure to be applied under the situation, where the required condition for the incident angle of the plane reference wave is not satisfied, is given.

2. Theoretical analysis of the single-shot scheme

Figure 1 (a) shows a schematic diagram of optical geometry for recoding holograms in the method proposed by (Toge et al., 2008). It is just a off-axis geometry of digital holography where θ is an incident angle of the plane reference wave, λ the wavelength of the light in the holographic recording. We assume the interval of photodiodes in an image sensor is d. If the incident angle, θ, of the reference plane wave satisfies the equation given by

$$\frac{\lambda}{\sin\theta} = 3d \quad \text{or} \quad \sin\theta = \lambda/3d \tag{1}$$

phase of the reference wave along a surface of the image sensor changes periodically with spatial interval of $3d$. Therefore, the single-shot fringe pattern –hologram– recorded by the image sensor can be considered as a sum of three phase-shifted sub-holograms with constant phase difference of $2\pi/3$ in the reference waves as shown in Fig. 1 (b). Therefore, after interpolating the missing data point in each sub-hologram, complex amplitude of the object wave at the image sensor can be reconstructed by using three-step algorithm of phase-shifting interferometry (Schreiber et al., 2007).

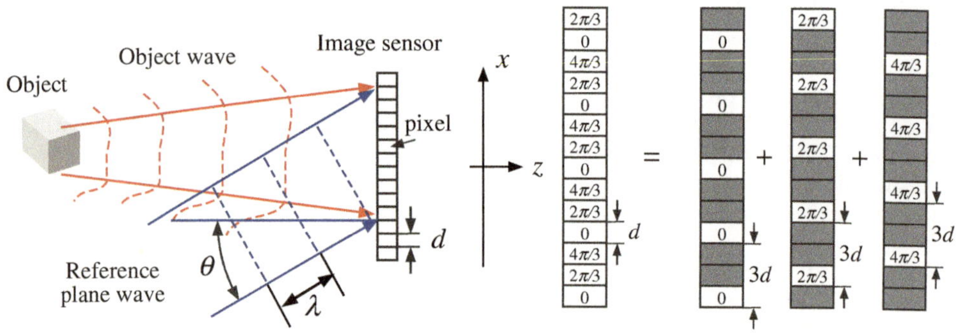

(a) Recording geometry

(b) resultant phase distributions at the image sensor under the condition of $\lambda/\sin\theta = 3d$

Fig. 1. Schematic diagrams of recording geometry (a) and resultant phase distributions at the image sensor under the condition of $\lambda/\sin\theta = 3d$ (b).

2.1 Hologram intensity distribution and its spectral distributions

Now let us consider mathematical expression of interference fringe on the image sensor and its sampled version –digital hologram intensity– obtained by sampling detection by the image sensor. If we represent object and reference waves at the image sensor plane by

$$U_O(x,y) = a(x,y)\exp\{i\phi(x,y)\}, \tag{2}$$

$$U_R(x,y) = A\exp\{i2\pi\xi_0 x\}, \tag{3}$$

respectively, the hologram intensity, $I(x,y)$, at the image sensor is given by

$$\begin{aligned} I(x,y) &= |U_R(x,y)+U_O(x,y)|^2 \\ &= A^2 + a^2(x,y) + 2Aa(x,y)\cos\{2\pi\xi_0 x + \phi(x,y)\} \\ &= b(x,y) + c(x,y)\exp(i2\pi\xi_0 x) + c^*(x,y)\exp(-i2\pi\xi_0 x) \end{aligned} \tag{4}$$

where

$$b(x,y) = A^2 + a^2(x,y), \tag{5}$$

$$c(x,y) = \frac{1}{2}Aa(x,y)\exp\{i\phi(x,y)\}, \tag{6}$$

$$\xi_0 = \frac{\sin\theta}{\lambda}, \tag{7}$$

respectively. ξ_0 means a carrier frequency in the interferogram introduced by the oblique incidence of the plane reference wave. One-dimensional Fourier transform of $I(x,y)$ with regard only to x axis gives

$$\tilde{I}(\xi,y) \equiv \mathcal{F}_x[I(x,y)] = B(\xi,y) + C(\xi-\xi_0,y) + C^*(\xi+\xi_0,y) \tag{8}$$

where $\tilde{I}(\xi,y) = \mathcal{F}_x[I(x,y)]$, $B(\xi,y) = \mathcal{F}_x[B(x,y)]$, and $C(\xi,y) = \mathcal{F}_x[c(x,y)]$, respectively. From this expression, it is clear that information of the object wave is mainly stored in the term $C(\xi,y)$ shifted by the carrier frequency $\xi_0 = \sin\theta/\lambda$ in the spectral domain. Figure 2(a) shows its schematic representation.

2.2 Image sensor as a sampling array and its spectral distributions

In digital holography, hologram intensity distribution is detected by an image sensor where the detection of light intensity is performed as spatial sampling by the photodiode in the image sensor. Therefore image sensor can be considered as the sampling array with interval same as the pixel interval. Therefore a mathematical expression for this sampling array can given in ideal case (Gaskill, 1978) by

$$s(x,y) = \sum_m\sum_n \delta(x-nd)\delta(y-md), \tag{9}$$

and its one-dimensional Fourier transform is given by

$$\begin{aligned} \tilde{S}(\xi,y) = \mathcal{F}_x[s(x,y)] &= \sum_m\sum_n \exp(2\pi nd\xi)\,\delta(y-md) = \sum_m\sum_n \exp\left(\pi\frac{n}{\xi_N}\xi\right)\delta(y-md), \\ &= \sum_m\sum_n \delta\left(\xi-\frac{n}{d}\right)\delta(y-md) = \sum_m\sum_n \delta(\xi-n2\xi_N)\,\delta(y-md), \end{aligned} \tag{10}$$

where $\xi_N = 1/2d$ is the Nyquist frequency of the image sensor. Figure 2(b) shows its schematic representation.

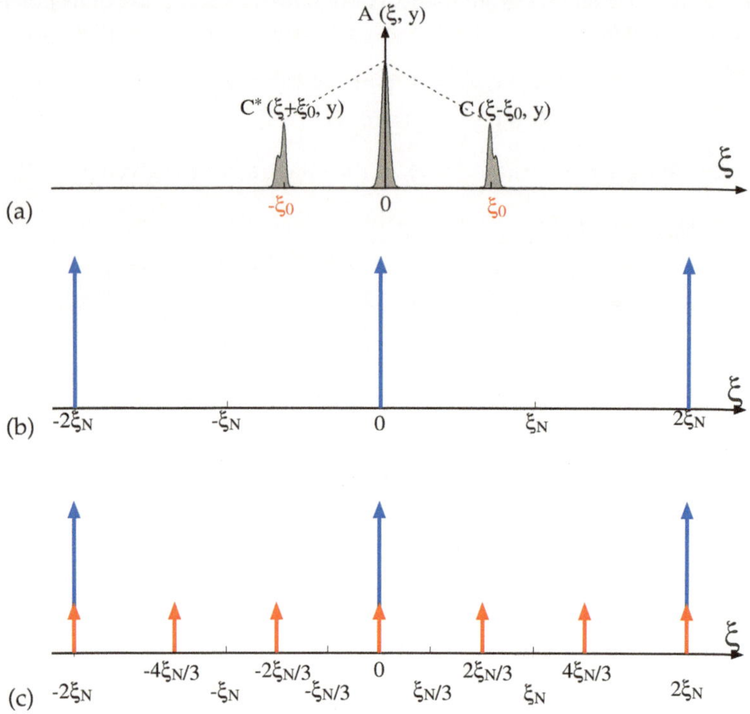

Fig. 2. Spectral distribution of interference fringes and pixel arrays of the image sensor. (a) spectral distribution of the interference fringe at the image sensor, (b) spectral distribution of a pixel array of the image sensor with spatial interval of d, and (c) that of a pixel array with the spatial interval of $3d$, respectively. Here $\xi_N = 1/2d$ is Nyquist frequency of the original pixel array of the image sensor.

Since the original sampling array of the image sensor with the interval of d in x direction can be recognized as a sum of sparse sampling arrays with an extended interval of $3d$ and spatial shift of detecting point by d, Eqs.(9) and (10) can be rewritten as

$$s(x,y) = \sum_{j=0}^{2} s^{(j)}(x,y) = \sum_{j=0}^{2} \sum_{n'} \sum_{m} \delta(x - n'3d - jd)\delta(y - md), \tag{11}$$

and

$$\tilde{S}(\xi,y) = \sum_{j=0}^{2} \tilde{S}^{(j)}(\xi,y) = \frac{1}{3d} \sum_{j=0}^{2} \sum_{n'} \sum_{m} \exp\left[-i2\pi n' \frac{j}{3}\right] \delta\left(\xi - \frac{n'}{3d}\right) \delta\left(y - md\right), \tag{12}$$

$$= \frac{2\xi_N}{3} \sum_{j=0}^{2} \sum_{n'} \sum_{m} \exp\left[-i2\pi n' \frac{j}{3}\right] \delta\left(\xi - n' \frac{2\xi_N}{3}\right) \delta\left(y - md\right), \tag{13}$$

respectively. Schematic representation of absolute value of the spectral distribution $\tilde{S}^{(j)}$ is shown in Fig. 2(c). Please note that, in addition to the widen interval of 3d, the shift of the detecting point jd is introduced in the spatial domain representation of Eq. (11). These two differences result in reduced interval of spectral array with $1/3d = 2\xi_N/3$ and new phase term whose phase value depends on the number j and spectral order n' in the spectral domain representation of Eqs. (12) and (13). For the sub-detector array $s^{(0)}(x,y)$, the additional phase for spectral array is 0 for all impulsive spectrum of the order n' since $j = 0$. In contrast to this, for the sub-detector array $s^{(1)}(x,y)$ and $s^{(2)}(x,y)$ result in the impulsive spectral arrays $\tilde{S}^{(1)}(\xi,y)$ and $\tilde{S}^{(2)}(\xi,y)$ with the phase of $-(2\pi/3)n'$ and $-(4\pi/3)n'$, respectively. As will be shown in the next subsection this phase term plays very important role in the method for single-shot phase-shifting digital holography (Harada et al., 2008; 2009).

2.3 Digital holograms detected by the image sensor and its spectral distributions

Let us consider the digital hologram detected by the image sensor of Eq. (9) and its sub-holograms which correspond to the hologram considered to be detected by individual sparse sampling array, $s^{(j)}(x,y)$, given by Eq. (11). Mathematical expression of the digital hologram, $I_s(x,y)$, detected by the image sensor is simply given by a product of the hologram intensity, $I(x,y)$, given by Eq. (4), and the sampling array, $s(x,y)$, given by Eq. (9). Therefore, spectral expression of the digital hologram, $\tilde{I}(\xi,y)$ can be given by the convolution integral between their Fourier spectra as follows:

$$\tilde{I}_s(\xi,y) = \tilde{I}(\xi,y) *_\xi \tilde{S}(\xi,y), \tag{14}$$

where $*_\xi$ stands the convolution integral with regard to the variable ξ.
In the same manner and with help of linearity shown in Eq. (11), mathematical expressions for the digital sub-holograms, $I_s^{(j)}(x,y)$ $(i = 0,1,2)$, and their one-dimensional Fourier transform, $\tilde{I}_s^{(j)}(\xi,y)$ are given by

$$I_s^{(j)}(x,y) = I(x,y) \cdot s^{(j)}(x,y), \tag{15}$$

$$\tilde{I}_s^{(j)}(\xi,y) = \tilde{I}(\xi,y) *_\xi \tilde{S}^{(j)}(\xi,y). \tag{16}$$

By substituting Eqs. (8) and (13) into Eq. (16), we obtain important expression for spectral distribution of the sub-holograms as follows:

$$\tilde{I}_s^{(j)}(\xi,y) = \frac{2\xi_N}{3}\sum_{n'}\exp\left[-i2\pi n'\frac{j}{3}\right]\left[B\left(\xi - n'\frac{2\xi_N}{3},y\right) + C\left(\xi - n'\frac{2\xi_N}{3} - \xi_0,y\right)\right.$$
$$\left.+C^*\left(\xi - n'\frac{2\xi_N}{3} + \xi_0,y\right)\right]\sum_m \delta(y - md) \quad (i = 0,1,2) \tag{17}$$

This equation implies that spectral distribution of the hologram intensity is replicated at the spatial frequency of $n'2\xi_N/3$ and modulated in phase by $-(2\pi n'/3)j$ depending on the replication order n'. According to the interpolating procedure for missing data points in the method Toge et al. (2008), we simply observe spectra falling into the Nyquist base-band of the sub-sampling array, i.e., $|\xi| < \xi_N/3$. Now we are a position to investigate the properties of the spectrum for various condition of the reference plane wave.

2.4 Considerations

In the present subsection, we investigate effects of incident angle of the reference plane wave in the single-shot phase-shifting digital holography from spectral point of view based on Eq. (17).

First case is the case where Eq. (1) is satisfied. Corresponding spectral distribution is shown in Fig. 3(a). In this case, spectral components of the hologram falling into the Nyquist base-band, $|\xi| \leq \xi_N/3$, are found to be B of $n' = 0$, C^* of $n' = +1$, and C of $n' = -1$, and their locations are all at the origin of spectral coordinate. Therefore, the three sub hologram obtained in this case are give by

$$\overline{I_s^{(0)}}(\xi, y) = \frac{2\xi_N}{3}\left[B(\xi, y) + C(\xi, y) + C^*(\xi, y)\right]\sum_m \delta(y - md), \quad (18)$$

$$\overline{I_s^{(1)}}(\xi, y) = \frac{2\xi_N}{3}\left[B(\xi, y) + C(\xi, y)\exp\left(i\frac{2\pi}{3}\right) + C^*(\xi, y)\exp\left(-i\frac{2\pi}{3}\right)\right]\sum_m \delta(y - md), \quad (19)$$

$$\overline{I_s^{(2)}}(\xi, y) = \frac{2\xi_N}{3}\left[B(\xi, y) + C(\xi, y)\exp\left(i\frac{4\pi}{3}\right) + C^*(\xi, y)\exp\left(-i\frac{4\pi}{3}\right)\right]\sum_m \delta(y - md), \quad (20)$$

where symbol $\overline{\cdots}$ denotes filter out only low-frequency component less than $\xi = \xi_N/3$. This is, in fact, performed by interpolating missing data due to sparse sampling in the space domain. As shown in Eqs.(18)—(20), important component C in the each sub hologram has the relative phase difference of $2\pi/3$, which corresponds relative phase shift in the space domain representation of the sub holograms. Therefore we are able to simply apply the three-step algorithm in phase-shifting interferometry Schreiber et al. (2007) for calculating complex amplitude of the object wave at the image sensors, *i.e.*, amplitude $a(x, y)$ and phase $\phi_W(x, y)$ (subscript $_W$ denotes *wrapped* within $-\pi$ to $+\pi$),

$$a(x, y) = \frac{1}{A}\left[\left\{\overline{I_s^{(1)}}(x, y) - \overline{I_s^{(2)}}(x, y)\right\}^2 + \left\{2\overline{I_s^{(0)}}(x, y) - \overline{I_s^{(1)}}(x, y) - \overline{I_s^{(2)}}(x, y)\right\}^2\right]^{1/2}, \quad (21)$$

$$\phi_W(x, y) = \tan^{-1}\left[\frac{\sqrt{3}\left\{\overline{I_s^{(1)}}(x, y) - \overline{I_s^{(2)}}(x, y)\right\}}{2\overline{I_s^{(0)}}(x, y) - \overline{I_s^{(1)}}(x, y) - \overline{I_s^{(2)}}(x, y)}\right]. \quad (22)$$

Next, we consider the case where Eq.(1) is not satisfied. Figure 3(b) shows the case where $\sin\theta > \lambda/3d$. Spectral components of the hologram falling into the Nyquist base-band, $|\xi| \leq \xi_N/3$, are the same as the previous case but their locations are different from the origin. Taking account this fact, the three sub holograms obtained in this case are given by

$$\overline{I_s^{(0)}}(\xi, y) = \frac{2\xi_N}{3}\left[B(\xi, y) + C\left\{\xi - \left(\xi_0 - \frac{2\xi_N}{3}\right), y\right\} + C^*\left\{\xi + \left(\xi_0 - \frac{2\xi_N}{3}\right), y\right\}\right] \times \sum_m \delta(y - md), \quad (23)$$

$$\overline{I_s^{(1)}}(\xi,y) = \frac{2\xi_N}{3}\left[B(\xi,y) + C\left\{\xi - \left(\xi_0 - \frac{2\xi_N}{3}\right), y\right\}\exp\left(i\frac{2\pi}{3}\right)\right.$$
$$\left. + C^*\left\{\xi + \left(\xi_0 - \frac{2\xi_N}{3}\right), y\right\}\exp\left(-i\frac{2\pi}{3}\right)\right]\sum_m \delta(y - md), \quad (24)$$
$$\overline{I_s^{(2)}}(\xi,y) = \frac{2\xi_N}{3}\left[B(\xi,y) + C\left\{\xi - \left(\xi_0 - \frac{2\xi_N}{3}\right), y\right\}\exp\left(i\frac{4\pi}{3}\right)\right.$$
$$\left. + C^*\left\{\xi + \left(\xi_0 - \frac{2\xi_N}{3}\right), y\right\}\exp\left(-i\frac{4\pi}{3}\right)\right]\sum_m \delta(y - md). \quad (25)$$

As shown in Eqs. (23)–(25), phase-shift relationship between these three sub holograms still remains but the components $C(\xi, y)$ and $C^*(\xi, y)$ are modulated by the spatial linear carrier with spatial frequency of $\xi_0 - 2\xi_N/3$. It comes from the misalignment of the reference plane wave. Therefore, simple application of the three-step algorithm given by Eqs. (21) and (22) will lead the wrong result. For the case of the color digital holography based on the single-shot method proposed by Sato *et al.* Toge et al. (2008), the angle of the incident satisfying Eq. (1) is inevitably different with different color of the light. Therefore, in such case, one must align the optical components for each color of the reference waves independently. This suggests that the color digital holography by Sato *et al.* Toge et al. (2008) seems to be done with very careful and technical procedures. The simple way to obtain the correct complex amplitude of the object wave at the image sensor without any careful alignment is to compensate the phase due to linear carrier with the frequency of $\xi_0 - 2\xi_N/3$ after unwrapping the phase obtained by using Eq. (22). More concretely, correct phase distribution is obtained by subtracting the following phase:

$$2\pi\left[\xi_0 - \frac{2\xi_N}{3}\right] nd \qquad (n = 0, 1, 2, \cdots). \quad (26)$$

from the unwrapped phase at each pixel. For linear-phase compensation, we have to measure additional hologram without object and derive the frequency of the linear carrier before or after the measurement of object wave.

Another extreme case where Eq. (1) is not satisfied is shown in Fig. 3(c). In this case the incident angle, θ, is too small for components Cs of the higher order of n' to fall into the Nyquist base-band, $|\xi| \leq \xi_N/3$. Therefore, there is no phase-shifting relationship between three sub holograms obtained from single off-axis hologram, and therefore, reconstruction of object waves using Eqs. (21) and (22) will be in vain both amplitude and phase. One solution to derive the correct complex amplitude in the present case is to use of the Fourier transform technique Takeda et al. (1982).

3. Verification by computer simulations

In order to demonstrate the present theoretical analysis, computer simulations are performed for each case discussed in the previous section. The object wave at the image sensor considered here is given by

$$U_O(x,y) = \exp\left[i2\pi(x^2 + y^2)^2\right] \quad (|x|, |y| \leq 1[\text{mm}]), \quad (27)$$
$$(28)$$

i.e., the phase object is assumed. Total number of pixels of the image sensor in the horizontal and vertical directions is 256. Therefore pixel size, d, is 7.8μm. Figures 4 (a) and (b) show a

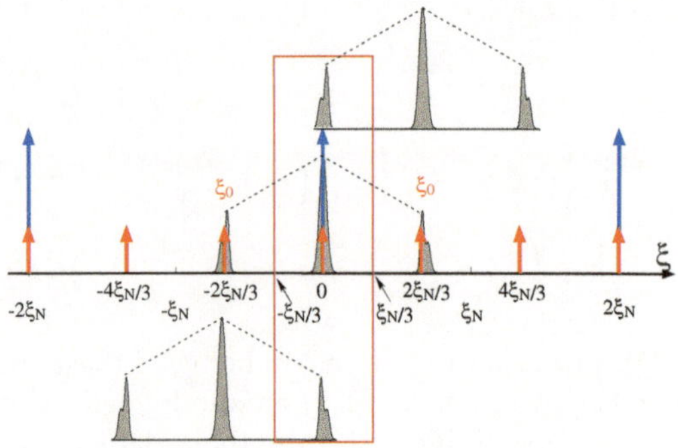

(a) $\sin\theta = \lambda/3d,\ \ \xi_0 = 1/3d = 2\xi_N/3$

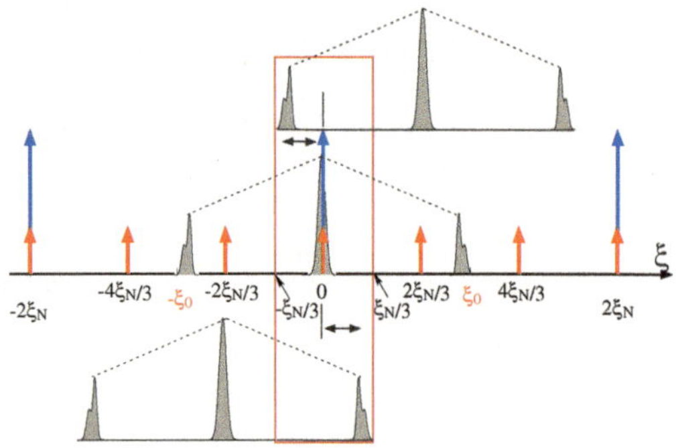

(b) $\lambda/2d > \sin\theta > \lambda/3d,\ \ 1/2d = \xi_N > \xi_0 > 1/3d = 2\xi_N/3$

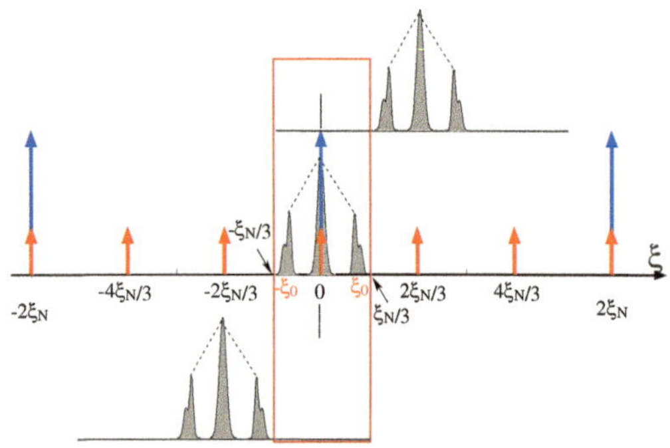

(c) $\sin\theta < \lambda/6d,\ \ \xi_0 < 1/6d = \xi_N/3$

Fig. 3. Spectral distributions of resultant sub-holograms for various conditions of the incident angle of the reference plane wave. (a) $\sin\theta = \lambda/3d$ and $\xi_0 = 1/3d = 2\xi_N/3$, (b) $\sin\theta > \lambda/3d$ and $\xi_0 > 1/3d = 2\xi_N/3$, and (c) $\sin\theta < \lambda/6d$ and $\xi_0 < 1/6d = \xi_N/3$, respectively.

phase distribution of the object wave under test in gray level and its intensity distribution of a hologram recorded by in-line geometry, respectively.

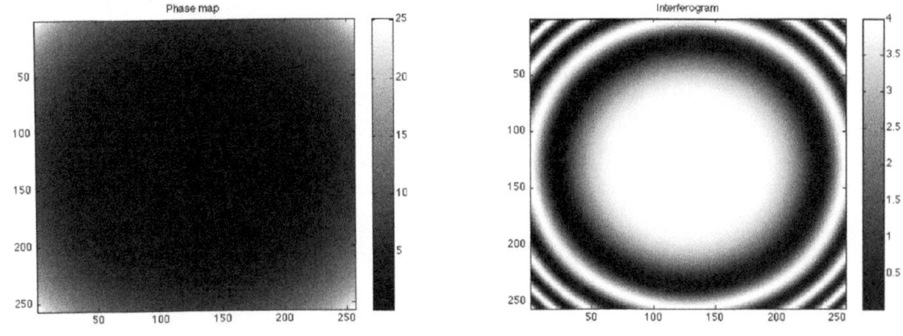

(a) (b)

Fig. 4. Object wave at the image sensor investigated in computer simulation: (a) phase distribution in gray level and (b) intensity distribution of a hologram obtained by the in-line geometry.

Figures 5(a)–(c) show intensity distributions of holograms detected by the image sensor in the off-axis recording geometry shown in Fig. 1 (a) with an angle of (a) $\sin\theta = \lambda/3d$, (b) $\sin\theta = 1.4\lambda/3d > \lambda/3d$, and (c) $\sin\theta = 0.8\lambda/6d < \lambda/6d$, respectively. Corresponding spatial carrier frequency introduced by tilted reference plane waves are (a) $\xi_0 = 1/3d = 2\xi_N/3$, (b) $\xi_0 = 1.4/3d = 2.8\xi_N/3$, and (c) $\xi_0 = 0.8/6d = 0.8\xi_N/3$, respectively. Corresponding spectral relations are schematically given in Figs. 3(a), (b), and (c), respectively. In the case of the Figs.5 (a) and (b), where side-band spectra of the higher order ($n' = \pm 1$) falls in to the Nyquist base-band of $|\xi| \leq \xi_N/3 = 1/6d$ [see Figs.3(a) and (b)], intensity distributions of the detected holograms contain aliased or moiré fringes. In contrast to them, Fig. 5 (c) shows clear interference fringes with linear carrier because there are no higher order side-band spectra in the Nyquist base-band [see Fig. 3(c)], and therefore, no moiré fringes.

Figures 5(a1)–(a3) denote sub-holograms obtained from the Fig. 5(a) by sparsely sampling with the interval of $3d$. Each of them shows similar coarse fringe pattern those obtained in the in-line geometry as shown in Fig. 4(b). Furthermore, their coarse fringes are shifted in their phase $2\pi/3$ each other. This fact confirms that phase distribution of the object wave at the image sensor can be evaluated directly by using Eq.(22) after interpolating process.

Figures 5(b1)–(b3) show sub-holograms obtained from the Fig. 5(b) by using the same manner. The coarse fringes in these figures are different shape from those of the in-line holograms shown in Figs. 5(a1)–(a3) since there are carrier fringes with a spatial frequency of $\xi_0' = |\xi_0 - 2\xi_N/3| = |2.8\xi_N/3 - 2\xi_N/3| = 0.8\xi_N/3$ due to large amount of tilt in the reference plane wave. However, it must be noted that there is the phase-shift relation between these coarse fringes with the phase value of $2\pi/3$ each other. Therefore, after interpolating process we obtain three phase-shift fringes which are available to calculate a phase distribution where the phase of object wave and linear phase of $\xi_0' \, nd = (0.6\xi_N/3)nd$ $(n = 1, 2, 3, \ldots)$ are added. Final phase distribution of the object wave at the image sensor can be obtained by subtracting the spatial linear phase $\xi_0' \, nd = (0.6\xi_N/3)nd$ $(n = 1, 2, 3, \ldots)$. This result supports our derived

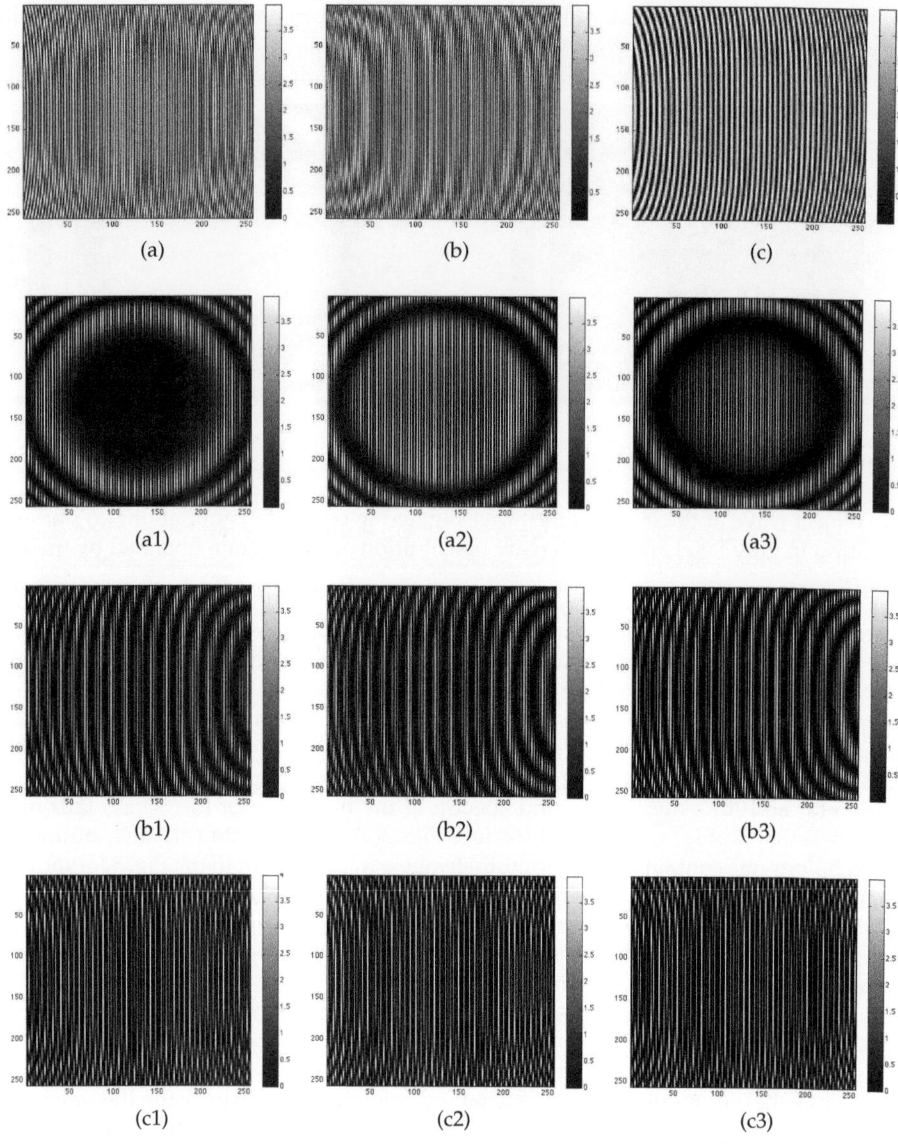

Fig. 5. Typical results of the computer simulations: (a) interference fringes, $I_s(x,y)$, for $\sin\theta = \lambda/3d$ and $\xi_0 = 1/3d = 2\xi_N/3$, (b) interference fringe, $I_s(x,y)$, for $\sin\theta = 1.4\lambda/3d > \lambda/3d$ and $\xi_0 = 1.4/3d = 2.8\xi_N/3$, (c) interference fringe, $I_s(x,y)$, for $\sin\theta = 0.8\lambda/6d < \lambda/6d$ and $\xi_0 = 0.8/6d = 0.8\xi_N/3$, (a1)–(a3) corresponding sub-holograms $I_s^{(0)}, I_s^{(1)}, I_s^{(2)}$ for $\sin\theta = \lambda/3d$, (b1)–(b3) corresponding sub-holograms $I_s^{(0)}, I_s^{(1)}, I_s^{(2)}$ for $\sin\theta = 1.4\lambda/3d > \lambda/3d$, and (c1)–(c3) corresponding sub-holograms $I_s^{(0)}, I_s^{(1)}, I_s^{(2)}$ for $\sin\theta = 0.8\lambda/6d < \lambda/6d$.

equations and procedures, *i.e.*, Eqs.(23)–(25) and compensation of the linear phase given by Eq. (26), for the case where the condition Eq. (1) is not satisfied.
Figures 5(c1)–(c3) show sub-holograms from the Fig. 5(c). In contrast to previous two cases, there are no phase-shifting relation between each sub-hologram. This result comes from the fact that there are no aliasing, in other word, no moiré fringe in the obtained hologram shown in Fig. 5(c). As predicted by our theoretical analysis we are not able not derived the phase map of the object wave at the image sensor by using the phase-shifting algorithm in this case. It will be done by using Fourier transform method Takeda et al. (1982) with the single-shot hologram of Fig. 5(c).

4. Conclusions

We have investigated theoretically tolerance of the incident angle of plane reference wave in space-division multiplexed single-shot phase-shifting digital holography presented by Sato *et al* Toge et al. (2008). It is found from our analysis that the rigorous alignment of the incident angle of the plane reference wave is very important requirement to reconstruct the correct complex amplitude of the object wave by direct application of the phase-shifting algorithm. It is also found that even if the incident angle of the plane reference wave departs from the condition given by Eq. (1), we have phase-shifted sub holograms from the single digital hologram by sparsely sampling. However, the direct application of the phase-shifting algorithm for these phase-shifted holograms results in wrong complex amplitude especially in phase distribution. We described the way to obtain correct phase distribution in this case from our theoretical analysis. That is to subtract the linear phase distribution introduced by misalignment of the reference plane wave. This solution was first derived and brought to light by our theoretical analysis within the authors knowledge, and widen field of the single-shot phase-shifting digital holography using off-axis plane wave illuminations.

5. References

Awatsuji, Y.; Sasada, M.; Kubota, T. (2004). Parallel quasi-phase-shifting digital holography, *Applied Physics Letters*, Vol.85, No.6, 1069-1071.

Gaskill, J. (1978). *Linear Systems, Fourier Transforms, and Optics*, John Wiley & Sons, ISBN:0-471-29288-5, New York.

Harada, Y.; Goto, H.; Iwaki, T.; Wakaguri, S. (2008). Single-shot phase-shifting digital holography based on the spatial carrier interferometry, *Extended Abstract on Optics & Photonics Japan 2008*, Optical Society of Japan (Japan Society of Applied Physics), Tsukuba, Japan, pp.284-285. (in Japanese).

Harada, Y.; Goto, H.; Wan, A. (2009). Spatially-multiplexed single-shot phase-shifting digital holography based on the spatial carrier interferometry, *Holographic Display Artists and Engineers Club (HODIC) Circular*, Optical Society of Japan (Japan Society of Applied Physics), Vol.20, No. 29, 35-44. (in Japanese).

Kreis, T. (2005). *Handbook of Holographic Interferometry*, Wiley VCH, ISBN:3-527-40546-1, Weinheim.

Nomura, T.; Murata, S.; Nitanai, E; Numata, T. (2006). Phase-shifting digital holography with a phase difference between orthogonal polarization, *Applied Optics*, Vol.45, No. 4873–4877.

Schnars, U. & Jueptner, W. (1978). *Digital Holography*, Springer, ISBN:3-540-21934-x, Berlin.

Schreiber, H.; Bruning, J.H.; Greivenkamp, J.E. (2007). Phase Shifting Interferometry, *in* Maralaca, D., ed., *Optical Shop Testing, Third Edition*, John Wiley & Sons, ISBN:978-0-471-48404-2, Hoboken, pp.547-666.

Takeda, M.; Ina, H.; Kobayashi, S. (1982). Fourier-transform method of fringe-pattern analysis for computer-based topography and interferometry, *Journal of Optical Society of America*, Vol.72, No.1, 156-160.

Toge, H.; Fujiwara, H.; Sato, K. (2008). One-shot digital holography for recording color 3-d images, *Proceedings of SPIE*, Vol.69120, 69120U-1 - 69120U-7.

Wyant, J.C. (2003). Dynamic Interferometry, *Optics & Photonics News* Vol.14, No.4, 36-41.

Yamaguchi, I. & Zhang, T. (1997). Phase-shifting digital holography, *Optics Letters*, Vol.22, No.16, 1268-1270.

6

Multi-Channel Adaptive Interferometers Based on Dynamic Hologram Multiplexing

Roman Romashko and Yuri Kulchin
Institute of Automation and Control Processes FEB RAS
Russia

1. Introduction

Optical interferometry was always considered as one of the most flexible and sensitive techniques for measuring mechanical vibrations (Hariharan, 1990; Osterberg, 1932). The number of applications of this method tremendously increased after the discovery of lasers (in 1960) and the development of low-loss optical fibres (in 1966). A laser as a source of coherent radiation assures high sensitivity of an interferometer while an optical fibre guarantees compactness, light weight, immunity to electromagnetic influence, and capability of the measuring system to operate in hazardous conditions (high temperature, radiation, etc.). As known, the phase of a light field cannot be directly detected. It is an interferometer, which provides the detection of the phase difference by combining the signal wave (which bears information about the measurand) with the coherent reference wave and measuring changes of the resultant light flux by a photodetector. The sensitivity of an interferometer to phase difference is limited by shot noise of the photoelectrons and can be extremely high: the theoretical minimum detectable displacement (corresponding to the phase difference) is $1.1{\cdot}10^{-16}$ m$\cdot$Hz$^{-1/2}$ for 10 mW of detected laser power at the wavelength of 500 nm (Forward, 1978; Wagner & Spicer, 1987). However, two main problems should be overcome to achieve such high sensitivity in non-laboratory conditions. The first is the necessity of precise adjustment of the object and the reference wavefronts interfering at the photodetector. This requirement makes difficult, in particular, the use of multimode optical fibres in interferometric sensors. The second is the need to keep constant the average phase shift between the interfering wavefronts. This is more fundamental than the first one because the induced phase shift (which is assumed to be proportional to influence of the measurand on the optical paths) is nonlinearly transferred into the light-flux change at the detector via the familiar cosine interference function.

A simple and elegant solution of both these problems was achieved when a conventional beam-splitter, which serves for combining the reference and the object waves, was replaced by a dynamic hologram continuously recorded in a photorefractive crystal (PRC). Since the configuration in this case involves only two waves that interfere inside the crystal, it was called two-wave mixing, TWM (Huignard & Marrakchi, 1981). Considering that the photorefractive dynamic hologram adapts not only the signal wavefront to the reference one but it is also self-adapted to slow temporal variations of the phase difference, this type of optical system is referred to as adaptive interferometry (Stepanov, 1991). The idea of using photorefractive crystals in an adaptive interferometer with multimode optical fibres was first proposed in 1980 by Hall et al. (Hall, et al., 1980). It was pointed out in this early paper that the physical mechanism of the dynamic grating formation in photorefractive crystals strongly affects phase-to-intensity transformation. The highest sensitivity to small phase excursions is achieved in the linear mode of phase detection when spatial variations of the refractive index inside the crystal are either non-shifted or shifted by a half of the grating period with respect to the interference fringes (Hall, et al., 1980). Such spatial shift occurs when the hologram is recorded in a photorefractive crystal under strong DC-field, in the so called drift regime (Young, et al., 1974). The main disadvantage of this approach is screening of the external electric field which leads to serious suppression of coupling of interfering beams in the illuminated part of the crystal. Another technical problem is overheating of the crystal under strong DC electric field. It is especially serious for fast photorefractive crystals because of their high photoconductivity. Strong overheating may even lead to the breakdown of the crystal. To prevent this DC-field should be applied to the crystal just during a short time (typically tens of milliseconds) followed by the relaxation period (typically tens of seconds). Thus the measurement can be done only in the pulse regime, which could be unacceptable for certain applications. Moreover, realization of this regime requires use of special synchronizing electronics that also complicates the measuring system as a whole. Alternatively, holographic recording in the diffusion regime (without external electric field) leads to the less sensitive, quadratic regime of phase demodulation (Kamshilin, et al., 1986). However, as it was shown later by Kamshilin and Grachev, the linear phase-to-intensity transformation can be achieved even in the diffusion regime of the hologram recording if the interfering waves have different polarization states (Kamshilin & Grachev, 2002). The last approach is based on vectorial-wave mixing (VWM) in photorefractive crystals for which the theory was developed by Sturman et al. (Sturman, et al., 1999). It is worth noting that an adaptive interferometer based on the technique proposed in (Kamshilin & Grachev, 2002) operates in so called enhanced diffusion mode when a strong AC-electric field is applied to a crystal for enhancing wave coupling. Further development of VWM-based approach has allowed one to find solutions which it made it possible:

- to avoid use of any external electric field due to recording the hologram in pure diffusion mode in reflection geometry of wave coupling at high spatial frequencies (Di Girolamo, et al., 2007a; Romashko, et al., 2005a);
- to use an object wave with arbitrary polarization (e.g. completely depolarized) without any polarization filtering, due to recording the hologram in orthogonal geometry of VWM (Di Girolamo, et al., 2010);
- to reduce significantly light intensity used for recording a dynamic hologram without worsening interferometer sensitivity and response time (Romashko, et al., 2007).

These solutions have allowed to significantly simplify the scheme of adaptive interferometer (by removing high-voltage suppliers, synchronizing feedback loops, etc.), improve performance stability, reduce optical loses, noise and energy-consumption. As a result, adaptive interferometers become more promising tools for different practical applications.

It is worth noting that nowadays the adaptive interferometers are intensively developed and find new and new application areas. First of all is non-destructive testing of materials and elements of technical constructions. In this area adaptive interferometry becomes one of the most promising techniques due to its possibility to operate reliably in an industrial environment keeping the sensitivity at the high level approaching the classical homodyne detection limit. More specifically, adaptive interferometers are used in so-called laser ultrasound systems for non-destructive remote inspection of internal defects inside materials and constructions (Dewhurst & Shan, 1999). In these systems an acoustic wave (a sound pulse) generated by very short and powerful laser pulse propagates through a sample under the test and produces very week vibrations of the backside surface with the tiny amplitude of the order of 1 nm or even smaller. These vibrations (if they are detected by an adaptive interferometer) allow quality estimations of the tested object along the path of the sound pulse.

Another area of applications of phase demodulators based on photorefractive dynamic holograms is stabilization of fibre-optical interferometric sensors eliminating all unwanted influences of the environment (e.g., temperature changes) (Di Girolamo, et al., 2007a; Fomitchov, et al., 2002; Kamshilin, et al., 1995; Kamshilin, et al., 1998; Qiao, et al., 2006; Romashko, et al., 2005b). Adaptive interferometers find also application for real-time analysis of deformations and vibrations of technical constructions (such as space reflectors (Pauliat, et al., 2006) or even artworks (Thizy, et al., 2007)), for molecular recognition (Peng, et al., 2007), for acousto-optical imaging of biological tissues (Delaye, et al., 2005), for refractive index measurement (Lichtenberg, et al., 2005), in optical coherence-domain reflectometry systems (Peng, et al., 2003), and many others.

Among those applications there is a class of practical problems where several simultaneous measurements are required (Kulchin, 2001; Peiponen, et al., 2009). Thereby a development of multichannel adaptive system is a topical problem.

Taking into account that a key element of an adaptive interferometer is a dynamic hologram (DH) recorded in PRC, two possible strategies for development of multichannel systems can be considered. The first is based on using of several crystals, one for each channel. The advantage of this extensive approach is a possibility of providing complete independency of the system channels performance. However it leads to system complexity (due to the large number of reference beams, optical elements, etc.), disproportional increase of system cost, increase of energy consumption, reduced reliability and so on. An alternative strategy for development of a multi-channel adaptive interferometer consists in multiplexing of DHs in single PRC. Note that multiple holograms recording in same volume of photosensitive material is widely used for data storage in holography memory systems (Peiponen, et al., 2009; Steckman, et al., 2001), for correlation processing and pattern recognition (Feng, et al., 2000; Wen & Yang, 1997), in multichannel optical communication systems (An, et al., 2001; Petrov, et al., 2001), in multichannel measurement systems (Andersen, et al., 2009; Kujawinska & Robinson, 1988) and many others.

It is worth noting that static (or permanent) holograms can be multiplexed in a photosensitive material one by one. On contrary, a distinctive feature of both recording and reading out dynamic holograms is that the hologram disappears if light waves are switched off. Therefore a multiplexing of dynamic holograms is possible only in case of their simultaneous recording. This could leads to additional cross-talk between channels due to pair wise interaction of two arbitrary signal waves in the PRC. Moreover, an overlap of light fields from neighboring channels in a crystal volume could affect the parameters of the holograms and, as a consequence, the sensitivity of particular holographic channels. All these circumstances must be taken into account in the development of multichannel adaptive interferometric system.

All approaches for multiple hologram recording can be grouped into three classes: spatial, angular and spectral multiplexing techniques. In the first approach, holograms are recorded in different parts of the same crystal. This technique allows one to provide maximal independence of channel performance; the cross-talk is practically precluded. However, the number of multiplexed holograms (and therefore number of channels) is limited by the crystal size, and in most cases does not exceed a few tens (Kulchin, et al., 2000b). Moreover, an equivalent number of individual reference beams is required for recording of each hologram in this case, which also leads to complexity of the measurement system and brings its efficiency down.

Angular multiplexing of holograms allows significant increase in the number of channels formed in a single crystal. In this case the same volume of crystal can be used for recording several holograms, signal light beams can partially or even completely overlap in a crystal and a common reference beam can be used for all holograms. However cross-talk between channels becomes more probable in such a scheme. This defines one of the important problems which should be solved. It is necessary to find conditions which preclude a cross-talk between angular multiplexed holograms or the cross-talk level will be below the inherent noise of the system.

The spectral multiplexing approach is based on using different wavelengths for dynamic hologram recording. This approach naturally fits with the WDM-technique of fiber-optical sensors (especially FBG) multiplexing. As a result effective multichannel measurement systems can be created by combined these two principles. However a realization of the spectral multiplexing approach requires keeping a number of peculiarities. In particular, the spectral working range should match the spectral sensitivity of a photosensitive material (e.g. PRC). Moreover, it is necessary to take into account the possible appearance of cross-talk due to overlap of multiplexed channel spectra. The last requirement together with the first one constrains the number of channels which could be realized in a multichannel system based on spectral multiplexing.

Below in the following Sections 4-6 we consider more detailed practical realizations of multichannel adaptive interferometers and measurement systems based on the above listed principles of dynamic hologram multiplexing in a photorefractive crystal. First we give briefly the basics of wave mixing in photorefractive crystal (Section 2) and introduce a relative detection limit as a parameter which characterizes adaptive interferometer sensitivity (Section 3).

2. Two-wave mixing in photorefractive crystal: Vectorial model

The photorefractive (PR) effect known as light induced change of material refractive index was discovered in Bell Laboratory in 1966 (Ashkin, et al., 1966). In order to be photorefractive a material should have photoconductive and electro-optic properties. Then, in such a material, photo-induced charges migrate in the presence of light from regions of high optical intensity to regions of low optical intensity and eventually reach a static charge distribution. This charge distribution creates a strong, static electric field (so called space-charge field), which in turn alters the index of refraction of the material by the first-order electro-optic (Pockels) effect. Since its discovery in $LiNbO_3$ and $LiTaO_3$ the PR effect was observed in a great number of materials, including dielectrics, semiconductors, liquid crystals, organic polymers and even in liquids and gases.

The first theoretical model of the PR effect was proposed by Kukhtarev etal. in 1979 (Kukhtarev, et al., 1979). Later in 1999, Sturman et.al. developed more general and more rigorous theoretical model of two-wave mixing in PR crystal which takes into account a vectorial nature of light waves, anisotropic properties of photorefractive materials and anisotropy of light diffraction at a dynamic holographic grating (Sturman, et al., 1999).
After decades of studying the PR effect a number of fundamental monographs devoted to wave-mixing in PR materials were written (Petrov, et al., 1983; Petrov, et al., 1991; Solymar, et al., 1996; Sturman & Fridkin, 1992). These books can be recommended to the reader interested in detailed study of PR effect. In this Chapter we give just the main points of the theoretical model of vectorial wave mixing in PR crystal of cubic symmetry (Sturman, et al., 1999) which can help better understanding of further material.
Consider two coherent waves with vectorial amplitudes $\mathbf{A}_1$ and $\mathbf{A}_2$ entering a PR crystal of cubic symmetry under an external electric field $\mathbf{E}_0$ through its (xy)-face (Fig.1). In paraxial approximation valid for small angles between wave vectors $\mathbf{k}_1$ and $\mathbf{k}_2$ and axis z the wave amplitudes can be considered as 2D-vectors with x- and y-components.

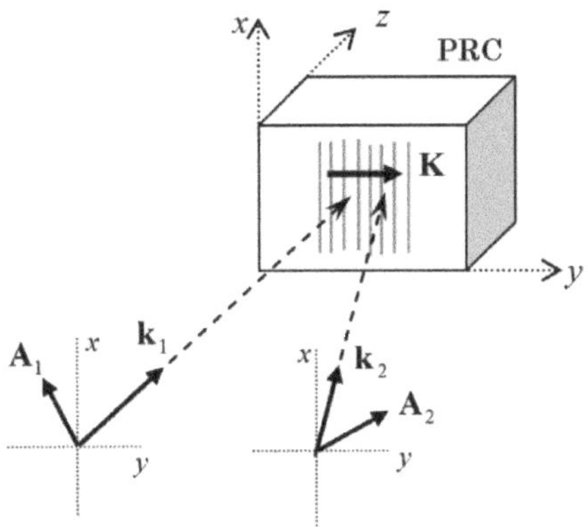

Fig. 1. Geometry of vectorial wave mixing in a photorefractive crystal

The waves' interference will result in the appearance of space charge field $\mathbf{E}_K$ which alters the crystal refractive index and forms a dynamic holographic grating with wave vector $\mathbf{K} = \mathbf{k}_1 - \mathbf{k}_2$ in plane (xy). Diffraction of mixed waves at the dynamic grating provides their coupling which results in change of wave amplitudes at the crystal output. This process is described by the system of coupled-wave equations:

$$\begin{cases} \left(\dfrac{\partial}{\partial z} + \dfrac{\alpha}{2} - i\hat{\mathbf{G}}\right)\mathbf{A}_1 = iE_K \hat{\mathbf{V}} \mathbf{A}_2, \\ \left(g\dfrac{\partial}{\partial z} + \dfrac{\alpha}{2} - i\hat{\mathbf{G}}\right)\mathbf{A}_2 = iE_K^* \hat{\mathbf{V}} \mathbf{A}_1, \end{cases} \tag{1}$$

where g is the parameter which takes into account relative direction of wave propagation in the crystal (g = +1 in transmission geometry where waves propagate in same direction, and g = -1 in reflection geometry where waves propagate in opposite directions); α is the light absorption coefficient; E_K is the amplitude of space-charge field $\mathbf{E_K}$. Using the conventional one-trap-one-band model for the charge transfer (Petrov, et al., 1991) and assuming provisionally that the applied field is parallel to the grating vector the space charge-field can be found, within the linear approximation in the contrast of the interference pattern, as

$$E_K = im\frac{(E_D + iE_0)E_q}{E_D + E_q + iE_0}, \tag{2}$$

where $m = 2\mathbf{A}_1\mathbf{A}_2^* / I_0$ is the contrast of the interference pattern, I_0 is the total light intensity, E_0 is the amplitude of external electric field applied to the crystal, while E_D and E_q are the characteristic fields,

$$E_D = \frac{Kk_BT}{e} \text{ (diffusion electric field)}, \; E_q = \frac{eN_t}{\varepsilon\varepsilon_0 K} \text{ (traps saturation field)}, \tag{3}$$

where e is the elementary charge, k_B is the Boltzmann constant, T is the absolute temperature, N_t is the effective concentration of photorefractive centers (traps), $\varepsilon\varepsilon_0$ is the dielectric constant. The matrices $\hat{\mathbf{G}}$ and $\hat{\mathbf{V}}$ are given by

$$G_{ij} = sE_0H_{ij}^{(0)} + i\rho\delta_{ijz}, \qquad V_{ij} = sH_{ij}, \tag{4}$$

here $s = -\pi n_0^3 r_{41} / \lambda$ is a material parameter; n_0 is the refractive index; r_{41} is the electro-optic coefficient; λ is the wavelength; ρ is the rotatory power; δ_{ijz} is the unit antisymmetric third rank tensor; $E_0 = |\mathbf{E_0}|$; $E_K = |\mathbf{E_K}|$. The Latin subscripts denote Cartesian components, and indices i and j assume independently the values x and y.

The tensors $\hat{\mathbf{H}}^{(0)}$ and $\hat{\mathbf{H}}$ responsible for electro-optic contribution to variation of crystal dielectric permittivity caused by external electric field $\mathbf{E_0}$ and internal space-charge field $\mathbf{E_K}$, are respectively:

$$H_{ij}^{(0)} = \frac{r_{ijl}E_{0l}}{r_{41}E_0}, \qquad H_{ij} = \frac{r_{ijl}E_{Kl}}{r_{41}E_K}, \tag{5}$$

where r_{ijl} is the electro-optic tensor. Taking into account that the space-charge field appears due to interference of mixed waves as well as a form of coupled-wave equations (1), one can conclude that tensor $\hat{\mathbf{H}}$ has the character of the wave-coupling matrix. The last becomes more clear in the case of waves mixing in non-gyrotropic crystal ($\rho = 0$) without application of external electric field ($E_0 = 0$) when the system (1) is reduced to the following

$$\begin{cases} \left(\frac{\partial}{\partial z} + \frac{\alpha}{2}\right)\mathbf{A}_1 = smE_K\hat{\mathbf{H}}\mathbf{A}_2, \\ \left(g\frac{\partial}{\partial z} + \frac{\alpha}{2}\right)\mathbf{A}_2 = -smE_K\hat{\mathbf{H}}\mathbf{A}_1. \end{cases} \tag{6}$$

As seen from Eqs.(6) in order for the waves to couple they must interfere ($m \neq 0$) with matrix $\hat{\mathbf{H}}$ being non-zero.

By solving Eqs.(1), one can find amplitudes of mixed waves at the output of the PR crystal. Note that a particular solution depends on wave mixing geometry which, in its turn, is determined by the orientation of holographic grating vector, **K**, and the directions of wave propagating (vectors $\mathbf{k}_1$ and $\mathbf{k}_2$) with respect to crystallographic axes of the crystal (Romashko, et al., 2010; Sturman, et al., 1999).

It is worth noting that the particular type of the matrix $\hat{\mathbf{H}}$ is closely related to the type of light diffraction at a dynamic hologram. Thus, in the case of pure isotropic diffraction which preserves the polarization of the diffracted light, the coupling matrix is proportional to the unit matrix, $\hat{\mathbf{H}} = const\hat{\mathbf{1}}$. In all other cases the diffraction is anisotropic and is accompanied by change of diffracted light polarization state. Particularly if the polarization plane of diffracted wave is rotated by 90 degrees with respect to the incident light the matrix is $\hat{\mathbf{H}} = \begin{pmatrix} 0 & 1 \\ 1 & 0 \end{pmatrix}$. The last case is considered as most efficient for providing a linear regime of phase demodulation in the adaptive interferometer based on a dynamic hologram recorded in PR crystal in diffusion mode, i.e. without application of external electric field (Kamshilin, et al., 2009).

3. Adaptive ineterferometer sensitivity

Since in any interferometer a physical parameter, which has to be measured, is finally encoded in the phase modulation of the light wave, its sensitivity to small phase excursions is a parameter serving as a primary criterion for comparison of different systems. It is not only the configuration of the optical scheme but also its particular realisation (the output laser power, the generated wavelength, the type of the crystal, etc.) which determines the sensitivity of an adaptive interferometer. Comparison of different adaptive interferometers is usually done by estimation the extent to which their sensitivity is worse than that of the classical lossless interferometer (Delaye, et al., 1997; de Montmorillon, et al., 1997). It was shown in early papers devoted to analysis of a classical interferometer that its sensitivity can be extremely high if the available light power is not limited and measurements of the phase modulation is carried out within a very narrow frequency band (Bershtein, 1954; Forward, 1978). Evidently, the minimum detectable phase difference $\Delta\phi^{min}$ is defined by the noise level of the measuring system. There are several sources of the noise in an optical interferometer: laser noise, thermal and shot noise of the photodetector, and noise of amplifying electronics. When the light power arriving at the photo-detector is high enough, shot noise (which is proportional to the square root of the received light power) of the photo-excited charge carriers prevails over all the other noise levels (Bershtein, 1954; Wagner & Spicer, 1987). The shot noise is the instability of the photodetector current caused by statistical fluctuation of the number of received photons. Its level is primarily defined by the average number of photons and it is given as:

$$\left\langle i_{shot}^2 \right\rangle = 2\frac{e^2\eta}{h\nu}\left\langle P_D \right\rangle \Delta f \, , \tag{7}$$

where e is the electron's charge, η is the quantum efficiency of the photo-detector, $h\nu$ is the energy of the photon, and Δf is the frequency bandwidth of the detection electronics.

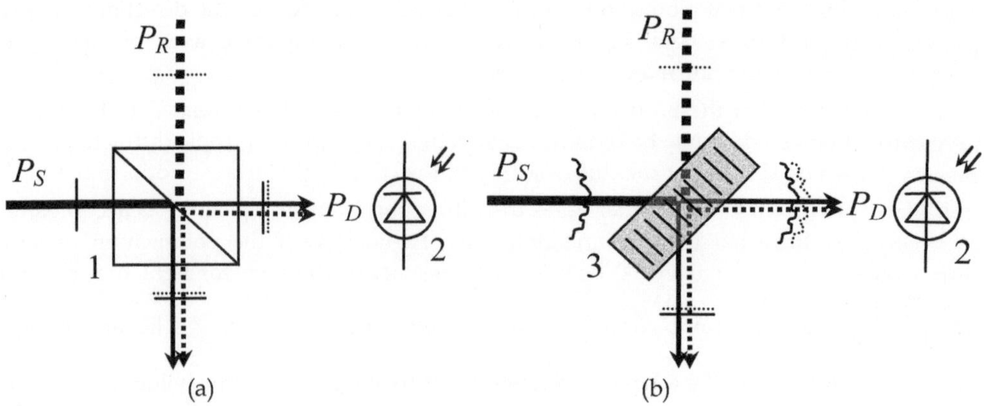

Fig. 2. Combining the signal (P_S) and the reference (P_R) waves using the conventional beam-combiner in the classical interferometer (a) and the dynamic hologram of an adaptive interferometer (b): 1 – the beam combiner; 2 – the photodetector; 3 – the photorefractive crystal

Let us now calculate the signal-to-noise ratio for phase demodulation in the classical interferometer, which is schematically shown in Fig. 2(a). As is known, the highest sensitivity of the interferometer is achieved at the quadrature condition, when the average phase difference between the interfering beams is equal to $\pi/2$ (Osterberg, 1932). When we introduce a small phase difference, $\Delta\phi$, the light power arriving at the photo-detector is

$$P_D = \frac{1}{2}\left(P_S + P_R\right) + \sqrt{P_S P_R}\sin\Delta\phi \approx P_S\frac{1+\beta}{2} + P_S\sqrt{\beta}\cdot\Delta\phi\,, \tag{8}$$

where P_R and P_S are the power of the reference and object wave, respectively, β is their ratio. Here we neglected all optical losses and assumed that the cross-sections of the interfering beam are the same. The informative signal in the interferometer is the electric current of the photodetector proportional to $\Delta\phi$. Assuming linear response of the photodetector to the incoming light power P_D, the signal is directly proportional to the second term in Eq. (8), while the first term is responsible for the shot noise given by Eq.(7). Therefore, the signal-to-noise ratio achieved in the classical interferometer is

$$SNR_C = \frac{\Delta\phi}{Q}\sqrt{\frac{2P_S}{1+\beta^{-1}}}\,, \quad \text{with} \quad Q = \sqrt{\frac{4h\nu\Delta f}{\eta}}\,. \tag{9}$$

The minimum detectable phase difference (the classical homodyne detection limit), $\Delta\phi_C^{min}$, is usually defined as the phase shift, which leads to $SNR = 1$. For large beam intensity ratio ($\beta >> 1$) Eq. (9) yields

$$\Delta\phi_C^{min} = \sqrt{\frac{2h\nu\Delta f}{\eta P_S}}\,. \tag{10}$$

Performance of any adaptive interferometer, which is schematically shown in Fig. 2(b), is conveniently compared with the performance of the classical interferometer by introducing the parameter

$$\delta_{rel} = \frac{\Delta\phi_A^{min}}{\Delta\phi_C^{min}} = \frac{SNR_C}{SNR_A}, \tag{11}$$

which can be called the relative detection limit. In Eq. (11) $\Delta\phi_A^{min}$ is the minimal detectable phase difference in the adaptive interferometer and SNR_A is the signal-to-noise ratio achieved in the adaptive interferometer by introducing the phase shift of $\Delta\phi$. On the one hand, the relative detection limit, δ_{rel}, is the ratio of the detection limits of both interferometers. On the other hand, it shows how the response of the classical interferometer to a known phase difference $\Delta\phi$ is larger than that of the adaptive interferometer.

Correct comparison of the adaptive and classical interferometers can be carried out if we use the same power of the object beam, the same photodetector and electronic circuit, and the same phase difference between the interfering beams. If the performance of the adaptive interferometer is analytically described, the relative detection limit can be calculated theoretically. Alternatively, δ_{rel} can be measured in the experiment as follows. Suppose that introduction of the small phase difference of $\Delta\phi$ in the adaptive interferometer results in a change of the light power at the photodetector by ΔP_D but the mean power, $<P_{D0}>$, remains the same. In this case the signal-to-noise ratio of the adaptive interferometer is

$$SNR_A = Q^{-1}\frac{\Delta P_D}{\sqrt{P_{D0}}}. \tag{12}$$

The light power P_{D0} is certainly related to the incident power P_S as $P_{D0} = P_S\Im$, where $\Im$ is the system transmission, which takes into account the optical losses in the crystal and all other optical elements ($\Im \le 1$), which is readily mbasured in the experiment. By introducing Eqs. (9) and (12) into Eq. (11), we get for $\beta >> 1$:

$$\delta_{rel} = \sqrt{\frac{2}{\Im}}\frac{P_{D0}}{\Delta P_D}\Delta\phi. \tag{13}$$

The relative detection limit is always larger than unity and it increases for higher optical losses (when $\Im$ decreases). The ratio $\Delta P_D/P_{D0}$ is readily measured because it is equal to the ratio of the modulated part of the photodetector current to its non-modulated part (i.e., to visibility of an oscilloscope trace when the small phase excursion of $\Delta\phi$ is applied to interferometer. Other parameters of equation (13) can be also readily estimated in the experiment. Therefore, Eq. (13) provides easy way to measure the relative detection limit δ_{rel}. This parameter is very convenient for comparison of different adaptive interferometers. Moreover, it also allows us to estimate the minimal detectable phase shift $\Delta\phi_A^{min}$ for the particular adaptive interferometer as $\Delta\phi_A^{min} = \delta_{rel}\Delta\phi_C^{min}$.

In following Sections we consider different schemes for multichannel adaptive interferometers based on spatial, angular and spectral multiplexing of dynamic holograms in a photorefractive crystal.

4. Spatial multiplexing of dynamic holograms

A multichannel adaptive interferometric demodulator proposed in (Kulchin, et al., 2000b) is built using the spatial multiplexing of dynamic holograms which are recorded in $Bi_{12}TiO_{20}$ crystal in its different parts (Fig.3). The holographic demodulation channels operate on the basis of the so called fanning effect (Voronov, et al., 1980) which appears as self-diffraction of light wave in photorefractive media. The nature of this effect consists in energy transfer between light beam entered the photorefractive crystal and beams scattered by crystal defects and inhomogeneities (Feinberg, 1982). Interference of initially weak scattered beams with the injected beam forms a set of dynamic holographic gratings. Diffraction of the input beam by these holograms leads to enhancement of those scattered waves which satisfy the phase-matching condition (Cronin-Golomb & Yariv, 1985). Due to holographic nature of fanning effect, any fast modulation of the phase of the injected light wave results in a variation of scattered light intensity; so the demodulation signal appears. Figure 4 illustrates light intensity distribution (fanning light) at the output of crystal when two holographic demodulation channels are formed. The channels optical beams are so close in a crystal that fanning waves are completely overlapped in the near field (Fig.4(a)) and can be separated only in the far field (Fig.4(b)), where they can be independently measured.

The authors show that demodulation signal in a channel decreases if the distance between light beams associated with multiplexed channels becomes below the lateral dimension of a channel (beam diameter) d (Fig.5). Such depletion is caused by reduction of interference pattern contrast m in a channel when light from another channel gets inside it. However this depletion is not critical – the output signal level reduces only by 2.5 dB even if optical fields from multiplexed channels completely overlap ($x = 0$).

Both the peculiarity and the advantage of holographic channel performance on the base of fanning effect are related to the fact that holograms are formed in a crystal without an external reference beam – the light waves scattered at the crystal defects act as a reference waves. Therefore radiation in a particular channel can be mutually non-coherent with other channels. Due to this, overlap of light waves from different channels does not lead to formation of cross-holograms and, as a sequence, does not produce cross-talk between channels. Figure 5 shows dependency of demodulation signal in 1st channel tuned to the

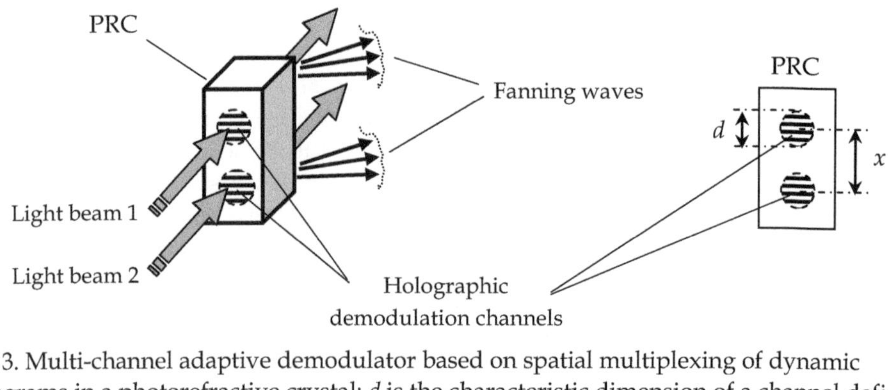

Fig. 3. Multi-channel adaptive demodulator based on spatial multiplexing of dynamic holograms in a photorefractive crystal: d is the characteristic dimension of a channel defined by light beam spot diameter, x is the distance between channel centers

frequency of 2nd channel on relative distance between channels. As seen this signal does not depend on channels overlapping ratio, while its level does not exceed the inherent noise level of a channel (- 11 dB).

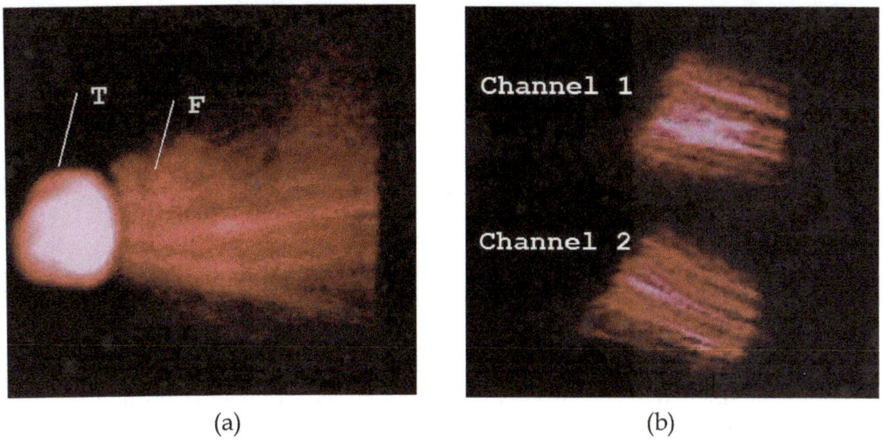

Fig. 4. Photographs of light field at the output of PR crystal in which two holographic channels are created: (a) near field image (T -beam transmitted through the crystal; F - fanning waves), (b) far field image of fanning waves (transmitted beam is blocked)

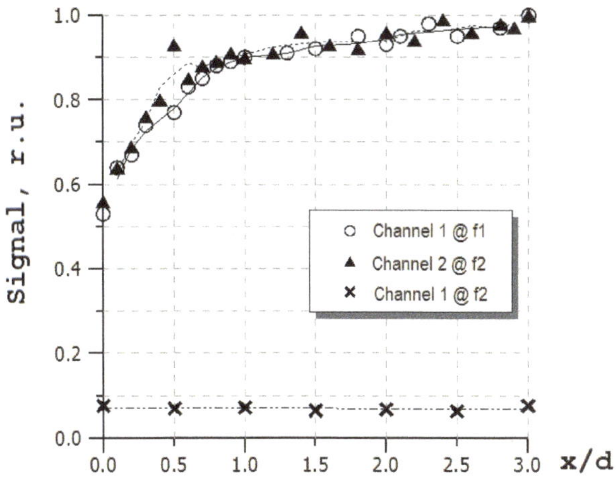

Fig. 5. Dependency of demodulation signals in two channels multiplexed in PR crystal on the relative distance between them. Channels provide processing of optical signals from two fiber-optical sensors detecting vibration at two different frequencies f_1 and f_2. The signal level in 1st channel tuned the frequency of 2nd channel does not exceed noise level of detection electronic equipment, which demonstrates the practical absence of cross-talk between channels

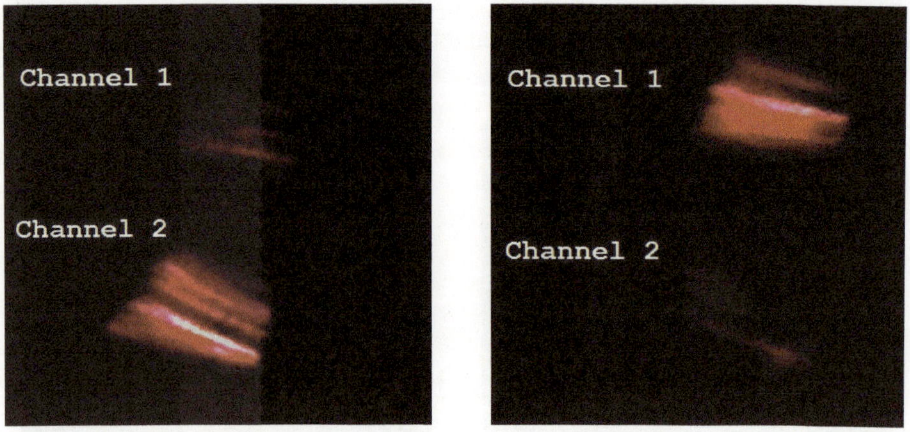

(a) (b)

Fig. 6. Photographs of fanning waves at the output of two-channel holographic demodulator formed in PR crystal when a mechanical impact under measurement is applied to a sensor and is detected in the 1st channel (a) and in the 2nd channel (b)

Figure 6 shows photographs recorded in the far field of fanning waves intensity distribution for the case of two holographic channels multiplexed in BTO crystal. The distance between channels in a crystal is equal to a channel radius (x/d = ½), providing partial overlap of channel light fields. The radiation entering the crystal is coming from two independently operated fiber-optical sensors - single-fiber multimode interferometers which have speckled waves at their outputs (Kulchin, 2001). When one of sensors remained quiet, the other one was exposed to a sudden impact with amplitude sufficient for rearrangement of speckle pattern and, as consequence, to diminishing of fanning wave intensity in the corresponded channeling (Fig.6). At the same time the level of fanning wave intensity in the rest channel remained constant. This result demonstrates the independence of fanning-effect-based channels operation even in the case of significant overlap of channel light fields in the bulk photorefractive crystal.

The two-channel adaptive fiber-optical measurement system based on the considered multichannel holographic demodulator was tested for monitoring of cracks formation in an object under increasing mechanical stress (Kulchin, et al., 2000a). Two fiber-optical sensors based on multimode fibers were first embedded into the structure of the test object when it was created (see Fig.7). Mechanical stress caused by gradually increasing pressure applied to the object led to initiation of cracks whose appearance is accompanied by generation of sound pulses. In their turn sound pulses propagating in the object affect the optical fibers and modulate the phase of optical radiation guided by fibers. Phase demodulation is performed by means of dynamic holograms multiplexed in PRC. Figure 8 shows oscilloscope traces of demodulated signals in two channels.

It is worth noting that the adaptive property of a holographic demodulator formed in PRC allows not only protection of a measurement system from the influence of the environment (temperature drift, pressure variation, etc.) but also makes it insensitive to slow deformation of the object, providing a detection of only those moments in time when cracks appear.

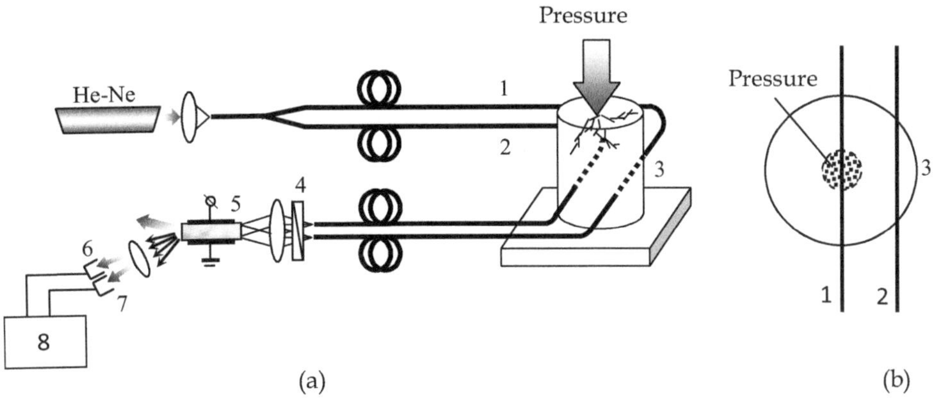

Fig. 7. Two-channel adaptive fiber-optical measurement system (a) and layout of sensing optical fibers in the object under the test – top view (b): 1, 2 – optical fibers; 3 – test object; 4 – polarizer; 5 – PR crystal $Bi_{12}TiO_{20}$; 6, 7 – channels photodetectors; 8 – recording system

As seen from the signal traces, crack frequency increases with growing of object deformation. It is more typical for the center part of the object where external pressure is applied and where the fiber #1 is placed (see Fig.7(b)). The object being stressed was broken at 325th second. The fiber placed in the center was broken together with the object. As a result the signal disappeared in the first channel. The second fiber remaining unbroken still detects weak signals indicating cracks which appear now in the object fragments. Start from 370th second, the press machine which creates mechanical stress in the object powders its central part thus stopping cracks formation process.

As seen from the oscilloscope traces, both channels operate independently without any detectable cross-talk.

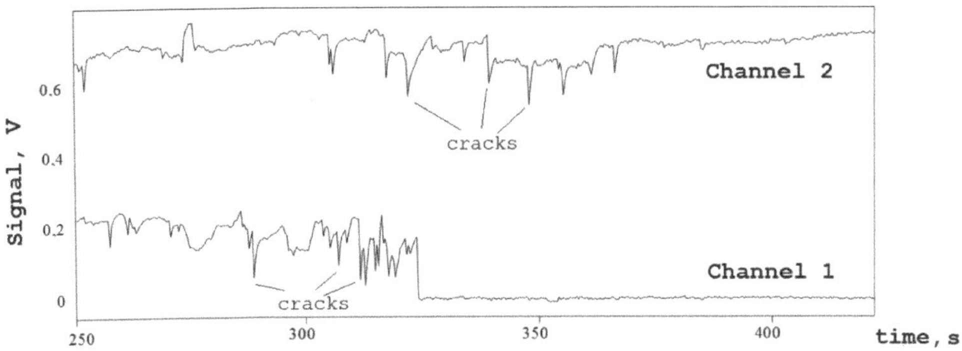

Fig. 8. Oscilloscope traces of photodetector signals in two channels of an adaptive fiber-optical measurement system which monitors crack the formation process in a solid state object

5. Angular multiplexing of dynamic holograms

The authors of paper (Fomitchov, et al., 2002) realized a multi-channel measurement system on the basis of angular multiplexing of dynamic holograms in a photorefractive crystal. The scheme of the multichannel holographic demodulator is shown in Fig.9. Several signal waves coming from fiber-optical sensors which transform a measurand (ultrasound waves) to light phase modulation are directed to the photorefractive crystal $Bi_{12}SiO_{20}$ where they are mixed with a common reference wave. Pair wise interference of all signal waves with the reference one forms a set of main dynamic holograms which provide phase demodulation of the corresponding signal waves.

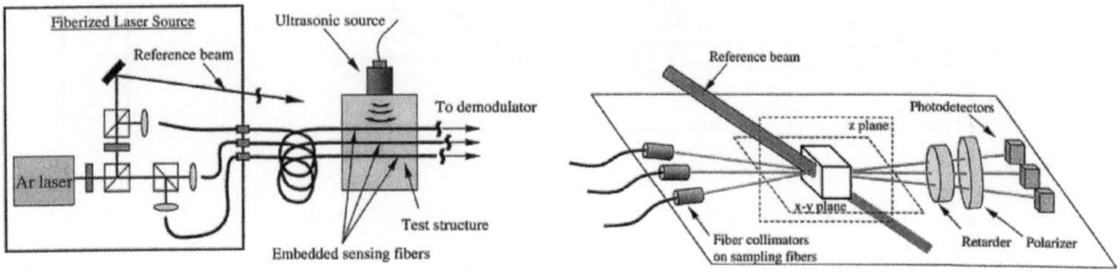

(a) (b)

Fig. 9. Optical schematic of laser source and typical arrangement of sensing fibers (a) and three-channel adaptive demodulator based on angular multiplexing of dynamic holograms in BSO crystal (b)

Use of a common reference beam for all multiplexed holograms implies that all signal waves are mutually coherent. In this case, overlap of signal beams could lead to formation of cross holograms and, as a sequence, to appearance of cross-talk between multiplexed holographic channels. In order to minimize the effect of cross holograms and to reduce a cross-talk the following approach is used. As seen from Fig.9 the holograms are recorded in so called transmission geometry – all mixed beams propagate in the crystal in same direction at small angles to each other. Having large spatial period Λ (or small wave number, $K = 2\pi / \Lambda$) the holographic grating has low diffraction efficiency which is defined by space charge field E_K proportional to K in accordance with Eqs.(2)-(3). In order to enhance the wave coupling efficiency in transmission holograms a strong external electric ac field with amplitude $E_0 = 6$ kV/cm and frequency 3000 Hz was applied to the PRC. Meanwhile the geometry of beams propagating in the crystal has provided mutual orthogonality of grating vectors of main and cross-holograms (see Fig.10). Indeed, the incidence plane (xy) of signal beams forming cross-holograms is orthogonal to the incidence plane (yz) of the reference beam which forms the main holograms. Thus, an external electric field applied to the crystal along grating vectors of main holograms has provided enhancement of their diffraction efficiency (see Eq.(2)), while cross-holograms remained unamplified. Moreover the angle between any pair of signal beams is much smaller than that between any signal and reference beams. This makes the cross-holograms weaker in comparison with the main holograms, thus reducing the cross-talk between channels

The three-channel adaptive measurement system based on the above considered approach of dynamic holograms angular multiplexing was applied to the detection of ultra-sound pulse propagation in a composite structure material (CSM). Three sensing fibers were embedded between CSM layers as shown in Fig.11(a). A 5-MHz contact transducer was used to launch bulk waves into the composite. The ultrasonic wave packet was detected by the sensing fibers, which were located at different depths within the composite.

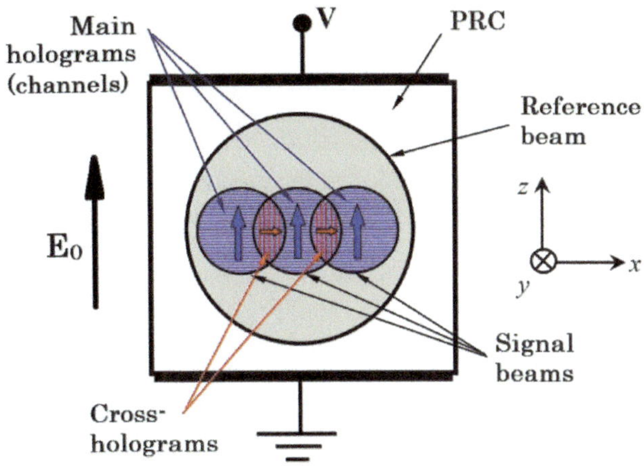

Fig. 10. Geometry of multi-wave mixing in PRC providing angular multiplexing of transmission dynamic holograms: all signal waves belong to horizontal plane (xy) and propagate in direction close to axis y; the reference wave belongs to plane (yz) and makes the angle of 5° with plane (xy)

The normalized ultrasonic signals are shown in Fig. 11(b). From the measured signals it can be seen that the bulk longitudinal waves are clearly detected by all three channels with good sensitivity. The longitudinal wave arrives at different times at different sensor locations, providing information about the wave speed (related to material properties of the composite). The measured longitudinal velocity of 2480 m/s agrees well with data obtained for the same composite material by a different technique (Fomitchov, et al., 2001). Furthermore, the absence of any spurious signals in sensors 2 and 3 at the time of arrival of the ultrasonic wave packet at sensor 1 is an indication of the absence of cross-talk in the multichannel adaptive system based on dynamic holograms multiplexed in a PR crystal.

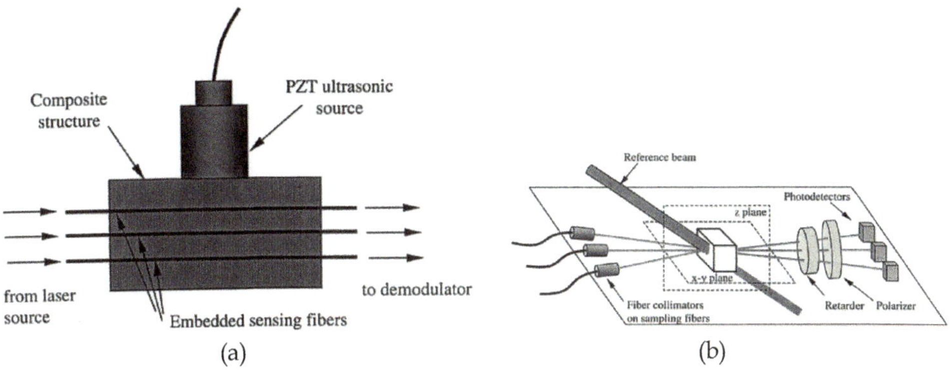

Fig. 11. (a) Experimental arrangement used to detect bulk ultrasonic waves in a composite sandwich structure, (b) detected ultrasonic signals for all three channels

In spite of such a satisfactory performance the system proposed in (Fomitchov, et al., 2002) has a major drawback. This is the necessity of using strong external electric field which leads to the technical problems common to all systems based on drift recorded dynamic holograms mentioned in the Introduction. Moreover, in order to realize a linear regime of phase demodulation the retarder plate installed after the crystal is used together with polarizer which introduces optical losses and produces polarization noise (Di Girolamo, et al., 2007b). In its turn, it was shown in the paper (Di Girolamo, et al., 2007a) that high efficiency of dynamic hologram (and, as sequence, high sensitivity of adaptive interferometer) can be achieved even without any external electric field if the hologram is recorded in reflection geometry at high spatial frequencies, *K*. Moreover, the linear mode of phase demodulation can be realized in this case also without any polarization filtering - due to VWM of waves with different polarization states (e.g., linear and elliptical) in a geometry which supports anisotropic diffraction of light (see Section 2). As was shown further in our paper (Di Girolamo, et al., 2008) a multiplexing of diffusion dynamic holograms of the reflection type can became a basis for an efficient multi-channel adaptive measurement system. Consider this approach in more details.

The geometry of reflection holograms multiplexing in a PR crystal of cubic symmetry is shown in Fig.12. Two signal waves S_1 and S_2 propagate in the crystal along its principal crystallographic axis [001] and mix with the reference wave R propagating in the opposite direction toward to them. As in the previous case, all mixed waves are mutually coherent. Thus their pair wise interference can lead in general to formation of dynamic holograms of two kinds - (i) the main DHs recorded by a reference and a signal beam and (ii) cross holograms recorded by pair of signal beams. The main holograms form demodulation channels while the presence of cross-holograms can become a source of cross-talk between channels. Moreover, the multiple recording of DHs in a single crystal can lead in general to reduction of diffraction efficiency of a particular dynamic hologram and, as sequence, to depletion of the sensitivity in particular channel. Both the rate of sensitivity decrease and the cross-talk level determine the performance of a multi-channel holographic demodulator. Consider this in more details.

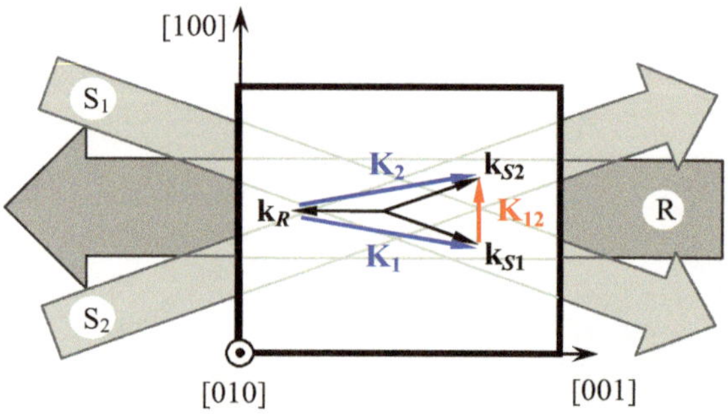

Fig. 12. The geometry of multiplexing of two reflection dynamic holograms in a PR crystal: $\mathbf{k}_R, \mathbf{k}_{S1}, \mathbf{k}_{S2}$ are wave vectors of reference (R) and signal (S_1, S_2) light waves, respectively; $\mathbf{K}_1, \mathbf{K}_2$ are grating vectors of main holograms; $\mathbf{K}_{12}$ is the grating vector of cross-hologram

Recall that two-wave mixing in PR crystal is described by system of coupled-wave equations (1) or (6) where $\hat{\mathbf{H}}$ is the coupling matrix whose components are determined by the orientations of the grating vector and wave propagation directions with respect to the crystallographic axis (see Eq.(5)). One can show that for the geometry depicted in Fig.12 the matrix $\hat{\mathbf{H}}$ has non-zero components ($\hat{\mathbf{H}} = \begin{pmatrix} 0 & 1 \\ 1 & 0 \end{pmatrix}$) only for counter propagating waves coupled at hologram by the grating vector parallel to the principal crystallographic axis [001]. As seen in Fig.12 this requirement is satisfied for any pair of signal and reference waves.

On the contrary, the grating vector $\mathbf{K}_{12}$ of the cross-hologram formed by any pair of two signal waves is orthogonal to the axis [001]. By using Eq.(5) one can show that all components of the coupling matrix $\hat{\mathbf{H}}$ in this case will be zero, $\hat{\mathbf{H}} = \begin{pmatrix} 0 & 0 \\ 0 & 0 \end{pmatrix}$. This means that, in such geometry, there is no coupling between signal waves even if they completely overlap, and hence there is no cross-talk between channels associated with these signal waves.

However the situation changes if the orthogonality of a cross-hologram grating vector $\mathbf{K}_{12}$ to direction [001] is violated (this becomes true if signal beams enter the crystal at different angles). Thus the efficiency of two signal waves coupling in this case will depends on the [001]-axis component of the vector $\mathbf{K}_{12}$. Taking into account that the efficiency of waves mixing at the main hologram (and hence the level of demodulation signal in a channel) is determined by wave number K (i.e. K_1 or K_2) in accordance with Eq.(3), one can show that the cross-talk level can be estimated with good accuracy by following

$$C \approx 20\log\left(h\frac{|\sin\Theta|}{2}\right), \tag{14}$$

where Θ is the angle between vector $\mathbf{K}_{12}$ and normal to axis [001], $h \le 1$ is the coefficient which takes into account degree of signal beams overlap in a crystal.

Figure 13 shows the dependency $C(\Theta)$ from which it is seen that violation of orthogonality $\mathbf{K}_{12} \perp [001]$ leads to a rise in cross-talk. However, its level remains below the inherent noise of a channel (–30 dB) over a wide range of angles Θ. The last provides wide angular aperture for entry of signal beams to a crystal (Fig.14). Thus, for the CdTe crystal having refractive index ~2,8 the aperture within which a cross-talk is practically absent can reach 60 degrees. Moreover the signal beams can belong to different planes (e.g., spaced in fantail way) in contrast to the previous multiplexing technique where they lay in same plane, see Fig.9(b). The wide angular range of signal beams input makes the formation of holographic channels easy and increases their potential number.

As it was shown in the paper (Di Girolamo, et al., 2008) another possible reason for cross-talk in the geometry of reflection holograms multiplexing is related to residual mechanical stresses or defects in PR crystal which locally disturb its symmetry and partially enable interaction of signal beams even in forbidden geometry (when $\mathbf{K}_{12} \perp [001]$). In this case matrix $\hat{\mathbf{H}}$ can have non-zero components. However the cross-talk level still does not exceed channel inherent noise. A low level of cross-talk is obtained by low spatial frequency of cross hologram grating K_{12} which is much smaller than that of main holograms, K_1 and K_2. It worth noting that cross-talk related to crystal defects/stresses can be excluded (or at least minimized) by using crystals of better quality.

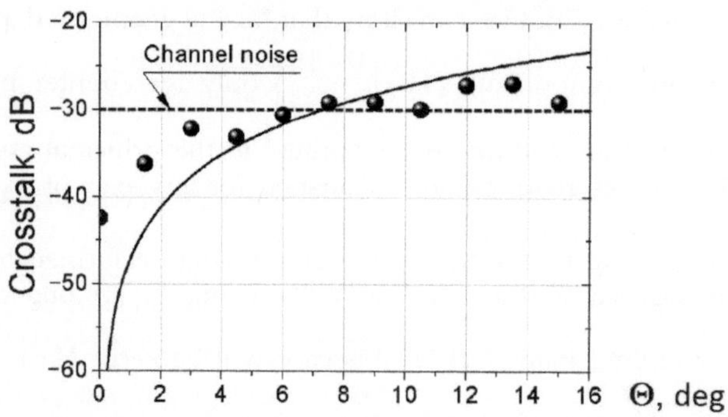

Fig. 13. Cross-talk between channels as function of angle between cross-hologram grating vector and normal to the crystal axis [001]: line represents theoretical dependency calculated in accordance with Eq.(14); markers are experimental data

As seen from Eqs.(2), (6), the efficiency of wave coupling at a dynamic hologram (and hence demodulation signal amplitude) is defined by the interference pattern contrast $m = 2\mathbf{A}_1\mathbf{A}_2^* / I_0$. However this expression is valid only for the two-wave mixing case. Introduction of additional signal waves (related with other channels) changes the contrast.

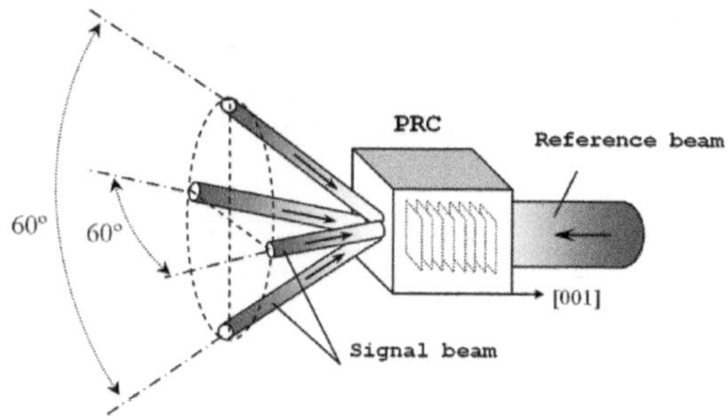

Fig. 14. Spatial arrangement of beams providing multiple recording of reflection dynamic holograms in a photorefractive crystal without noticeable crosstalk

In spite of mutual coherence of all light beams crossed in the crystal, the change of contrast is related mostly to increase of total light intensity. High angular selectivity of volume holograms allows one to exclude contributions to the variable part of the contrast from additional interference patterns having spatial frequencies and orientation different from those of main hologram. Thus the contrast of the interference pattern in the one of N channels can be estimated by

$$m(N) = \frac{2\sqrt{I_S I_R}}{I_R + I_S + h\sum_{j=1}^{N-1} I_j}, \tag{15}$$

where I_R, I_S and I_j are the intensities of reference beam, signal beam of a particular channel and signal beam of the j-th additional channel, respectively.

It is worth noting that maximum sensitivity of an adaptive interferometer based on a dynamic hologram is achieved at a certain ratio of mixed beam intensities (Di Girolamo, et al., 2008). Thus, taking into account that a common reference beam is used in the considered geometry of hologram multiplexing the optimized multichannel system will have signal beams of equal intensity, $I_j = I_S, (j = 1, N)$. Then, Eq.(15) for contrast of interference pattern in one of N multiplexed channels can be reduced with good accuracy to the following

$$m(N) \approx \frac{2\sqrt{I_S I_R}}{I_R + hNI_S}. \tag{16}$$

By solving the system of coupled wave equations (1) talking into account Eq.(16) one can analyze one channel performance under conditions of decreasing number of channels. Figure 15 shows dependency of the relative detection limit δ_{rel} on multiplexed channels number N. As seen, increase of N is accompanied by increase of detection limit (equivalent to reduction of sensitivity, see Eq.(11)) in each channel. However this increase is not critical and depends on the crystal sample. Thus, formation of 10 channels in one sample of CdTe crystal leads to doubling of δ_{rel} while for another sample of same crystal this increase amounts to just 23%. Taking into account the extremely high sensitivity of an adaptive interferometer which approaches that of classical interferometer (Kamshilin, et al., 2009), one can conclude that even a reduction of sensitivity by a factor of two can be considered as not significant and remains acceptable in most practical applications. Additionally mutual influence of multiplexed channels can be diminished by reduction of the degree of signal beams overlap h (see Eq.(15)) using their tighter focusing.

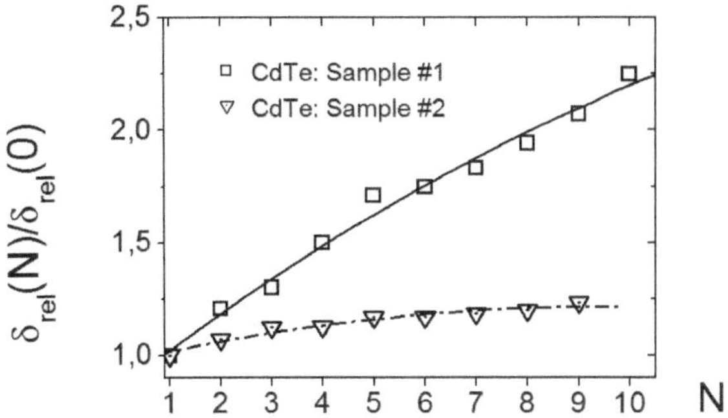

Fig. 15. Dependence of relative detection limit in one channel of holographic multi-channel system on number of multiplexed channels

The scheme of a two-channel adaptive interferometer based on angular multiplexing of reflection dynamic holograms in a photorefractive crystal CdTe in the above-considered geometry is shown in Fig. 16. Two signal beams coming from two sensing optical fibers are focused by lenses into the crystal where they are mixed with a reference beam propagating in the opposite direction along the crystallographic axis [001]. Calibrated piezoelectric transducers attached to the fibers simulate the action of measurand and introduce phase modulation with amplitudes 0.30 and 0.15 rad in the first and the second channels, respectively. This is equivalent to a detection of mechanical vibration with amplitude of 25 and 13 nm (at wavelength 1.06 μm), respectively. Oscilloscope traces of detected signals in two channels of the adaptive interferometer are shown in Fig.17. As seen the measurement system provides good SNR in both channels in spite of the weak level of measured vibration and wide-band detection mode (Δf = 50 MHz). Moreover, the two channels operate independently without any detectable cross-talk.

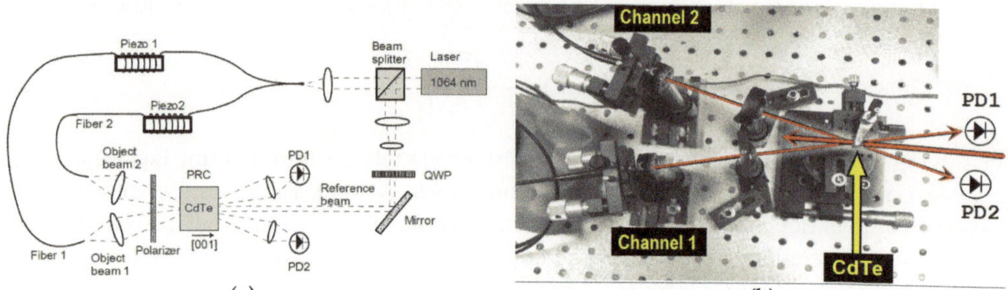

(a) (b)

Fig. 16. Layout (a) and partial photo (b) of two-channel adaptive interferometer based on dynamic reflection holograms angularly multiplexed in PR crystal: PD1 and PD2 are photodetectors which register signal beam intensities in 1st and 2nd channels

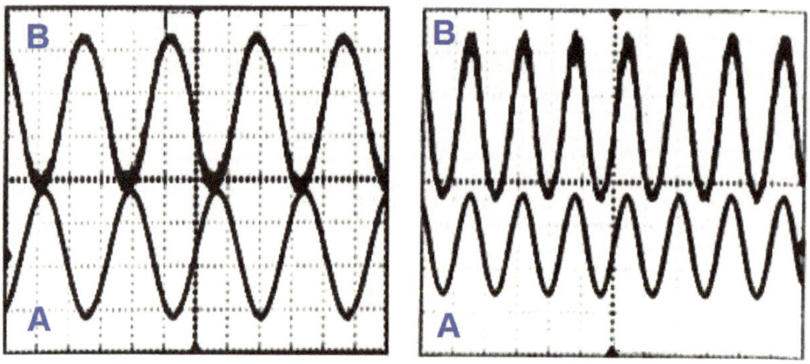

Fig. 17. Snapshots of oscilloscope traces demonstrating the operation of a two-channel adaptive measurement system: the first channel (left) and the second channel (right); (A) modulation signal which excites vibration in sensing fibers, (B) signal detected in a channel

6. Spectral multiplexing of dynamic holograms

Spectral (or wavelength division) multiplexing is one of effective techniques for building multichannel optical and fiber-optical measurement systems. In particular, this approach is highly demanded if fiber Bragg gratings (FBG) are used as sensors (Grattan & Meggitt, 1998). FBG sensors are an increasingly important emerging technology in the area of intelligent structural health monitoring (SHM) of civil, mechanical, naval, and aerospace structures (Claus, 1992; Culshaw & Dakin, 1996). A large number of FBG sensors can be easily written in a single fiber. A measurand (for SHM applications, FBG sensors are typically used to monitor static or dynamic strains) causes spectral shifts in the FBG sensor transmittivity (equivalently reflectivity). The spectral shifts can be monitored in several ways with an appropriate demodulator. For spectrally encoded FBG sensors, extant demodulation schemes can be classified into three categories: scanning type, spectrometry based, and interferometry based. Scanning-type techniques include Fabry–Perot scanning filters (Kersey, et al., 1993), acoustooptic tunable filters (Xu, et al., 1993), and tunable laser sources (Fomitchov & Krishnaswamy, 2003). All these scanning-type techniques suffer from the fact that at any instant only one FBG sensor can be interrogated. Such approaches are not applicable if all the sensors have to be interrogated *simultaneously* for the purpose of monitoring impact signals and acoustic emissions (Perez, et al., 2001). Spectrometric methods (Davis & Kersey, 1995) suffer from low sensitivity and are not suitable for dynamic measurements if several sensors have to be active at all times. Interferometric methods such as the Mach–Zehnder interferometer (Kersey, et al., 1992) are ideally suited to monitor dynamic strains; however, they require electronic feedback to actively compensate for any quasi-static drift to maintain the interferometer at quadrature. This makes the cost of multiplexing high since each sensor requires its own feedback system. Therefore, a cost-effective multichannel demodulator capable of providing not only simultaneous interrogation of FBG-sensor arrays but also stable performance in unstable environment is required.

In the paper (Qiao, et al., 2006) authors show that such a demodulator can be developed on the basis of dynamic holograms formed in a photorefractive crystal. Consider this in more details.

Due to its adaptive properties a PRC-based demodulator can compensate all slow (including quasi-static) impacts on sensors and provide stable detection of high-frequency or pulse measurands (e.g., vibration, ultra-sonic waves, acoustic emission, etc.) without using any feedback loops or external stabilizing electronic circuits. As will be considered below, the demodulator based on two-wave mixing in a PR crystal can provide both spectral decoding of FBG-sensor signals and their effective spectral demultiplexing.

The spectral shift induced by strain, ε_F, and temperature change, ΔT, in a Bragg-grating sensor can be written as (Murray, et al., 2000):

$$\Delta\lambda_B = \lambda_B \left\{ 1 - \frac{n_{\text{eff}}^2}{2}\left[p_{12} - u(p_{11} + p_{12})\right]\varepsilon_F + (\alpha_T + \alpha_N)\Delta T \right\}, \tag{17}$$

where, λ_B is the center wavelength of the FBG sensor, $\Delta\lambda_B$, is the wavelength shift caused by strain or temperature, n_{eff} is the effective refractive index of the fiber, p_{ij} $(i, j = 1, 2)$ are the components of the strainoptic tensor, u is Poisson's ratio, α_T is the thermal expansion coefficient, and α_N is the thermo-optic coefficient of the fiber. For typical Bragg sensors at 1550 nm, it has been estimated that one microstrain (1×10^{-6}) will lead to a 1.2 pm change in wavelength, and a 1 °C change in temperature will lead to about 13 pm change in wavelength (Othonos & Kalli, 1999). This fact emphasizes again the necessity of compensation for environment effects in the measurement system.

The adaptive interferometer based on dynamic holograms provides transformation of transient phase shift to intensity modulation of signal beam, while the signal of FBG-sensor is encrypted in a shift of central wavelengths in the reflection spectra. At the same time the adaptive interferometer also can be configured as a *spectral* demodulator. Indeed, a spectral shift $\Delta\lambda_B$ of the signal *and* reference beams can be effectively converted into a relative phase shift $\Delta\phi$ *between* the beams due to travel through unbalanced optical paths (Othonos & Kalli, 1999):

$$\Delta\phi(t) = \frac{2\pi b}{\lambda_B^2}\Delta\lambda_B, \tag{18}$$

where b is the optical path difference (OPD). It is noteworthy that a similar principle of spectra change transformation to phase modulation is used as well in Mach-Zehnder and other unbalanced interferometers (Kersey, et al., 1992).

From Eq. (18), it appears that the greater the OPD, the larger the equivalent phase shift, and therefore the stronger the interference signal should be. However, typically broadband light sources are used to illuminate the FBG sensors, and the FBG reflection spectrum typically has a finite line width of the order of 0.1–0.4 nm. This implies that coherence of the two interfering beams needs to be taken into account both in the photorefractive grating creating process, and in the subsequent interference between the transmitted signal and the diffracted reference beams. The fringe visibility due to the interference of two beams of finite spectral width Δk is given by

$$m(b) = \frac{2\sqrt{\beta}}{\beta+1}\exp\left\{-\frac{\Delta k^2 b^2}{16\ln 2}\right\}, \tag{19}$$

where β is the intensity ratio of reference and signal beams. The factor before *exp*-function in Eq.(19) is nothing other than interference fringe contrast m defining space charge field in PR crystal (see Eq.(2)). Incorporating the degradation in fringe visibility due to low coherence in the TWM analysis (by solving the coupled-wave equations) leads to the following expression for spectral-shift demodulated interference signal:

$$\Delta I_S \propto \left[\exp(\gamma' L)\sin(\gamma'' L)\right]\exp\left\{-\frac{\Delta k^2 b^2}{16\ln 2}\right\}\frac{b}{\lambda_B^2}\Delta\lambda_B, \tag{20}$$

where $\gamma = \gamma' + i\gamma''$ is the TWM complex gain and L is the crystal length in the beam propagation direction.

The effect of decreased fringe visibility on the photorefractive grating formation process is not a significant factor, since the preferred mode of two-wave mixing in PRCs is in the low intensity modulation regime (this avoids the creation of higher order index gratings that could lead to cross talk when the system is multiplexed). Furthermore, the system is operated at near quadrature, when $\gamma'' L \approx \pi/2$ (Solymar, et al., 1996), and since the TWM energy gain, γ', is typically very small, the variations in the first two terms with the OPD are not significant. The dependence of the interference signal on the OPD is therefore predominantly given by

$$\Delta I_S \propto \exp\left\{-\frac{\Delta k^2 b^2}{16\ln 2}\right\}\frac{b}{\lambda_B^2}\Delta\lambda_B. \tag{21}$$

As seen from Eq.(21), the wavelength demodulation signal is a strong function of the OPD d with the exponential decay arising from loss of coherence, and the linear increase arising from the increase with OPD in the phase shift due to spectral change. Figure 18 is a plot of the signal amplitude versus the OPD for different linewidth FBG sensors. As indicated in Fig. 18, when the OPD equals zero, there is no wavelength-demodulated signal detected, and as the OPD increases the signal amplitude increases to a maximum beyond which the signal starts to drop due to decreasing fringe visibility. For each given line width of the FBG sensor, there exists an optimum value of the OPD that maximizes the wavelength-demodulated signal. It is also clear from Fig. 18 that the narrower the linewidth, the larger the optimum OPD, and therefore the larger the demodulated signal. However, it is not always better to use a narrower linewidth FBG sensor. First, a narrower linewidth FBG sensor is usually longer in length (Othonos & Kalli, 1999), which decreases the highest frequency to which the FBG can respond (Coppola, et al., 2001). Second, a larger OPD decreases the dynamic range that the FBG sensor can measure.

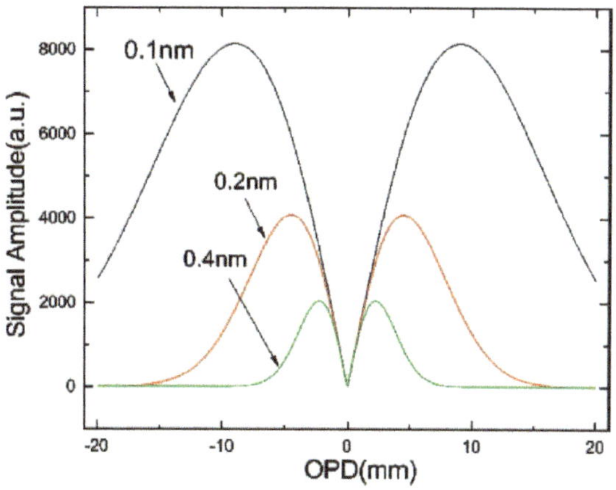

Fig. 18. Wavelength demodulated signal amplitude as a function of the OPD for different linewidth FBG sensors

The scheme of the measurement system based on FBG sensor and photorefractive dynamic hologram proposed in (Qiao, et al., 2006) is shown in Fig. 19. The FBG sensor is illuminated by a broadband amplified spontaneous emission (ASE) source in the C band (1530 to 1570 nm), and the reflected light is coupled by a circulator into an erbium-doped fiber amplifier (EDFA) that works in saturation mode (output 500 mW). The amplified light is split using a 1×2 coupler (splitting ratio 95/5) into reference (95%) and signal (5%) beams that travel unbalanced optical paths to the InP:Fe photorefractive crystal. The InP:Fe crystal is oriented for two-wave mixing in the direct detection configuration (Delaye, et al., 1997), in which both beams enter the crystal by the $(\bar{1}10)$ face. An external dc field (6 kV/cm) was applied along [001] direction to provide the quadrature condition of phase detection. The authors note that to apply a continuous dc field across the InP:Fe crystal, a Peltier cooler was used to prevent electrical breakdown due to crystal overheating.
In the work (Qiao, et al., 2006) a FBG sensor centered at 1552 nm with a linewidth of 0.1 nm, length of 10 mm, and reflectivity of 50% was glued onto a piezoelectric transducer (PZT) stretcher that was used to induce known dynamic strains (with amplitude 10 μstrain at frequency 10 kHz). The light reflected from the FBG sensor undergoes spectral shift due to strain-induced changes in the Bragg reflectivity. The sensor by itself is sensitive to both quasi-static and dynamic strains, and is also subject to temperature drift. However the TWM

demodulator system will automatically compensate for quasi-static drifts and track only the dynamic strains.

Oscilloscope traces of demodulated signal obtained at different values of the optical path difference are shown in Fig. 20. As seen, an intermittent dc field is applied from 1 to 6 ms with respect to a reference trigger, and the photorefractive grating initially builds up. The dynamic strain is applied as a tone burst starting from 2 to 6 ms. When the OPD equals to zero, although the TWM energy gain is at its maximum, there is no detected wavelength demodulated signal because there is no OPD to convert the wavelength shift into phase shift. As the OPD increases, the wavelength demodulated signal starts to appear.

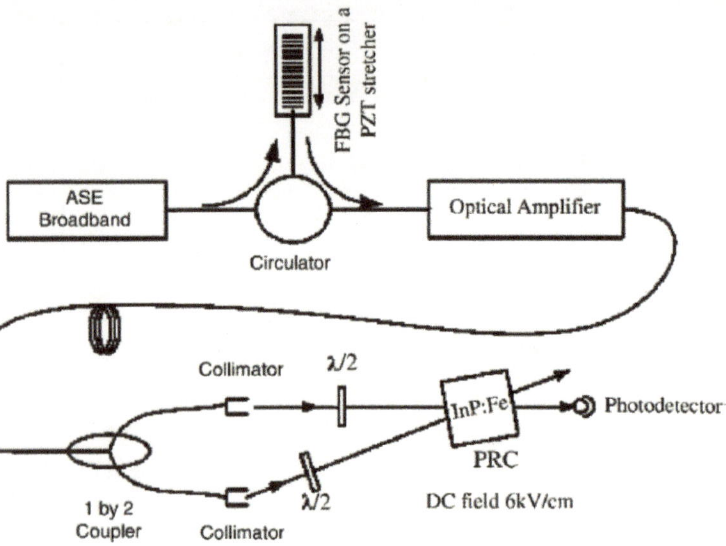

Fig. 19. Experimental configuration of the FBG sensor and the TWM wavelength demodulator

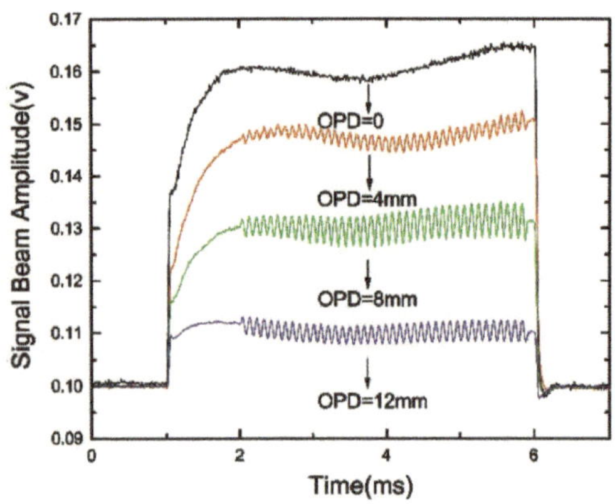

Fig. 20. Wavelength demodulated signal at different values of OPD

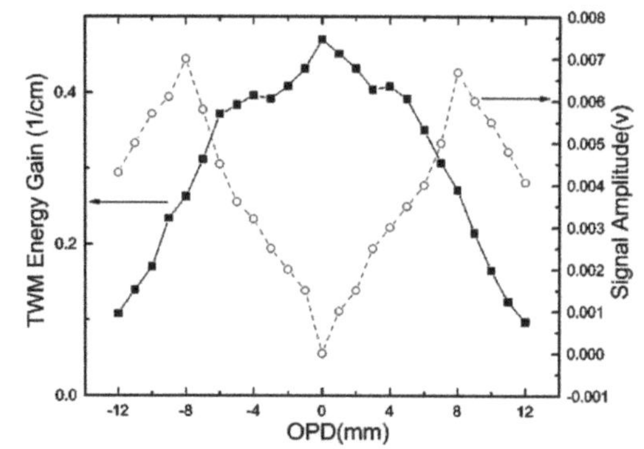

Fig. 21. Plot of wavelength demodulated signal amplitude and TWM gain versus OPD

Figure 21 is a plot of the wavelength demodulated signal amplitude and TWM energy gain versus the OPD. The optimum OPD that maximizes the wavelength demodulated signal for the 0.1 nm linewidth FBG sensor is found to be 8 mm. For an OPD of 8 mm, Eq. (18) indicates a wavelength-to-phase shift conversion sensitivity of about 21 radians per nanometer wavelength shift at 1550 nm wavelength. This translates to 0.0252 radian/με. Such phase shifts are readily detectable by the TWM interferometer. Also note that the trend of the signal amplitude curve is similar to that of the theoretical curve shown in Fig. 18. Also, as was mentioned above, the TWM energy gain γ' is indeed a small number and experimentally is found to vary from 0.47 to 0.1 cm^{-1} as the OPD changes from 0 to 12 mm. This causes the first exponential term in Eq. (20) to vary from 1.6 to 1.1, which is much smaller than the change in the second exponential term that decreases from 1 to 0.03.

The minimum detectable strain with the setup of adaptive interferometer proposed and realized in (Qiao, et al., 2006) was measured to be 0.25 με, corresponding to a spectral shift of 0.3 pm. The minimum detectable spectral shift, which is limited by the ASE source and EDFA intensity noise, can be further improved by using balanced photodetection to cancel the intensity noise.

Let us consider now how the multi-channel measurement system can be created on the basis of decoding of spectral signals from FBG sensors in PR crystal. The general principle is following (Qiao, et al., 2006). Optical signals from FBG sensors with distinct spectral reflectivities and center wavelength separation, $\Delta\lambda_C$, propagate in common optical path and enter a single PR crystal where they record a set of dynamic holograms which in its turn provides demodulation of spectrally encoded signals. Then the signals coming from different FBGs are separated by means of band-drop filters widely used in optical communication systems.

The distance between FBG central wavelengths $\Delta\lambda_C$ is chosen to be sufficiently large so that stationary optical interference between the multiple channels cannot occur. In this case, inside the PRC each channel creates its own index grating of different grating pitch. The change in the index grating pitch, $\Delta\Lambda$, is related to the channel separation and the signal and reference beam angle θ:

$$\Delta\Lambda = \frac{\Delta\lambda_C}{2\sin(\theta/2)}. \tag{22}$$

It is worth noting that the FBG channel spacing $\Delta\lambda_C$ should be neither too small to avoid cross talk due to closely packed index gratings, nor too large to maximize the number of sensors that can be used within the limited bandwidth of the light source and that of spectral sensitivity of the PR crystal. Thus, for the C band (1530–1570 nm), the channel separation $\Delta\lambda_C$ was 4 nm, which, according to Eq.(22), will give rise to an index grating pitch shift of 76 nm at a beam angle of 3°. In the context of holographic storage elements using PRCs, it has been shown that multiple gratings can be written in a single PRC with negligible cross talk if the grating pitches were to differ by as little as 0.03 nm (Kume, et al., 1998). Thus, a 4 nm channel spacing allows 10 channels with very small cross talk in the C band. In practice, this can be increased by using narrower channel spacing, at the expense of decreased dynamic range.

The experimental configuration of a four-channel TWM wavelength demodulator demonstrated in (Qiao, et al., 2006) is shown in Fig. 22. Four 0.1 nm linewidth FBG sensors are connected in series and are centered at 1548, 1552, 1556, and 1560 nm, respectively. The experimental configuration is similar to that of the single channel configuration shown in Fig. 19 except that after the PRC, there are four band-drop filters which transmit a certain band (i.e., from 1546.8 to 1549.4 nm) and reflect all the other wavelengths. The band-drop filters therefore decouple the TWM-demodulator signals from the various FBG sensors prior to photodetection. The width of the band-drop filter should be chosen to be wider than the expected quasi-static phase shifts of the FBG sensors, and to be slightly narrower than the FBG channel spacing $\Delta\lambda_C$.

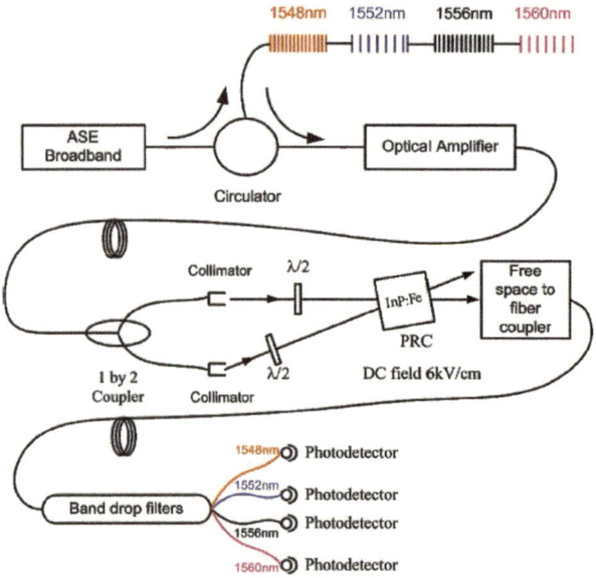

Fig. 22. Experimental configuration of the four-channel adaptive measurement system based on spectral multiplexing of FBG sensors and TWM demodulation of their signals at dynamic holograms in PRC

Performance of the four-channel adaptive measurement system is illustrated by Fig.23 which shows oscilloscope traces of demodulation signal in each channel of the system.

Dynamic strains with equal amplitude (5 με) and different frequencies were simultaneously applied to all FBG sensors: 10 kHz on FBG sensor 1 (1548 nm), 5 kHz on FBG sensor 2 (1552 nm), 2 kHz on FBG sensor 3 (1556 nm), and 20 kHz on FBG sensor 4 (1560 nm). Figure 23(a) shows that the four channels can be demodulated simultaneously. Note that the low frequency fluctuations seen in the signals are due to environmental noise-induced *intensity* fluctuations as explained earlier, and these can be removed using a balanced detection scheme if necessary. Also note that although the applied signal amplitudes for all four channels are the same, the demodulated signal amplitude for each channel is slightly different. This is mainly due to each channel having different optical intensities (due to nonuniform EDFA gain). In practice, this can either be precalibrated, or a gain-flattened EDFA can be used.

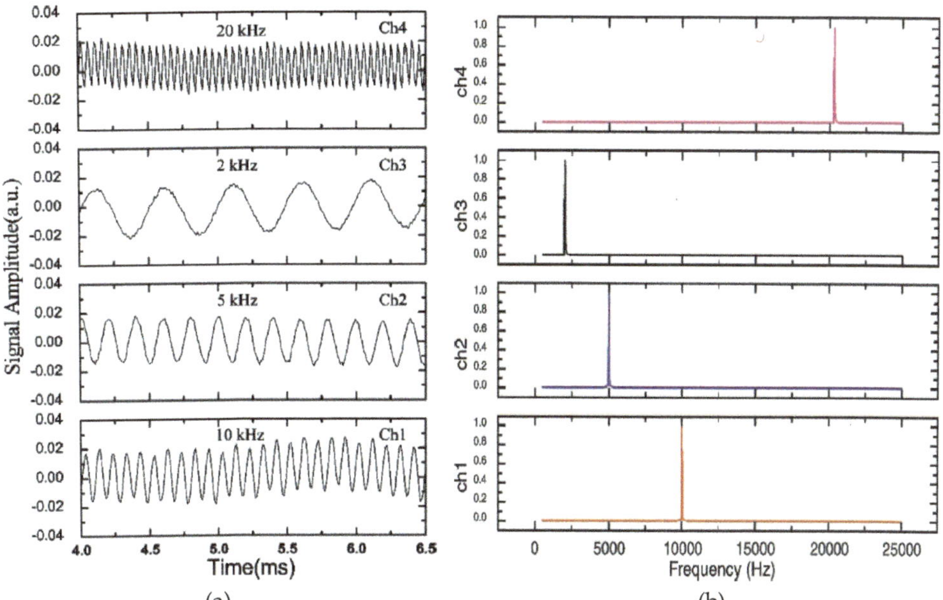

Fig. 23. Output demodulated signals (a) and their Fourier spectra (b) in four channels of adaptive measurement system based on FBG sensors and dynamic holograms spectrally multiplexed in PR crystal

The cross talk between these four channels can be inferred from the Fourier spectra of the signals shown in Fig.23(b). If there were cross talk, we would expect all the frequency components, namely 2, 5, 10, and 20 kHz, to show up in the signal spectrum of each channel. However, in the spectrum of channel 1, there is only the expected 10 kHz component, and no other frequency components are observed. This is true also for the other three channels. It is safe to conclude that the cross talk between these four channels is at most comparable to the noise level in the Fourier spectrum, which is at least 30 dB below the signal level in this particular case.

To conclude this Section we note that, apart from band-drop filters with fixed central wavelengths used in the paper (Qiao, et al., 2006), other WDM-demultiplexers including

those of scanning type can be used for spectral division of channels. Note as well that such a demultiplexer can be also built on the basis of dynamic holographic grating recorded in a PR crystal (Hukriede, et al., 2003; Petrov, et al., 2001; Petrov, et al., 2003; Runde, et al., 2005).
The scheme of such photorefractive band-drop filter proposed in (Petrov, et al., 2003) is shown in Fig. 24. Two coherent light beams A and B (λ_w = 532 nm) making the angle 2θ enter the photorefractive crystal $BaTiO_3$:Co where they form a dynamic holographic grating with the index pitch $\Lambda_F = \lambda_w/(2\sin\theta)$. Radiation to be spectrally filtered enters the crystal along the wave vector of holographic grating $\mathbf{K_F}$ (see Fig. 24) which acts as a Bragg mirror with central wavelength

$$\lambda_B = \frac{n_C \lambda_w}{\sin\theta}, \tag{23}$$

where n_C is the refractive index of crystal.
Figure 25 shows typical transmission spectrum of band-drop filter based on a holographic grating formed in PRC. As seen, the filter has high Q-factor due to narrow line width ~0.1 nm.

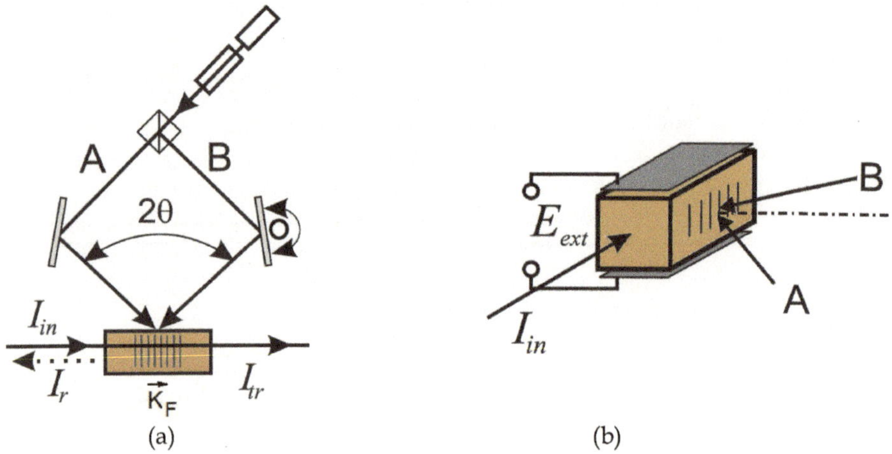

Fig. 24. Geometry of dynamic band-drop Bragg filter recording in PR crystal (a) and orientation of external electric field applied to a crystal (b)

The adaptive property of a dynamic grating formed in PRC allows one to perform fine tuning of the filter parameter thus providing on-line switching between different FBG sensors. Thus, change of the angle θ between recording beams leads to erasure of the previous grating and recording of a new one with new pitch corresponded to new central wavelength λ_B. As seen from Fig. 26 which shows experimental dependence of central wavelength on the angle θ, the change of the latter leads to shift of λ_B by 80 nm (from 1480 to 1560 nm). Authors (Petrov, et al., 2003) note that shift of the Bragg wavelength has no observable effect on the form of the transfer function and the diffraction efficiency. The speed of tuning was determined by the recording time of the dynamic hologram and was about 0.5 s.

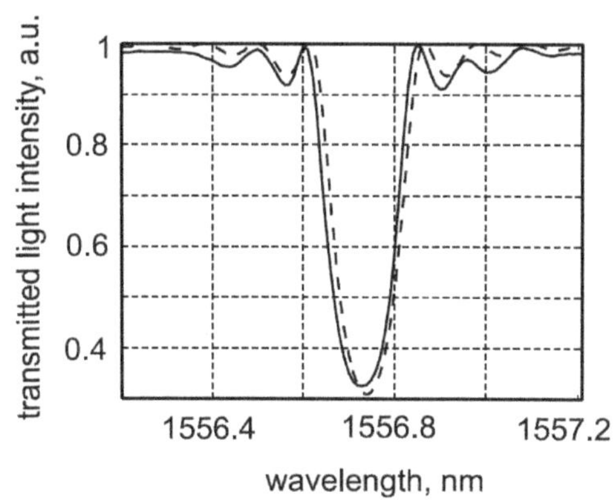

Fig. 25. The transfer function of the simple dynamic Bragg grating

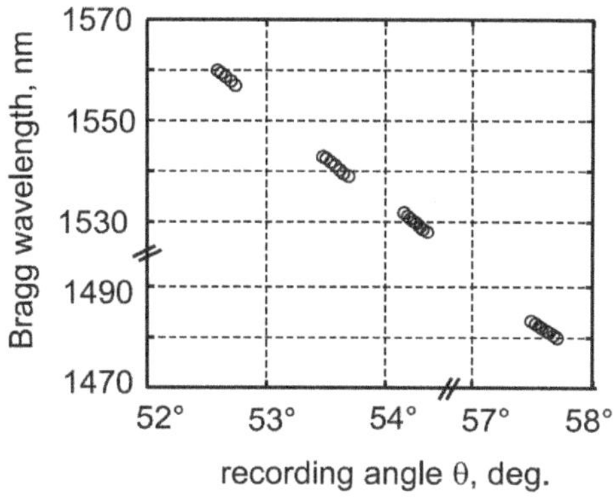

Fig. 26. The wavelength position of the filter transfer function versus the angle θ between the writing beams

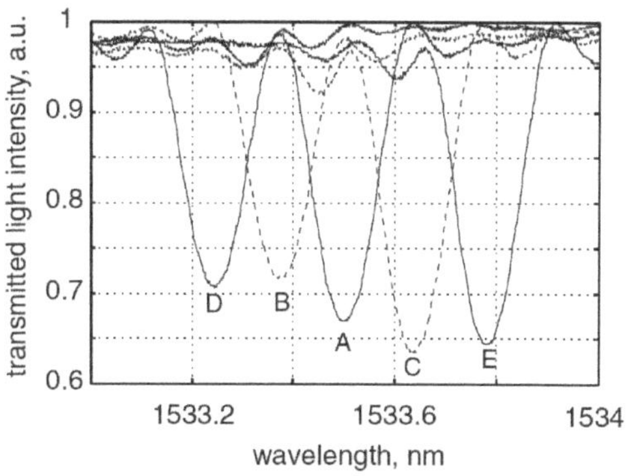

Fig. 27. The wavelength position of the filter transfer function versus the external electric field: A, $E_{ext} = 0$; B, $E_{ext} = -370$ V cm^{-1}; C, $E_{ext} = +389$ V cm^{-1}; D, $E_{ext} = -614$ V cm^{-1}; E, $E_{ext} = +653$ V cm^{-1}

The central wavelength of the filter based on a photorefractive grating can be additionally tuned electrically by application of external electric field to the photorefractive crystal. Figure 27 demonstrates such electrical tuning of the filter. This is an example of fast tuning via the electro-optically induced variations in the average refractive index. One can see that the application of voltages ranging from −614 to +655 V cm^{-1} provides tuning in the 0.55 nm range. The speed of tuning reached 2.2 nm μs^{-1} and was limited only by the available equipment (e.g. switching time of the power source used).

7. Conclusion

In this Chapter we considered different approaches to development of multi-channel adaptive measurement systems based on multiplexing of dynamic holograms in a photorefractive crystal. We show that such systems can provide (i) high sensitivity to detection of ultra-small physical quantities (close to the classical homodyne detection limit), (ii) cross-talk free performance, and (iii) adaptive properties which cancel uncontrollable environmental influence on the system. For development of such systems different approaches to dynamic holograms multiplexing, including spatial, angular, spectral and their combinations, can be effectively used. Moreover, dynamic gratings recorded in a photorefractive crystal can also be used for development of such elements of multi-channel systems as spectral tunable filters which can provide effective demultiplexing of signals with different wavelengths.

8. References

An, J. W.; Kim, N. and Lee, K. W. (2001). Volume holographic wavelength demultiplexer based on rotation multiplexing in the 90-deg geometry. *Opt. Comm.*, Vol.197, pp.247-254, ISSN 0030-4018

Andersen, G. P.; Dussan, L. and Chen, K. (2009). Holographic wavefront sensor. *Opt. Eng.*, Vol.48, No.8, p.085801, ISSN 0091-3286

Ashkin, A.; Boyd, G. D.; Dziedzic, J. M.; Smith, R. G.; Ballman, A. A.; Levinstein, J. J. and Nassau, K. (1966). Optically-induced refractive index inhomogeneities in $LiNbO_3$ and $LiTaO_3$. *Appl. Phys. Lett.*, Vol.9, No.1, pp.72-74, ISSN 0003-6951

Bershtein, I. L. (1954). Measurement of extremely small changes of path difference of two light oscillations. *Dokl. Acad. Nauk USSR*, Vol.94, No.4, pp.655-658

Claus, R. O. (Ed). (1992). *Fiber Optic Sensor Based Smart Materials and Structures*. Institute of Physics Publishing, ISBN 0-7503-0248-8, Bristol

Coppola, G.; Minardo, A.; Cusano, A.; Breglio, G.; Zeni, G.; Cutolo, A.; Calabro, A.; Giordano, M. and Nicolais, L. (2001). Analysis of feasibility on the use of fiber Bragg grating sensors as ultrasound detectors, *Proceedings of SPIE "Smart Structures and Materials 2001"*, Vol.4328, Febraury, 2001

Cronin-Golomb, M. and Yariv, A. (1985). Optical limiters using photorefractive nonlinearities. *J. Appl. Phys.*, Vol.57, No.11, pp.4906-4910, ISSN ISSN 0021-8979

Culshaw, B. and Dakin, J. (Eds). (1996). *Optical Fiber Sensors. Applications, Analysis, and Future Trends*. Artech House, London

Davis, M. A. and Kersey, A. D. (1995). Application of a fiber Fourier transform spectrometer to the detection of wavelength encoded signals from Bragg-grating sensors. *J. Lightwave Technol.*, Vol.13, pp.1289-1295, ISSN 0733-8724

Delaye, P.; Blouin, A.; Drolet, D.; de Montmorillon, L.-A.; Roosen, G. and Monchalin, J.-P. (1997). Detection of ultrasonic motion of a scattering surface using photorefractive InP:Fe under an applied dc field. *J. Opt. Soc. Am. B*, Vol.14, No.7, pp.1723-1734, ISSN 1520-8540

Delaye, P.; Roosen, G.; Ramaz, F.; Forget, B. C.; Atlan, M.; Boccara, A. C. and Gross, M. (2005). Photorefractive two wave mixing detection for acousto-optical imaging of biological thick tissues, *Proceedings of 10th Int. Conf. on Photorefractive Effects, Materials, and Devices PR-05*, Vol.99, Sanya, Hainan, China, 19-23 July, 2005

de Montmorillon, L.-A.; Delaye, P.; Launay, J.-C. and Roosen, G. (1997). Novel theoretical aspects on photorefractive ultrasonic detection and implementation of a sensor with an optimum sensitivity. *J. Appl. Phys.*, Vol.82, No.12, pp.5913-5922, ISSN 1089-7550

Dewhurst, R. J. and Shan, Q. (1999). Optical remote measurement of ultrasound. *Meas. Sci. Technol.*, Vol.10, pp.R139-R168 ISSN 0957-0233

Di Girolamo, S.; Kamshilin, A. A.; Romashko, R. V.; Kulchin, Y. N. and Launay, J.-C. (2007a). Fast adaptive interferometer on dynamic reflection hologram in CdTe:V. *Opt. Express*, Vol.15, No.2, pp.545-555, ISSN 1094-4087

Di Girolamo, S.; Kamshilin, A. A.; Romashko, R. V.; Kulchin, Y. N. and Launay, J.-C. (2007b). Sensing of multimode-fiber strain by a dynamic photorefractive hologram. *Opt. Lett.*, Vol.32, No.13, pp.1821-1823, ISSN 0146-9592

Di Girolamo, S.; Romashko, R. V.; Kulchin, Y. N.; Launay, J.-C. and Kamshilin, A. A. (2008). Fiber sensors multiplexing using vectorial wave mixing in a photorefractive crystal. *Opt. Express*, Vol.16, No.22, pp.18041-18049, ISSN 1094-4087

Di Girolamo, S.; Romashko, R. V.; Kulchin, Y. N. and Kamshilin, A. A. (2010). Orthogonal geometry of wave interaction in a photorefractive crystal for linear phase demodulation. *Opt. Comm.*, Vol.283, pp.128-131, ISSN 0030-4018

Feinberg, J. (1982). Asymmetric self-defocusing of an optical beam from the photorefractive effect. *J. Opt. Soc. Am.*, Vol.72, No.1, pp.46-51

Feng, W.; Yan, Y.; Jin, G.; Wu, M. and He, Q. (2000). Multiplexing of volume holographic wavelet correlation processor. *Opt. Comm.*, Vol.176, pp.49-59, ISSN 0030-4018

Fomitchov, P. A.; Kim, Y. K.; Kromin, A.; Krishnaswamy, S.; Achenbach, J. D. and Daniel, I. M. (2001). Distributed photoacoustic system for cure monitoring of composites, *Proceedings of SPIE "Advanced Nondestructive Evaluation for Structural and Biological Health Monitoring"*, Vol.4335, July, 2001

Fomitchov, P. A.; Murray, T. W. and Krishnaswamy, S. (2002). Intrinsic fiber-optic ultrasonic sensor array using multiplexed two-wave mixing interferometry. *Appl. Opt.*, Vol.41, No.7, pp.1262-1266, ISSN 1559-128X

Fomitchov, P. A. and Krishnaswamy, S. (2003). Response of a fiber Bragg-grating ultrasound sensor. *Opt. Eng.*, Vol.42, pp.956-963, ISSN 0091-3286

Forward, R. L. (1978). Wideband laser-interferometer graviational-radiation experiment. *Phys. Rev.*, Vol.17, No.2, pp.379-390

Grattan, K. T. V. and Meggitt, B. T. (Eds). (1998). *Optical Fiber Sensor Technology: Devices and Technology*. Chapman and Hall, ISBN 0-412-78290-1, London

Hall, T. J., Fiddy, M. A. and Ner, M. S. (1980). Detector for an optical-fiber acoustic sensor using dynamic holographic interferometry. *Opt. Lett.*, Vol.5, No.11, pp.485-487, ISSN 0146-9592

Hariharan, P. (1990). Optical interferometry. *Rep. Prog. Phys.*, Vol.54, No.3, pp.339-390, ISSN 0034-4885

Huignard, J. P. and Marrakchi, A. (1981). Two wave mixing and energy transfer in $Bi_{12}SiO_{20}$ crystals: application to image amplification and vibration analysis. *Opt. Lett.*, Vol.6, No.12, pp.622-624, ISSN 0146-9592

Hukriede, J.; Runde, D. and Kip, D. (2003). Fabication and application of holographic Bragg gratings in photorefractive lithium niobate channel waveguides. *J. Phys. D: Appl. Phys.*, Vol.36, pp.R1-R16, ISSN 1361-6463

Kamshilin, A. A.; Frejlich, J. and Cescato, L. (1986). Photorefractive crystals for the stabilization of the holographic setup. *Appl. Opt.*, Vol.25, No.14, pp.2375-2381, ISSN 1559-128X

Kamshilin, A. A.; Jaaskelainen, T.; Khomenko, A. V. and Garcia-Weidner, A. (1995). Multimode fiber-optic sensor using photorefractive double phase conjugator. *Appl. Phys. Lett.*, Vol.67, No.18, pp.2585-2587, ISSN 0003-6951

Kamshilin, A. A.; Jaaskelainen, T. and Kulchin, Y. N. (1998). Adaptive correlation filter for stabilization of interference-fiber-optic sensors. *Appl. Phys. Lett.*, Vol.73, No.6, pp.705-707, ISSN 0003-6951

Kamshilin, A. A. and Grachev, A. I. (2002). Adaptive interferometer based on wave mixing in a photorefractive crystal under alternating electric field. *Appl. Phys. Lett.*, Vol.81, No.16, pp.2923-2925, ISSN 0003-6951

Kamshilin, A. A.; Romashko, R. V. and Kulchin, Y. N. (2009). Adaptive interferometry with photorefractive crystals. *J. Appl. Phys.*, Vol.105, pp.031101(11), ISSN 0021-8979

Kersey, A. D.; Berkoff, T. A. and Morey, W. W. (1992). High resolution fiber-grating based strain sensor with interferometric wavelength shift detection. *Electron. Lett.*, Vol.28, pp.236-238, ISSN 0013-5194

Kersey, A. D.; Berkoff, T. A. and Morey, W. W. (1993). Multiplexed fiber Bragg grating strain-sensor system with a Fabry–Perot wavelength filter. *Opt. Lett.*, Vol.18, pp.1370-1372, ISSN 0146-9592

Kujawinska, M. and Robinson, D. W. (1988). Multichannel phase-stepped holographic interferometry. *Appl. Opt.*, Vol.27, No.2, pp.312-320, ISSN 1559-128X

Kukhtarev, N. V.; Markov, V. B.; Odulov, S. G.; Soskin, M. S. and Vinetskii, V. L. (1979). Holographic storage in electrooptic crystals. I. Steady state. *Ferroelectrics*, Vol.22, pp.949-960, ISSN 0015-0193

Kulchin, Y. N.; Romashko, R. V. and Piskunov, E. N. (2000a). Multi-channel adaptive fiber-optical system for monitoring of fast processes in solid state, *Proceedings of SPIE "Asia-Pacific Conference on Fundamental Problems of Opto- and Microelectronics APCOM-2000"*, Vol.4513, Vladivostok, Russia, September, 2000

Kulchin, Y. N.; Romashko, R. V.; Piskunov, E. N. and Kamshilin, A. A. (2000b). A multichannel correlation filter based on a photorefractive crystal for the processing of varying speckle fields. *Tech. Phys. Lett.*, Vol.26, No.6, pp.505-507, ISSN 1063-7850

Kulchin, Y. N. (2001). *Distributed Fiber-Optical Measurement Systems*, Fizmatlit, ISBN 978-5-9221-0072-4, Moscow, Russia

Kume, T.; Nonaka, K.; Yamamoto, M. and Yagi, S. (1998). Wavelength-multiplexed holographic data storage by use of reflection geometry with a cerium-doped strontium bariumniobate single-crystal structure and a tunable laser diode. *Appl. Opt.*, Vol.37, pp.334-339, ISSN 1559-128X

Lichtenberg, S.; Heinisch, S.; Petrov, V. M.; Petter, J. and Tschudi, T. (2005). Refractive index measurement using an adaptive interferometer based on phase-shift keying, *Proceedings of 10th Int. Conf. on Photorefractive Effects, Materials, and Devices PR-05*, Vol.99, Sanya, Hainan, China, 19-23 July, 2005

Murray, T. W.; Tuovinen, H. and Krishnaswamy, S. (2000). Adaptive optical array receivers for detection of surface acoustic waves. *Appl. Opt.*, Vol.39, No.19, pp.3276-3284, ISSN 1559-128X

Osterberg, H. (1932). An interferometer method of studying the vibrations of an oscillating quartz plate. *J. Opt. Soc. Am.*, Vol.22, No.1, pp.19-35

Othonos, A. and Kalli, K. (1999). *Fiber Bragg Gratings: Fundamentals and Applications in Telecommunications and Sensing*, Artech House Publishers, ISBN 978-0890063446

Pauliat, G.; Roosen, G.; Georges, M. P. and Pedrini, G. (2006). Passive introduction of carrier fringes in real-time photorefractive interferometers for single interferogram analysis. *J. Europ. Opt. Soc.*, Vol.1, p.06024, ISSN 1990-2573

Peiponen, K. E.; Myllylä, R. and Priezzhev, A. V. (2009). *Optical Measurement Techniques*, Springer, ISBN 978-3-540-71926-7, Berlin

Peng, L.; Varma, M. M.; Regnier, F. E. and Nolte, D. D. (2007). Adaptive interferometry of protein on a BioCD. *Appl. Opt.*, Vol.46, No.22, pp.5384-5395, ISSN 1559-128X

Peng, L.; Yu, P.; Nolte, D. D. and Melloch, M. R. (2003). High-speed adaptive interferometer for optical coherence-domain reflectometry through turbid media. *Opt.Lett.*, Vol.28, No.6, pp.396-398, ISSN 0146-9592

Perez, I.; Cui, H. and Udd, E. (2001). *Proceedings of SPIE "Smart Structures and Materials 2001"*, Vol.4328, Febraury, 2001

Petrov, M. P.; Stepanov, S. I. and Khomenko, A. V. (1983). *Photosensitive Electrooptic Materials in Holography and Optical Information Processing*, Nauka, Leningrad

Petrov, M. P.; Stepanov, S. I. and Khomenko, A. V. (1991). *Photorefractive Crystals in Coherent Optical Systems*, Springer-Verlag, ISBN 3-540-52603-X, Berlin, Germany

Petrov, V. M.; Denz, C.; Shamray, A. V.; Petrov, M. P. and Tschudi, T. (2001). Electrically controlled volume $LiNbO_3$ holograms for wavelength demultiplexing systems. *Optical Materials*, Vol.18, No.1, pp.191-194, ISSN 0925-3467

Petrov, V. M.; Lichtenberg, S.; Petter, J.; Tschudi, T.; Chamray, A. V.; Bryskin, V. V. and Petrov, M. P. (2003). Optical on-line controllable filters based on photorefractive crystals. *J. Opt. A: Pure Appl. Opt.*, Vol.5, pp.S471-S476, ISSN 1464-4258

Qiao, Y.; Zhou, Y. and Krishnaswamy, S. (2006). Adaptive demodulation of dynamic signals from fiber Bragg gratings using two-wave mixing technology. *Appl. Opt.*, Vol.45, No.21, pp.5132-5142, ISSN 1559-128X

Romashko, R. V.; Kulchin, Y. N. and Kamshilin, A. A. (2005a). Linear phase demodulation via reflection photorefractive holograms, *Proceedings of 10th Int. Conf. on Photorefractive Effects, Materials, and Devices PR-05*, Vol.99, Sanya, Hainan, China, 19-23 July, 2005

Romashko, R. V.; Kulchin, Y. N.; Shandarov, S. M.; Kargin, Y. F. and Volkov, V. V. (2005b). Adaptive correlation filter based on dynamic reflection hologram formed in photorefractive $Bi_{12}TiO_{20}$ crystal. *Opt. Rev.*, Vol.12, No.1, pp.59-60, ISSN 1340-6000

Romashko, R. V.; Kulchin, Y. N. and Kamshilin, A. A. (2007). Optimal geometry for fast and efficient hologram recording in photorefractive crystal. *Opt. Rev.*, Vol.14, No.4, pp.176-179, ISSN 1340-6000

Romashko, R. V.; Di Giloramo, S.; Kulchin, Y. N. and Kamshilin, A. A. (2010). Photorefractive vectorial wave mixing in different geometries. *J. Opt. Soc. Am. B*, Vol.27, No.2, pp.311-317, ISSN 1520-8540

Runde, D.; Breuer, S. and Kip, D. (2005). Holographic refelection filters in photorefractive LiNbO3 channel waveguides for applications as add/drop multiplexers, *Proceedings of 10th Int. Conf. on Photorefractive Effects, Materials, and Devices PR-05 (OSA Trends in Optics and Photonics)*, Vol.99, Sanya, Hainan, China, June, 2005

Solymar, L.; Webb, D. J. and Grunnet-Jepsen, A. (1996). *The Physics and Application of Photorefractive Materials*, Clarendon Press, Oxford

Steckman, G. J.; Pu, A. and Psaltis, D. (2001). Storage density of shift-multiplexed holographic memory. *Appl. Opt.*, Vol.40, No.20, pp.3387-3394, ISSN 1559-128X

Stepanov, S. I. (1991). Adaptive interferometry: a new area of applications of photorefractive crystals, In: *International Trends in Optics*, J.W. Goodman, (Ed.), pp.125-140, Academic Press, Inc., ISBN 0-12-289690-4, New York, London

Sturman, B. I. and Fridkin, V. M. (1992). *The Photovoltaic and Photorefractive Effect in Noncentrosymmetric Materials*, Gordon and Breach, ISBN 978-2-88124-498-8, New York, USA

Sturman, B. I.; Podivilov, E. V.; Ringhofer, K. H.; Shamonina, E.; Kamenov, V. P.; Nippolainen, E.; Prokofiev, V. V. and Kamshilin, A. A. (1999). Theory of photorefractive vectorial wave coupling in cubic crystals. *Phys. Rev. E*, Vol.60, No.3, pp.3332-3352, ISSN 0031-9007

Thizy, C.; Georges, M. P.; Kouloumpi, E.; Green, T.; Hackney, S. and Tornari, V. (2007). Photorefractive holographic interferometry for movable artwork assessment, *Proceedings of OSA Topical Meeting "Controlling light with light: Photorefractive Effects, Photosensitivity, Fiber Gratings, Photonic Materials and More"*, October, 2007

Voronov, V. V.; Dorosh, I. R.; Kuzminov, Y. S. and Tkachenko, N. V. (1980). Photoinduced light scattering in cerium-doped barium strontium niobate crystals. *Sov. J. Quantum Electron.*, Vol.10, No.11, pp.1346-1349, ISSN 0049-1748

Wagner, J. W. and Spicer, J. B. (1987). Theoretical noise-limited sensitivity of classical interferometry. *J. Opt. Soc. Am. B*, Vol.4, No.8, pp.1316-1326, ISSN 1520-8540

Wen, Z. and Yang, X. (1997). Multichannel photorefractive correlator for rotation-invariant optical pattern recognition. *Opt. Comm.*, Vol.135, pp.212-216, ISSN 0030-4018

Xu, M., Geiger, G. H. and Dakin, J. P. (1993). Modeling and performance nalysis of a fiber Bragg grating interrogation system using an acousto-optic tunable filter. *J. Lightwave Technol.*, Vol.14, No.391, p.396, ISSN 0733-8724

Young, L., Wong, W. K., Thewalt, M. L. and Cornish, W. D. (1974). Theory of formation of phase holograms in lithium niobate. *Appl. Phys. Lett.*, Vol.24, No.6, pp.264-265, ISSN 0003-6951

7

Infrared Holography for Wavefront Reconstruction and Interferometric Metrology

Sergio De Nicola, Andrea Geltrude,
Massimiliano Locatelli, Kais Al-Naimee,
Riccardo Meucci and F.Tito Arecchi
Istituto Nazionale di Ottica – CNR Firenze
Italy

1. Introduction

Long wavelength interferometry has been widely applied in different fields, such as infrared optics, infrared transmitting materials, high-reflective multilayer dielectric coatings for high-power laser systems. In optical metrology, long-wave interferometers are also employed for shape measurement of reflective rough surfaces and for testing optical systems that requires deep asphericss. An advantage of using longer wavelength is that the aspheric departure from the best fit reference-sphere, in unit of the probing wavelength, is reduced at longer wavelength, thus allowing one to obtain an interferogram of the deep aspheric under test. This leads to extension of the unambiguous distance measurement range, a well-known problem in interferometric metrology, where the essential difficulty relates to the interferometric fringe order, which cannot be determined unambiguosly from a single measurement of phase interference. This problem is also of particular significance in digital holographic and interferomety based applications where digital processing of the recorded interferogram makes it possible to extract quantitatively amplitude and phase of the numerically reconstructed wavefront (Cuche el al., 1999a, 1999b; Vest, 1979; Yaroslavsky & Eden, 1996). Digital holography based imaging techniques provide real time capabilities to record also three-dimensional objects using interference between an object wave and a reference wave captured by an image sensor such a CCD sensor. Three-dimensional 3D information of the object can be obtained from the numerical reconstruction of a single digitally recorded hologram, since the information about the optically interfering waves is stored in the form of matrices. The numerical reconstruction process offers many more possibilities than conventional optical processing (Goodman & Lawrence, 1967; Stetson & Powell, 1966; Stetson & Brohinsky, 1985). For example, it is possible to numerically focus on any section of the three dimensional volume object without mechanical focusing adjustment (Grilli et al., 2001; Lai et al., 2000), correct optical components defects such as lens aberrations or compensate the limited depth of field of an high magnification microscope objective. Full digital processing of holograms requires high spatial resolution sensor arrays with demanding capabilities for imaging applications and non destructive testing. In this regard, several new recording materials and optoelectronic sensors have been devised for

recording holograms beyond the visible wavelengths, in the infrared. Coherent imaging based on digital processing of infrared holograms presents the great advantage of providing quantitative data from the recorded interference pattern (Lei et al., 2001a, 2001b; Leith & Upatnieks, 1965; Nilsson & Carlsson, 2000; Seebacker et al., 2001) but, the lower spatial resolution of the infrared sensor arrays compared to those working in the visible region, appears to be a major limiting factor to accurate numerical wavefront reconstruction for the digitized interferogram.
In this chapter we will describe potential applications of digital processing methods for whole optical wavefront reconstruction of infrared recorded holograms. The basic principle of numerical reconstruction of digitized holograms will be reviewed and we will discuss numerical methods to compensate the loss of spatial resolution at longer wavelength and the limited spatial resolution of the recording array We will present several examples to show that infrared digital holography can be exploited as a high accuracy technique for testing both reflective and transmissive objects. In order to demonstrate the feasibility of digital holography in the infrared region we will describe a two beam interferometric technique for recording holograms, both in reflection and transmission type geometry which employs a pyroelectric sensor array. The availability of pyroelectric sensor arrays for recording digital infrared holograms offers the advantage of a compact and efficient interferometric set-up for recording infrared hologram. We will also present application of digital holographic processing methods for numerical reconstruction of wavefronts of optical beams, such as Laguerre-Gauss beams possessing wavefront singularities in the infrared region. In fact, this technique can be usefully employed for characterizing the vorticity of infrared beams of potential use in optical telecommunication applications, where the preservation of the purity of the mode along the propagation direction is a problem of crucial importance. We will present examples of characterization of the vortex signature of infrared Laguerre-Gauss beam through the numerical reconstruction of the singular phase of the beam from the digitized hologram.

2. Theory and principle of operation of digital holography

The principle of the optical recording and reconstruction in classical holography is illustrated in Fig.1. The reference beam R interferes at the plane of the holographic plate at off-axis angle θ with respect to the object beam O. In this set-up the reconstructed image is spatially separated from the zero-order diffraction and the second image, the so-called 'twin image'. These three diffraction orders propagate along different directions and can be observed separately, leading to a significant improvement compared to the original in-line configuration originally developed by Gabor (Gabor, 1948; Goodman, 1996) where the zero-order and the two conjugate images overlap.
The intensity distribution $I(x,y)$ across the x-y holographic recording plane can be written as the modulus squared of the complex superposition $O(x,y)+R(x,y)$, namely

$$I(x,y)=|O(x,y)+R(x,y)|^2=|R(x,y)|^2+|O(x,y)|^2+R^*(x,y)O(x,y)+R(x,y)O^*(x,y) \quad (1)$$

where the symbol * denotes complex conjugation, $O(x,y)=|O(x,y)|\exp[i\varphi_O(x,y)]$ is the complex amplitude of the object wave with real amplitude $|O(x,y)|$ and phase $\varphi_o(x,y)$ and

$R(x,y)=|R(x,y)|\exp[i\varphi_R(x,y)]$ is the complex amplitude of the reference wave with real amplitude $|R(x,y)|$ and phase $\varphi_R(x,y)$. Eq. (1) can be modified to the form

$$I(x,y)=\left(|R(x,y)|^2+|O(x,y)|^2\right)+2|R(x,y)||O(x,y)|\cos\left(\varphi_o(x,y)-\varphi_R(x,y)\right) \tag{1a}$$

which shows explicitly that the recorded hologram contains a term with an amplitude and phase modulated spatial carrier. For the numerical reconstruction of the recorded hologram, the interference pattern $I(x,y)$ is illuminated by the reference wave $R(x,y)$, i.e , we have

$$R(x,y)I(x,y)=R(x,y)|R(x,y)|^2+R(x,y)|O(x,y)|^2+|R(x,y)|^2 O(x,y)+R^2(x,y)O^*(x,y) \tag{2}$$

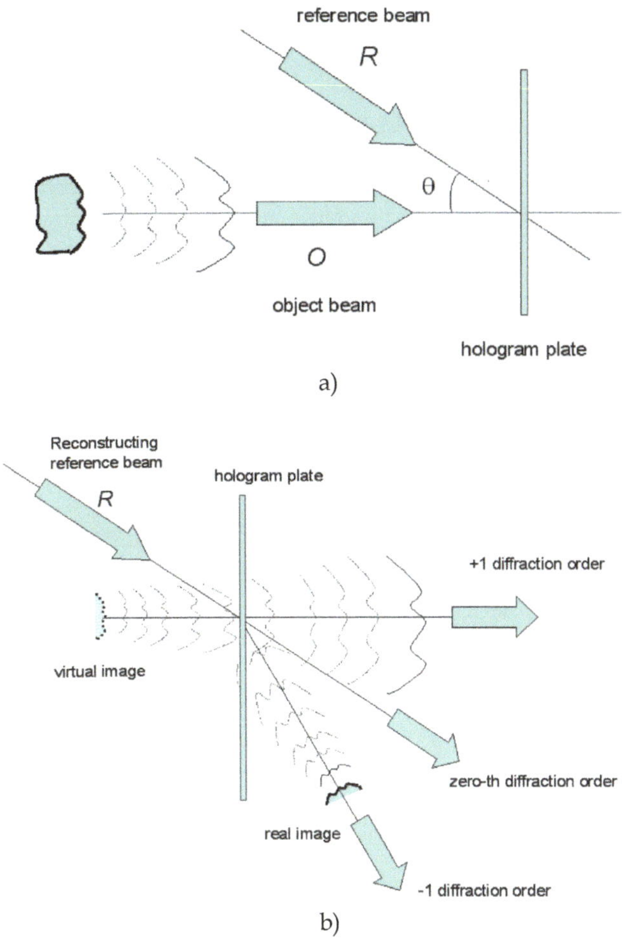

Fig. 1. Optical configuration for recording (a) and for reconstruction (b) of off-axis holograms.

The first term on the right side of this equation is proportional to the reference wave field, the second one is a spatially varying "cloud" surrounding the first term. These two terms constitute the zero-order of diffraction or DC term. The third term represents, apart for a constant factor, an exact replica of the original wavefront $O(x,y)=|O(x,y)|\exp(i\varphi_o(x,y))$ and for this reason it is called virtual image, or simply image of the object. The last term is another copy, the so called twin image of the original object wave or real image.
In holography the hologram can be regarded as an amplitude transmittance that diffracts the reference wave. In DH the object wave field is determined through the numerical calculation of the optical field propagation of $R(x,y)h(x,y)$ from the holographic plane back to the object plane ξ,η . The numerical reconstruction of a digitally recorded hologram follows the scalar diffraction theory in the Fresnel approximation of the Rayleigh-Sommerfield diffraction integral. The reconstructed diffracted field $Q(\xi,\eta)$ in the reconstruction plane ξ,η at distance d from the hologram plane can be written in the paraxial approximation in the following form

$$Q(\xi,\eta)=\frac{1}{i\lambda d}\exp\left(i\frac{2\pi}{\lambda}d\right)\int_{-\infty}^{\infty}\int_{-\infty}^{\infty}R(x,y)I(x,y)\exp\left[i\frac{\pi}{\lambda d}\left((x-\xi)^2+(y-\xi)^2\right)\right]dx\,dy \qquad (3)$$

Eq.(3) is the starting point for reconstructing numerically the digitized hologram in the paraxial approximation, where the x and y values and the corresponding ξ and η values in the reconstruction plane are small compared to the distance d (see Fig.2).

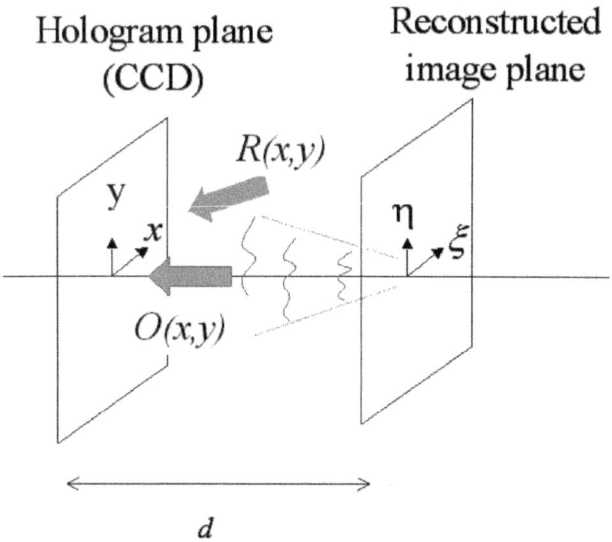

Fig. 2. Optical set-up in off-axis digital holography

Once the complex field $Q(\xi,\eta)$ has been calculated at distance d, the intensity $I(x,y;d)$ and phase distribution $\varphi(x,y;d)$ of the reconstructed image can be determined by the following relations

$$I(x,y;d) = |Q(x,y)|^2 \tag{4a}$$

$$\varphi(x,y;d) = \arctan\frac{\mathrm{Im}[Q(x,y)]}{\mathrm{Re}[Q(x,y)]} \tag{4b}$$

Eq. (4b) provides phase values wrapped in the interval $[\pi,\pi]$. A well-known unwrapping procedure can be employed to convert the phase modulo-2π into a continuous phase distribution in order to obtain a smooth phase image.

2.1 Reconstruction methods

In the following we will describe the main methods for numerical reconstruction of holograms (Kreis & Jueptner, 1997; Kronrod et al., 1972; Schnars & Jueptner, 1994, 2002).

2.1.1 Fresnel transformation method (FTM)

The convolution integral given by Eq.(3) can be manipulated to obtain the reconstructed diffracted field $Q(\xi,\eta)$ in terms of the so-called Fresnel transformation of the hologram function. Indeed it results that

$$\begin{gathered} Q(\xi,\eta) = \frac{1}{i\lambda d}\exp\left(i\frac{2\pi}{\lambda}d\right)\exp\left[i\frac{\pi}{\lambda d}\left(\xi^2+\eta^2\right)\right] \\ \int_{-\infty}^{\infty}\int_{-\infty}^{\infty} R(x,y)I(x,y)\exp\left[i\frac{\pi}{\lambda d}\left(x^2+y^2\right)\right]\exp\left[-i\frac{2\pi}{\lambda d}(\xi x + y\eta)\right]dx\,dy \end{gathered} \tag{5}$$

Eq.(5) shows that the reconstruction field is determined essentially by the two-dimensional Fourier transformation of the multiplication of the hologram $I(x,y)$ by the reference wave $R(x,y)$ and the chirp phase function

$$w(x,y) = \exp\left[i\frac{\pi}{\lambda d}\left(x^2+y^2\right)\right] \tag{5a}$$

Eq. (5) can be written in terms of the Fourier integral

$$Q(\xi,\eta) = \frac{1}{i\lambda d}\exp\left(i\frac{2\pi}{\lambda}d\right)\exp\left[i\frac{\pi}{\lambda d}\left(\nu_\xi^2+\nu_\eta^2\right)\right]\left(F^{+1}\left[R(x,y)I(x,y)w(x,y)\right](\nu_\xi,\nu_\eta)\right) \tag{6}$$

where the direct (+1) or inverse (-1) continuous two dimensional Fourier transformations of the function $f(x,y)$ are defined, respectively, by

$$F^{\pm 1}\left[f(x,y)\right](\nu_\xi,\nu_\eta) = \int_{-\infty}^{\infty}\int_{-\infty}^{\infty} f(x,y)\exp\left[\mp i2\pi\left(\nu_\xi x + \nu_\eta y\right)\right]dxdy \tag{7}$$

In Eq. (7) ν_ξ and ν_η are the spatial frequencies (Kreis, 2002a, 2002b) corresponding to the spatial variables ξ and η in the reconstruction plane and they are related to the reconstruction wavelength λ and to the reconstruction distance *d* by the following relations

$$\nu_\xi = \frac{\xi}{\lambda d} \qquad \nu_\eta = \frac{\eta}{\lambda d} \tag{8}$$

With the off-axis geometry the object wave and the reference wave arrive in the hologram plane with separate directions and, according to the above equations, the different terms of the numerically reconstructed wavefront propagate along different directions, owing to their different spatial frequencies. In fact, if Eq.(2) is substituted into Eq. (5), it is clear that the reconstruction of the DC term, the virtual and the real image are essentially governed by the frequency content of the respective spectra at the reconstruction distance *d*, which ultimately impose restrictions on the spatial bandwidth of the object and reference beam. If the reference field is given by $R(x,y)=\sqrt{I_R}\exp\left[i\left(k_x x+k_y y\right)\right]$ where $I_R=\left|R(x,y)\right|^2$ is the intensity of the reference field and $\mathbf{k}=\left(k_x,k_y,k_z\right)$ is the corresponding wave vector, the three terms are separated in the Fourier domain corresponding to the reconstruction plane ξ,η at distance *d*. The zero-order is located around the origin while the image and the twin image are symmetrically centred on $\left(k_x/2\pi,k_y/2\pi\right)$ and $\left(-k_x/2\pi,-k_y/2\pi\right)$, respectively. To achieve good quality reconstruction in DH, the sampling theorem (Nyquist criterion) has to be fulfilled across the whole CCD array area. The criterion requires at least two pixels per fringe period and this implies that the maximum interference angle $\alpha_{\max}$ between the spherical wavelet from each point of the object and the reference wave field is determined by the pixel size Δx according to the relation

$$a_{\max}=\frac{\lambda}{2\Delta x} \tag{9}$$

Relation (9) expresses the fact that for recording an hologram by a CCD array with pixel spacing Δx at least two pixels per fringe are needed. For example in case of a camera with pixel size Δx = 9 μm, the maximum interference angle is $\alpha_{\max}\approx 1.7°$ for λ=532 nm. Mathematically, the two-dimensional spatial sampling $I(n\Delta x,m\Delta y)$ of the hologram $I(x,y)$ on a rectangular raster of $N\times M$ points can be described by the following relation

$$I\left(n\Delta x,m\Delta y\right)=I\left(x,y\right)rect\left(\frac{x}{N\Delta x},\frac{y}{M\Delta y}\right)\sum_{n=1}^{N}\sum_{m=1}^{M}\delta\left(x-n\Delta x,\,y-m\Delta y\right) \tag{10}$$

where $\delta(x,\,y)$ is the two-dimensional Dirac-delta function, *n* and *m* are integer numbers, $N\Delta x\times M\Delta y$ is the area of the digitized hologram and $rect(x,y)$ is equal to one if the point of coordinated (x,y) is inside the area of the digitized hologram, and is zero elsewhere. Δx and Δy in Eq.(10) are the distances between the neighboring pixels on the CCD array in the horizontal and vertical directions, respectively. If the whole CCD array has a finite width given by $N\Delta x\times M\Delta y$, where *N* and *M* are the pixel numbers in each directions, the discrete representation of the Fresnel reconstruction integral given by Eq. (5) can be written as

$$\begin{aligned}Q\left(r\Delta\nu_\xi,s\Delta\nu_\eta\right)=&\frac{1}{i\lambda d}\exp\left(i\frac{2\pi}{\lambda}d\right)\exp\left[i\pi\lambda d\left(r^2\Delta\nu_\xi^2+s^2\Delta\nu_\eta^2\right)\right]\\&\times\Delta x\Delta y\sum_{n=-N/2}^{N/2-1}\sum_{m=-M/2}^{M/2-1}I\left(n\Delta x,m\Delta y\right)R\left(n\Delta x,m\Delta y\right)w\left(n\Delta x,m\Delta y\right)\exp\left[i2\pi\left(\frac{rn}{N}+\frac{sm}{M}\right)\right]\end{aligned} \tag{11}$$

Eq. (11) can allows to compute a matrix of $N \times M$ complex numbers of the reconstructed field via the discrete two-dimensional fast Fourier transform algorithm. According to the theory of discrete Fourier transform, the sampling frequency intervals are given by $\Delta\nu_{\xi} = 1/N\Delta x$ and $\Delta\nu_{\eta} = 1/M\Delta y$ which, together with relations (8) allow us to determine the dimensions $\Delta\xi \times \Delta\eta$ of the reconstruction pixel, namely

$$\Delta\xi = \frac{\lambda d}{N\Delta x} \qquad \Delta\eta = \frac{\lambda d}{M\Delta y} \tag{12}$$

According to Eq. (12) the pixel width in the reconstructed plane is different from those of the digitized hologram and it scales inversely to the aperture of the optical system, i.e., to the side length $S = N\Delta x$ of the hologram (limiting the analysis to the x-direction for the sake of simplicity). This result is in agreement with the theory of diffraction which predicts that at a distance d from the hologram plane the developed diffraction pattern is characterized by the diameter $\lambda d/S$ of its Airy disk (or speckle diameter). Therefore the resolution of the reconstructed image (amplitude or phase image) is limited by the diffraction limit of the imaging system through the automatic scaling imposed by the Fresnel transform. Assuming that the hologram $I(x,y)$ has spatial frequencies smaller than those in the quadratic phase factor $w(x,y)$, the main problem when calculating Eq. (11) comes from an adequate sampling of the exponential function $w(x,y)$ inside the integral and of the global phase factor $\exp\left[i\pi\lambda d\left(r^2\Delta\nu_{\xi}^2 + s^2\Delta\nu_{\eta}^2\right)\right]$ multiplying the expression in Eq.(11). Assuming the sampling of $w(x,y)$ in the Nyquist limit, it is easy to obtain the approximate condition (limiting our analysis to one dimension only) that determines the range of distances d where the discrete Fresnel reconstruction algorithm (cfr. Eq (11)) gives good results, namely

$$d \geq d_c = \frac{N\Delta x^2}{\lambda} \tag{13}$$

The same argument can be applied to the global phase factor $\exp(i\pi\lambda d r^2\Delta\nu_{\xi}^2 + s^2\Delta\nu_{\eta}^2)$, which has too rapid variation with increasing spatial distance d, giving, this time, the condition $d \leq d_c$. A good reconstruction of both the amplitude and phase-contrast image is accomplished only if the equality $d = d_c$ is assumed, while, if only the amplitude-contrast reconstruction is considered, the less restrictive condition $d \geq d_c$ must hold. In fact, in this case the global phase factor is unessential for evaluating intensity distribution and the intensity profiles are of lower spatial frequency variation than the corresponding phase. Note that the size $\Delta\xi_c$ of the reconstruction pixel at distance $d = d_c$ in the Nyquist limit coincides with pixel size of the sampled hologram, i.e., $\Delta\xi_c = \lambda d_c/N\Delta x = \Delta x$. As an example, for *N=512* pixel, λ=632nm and pixel size $\Delta x = \Delta y = 11\mu\text{m}$, the Fresnel method is valid for distance greater than 98mm; for *N=1024* pixel, λ=532nm and $\Delta x = \Delta y = 6.7\mu\text{m}$, the distance has to be greater than 24.5mm. From Eq. (12) it can be easily deduced that the width $S_I = N\Delta\xi$ of the numerical reconstruction increases linearly with reconstruction distance d according to the scaling law

$$S_I = \frac{\lambda d N}{S} \tag{14}$$

where N and the width $S = N\Delta x$ of the hologram are input parameters in the reconstruction process. Nevertheless this result is only compatible with condition given by Eq.(13), derived from the appropriate sampling of reconstructed amplitude image. This means that maintaining the number N as a constant may lead to a badly sampled reconstructed image if the reconstruction distance does not satisfy the above requirements. By padding the recorded hologram with zeros in the border, after Fresnel diffraction the external part of the reconstructed hologram can be prevented from leaving the support matrix and entering the opposite side of the matrix because of aliasing. This is of particular importance in the numerical reconstruction of off-axis recorded holograms where the use of the pixel area of the recording array is less efficient.

2.1.2 Convolution transformation method

An alternative way of numerical reconstruction of holograms is through the calculation of the propagated angular spectrum, the so called "convolution approach" to digital holography. The reconstructed field, in the paraxial approximation, can be written in this case in the following form

$$Q\left((\nu_\xi,\nu_\eta)\right) = \exp\left(i\frac{2\pi}{\lambda}d\right)\mathrm{F}^{-1}\left\{(\exp\left[-i\pi\lambda d\left(\nu_\xi^2+\nu_\eta^2\right)\right]F^{+1}\left[R(x,y)h(x,y)\right](\nu_\xi,\nu_\eta)\right\} \quad (15)$$

where the Fourier transform of the chirp function $w(x,y)$ given by Eq. (4) has been used, namely

$$F^{+1}\left[w(x,y)\right](\nu_\xi,\nu_\eta) = id\lambda\,\exp\left[[-i\pi\lambda d\left((\nu_\xi^2+\nu_\eta^2)\right]\right)\right] \quad (16)$$

It can be shown that when the angular spectrum is used, the use of two Fourier transforms for computing Eq.(15), once for taking the Fourier transform of the hologram (multiplied by the reference wave) and another time for taking the inverse Fourier transform, leads to a cancellation of the scale factor between the input and output field to obtain that the pixel size of the reconstructed images equal to that of the sampled hologram, i.e., $\Delta\xi = \Delta x$ and $\Delta\eta = \Delta y$ and the actual sizes of h the input hologram and reconstructed image are identical $(S_I = S)$. We point out that, although Eq.(6) and Eq.(15) are formally equivalent, the different use of the DFT algorithm to perform the calculation of the same diffraction integral, makes the convolution-based algorithm valid for near distances $d \le d_c$. Both methods overlap at distance $d = d_c$. Clearly this method is computationally more expensive than the direct evaluation of the Fresnel integral, since it requires two Fourier transforms (one direct and one inverse) but it is advantageous for keeping constant the length scales of the reconstructed images for all distances satisfying the near-field approximation (Zhang et al., 2004). From the discrete complex values of the reconstructed field, the intensity $I(r,s;d)$ and phase distribution $\varphi(r,s;d)$ of the reconstructed image can be determined according to the following two equations

$$I(r,s;d) = \left|Q(r\Delta\xi, s\Delta\eta)\right|^2 \quad (17a)$$

$$\varphi(r,s;d) = \arctan\frac{\mathrm{Im}\left[Q(r\Delta\xi, s\Delta\eta)\right]}{\mathrm{Re}\left[Q(r\Delta\xi, s\Delta\eta)\right]} \quad (17b)$$

2.2 Digital holographic interferometry

The fringe pattern that results from the interference of the reference beam and an object beam carries phase information on the object under test and any change in its state gives rise to a corresponding modification in the phase information.

If the complex fields $Q(\xi,\eta;s_1)$ and $Q(\xi,\eta;s_2)$ are the numerical reconstructions of two holograms recorded at different states s_1 and s_2 of the object, the corresponding phase change $\Delta\phi(\xi,\eta;s_2-s_1)$ is given by

$$\Delta\phi(\xi,\eta;s_2-s_1)=Arg\left[Q(\xi,\eta;s_2)\right]-Arg\left[Q(\xi,\eta;s_1)\right] \tag{18}$$

by taking into account that

$$Arg\left[Q(\xi,\eta,s)\right]=\arctan\left[\frac{\mathrm{Im}(Q(\xi,\eta,s))}{\mathrm{Re}(Q(\xi,\eta,s))}\right] \tag{18a}$$

we can express the phase change in the following form

$$\Delta\phi(\xi,\eta;s_2-s_1)=\frac{\mathrm{Re}\left[Q(\xi,\eta;s_2)\right]\mathrm{Im}\left[Q(\xi,\eta;s_1)\right]-\mathrm{Re}\left[Q(\xi,\eta;s_2)\right]\mathrm{Im}\left[Q(\xi,\eta;s_1)\right]}{\mathrm{Re}\left[Q(\xi,\eta;s_2)\right]\mathrm{Re}\left[Q(\xi,\eta;s_1)\right]+\mathrm{Im}\left[Q(\xi,\eta;s_2)\right]\mathrm{Im}\left[Q(\xi,\eta;s_1)\right]} \tag{19}$$

For instance, in the case of deformation measurement, s_1 and s_2 are states of deformation of the object under investigation and the calculated interference phase provides information about the displacement field onto the surface in case of opaque objects or the full optical path variation occurring in transparent objects (Rastogi, 1994). Digital Holographic interferometry has also been applied to measure deformations of either large and very small objects, to investigate the refractive index changes (Dubois et al., 1999; Zito et al., 2009) and for comparing accurately the shapes of objects and for removing spherical aberrations introduced by high numerical aperture objective lenses employed in digital holographic microscopy applications where the paraxial approximations implicit in the Fresnel treatment often fails (Jueptner et al, 1987; Kato et al., 2003; Kim, 2000; Pedrini et al., 2001).

3. Infrared digital holography

In this section we present some applications of digital holography and of the previously described wavefront reconstruction technique in the infrared region as a high accuracy technique for testing reflective objects. The method makes use of a two-beam interferometric set-up (De Nicola et al., 2008) for recording digital off-axis interferograms of reflective objects at wavelength 10.6μm. A scheme of the set-up is shown in Fig. 3.

A conventional gas flowing CO_2 laser emitting on the P(20) line at λ=10.6 μm is employed as an infrared source. The laser cavity, 82cm long, is defined by a partially reflective flat mirror (R=95%) and an out-coupling mirror (R=90%) of 3m radius of curvature mounted upon a piezoelectric translator. The output laser beam is horizontally polarized by means of an intracavity ZnSe Brewster window. The system, pumped by an electric discharge of 10mA, when the laser threshold is approximately 9mA provides an output optical power in the range of 500-800mW. The spatial profile of the laser is set in the fundamental TEM_{00} Gaussian mode by means of an intra-cavity iris diaphragm. In this configuration, the laser

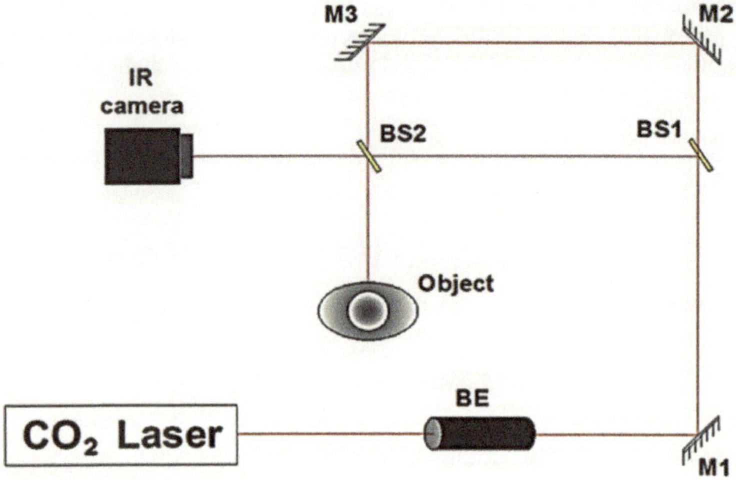

Fig. 3. Mach-Zehnder interferometer employed to record IR digital holograms of reflective samples: M1, M2 and M3, mirrors; BE beam expander; BS1 and BS2, beam splitters.

beam is characterized by a spot size of 6mm on the flat mirror and a divergence less than 2mrad. Figure 3 shows the DH optical set-up based on a Mach–Zehnder interferometer. The infrared beam is directed to a beam expander with magnification 2.5 × and 15 mm diameter of output beam. Two mirrors M1 and M2 and two beam splitters, BS1 (70T/30R at 45^0 angle of incidence) and BS2 (50T/50R at 45^0 angle of incidence), are used to form the interferometer. The beam splitters are ZnSe coated windows with a diameter 50mm each. The interferometer allows to record the interference patterns between the two beams, the reference beam and that reflected by the test object on the detection plane of a pyroelectric videocamera _Spiricon Pyrocam III, Model PY-III-C-A, which has a matrix of $N \times N = 124 \times 124$ pyroelectric sensor elements of LiTaO3, with square pixel size $85\mu m \times 85\mu m$ and center-to-center spacing of the pixels, the pixel pitch, $100\mu m \times 100\mu m$. The reference beam interferes with the object at a small angle $\alpha \leq \lambda/2\Delta\xi = 3°$, as required by the sampling theorem. The pyroelectric camera allows detection of CW infrared laser radiation by means of an internal chopper. It is connected to a personal computer to record digitized fringe patterns. The reflective objects used are two opaque aluminium blocks. The first one is a rectangle of size 20mm × 35mm that has letters inscribed. The letters are " UOR" and "XUO" (about 3mm × 4 mm each). The second one is a disc of radius 25.4mm which has inscribed a set of concentric circular tracks. The aluminium blocks are located at distance *d*=250mm from the array. The discrete finite form of Eq. (11) is obtained through the pixel pitch $\Delta\xi \times \Delta\eta = 100\mu m \times 100\mu m$ of the pyrocam array and the reconstructed object field $\psi(x,y;d)$ is obtained by applying the 2D fast Fourier transform (2D-FFT) algorithm to the discrete samples of the quantity $I(\xi,\eta)R(\xi,\eta)\exp\left[i\pi\left(\xi^2+\eta^2\right)/\lambda d\right]$. By the 2D-FFT algorithm, the size of the reconstruction pixel at distance d is $\Delta x \times \Delta y = d\lambda/N\Delta\xi \times d\lambda/N\Delta\eta$. For a reconstruction distance *d*= 250 mm , the resolution of the reconstructed field distribution is limited by the size $\Delta x \times \Delta y = 213\mu m \times 213\mu m$ of the reconstruction pixel. To compensate for the loss of resolution with increasing reconstruction distance *d*, the digitized hologram was padded with value of zero to achieve a larger array of $N^* \times N^* = 256 \times 256$ points and a reconstruction pixel of size 103μm ×103μm at a distance *d*= 250mm. This process increases the number of 2D-FFT points, changing artificially the period, while holding the spatial sampling rate of the digitized hologram fixed. In this manner, spatial spectral components of the reconstructed images, originally hidden from view can be shifted to points where they can be observed. However aliasing created by the use of a 2DFFT may result from spreading or leakage of the spectral components away from the correct frequency leading to undesirable modification of the reconstructed image. Aliasing occurs during numerical implementations if the reconstruction distance is less than the minimum object-to-hologram recordable distance $d_{\min} = N^* \Delta\xi^2 / \lambda$. Increasing the size N^* of the window function tends to increase $d_{\min}$. The condition $d > d_{\min}$ sets the maximum number of samples with value of zero at the end of the digitized holograms for a given hologram recording distance. In our case, for the infrared wavelength λ=10.6 μm, the condition is satisfied since we have $d_{\min}$ =241 mm for $N^* = 256$. The reconstruction was obtained with a reference beam of the form $R(\xi,\eta) = \exp\left[i\left(k_x\xi + k_y\eta\right)\right]$ where the two components k_x and k_y of the wave-vectors were adjusted to centre the image in the reconstruction plane. The determined phase values are wrapped (Yamaguchi & Zhang, 1997; Yamaguchi et al., 2002) in the interval $[-\pi,\pi]$. A well-known unwrapping procedure can be employed to convert the phase modulo-2π into a continuous phase distribution in order to obtain a smooth 3-D phase profile (Demoli et al., 2003; Gass et al., 2003; Schnars, 1994). The height distribution $h(x,y)$ of the object is the information to be retrieved. It is related to the reconstructed phase distribution $\varphi(\xi,\eta;d)$ by the simple relationship $h(x,y) = \lambda\varphi(x,y;d)/4\pi$. Fig.(4) shows the results of the reconstruction of the image of the letters inscribed in the first aluminium block. Amplitude images were reconstructed from the digitized infrared holograms according to method discussed in the previous section. In Fig.4(a), 4(c) the size of the reconstruction pixel is 213μm×213μm; Fig. 4(b), 4(d) shows the corresponding reconstructions with padding operations.

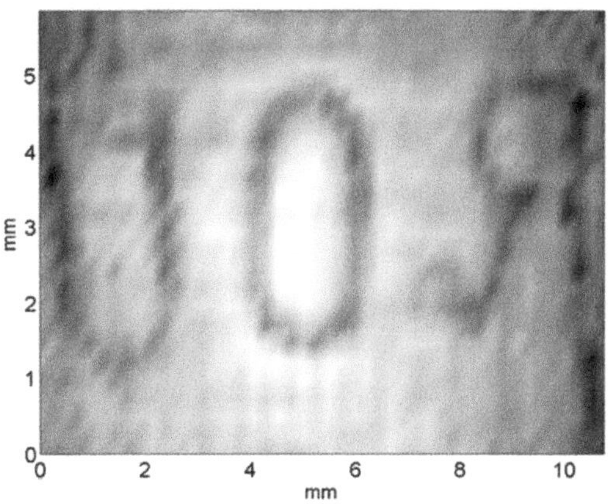

a)

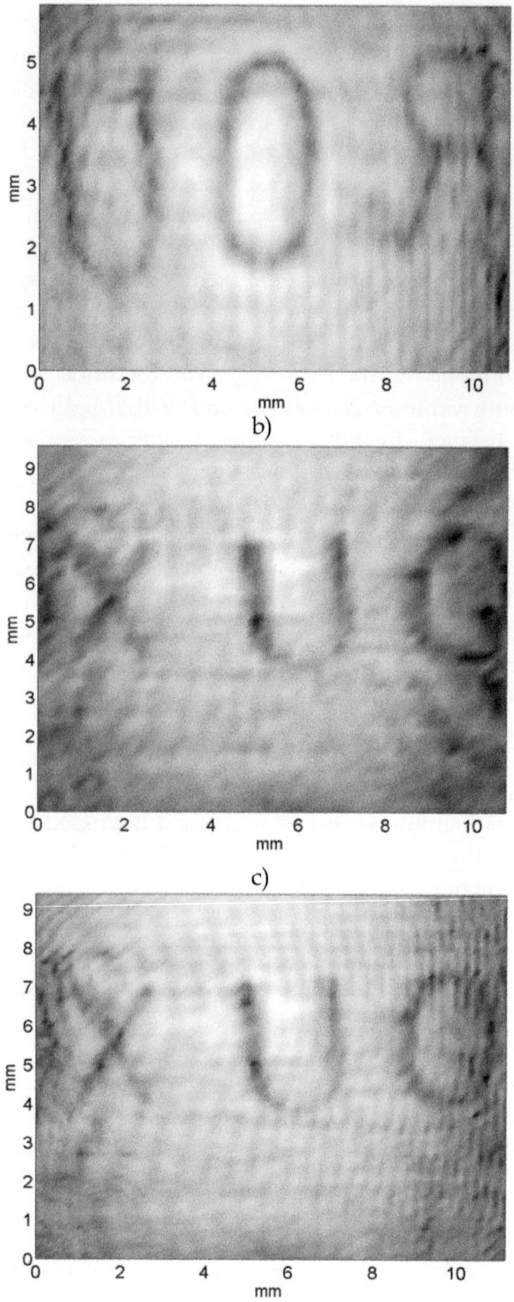

Fig. 4. Amplitude reconstruction of the letters "ROU" before padding operation (a), after zero padding (b); "XUO" before padding operation (c), after padding (d).

The size is 103μm×103μm, and the resolution is clearly improved, as it can be seen by comparing the reconstructed images shown Fig. 4(b), 4(d) with those of Fig. 4(a), 4(c).
The hologram of the aluminium disk shaped object is shown in Fig. 5. The concentric circular tracks superimposed on the interference fringes are clearly visible.

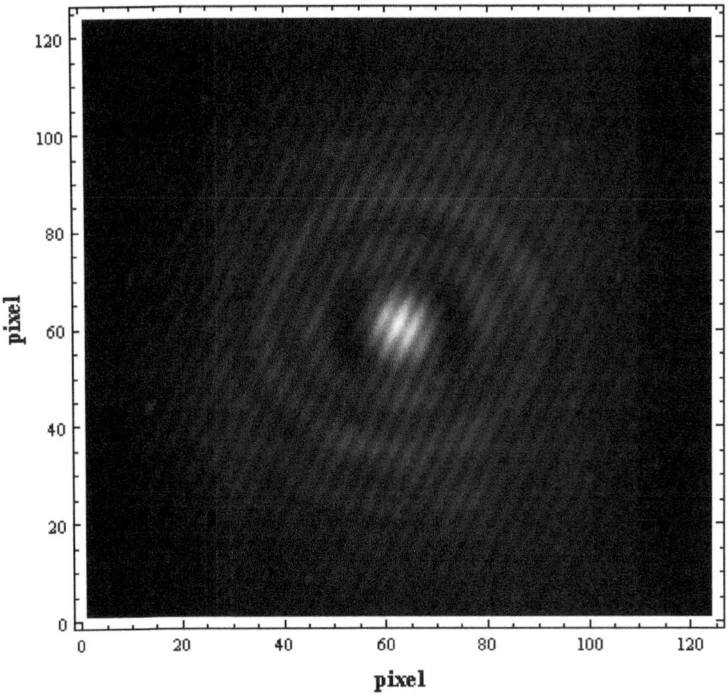

Fig. 5. Infrared hologram of an aluminium disk shaped object which has inscribed a set of concentric circular tracks.

Figure (6) shows the phase images reconstructed form the disk shaped object. Fig. 6(a), 6(c) and 6(e) display the phase distribution of the object reconstructed without padding operation and Fig. 6(b), 6(d) and 6(f) show the corresponding reconstructions with zero padding operation. It is noticeable in the 3D perspectives plots of the reconstructed phase shown in Fig. 6(d) and 6(f) that the circular-shaped tracks inscribed in the steel disk are better resolved with padding operation.

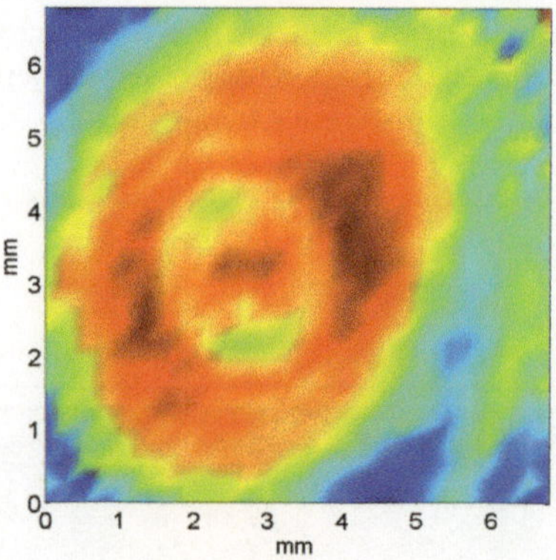

a)

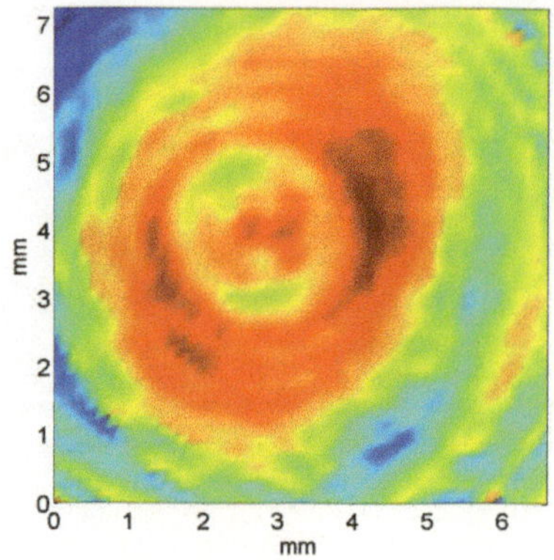

b)

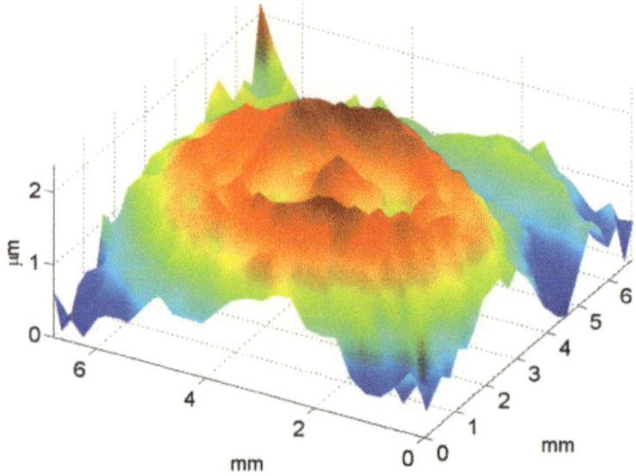

c)

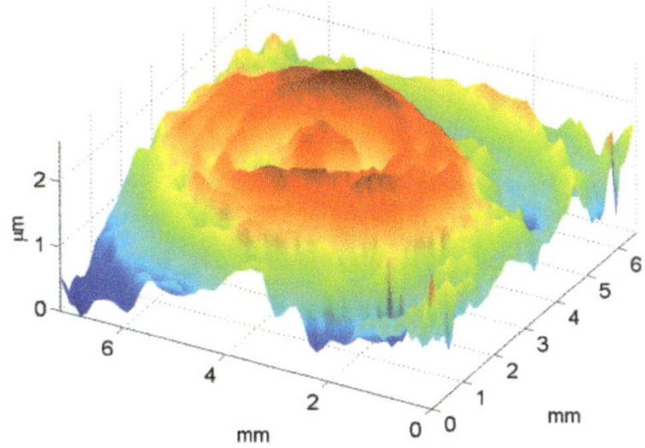

d)

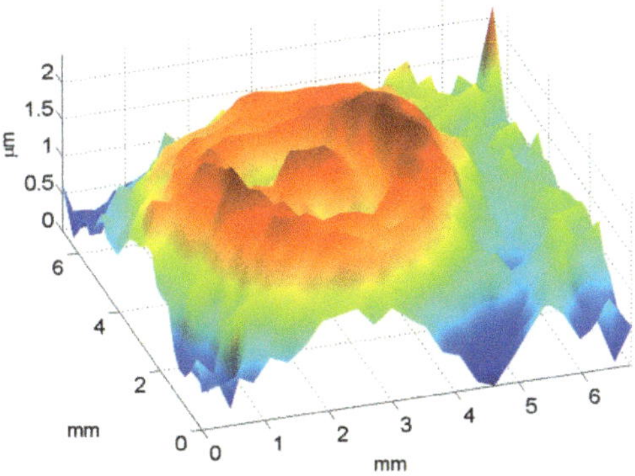

e)

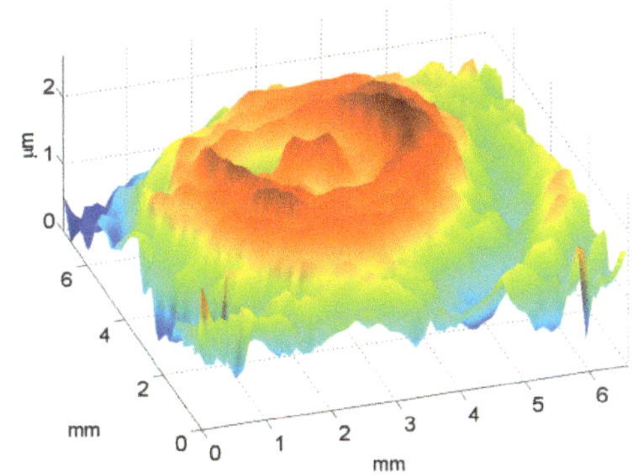

f)

Fig. 6. Phase images of aluminium disc:(a) two-dimensional reconstruction without padding operation; (b) 2D reconstruction with padding; (c), (e) two different 3D maps of the reconstructed surface profile without padding; (d), (f) with zero padding.

3.1 Large object investigation

Whenever large object recording and reconstruction is needed, digital holography with CO_2 laser (Allaria et al., 2003) has many advantages, with respect to digital holography in the visible region. First of all, a long wavelength radiation has a lower sensitivity to sub micrometric vibrations and this peculiarity provides a higher fringe visibility when large samples are investigated. A second advantage is related to the high output power of CO_2 lasers allowing to irradiate more efficiently and uniformly the surface of large samples (Pedrini et al.,1996). The last advantage derives directly from Eq. 9. Working with a longer wavelength, it is possible to use larger angles and, ultimately, smaller distances between the object and the recording device. In Fig. 7, a configuration to realize Infrared Digital Holography (IDH) of large objects is shown (Pelagotti et al., 2010). In this case we used a CW CO_2 laser emitting in the TEM_{00} mode at 10.6μm with 110W output power. Interferometric patterns where recorded by means of an ASi microbolometric infrared camera with a resolution of 640x480 pixels and 25μm pixel pitch. In this configuration the object is placed at the minimum distance allowed by Eq. 9. In this set up the laser beam is first divided by a ZnSe beam splitter (BS_1) which reflects 90% of the impinging radiation and transmits the remaining 10%. The transmitted beam, which constitutes the reference beam, after passing through a variable attenuator (VA), is enlarged by a ZnSe spherical lens of 1 inch focal length (L_2) in order to reach the thermocamera with a suitable low intensity. The object beam is enlarged by means of a spherical ZnSe lens (L_1) of 1 inch focal length and directly sent to the object under investigation.

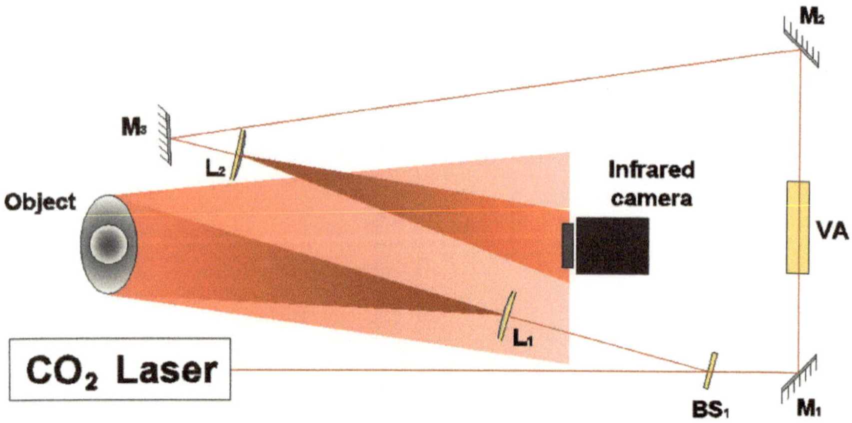

Fig. 7. Single beam setup: M_1, M_2 and M_3 are plane mirrors; BS_1 is a ZnSe 90/10 beam splitter; L_1 and L_2 are ZnSe spherical lenses 1in focal length; the object is a plastic mannequin 1.90cm high.

The speckle interferogram, whose visibility is controlled by means of the variable attenuator (VA), is collected by the recording device. With this configuration we recorded holograms of the bronze statue 30cm high shown in Fig. 8(a) (Paturzo et al., 2010). The hologram and its numerical amplitude reconstruction are shown in Fig 8(b), Fig 8(c), respectively.

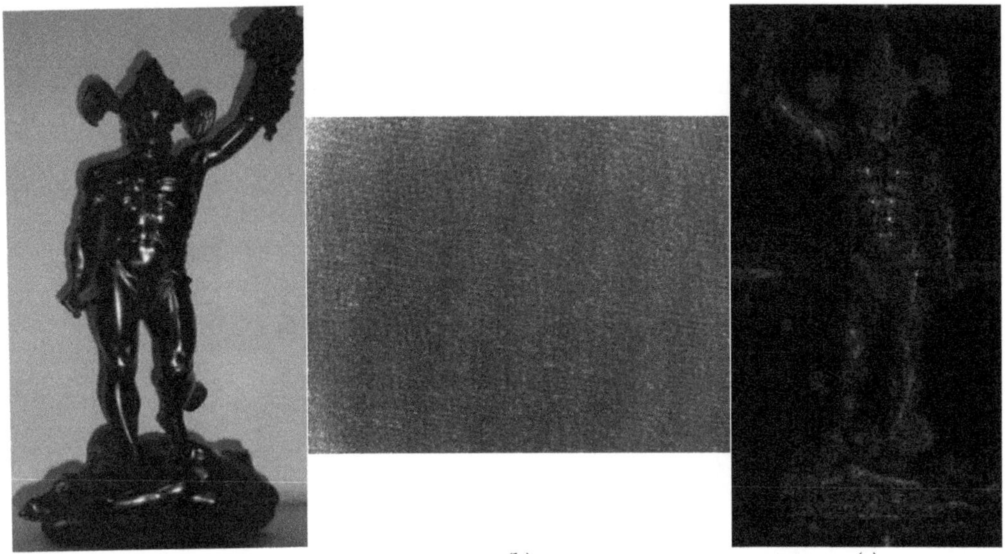

(a) (b) (c)

Fig. 8. (a) Photo of the Perseus statue; (b) acquired interferogram; (c) numerical reconstruction.

4. Digital holography for wavefront reconstruction of infrared Laguerre-Gaussian modes

In this section we present some applications of digital holography for analyzing the spatial distribution and signature vorticity of Laguerre-Gaussian (LG) modes. LG modes have been largely employed for interesting application such as optical trapping, rotational frequency shift and optical manipulation of micrometric systems (Allen et al., 1992, 1999; Arecchi et al., 1991). The concept of singularities in electromagnetic fields possessing an orbital angular momentum (Soskin & Vasnetov, 2001; Leach et al., 2002) has an extensive literature because of their interesting properties and potential applications. Laguerre-Gaussian beams are examples of optical fields, that possess wave-front singularities (optical vortices) of topological charge ℓ, where ℓ can take any integer value, that are related to the azimuthal angular dependence $\exp(i\ell\phi)$ in the transverse distribution of the optical field. The azimuthal mode index ℓ (orbital helicity) is physically related to OAM, $\ell\hbar$ per photon, of the LG modes. The LG modes of azimuthal index ℓ and radial index p are

$$\psi_{\ell p}(x,y,z)=\sqrt{\frac{2p!}{\pi(\ell+p)!}}\frac{1}{w(z)}\left(\frac{r\sqrt{2}}{w(z)}\right)^{\ell}\exp\left(-\frac{r^2}{w(z)^2}\right)\times$$
$$L_p^{\ell}\left(\frac{2r^2}{w(z)^2}\right)\exp\left(-\frac{ikr^2}{2R(z)}\right)\exp(-i\ell\phi)\times \tag{20}$$
$$\exp\left[-i(2p+\ell+1)\arctan\left(\frac{z}{z_r}\right)\right],$$

where r is the radial coordinate, $\phi = \arctan(y / x)$ the azimuthal coordinate and $L_p^\ell(x)$ are the generalized Laguerre polynomials with azimuthal and radial mode numbers ℓ and p, respectively. The production and characterization of free space Laguerre-Gaussian beams at comparatively longer wavelengths gives the possibility of using optical vortices for observing the transfer of angular momentum to relatively larger objects, since, for a fixed power, the angular momentum in the beam is proportional to the wavelength. Quantitative characterizations of the structure and signature of the vorticity of infrared LG beams can be obtained by using an interferometer for recording the interference pattern of the LG mode (the object beam) and an external plane-wave reference beam (De Nicola et al., 2010). The experimental set-up is shown in Fig. 9.

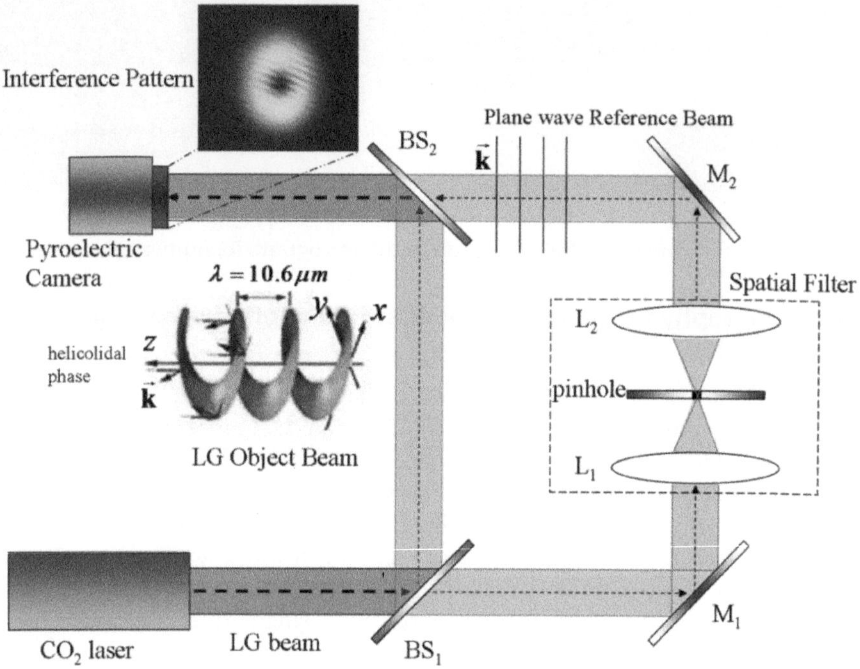

Fig. 9. Outline of the Mach-Zehnder interferometric setup. A CO_2 laser source operating at the wavelength λ=10.6μm produces an LG mode ψ_{10}, divided at the BS_1 (beam splitter) position into an object beam (LG mode) and a reference beam. The latter is a plane wave beam obtained after having spatially filtered and expanded the vortex field with the collimating lenses and pinhole system L1-P-L_2. M_1 and M_2 are mirrors. The object beam interferes with the reference beam at the beam splitter BS_2 position. The interference pattern is recorded with an IR pyroelectric camera.

A vortex optical field in the mid infrared range is produced by inserting a circular diaphragm of wide aperture (Fresnel number larger than one) inside the optical cavity of a CO_2 laser. This operation easily leads to laser emission in the doughnut mode TEM^*_{01} ,but the optical field sometimes presents impurities originating from the residual contributions of the fundamental mode or higher order cavity modes. The CO_2 laser source operates at the wavelength $\lambda=10.6\mu m$. The laser cavity is defined by a partially reflective flat mirror (R=95%) and an out-coupling mirror (R=90%) mounted upon a piezoelectric translator. As aforementioned, the LG mode is selected with a diaphragm and is horizontally polarized by means of an intracavity ZnSe Brewster window.

The interferometric configuration comprises two mirrors, indicated as M_1 and M_2, and two beam splitters, BS_1 (70T/30R at 45° angle of incidence) and BS_2 (50T/50R at 45° angle of incidence) as depicted in figure 9. The output laser beam is a linear superposition of TEM_{10} and TEM_{01} Hermite-Gaussian modes (the doughnut mode TEM^*_{01}) corresponding to $\ell=1$

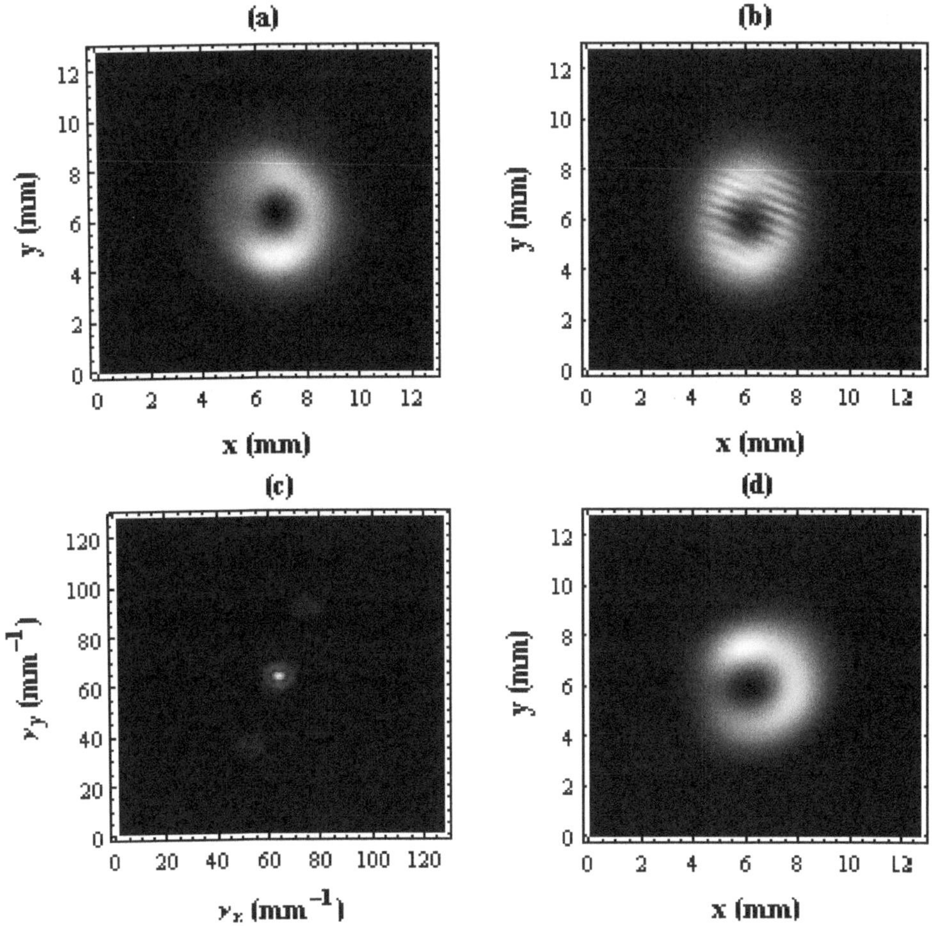

Fig. 10. Infrared Laguerre-Gaussian beam analysis based on Fresnel reconstruction of off-axis digital hologram: (a) intensity distribution of the LG mode ψ_{10} recorded at the IR camera plane; (b) digitized interference intensity profile between LG object beam and plane wave reference; (c) Fourier transform of the carrier modulated interference pattern; (d) reconstructed vortex beam amplitude

lenses (L_1 and L_2) of focal length f=3*cm* and a pinhole (P) of 250μm placed in between the two mirrors M_1 and M_2, as shown in Fig. 9. The two beam splitters (BS_1 and BS_2) are ZnSe and $p=0$, the lowest order vortex, with a beam waist w_o=2.4*mm* at the out-coupling mirror position. The LG mode ψ_{10} represents the object beam. The reference beam consists of a plane wave obtained with the spatial filter/expansion system L_1PL_2 that comprises two coated windows with a diameter of 50*mm*. The vortex output beam is split into reference and object beams at the BS_1 position. The interference pattern between the plane wave reference and the object beam is recorded on the detection plane of an internally chopped pyroelectric video camera (Spiricon Pyrocam III Model PY-III-C-A), having an infrared sensor of 124×124 pixel elements of $LiTaO_3$ with square size of 85μm and a pixel pitch Δx=100μm. The reference beam interferes with the object beam at a small angle ($\alpha \leq \lambda/(2\Delta x) \sim 3°$) as required by the sampling theorem. In Fig. 10(a) we can see the intensity distribution of the LG vortex beam ψ_{10} (ℓ=1 and p=0) with a dark central core, recorded by the pyroelectric array at distance d=52*cm* from the out-coupling mirror of the cavity. The digitized carrier modulated infrared interference pattern $I_{10}(x,y)$ in the (x, y)-plane of the camera is presented in Fig. 10(b).

Because of the off-axis recording geometry, the two-dimensional Fourier transform of the fringe pattern is characterized by three dominant diffraction orders, as shown in Fig. 10(c) where the amplitude of the Fourier transform has been represented. The full complex two-dimensional Fourier transform of the fringe pattern can be written in the form

$$\begin{aligned}\tilde{I}_{10}\left(\nu_x,\nu_y\right) = \tilde{I}_{dc}\left(\nu_x,\nu_y\right) + \tilde{\psi}_{10}\left(\nu_x - \nu_{xc}, \nu_y - \nu_{yc}\right) + \\ + \bar{\tilde{\psi}}_{10}\left(\nu_x + \nu_{xc}, \nu_y + \nu_{yc}\right),\end{aligned} \tag{21}$$

whereas

$$\tilde{I}_{10}\left(\nu_x,\nu_y\right) = \mathcal{F}\left\{I_{10}(x,y)\right\} = \iint I_{10}(x,y)\exp\left[2\pi i\left(\nu_x x + \nu_y y\right)\right]\mathrm{d}x\mathrm{d}y, \tag{22}$$

In eq. (21) $\tilde{I}_{dc}\left(\nu_x,\nu_y\right)$ is the contribution to the spectrum given by the low-frequency background illumination (Cuche et al., 2000; Liu et al., 2002; Onural, 2000), corresponding to a peak at low spatial frequencies; $\tilde{\psi}_{10}\left(\nu_x - \nu_{xc}, \nu_y - \nu_{yc}\right)$ and $\bar{\tilde{\psi}}_{10}\left(\nu_x + \nu_{xc}, \nu_y + \nu_{yc}\right)$ represent the Fourier spectrum of the vortex field and its complex conjugate, respectively, shifted by the carrier spatial frequencies ν_{xc} and ν_{yc} along the x- and y-direction, respectively. These two terms contribute to the two outer peaks of Fig. 10(c), and correspond to the positive and negative components of the vortex field spectrum. By taking the shifted inverse Fourier transform of $\tilde{\psi}_{10}\left(\nu_x - \nu_{xc}, \nu_y - \nu_{yc}\right)$, the amplitude of the vortex field can be determined, as shown in Fig 10(d), respectively.

The retrieved phase distribution in Fig. 11 shows a spiral profile with a discontinuity line typical of single topological charge vortex (White et al., 1991). We point out that the described method can be usefully employed for characterizing the vorticity of infrared beams of potential use in optical telecommunication applications, where the preservation of the purity of the mode along the propagation direction (Indebetouw, 1993) is a problem of crucial importance.

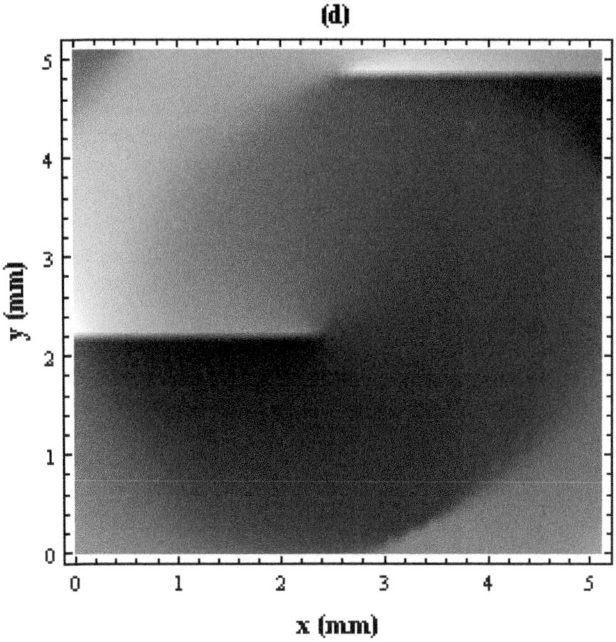

Fig. 11. Infrared Laguerre-Gaussian beam analysis based on Fresnel reconstruction of off-axis digital hologram: phase distribution of the lowest order vortex beam

5. Conclusions

In this chapter we have shown that the numerical reconstruction of a whole optical wavefield through digital holography can be successfully performed in the mid-infrared regime using pyroelectric and microbolometric sensors. Amplitude and phase reconstructions were obtained by back-Fresnel propagation from the hologram recording plane to the object plane. Digital holography is closely related to digital image processing and to the mathematical models of imaging. We have described methods for improving the accuracy of the reconstruction which allows us to compensate for the loss of resolution at longer wavelength and the low spatial resolution of the pyroelectric camera array. It is worth pointing out that the improved spatial resolution of digital holography in the mid infrared regime is a significant improvement in a number of biologically relevant measurements related to biological cell and tissue analysis, where electric potential or light induced phase changes are expected to play a significant role in the characterization of complex biological structures. Infrared digital holography has also been applied for large object investigation.

6. References

Allaria, E., Brugioni, S., De Nicola, S., Ferraro, P., Grilli, S. & Meucci, R. (2003). Digital Holography at 10.6 μm. *Optics Communications*, Vol. 215, pp. 257-262.

Allen, L., Beijersbergen, M.W., Spreeuw, R.J.C. and Woerdman, J.P. (1992). Orbital angular momentum of light and the transformation of Laguerre–Gaussian laser modes. *Phys. Rev. A*, Vol. 45, pp. 8185-8189.

Allen, M.L., Padgett, M.J. & Babiker, M. (1999). The orbital angular momentum of light. *Progress in Optics* (Vol. 39), Emil Wolf (Elsevier), pp. 291-372, Amsterdam, Netherlands.

Arecchi, F.T., Giacomelli, G., Ramazza, P. L. & Residori, S. (1991). Vortices and defects statistics in two dimensional optical chaos. *Phys. Rev. Lett.*, Vol. 67, pp. 3749-3751.

Cuche, E., Marquet, P. & Depeursinge, C. (1999a). Simultaneous amplitude-contrast and quantitative phase-contrast microscopy by numerical reconstruction of Fresnel off-axis holograms. *Appl. Opt.*, Vol. 38, pp. 6994-7001.

Cuche, E., Bevilacqua, F. & Depeursinge, C. (1999a). Digital Holography for quantitative phase-contrast imaging. *Opt. Lett.*, Vol. 24, pp. 291-293 .

Cuche, E., Marquet. P. & Depeursinge, C. (2000). Spatial filtering for zero-order and twin-image elimination in digital off-axis holography. *Appl. Opt.*, Vol. 39, pp. 4070–4075.

Demoli, N., Vukicevic, D. & Torzynski, M. (2003). Dynamic digital holographic interferometry with three wavelengths. *Opt. Express*, Vol. 11, pp. 767-774.

De Nicola, S., Ferraro, P., Grilli, S., Miccio, L., Meucci, R., Buah-Bassuah, P.K. & Arecchi, F.T. (2008). Infrared digital refelective-holographic 3D shape measurements. *Opt. Commun.*, Vol. 281, pp. 1445-1449.

De Nicola, S., Zito, G., Meucci, R., Buah-Bassuah, P.K. & Arecchi, F.T. (2010). Fourier projection method for measuring the two-point correlation of Laguerre-Gaussian modes. *J. of Optics*, Vol. 12, pp. 035404-035412.

Dubois, F., Joannes, L., Dupont, O., Dewandel, J.L. & Legros, J.C. (1999). An integrated optical set-up for fluid-physics experiments under microgravity conditions. *Meas. Sci. Technol.*, Vol. 10, pp. 934-945.

Gabor, D. (1948). A new microscopic principle. *Nature*, Vol. 161, pp. 777–778.

Gass, J., Dakoff, A. & Kim, M.K. (2003). Phase imaging without 2 π ambiguity by multi-wavelength digital holography. *Optics Letters*, Vol. 28, pp. 1141-1143.

Goodman, J.W. & Lawrence, R.W. (1967). Digital image formation from electronically detected holograms. *Appl. Phy. Lett.*, Vol. 11, pp. 77–79.

Goodman, J.W. (1996). *Introduction to Fourier Optics* (Ed. 2). McGraw-Hill, New York, USA.

Grilli, S., Ferraro, P., De Nicola, S., Finizio, A., Pierattini, G. & Meucci, R. (2001). Whole optical wavefields reconstruction by digital holography. *Opt. Exp.*, Vol. 9, pp. 294-302.

Indebetouw, G. (1993). Optical vortices and their propagation. *J. Mod. Opt.*, Vol. 40, pp. 73-87.

Jueptner, W.P., Werner, P., Kujawinska, M., Osten, W., Salbut, L.A. & Seebacher, S. (1987). Combined measurement of silicon microbeams by grating interferometry and digital holography. *Proceeding of SPIE International Conference on Applied Optical Metrology*, P. K. Rastogi and F. Gyimesi Eds., Vol. 3407, pp. 348-357.

Kato, J., Yamaguchi, I. & Matsumura, T. (2003). Multicolor digital holography with an achromatic phase shifter. *Opt. Lett.*, Vol. 27, pp. 1403-1405.

Kim, M. (2000). Tomographic three-dimensional imaging of a biological specimen using wavelength-scanning digital interference holography. *Opt. Express*, Vol. 7, pp. 305-310.

Kreis, T.M. & Jüptner, W. (1997). *Principles of Digital Holography*. Jüptner and Osten Eds., Fringe 97, Academic Verlag, pp. 253–363.

Kreis, T.M. (2002a). Frequency analysis of digital holography. *Opt. Eng.*, Vol. 41, pp. 771-778.

Kreis, T.M. (2002b). Frequency analysis of digital holography with reconstruction by convolution. *Opt. Eng.*, Vol. 41, pp. 1829-1839.

Kronrod, R.W., Merzlyakov, N.S. & Yaroslavskii, L.P. (1972). Reconstruction of a hologram with a computer. *Sov Phys Tech Phys.*, Vol. 17, pp. 333–334.

Lai, S., King, B. & Neifeld, N.A. (2000). Wavefront reconstruction by means of phase-shifting digital in-line holography. *Opt. Comm.*, Vol. 173, pp.155-160.

Leach, J., Padgett, M.J., Barnett, S.M., Franke-Arnold, S. & Courtial, J. (2002). Measuring the orbital angular momentum of a single photon. *Phys. Rev. Lett.*, Vol. 88, pp. 257901-04.

Lei, X., Xiaoyuan, P., Asundi, A.K. & Jianmin, M. (2001a). Hybrid holographic microscope for interferometric measurement of microstructures. *Opt. Eng.*, Vol. 40, pp. 2533-2539.

Lei, X., Xiaoyuan, P., Jianmin, M. & Asundi, A.K. (2001b). Studies of Digital Microscopic Holography with Applications to Microstructure Testing. *Appl. Opt.*, Vol. 40, pp. 5046-5052.

Leith, E. & Upatnieks, J. (1965). Microscopy by wavefront reconstruction. *J. Optical Society of America*, Vol. 55, pp. 569-570.

Liu, C., Li, Y., Cheng, X., Liu, Z., Bo, F. & Zhu, J. (2002). Elimination of zero-order diffraction in digital holography. *Opt. Eng.*, Vol. 41, pp. 2434–2437.

Nilsson, B. & Carlsson, T. (2000). Simultaneous measurement of shape and deformation using digital light-in-flight recording by holography. *Opt. Eng.*, Vol. 39, pp. 244-253.

Onural, L. (2000). Sampling of the diffraction field. *Appl. Opt.*, Vol. 39, pp. 5929-5935.

Paturzo, M., Pelagotti, A., Finizio, A., Miccio, L., Locatelli, M., Geltrude, A., Meucci, R. & Ferraro, P. (2010). Optical reconstruction of digital holograms recorded at 10μm: a route for 3D imaging at long IR wavelengths. *Optics Letters*, Vol. 35, pp. 2112-2114.

Pedrini, G., Tiziani, H.J. & Zoa, Y. (1996). Speckle size of digitally reconstructed wavefronts of diffusely scattering objects. *J. Mod. Opt.*, Vol. 43, pp. 395-407.

Pedrini, G., Schedin, S. & Tiziani, H.J. (2001). Aberration compensation in digital holographic reconstruction of microscopic objects. *J. Mod. Opt.*, Vol. 48, pp. 1035-1041.

Pelagotti, A., Locatelli, M., Geltrude, A., Poggi, P., Meucci, R., Paturzo, M., Miccio, L. & Ferraro, P. (2010a). Reliability of 3D imaging by digital holography at long IR wavelength. *Journal of Display Techonogy*, Vol. 6, pp. 465-471.

Rastogi, P.K. (1994). *Holographic Interferometry*. Springer Verlag, Berlin, Germany.

Schnars, U. (1994). Direct phase determination in hologram interferometry with use of digitally recorded holograms. *J. Opt. Soc. Am. A*, Vol. 11, pp. 2011-2015.

Schanrs, U. & Juptner, W. (1994). Direct recording of holograms by a CCD target and numerical reconstruction. *Appl. Opt.*, Vol. 33, pp. 179-181.

Schnars, U. & Juptner, W. (2002). Digital recording and numerical reconstruction of holograms. *Meas. Sci. Technol.*, Vol. 13, pp. R85–R101.

Seebacker, S., Osten, W., Baumbach, T. & Juptner, W. (2001). The determination of materials parameters of micro-components using digital holography. *Opt. Las. Eng.*, Vol. 36, pp. 103-126.

Soskin, M. S. & Vasnetov, M. V. (2001). Singular Optics. *Progress in Optics* (Vol. 42), Emil Wolf (Elsevier), pp. 219-276, Amsterdam, Netherlands.

Stetson, K.A. & Powell, R.L. (1966). Hologram Interferometry. *J. Optical Society of America*, Vol. 54, pp. 1161-1166.

Stetson, K.A. & Brohinsky, W.R. (1985). Electrooptic holography, its application to hologram interferometry. *App. Opt.*, Vol. 24, pp. 3631.

Vest, C.M. (1979). *Holographic Interferometry*. John Wiley, New York, USA.

White, A.G., Smith, C. P., Heckenberg, N.R., Rubinsztein-Dunlop, H., Mcduff, R. & Weiss, C. O. (1991). Interferometric measurements of phase singularities in the output of a visible laser. *J. Mod. Opt.*, Vol. 12, pp. 2531-41.

Yamaguchi, I. & Zhang, T. (1997). Phase-shifting digital holography. *Opt. Lett.*, Vol. 23. Pp. 1268-1270.

Yamaguchi, I., Matsumura, T. & Kato, J. (2002). Phase-shifting colour digital holography. *Optics Letters*, Vol. 27, pp. 1108-1110.

Yaroslavsky, L. & Eden, M. (1996). *Fundamentals of Digital Optics.* BirkhŠuser, Boston, USA.

Zhang, F., Yamaguchi, I. & Yaroslavsky, L.P. (2004). Algorithm for reconstruction of digital holograms with adjustable magnification. *Opt. Lett.*, Vol. 29, pp. 1668-1670.

Zito, G., Finizio, A. & De Nicola, S. (2009). Spatially resolved refractive index profiles of electrically switchable computer-generated holographic gratings. *Opt. Express*, Vol. 17, pp. 18843-18851.

8

Incoherent Holographic Interferometry

Kaige Wang
Department of Physics, Applied Optics Beijing Area Major Laboratory, Beijing Normal University, Beijing China

1. Introduction

The original meaning of coherence was attributed to the ability of light to exhibit interference phenomena. There are two types of coherence: temporal coherence and spatial coherence. Temporal coherence measures the ability of two relatively delayed beams to form interference fringe. Interference in a Michelson interferometer refers to temporal coherence. However, spatial coherence reflects the ability of a beam to interfere with a spatially shifted (but not delayed) version across the beam. Young's double-slit experiment is an example which concerns spatial coherence.

Since the first experimental realization of Young's double-slit interference, it has been known that the observation of interference-diffraction pattern of an object requires a spatially coherent source. The waves emitted from positions outside the coherent area possess independent irregular phases which may degrade the interference pattern. In the early days when coherent sources were unavailable, interference experiments were carried out with an extended thermal source restricted by a pinhole aperture, which can improve the spatial coherence. Holography is one of the most important applications of spatial interference. In the first holography experiment, Dennis Gabor stated that (Gabor, 1972) "The best compromise between coherence and intensity was offered by the high-pressure mercury lamp, which had a coherence length of only 0.1 millimeter, ... But in order to achieve spatial coherence, we had to illuminate, with one mercury line, a pinhole 3 microns in diameter. " The pinhole eventually reduced the power of the source and thus impeded the potential application of optical interferometric techniques such as holography. This barrier was overcome with the invention of the laser, whose intense and coherent beam was ideal for performing interference.

Coherent sources are obtainable within a certain range of optical frequencies. However, various holographic techniques have been developed using incoherent sources such as X-rays, electrons, and γ radiation. To improve the coherence one has to pay the cost of decreasing the intensity of the source, as already pointed out in relation to Gabor's experiment. A challenging question would be whether coherence is absolutely necessary in holography, or can we bypass the coherence requirement? As a matter of fact, in the early days of holography, a technique using incoherent illumination was first proposed by Mertz and Young (Mertz & Young, 1963), and then extended by Lohmann (Lohmann, 1965), Stroke and Restrick (Stroke & Restrick, 1965), and Cochran (Cochran, 1966). The strategy is based on the fact that each point of a spatially incoherent object produces, through interference of its wave fronts, a stationary interference pattern which uniquely encodes the position and intensity of the object point. However, the method has various weaknesses that significantly limit its application. The schemes proposed for incoherent holography can only record the intensity distribution of a fluorescent object. Moreover, each elementary fringe pattern is formed by two extremely tiny portions of the light, and the summation of many weak interference patterns results in a very large bias level in the hologram, much larger than that in coherent holography. Hence incoherent holography is appropriate only for objects with a low number of resolution elements (Goodman, 1996).

In this chapter we introduce an incoherent interference mechanism which seems to contradict our existing knowledge of the interference requirements. Consider a holographic interferometer in which an object is placed in one arm; the object wave then interferes with the reference wave in the other arm, resulting in an interference pattern which records the object information. To obtain well-defined interference fringes in an ordinary balanced interferometer where both arms have the same path length, the optical field illuminating the interferometer must have spatial coherence, that is, its transverse coherence length must be large enough. We find that the requirement of spatial coherence is due to the particular geometry of the balanced interferometer. If the two fields to be interfered travel through different lengths or different diffraction configurations, then spatial coherence is no longer necessary for spatial interference. Thus we have proposed several unbalanced interferometers where the two beams travel different path lengths or have different diffraction configurations (Zhang et al., 2009a, 2009b). These interferometers are capable of exhibiting interference using either coherent or spatially incoherent light, but their interference fringes are different, signifying that they have different origins.

The chapter is organized as follows. In Section 2 we first analyze why both temporal and spatial coherence conditions are necessary in an ordinary holographic interferometer. Then we propose an unbalanced holographic interferometer with different path lengths and demonstrate that it can exhibit interference using a light source with spatial incoherence. The setup is capable of performing holography. In Section 3 we further suggest several types of unbalanced interferometers which are able to realize incoherent interferometry. In particular, some of the schemes can reduce the requirement of temporal coherence and employ a true incoherent source such as a lamp with an extended illumination area. Moreover, we show that phase reversal diffraction can occur in the incoherent interferometer. Finally, a brief summary is given in Section 4.

2. What is incoherent interferometry?

We consider a holographic interferometer as shown in Fig. 1, in which an object field $E_o(x)$ and a reference field $E_r(x)$ are reflected from mirrors M_1 and M_2, respectively, to interfere at the recording plane RP. The intensity distribution of their superposed field is written as

$$I(x) = \langle |E_o(x) + E_r(x)|^2 \rangle = \langle |E_o(x)|^2 \rangle + \langle |E_r(x)|^2 \rangle + \{\langle E_r^*(x)E_o(x)\rangle + \text{c.c.}\}, \qquad (1)$$

where x is the transverse coordinate across the beam, and $\langle E_r^*(x)E_o(x)\rangle$ the interference term. In the interferometric scheme the two fields must come from the same source field $E_s(x)$. Let a transmitting object O with transmittance T(x) be close to the source, so the two fields can be expressed as

$$E_o(x,t) = \int T(x_1)h_o(x,x_1)E_s(x_1,t_1)dx_1, \qquad (2a)$$

$$E_r(x,t) = \int h_r(x,x_2)E_s(x_2,t_2)dx_2, \quad (2b)$$

where $h_o(x,x_1)$ and $h_r(x,x_2)$ are the impulse response functions for the object and reference paths, t_1 and t_2 are the times taken for the source fields to propagate to the recording plane through the object and reference paths, respectively. x_j $(j = 1,2)$ is the transverse coordinate in the source field. Hence the interference term is obtained to be

$$\langle E_r^*(x,t)E_o(x,t)\rangle = \int T(x_1)h_o(x,x_1)h_r^*(x,x_2)\langle E_s^*(x_2,t_2)E_s(x_1,t_1)\rangle dx_1dx_2. \quad (3)$$

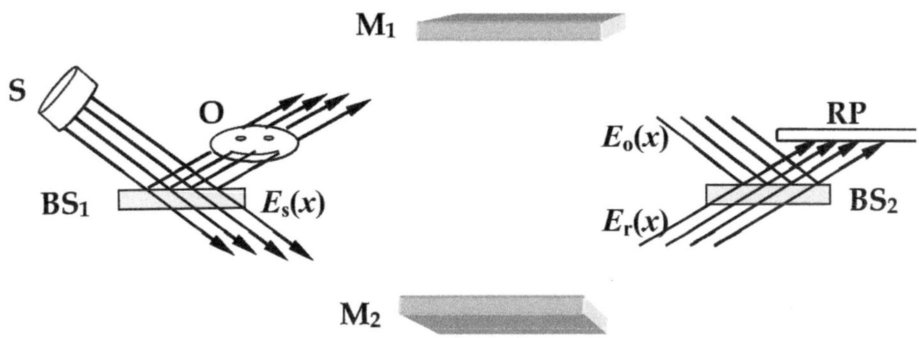

Fig. 1. A holographic interferometer consisting of two beamsplitters, BS_1 and BS_2, and two mirrors, M_1 and M_2; S is a source and O an object; RP is the recording plane.

The temporal coherence of the source field is characterized by the coherence time τ. If $|t_2 - t_1|$ is larger than τ, then $\langle E_s^*(x_2,t_2)E_s(x_1,t_1)\rangle = 0$ and interference never occurs. However, spatial coherence requires that the beam has a well-defined wave-front, such as a field emitted from a point source or a laser beam. When the source field satisfies both temporal and spatial coherence, then

$$\langle E_s^*(x_2,t_2)E_s(x_1,t_1)\rangle = E_s^*(x_2,t_2)E_s(x_1,t_1). \quad (4)$$

Thus coherent interferometry occurs in the form of

$$\langle E_r^*(x,t)E_o(x,t)\rangle = E_r^*(x,t)E_o(x,t), \quad (5)$$

which contains the object information. This is what has been known before.
We now consider the case when the source field satisfies temporal coherence (i.e. $|t_2 - t_1| < \tau$) but not spatial coherence. The first-order correlation function of the source field is written as

$$\langle E_s^*(x_2,t_2)E_s(x_1,t_1)\rangle = I_s e^{i\varphi}\delta(x_2 - x_1), \quad (6)$$

where φ is an arbitrary phase. For simplicity, the intensity distribution I_s is assumed to be homogeneous. Equation (6) describes complete spatial incoherence. The wave-front of the beam fluctuates randomly and any two positions across the beam are statistically independent. Substituting Eq. (6) into Eq. (3), we obtain

$$\langle E_r^*(x)E_o(x)\rangle = I_s e^{i\varphi} \int T(x_0)h_o(x,x_0)h_r^*(x,x_0)dx_0. \quad (7)$$

For a conventional balanced interferometer where the two optical paths have the same length, that is $h_o(x, x_0) = h_r(x, x_0)$, then $h_o(x, x_0)h_r^*(x, x_0)$ is x-independent. In the paraxial approximation, for example, the impulse response function for a propagation length z is given by

$$H(x, x_0; z) = \sqrt{\frac{k}{i2\pi z}}\exp(ikz)\exp\left[\frac{ik(x-x_0)^2}{2z}\right]$$

$$= \sqrt{\frac{k}{i2\pi z}}\exp(ikz)G(x - x_0, z), \quad (8a)$$

$$G(x, z) \equiv \exp\left[\frac{ikx^2}{2z}\right], \quad (8b)$$

where k is the wave number, so $H(x, x_0; z)H^*(x, x_0; z) = k/(2\pi z)$ is x-independent. Therefore, the interference term (7) yields

$$\langle E_r^*(x)E_o(x)\rangle \propto \int T(x_0)dx_0, \quad (9)$$

and the object information has been washed out. We can also prove that, no matter where the object is placed within the object path, Eq. (9) still holds provided the two paths of the interferometer have the same length. Perhaps this consequence brought about the misunderstanding that spatial interference in holography needs not only temporal but also spatial coherence. The following experiment will show that the condition of spatial coherence is not necessary if the interferometric scheme is modified.

In 2009 we proposed an incoherent interferometry setup as shown in Fig. 2 (Zhang et al., 2009a). In this unbalanced interferometer the object and reference arms have different lengths, $z_o = 16$cm and $z_r = 27$cm, respectively. The spatially incoherent light source is formed by passing a He-Ne laser beam of wavelength 632.8nm through a rotating ground glass disk. The object in the experiment is a double slit T of slit width $b = 125\mu m$ and spacing $d = 310\mu m$. The interference pattern can be recorded by either of two charge-coupled device (CCD) cameras.

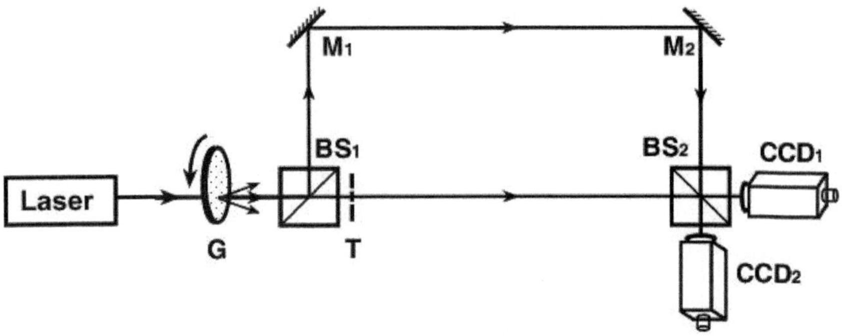

Fig. 2. Experimental scheme for an unbalanced interferometer where the two arms have different lengths. A laser and a rotating ground glass disk G form a spatially incoherent source; M_1 and M_2 are mirrors, BS_1 and BS_2 are beamsplitters, CCD_1 and CCD_2 are detectors, and T is a double-slit close to BS_1.

The laser beam has good temporal coherence which will not be affected by the scattering of the slowly rotating ground glass. The path difference between the two arms is much smaller than the longitudinal coherence length $c\tau$, where c is the light speed in vacuum and τ is the coherence time of the laser beam. However, the motion of the ground glass disk destroys the spatial coherence of the laser beam. This source emits what is known as pseudo-thermal light and satisfies Eq. (6) approximately. In contrast, a true thermal light source cannot satisfy the condition of temporal coherence in this setup.

The two-dimensional (2D) intensity patterns detected by CCD_1 are shown in Fig. 3. When the ground glass disk is moved step by step, the interference patterns are irregular speckles frame by frame and each frame is different, as shown in the two single-shot frames of Figs. 3(a) and 3(b). However, if we average over a number of exposures, as the number of frames increases, a well-defined interference pattern emerges gradually, as can be seen in Figs. 3(c)-3(g).

The experimental results tell us that, unlike the stationary pattern in coherent interferometry, the interference pattern here is irregularly fluctuating in time and the well-defined pattern can only be discerned after taking its statistical average. These striking features imply that incoherent interferometry originates from a different interference mechanism.

We can explain the experiment using Eq. (7). The interference term in this incoherent interferometer can be calculated as

$$\begin{aligned}\langle E_r^*(x)E_o(x)\rangle &= I_s e^{i\varphi} \int T(x_0)H(x,x_0;z_o)H^*(x,x_0;z_r)dx_0 \\ &\propto \exp[ik(z_o - z_r)] \int T(x_0)G(x-x_0,z_o)G^*(x-x_0,z_r)dx_0 \\ &= \exp[ik(z_o - z_r)] \int T(x_0)G(x-x_0,Z)dx_0, \end{aligned} \tag{10}$$

where function G defined by Eq. (8b) obeys

$$G(x,z_o)G^*(x,z_r) = G(x,Z), \tag{11}$$

when the effective diffraction length Z is defined by

$$\frac{1}{Z} = \frac{1}{z_o} - \frac{1}{z_r}. \tag{12}$$

Equation (10) defines a kind of incoherent interferometry, which represents the Fresnel diffraction integral of an object under the paraxial condition, and is the same as for coherent diffraction but with an effective object distance Z replacing the real one, z_o. Note that Z is negative for $z_o > z_r$, designating a phase reversal diffraction. We will discuss this issue in Subsection 3.3.

In the far-field limit, Eq. (10) presents the Fourier transform $\tilde{T}$ of object T:

$$\langle E_r^*(x)E_o(x)\rangle \propto \exp[ik(z_o - z_r)]\, G(x,Z)\tilde{T}\left(\frac{kx}{Z}\right). \tag{13}$$

For a double-slit of slit width b and spacing d, its Fourier transform reads

$$\tilde{T}(q) = \left(\frac{2b}{\sqrt{2\pi}}\right) \text{sinc}(qb/2)\cos(qd/2). \tag{14}$$

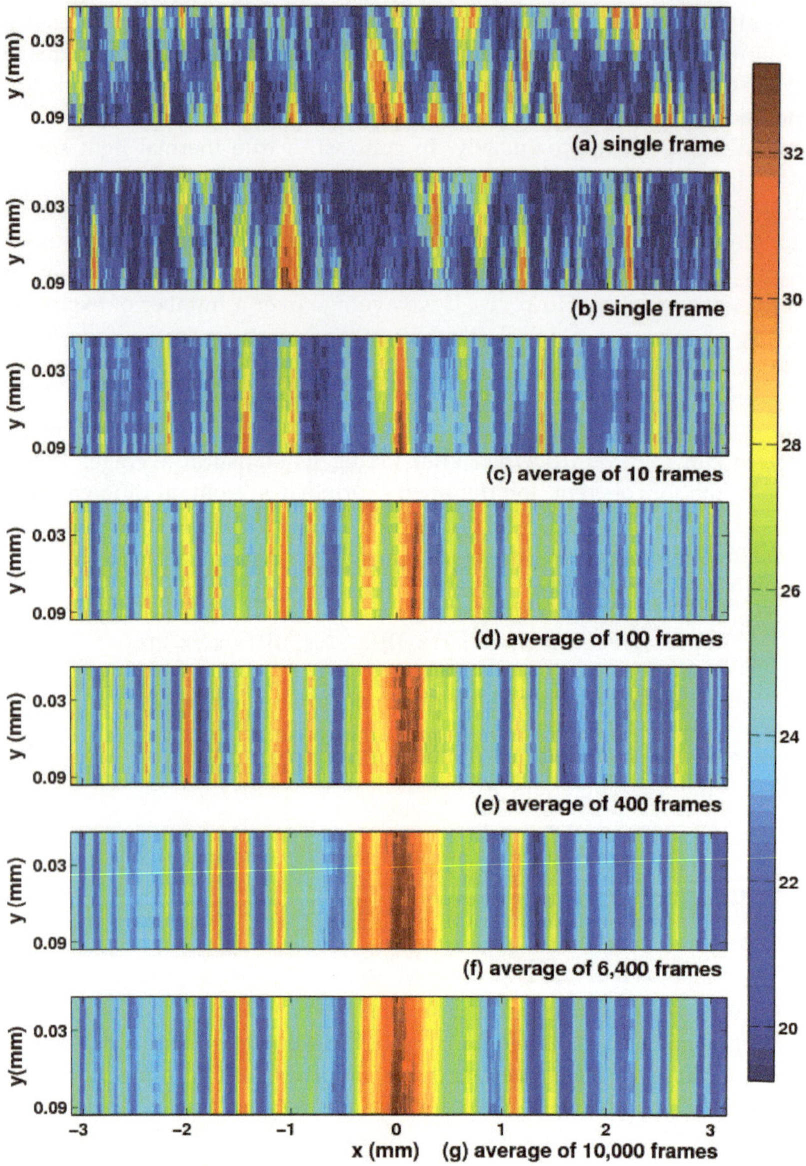

Fig. 3. Experimentally observed 2D interference patterns recorded by CCD_1 in the scheme of Fig. 2. Figs. (a) and (b) are individual single frames; (c), (d), (e), (f) and (g) are averaged over 10, 40, 400, 6400 and 10000 frames, respectively. The color-bar shows the relative intensity.

Besides the interference term, however, the output intensity distribution of the interferometer includes also the intensities of both the object and reference waves [see Eq. (1)]. In incoherent interferometry, these two intensities are homogeneously distributed.

In contrast, for the same interferometer driven by coherent light, Eq. (5) yields

$$\begin{aligned}\langle E_r^*(x)E_o(x)\rangle &\propto \exp[ik(z_o - z_r)] \int T(x_0)G(x - x_0, z_o)dx_0 \\ &\approx \exp[ik(z_o - z_r)]\, G(x, z_o)\tilde{T}\left(\frac{kx}{z_o}\right),\end{aligned} \tag{15}$$

where the last step is valid in the far-field limit. However, the intensity distribution of the reference waves is homogeneous whereas the intensity of the object wave exhibits the diffraction pattern

$$\langle |E_o(x)|^2\rangle \propto |\int T(x_0)G(x - x_0, z_o)dx_0|^2 \approx \left|\tilde{T}\left(\frac{kx}{z_o}\right)\right|^2. \tag{16}$$

In holography this pattern should be avoided by diminishing the intensity of the object wave.

To demonstrate the above theoretical explanation, Fig. 4 shows one-dimensional (1D) intensity patterns recorded by the two CCD cameras, where (a) and (b) are the interference fringes observed by CCD_1 and CCD_2, respectively. The left part shows the results when the interferometer is illuminated by spatially incoherent light. In our experimental scheme the effective diffraction length is calculated to be Z = 39.3cm. For a 50/50 beamsplitter BS_2, if one output field is $[E_r(x) + E_o(x)]/\sqrt{2}$, the other should be $[E_r(x) - E_o(x)]/\sqrt{2}$. We can see that the two interference patterns with a phase shift π are formed in the average of 10000 frames, which matches with the theoretical simulation of Eq. (13). Taking the difference and sum of the two output intensities gives, respectively, the net interference pattern and the intensity background, as shown in (c) and (d). As a matter of fact, the homogeneous intensity background verifies the spatial incoherence of the source.

To further confirm whether the interference pattern is related to the spatial incoherence, we compare it with the result obtained in the same interferometer using coherent light. For this we simply remove the ground glass disk in Fig. 2. The experimental results and theoretical simulation are shown in the right part of Fig. 4, where (a) and (b) show the stationary intensity patterns registered by CCD_1 and CCD_2, respectively. In the coherent case, these patterns contain the interference term and the diffraction pattern of the object. Again, taking the difference and sum of (a) and (b), we obtain the net interference pattern in (c) and the diffraction pattern of the object in (d), respectively. These experimental results are very different from the left part for incoherent interferometry, and agree well with Eqs. (15) and (16) with a diffraction length of z_o = 16cm for the coherent case.

The above theoretical analysis and experimental demonstration tell us that, in an unbalanced interferometer, spatially incoherent light is capable of performing holographic interference in a similar way to that of coherent light. When an object is illuminated by incoherent light, the coherent information is lost in the object wave itself, but can be reproduced in an unbalanced interferometric scheme. Physically, the mechanisms of the two types of interferometry are quite different. In the incoherent case, the interference pattern fluctuates irregularly in time, and a well-defined pattern can only be formed in the statistical summation. Moreover, the diffraction pattern of an object is dependent on the effective diffraction length, which is associated with the travel distances of both the object and reference waves.

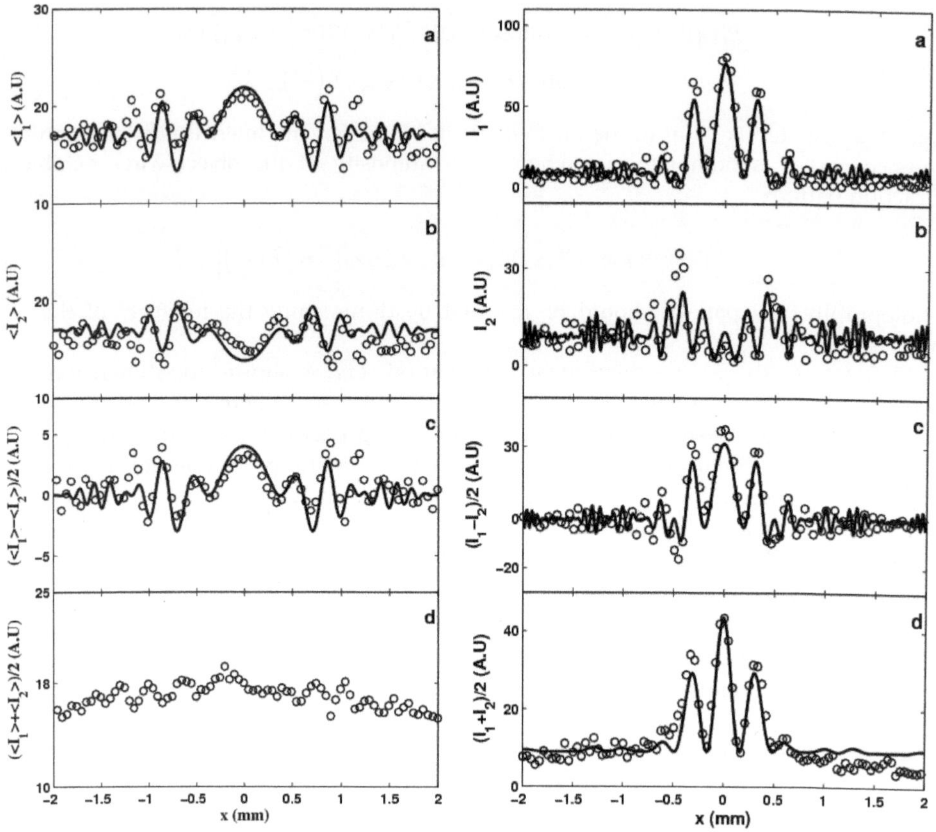

Fig. 4. Experimentally observed 1D interference patterns in the scheme of Fig. 2. Left and right parts are the patterns obtained using spatially incoherent and coherent light, respectively. (a) and (b) are the patterns registered by CCD_1 and CCD_2, respectively; (c) and (d) are their difference and summation, respectively. The patterns for the incoherent case are averaged over 10000 frames. Experimental data and theoretical simulation are given by open circles and solid lines, respectively.

3. Various schemes for incoherent interferometry

In the last section, we indicated that incoherent interferometry does not exist in a balanced interferometer and we proposed an unbalanced interferometer where its two arms have different lengths. Based on this attribute, in this section, we propose other unbalanced interferometers to perform incoherent interference.

3.1 Scheme I

As shown in Fig. 5, we modify the previous interferometer by inserting two lenses in the centre of the two arms (Zhang, July 2010). Since a lens can perform Fourier transformation

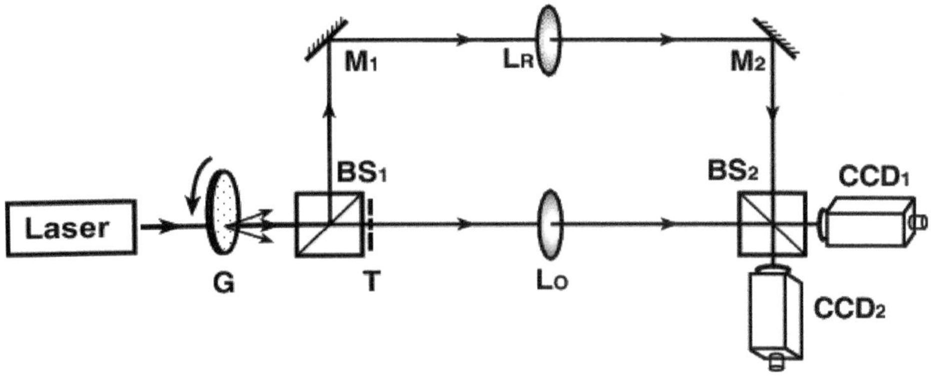

Fig. 5. Scheme I of incoherent interferometry, which is similar to the scheme of Fig. 2 but two lenses are inserted in the centre of the two arms. Both the input and output ports of the interferometer are located at the two focal planes of the lenses.

of a field between its two focal planes, we position both the input and output ports of the interferometer at the two foci of the lenses. The impulse response function between the two focal planes for a lens of focal length f is given by

$$h(x, x_0) = F(x, x_0; f) \equiv \sqrt{k/(i2\pi f)} \exp(i2kf) \exp(-ikxx_0/f), \tag{17}$$

where k is the wave number of the beam.

Let object $T(x_0)$ be close to the input port. In the incoherent regime, then Eq. (7) gives the interference term to be

$$\begin{aligned} \langle E_r^*(x) E_o(x) \rangle &\propto \int T(x_0) F(x, x_0; f_o) F^*(x, x_0; f_r) dx_0 \\ &\propto \exp[i2k(f_o - f_r)] \int T(x_0) \exp[-ikxx_0/f] dx_0 \\ &= \exp[i2k(f_o - f_r)]\, \tilde{T}[kx/f], \end{aligned} \tag{18}$$

where f_o and f_r are the focal lengths of the lenses in the object arm and reference arm, respectively. The effective focal length f is defined as

$$\frac{1}{f} = \frac{1}{f_o} - \frac{1}{f_r}. \tag{19}$$

In the experiment we take the two lenses to be of focal lengths $f_o = 7.5$cm and $f_r = 12$cm, and the double-slit as the object. The experimental results are plotted in Fig. 6. After eliminating the background, Fig. 6(c) shows the Fourier spatial spectrum of the double-slit. The theoretical curves are plotted by using Eqs. (18) and (19) with the effective focal length $f = 20$cm.

If a coherent plane wave drives the same interferometer, the interference term is obtained as

$$\langle E_r^*(x) E_o(x) \rangle \propto \exp[i2k(f_o - f_r)]\, \tilde{T}[kx/f_o] \delta(x), \tag{20}$$

where Dirac Delta function $\delta(x)$ comes from the focusing effect of the lens in the reference wave, and the Fourier transform of the object is governed by the focal length f_o of the object lens. However, this Fourier spectrum will not appear in the coherent interference term due to the modulation of Delta function.

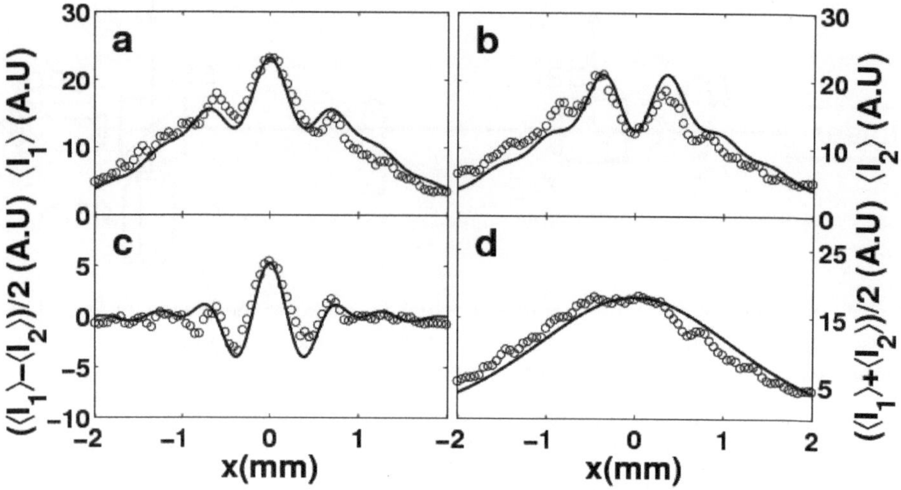

Fig. 6. Experimentally observed 1D interference patterns in Scheme I (Fig. 5). (a) and (b) are interference patterns (averaged over 10000 frames) registered by CCD_1 and CCD_2, respectively; (c) and (d) are their difference and summation, respectively. Experimental data and theoretical simulation are given by open circles and solid lines, respectively.

3.2 Scheme II

In the previous interferometric schemes, the two arms have different optical path lengths so they require the light source to have much better temporal coherence, which is possible only by using a laser beam. A quasi-monochromatic thermal light source with an extended area can be regarded as a spatially incoherent source with a short coherence time less than 0.1 nsec. Therefore a true thermal light source fails in these schemes. From an application point of view, we must look for other schemes if we wish to employ a true thermal light source. For this purpose, the two arms of the interferometer must have equal optical path lengths but different diffraction configurations. We propose the following schemes for a true thermal light source.

We first consider an interferometer in which the two arms have the same length, $2f_o$ (Zhang et al. 2009a). As shown in Fig. 7, a lens L_o of focal length $f_o = 19$cm is set in the centre of the object arm, while in the other arm the beam travels freely to beamsplitter BS_2. Hence the diffraction through the lens in the object arm has a different configuration from that for free propagation in the other arm.

Let object T(x) and the CCD cameras be placed at the two focal planes of lens L_o, which produces the Fourier transform of the object. In the coherent regime, the interference term given by Eq. (5) exhibits the Fourier spatial spectrum of the object T(x)

$$\langle E_r^*(x)E_o(x)\rangle \propto \widetilde{T}(kx/f_o). \tag{21}$$

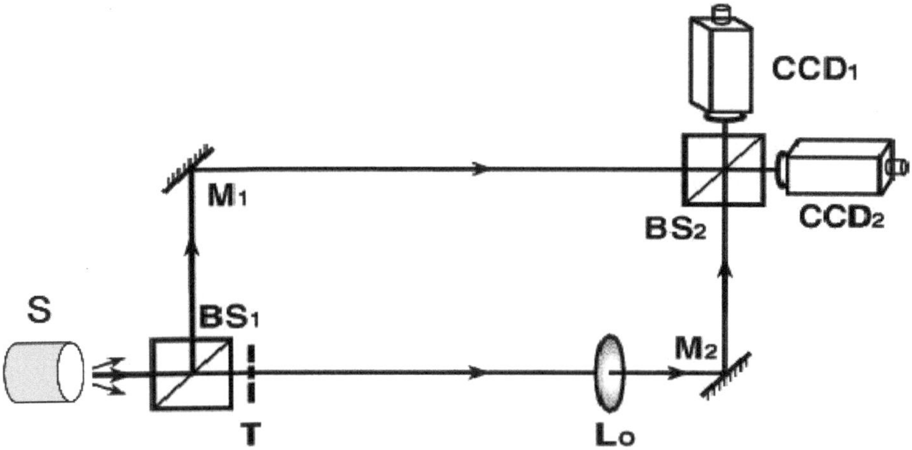

Fig. 7. Scheme II of incoherent interferometry, where S is a light source. A lens L_o of focal length f_o is set in the centre of the object arm, and both arms have the same length $2f_o$.

However, in the incoherent regime, Eq. (7) gives the interference term to be

$$\begin{aligned}\langle E_r^*(x)E_o(x)\rangle &\propto \int T(x_0)F(x,x_0;f_o)H^*(x,x_0;2f_o)dx_0 \\ &\propto \int T(x_0)G^*(x+x_0,2f_o)dx_0 \\ &\approx G^*(x,2f_o)\tilde{T}\left[\frac{kx}{2f_o}\right],\end{aligned} \tag{22}$$

where the last step is valid in the far-field limit. The equation shows the Fresnel diffraction pattern of the object with a diffraction length $2f_o$, different from the f_o in Eq. (21).

With the above experimental scheme, we have used four types of sources. Figure 8 shows 1D patterns using the same sources as in the scheme of Fig. 2; the left part corresponds to a He-Ne laser beam passing through a ground glass plate while the right part, the direct laser beam. We can see that the interference patterns in the left part are very similar to those in Fig. 4. As a matter of fact, these two incoherent interferometry schemes exhibit similar Fresnel diffraction with slightly different effective diffraction lengths, $Z = 39.3$cm and $2f_o = 38$cm. When the ground glass plate is removed, coherent interference patterns are registered by CCD_1 and CCD_2, as shown in the right part of Figs. 8(a) and 8(b), respectively. They are the spatial Fourier spectra of the double-slit, consisting of two parts, $\tilde{T}(kx/f_o) + c.c.$ and $|\tilde{T}(kx/f_o)|^2$. The former and the latter can be extracted by the difference and sum of (a) and (b), as shown in the right part of Figs. 8(c) and 8(d), respectively.

As the third source we use a Na lamp emitting true thermal light of wavelength 589.3 nm, with an illumination area of 10×10 mm², to replace the previous source in Fig. 2. In this case, the interference pattern directly appears on the CCD screen and it is not necessary to perform statistical averaging, as shown in Fig. 9(b). This is due to the fact that the response time of the CCD camera is much longer than the time scale of the thermal light fluctuations, so that averaging has already taken place in a single exposure. For comparison, Fig. 9(a) shows the 2D interference pattern corresponding to Fig. 8(a, left) for pseudo-thermal light. The two sets of fringes are similar, but have slightly different spacings, due to the different wavelengths of the two sources. Finally, we set a pinhole of diameter 0.36 mm after the lamp to dispel the spatial incoherence. With this point-like source, a different interference pattern, which has half the fringe spacing of that for the spatially incoherent source, appears on the CCD screen, as shown in Fig. 9(c). Consequently, the present scheme is favorable for realizing incoherent interferometry using a true incoherent source.

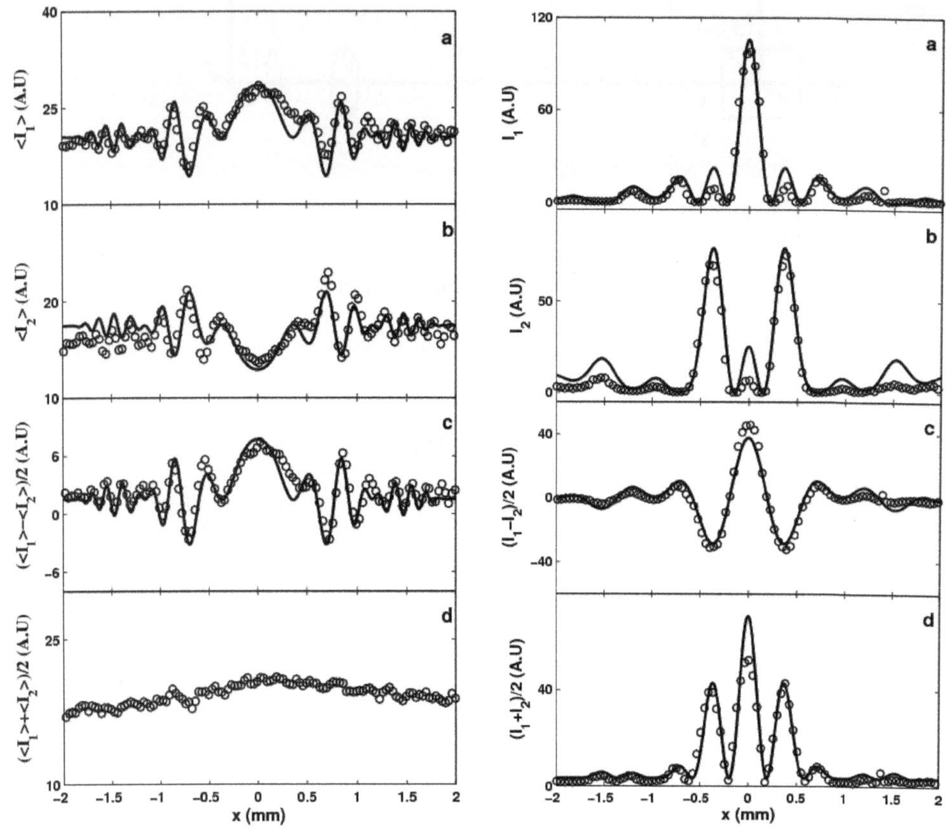

Fig. 8. Experimentally observed 1D interference patterns in the scheme of Fig. 7. Left and right parts are the patterns using spatially incoherent and coherent light, respectively. (a) and (b) are interference patterns (averaged over 10000 frames) registered by CCD_1 and CCD_2, while (c) and (d) are their difference and summation, respectively. Experimental data are given by open circles and theoretical simulation by solid lines.

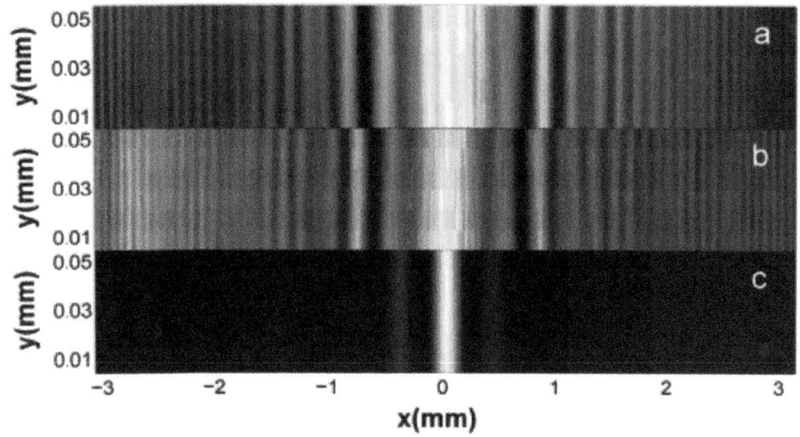

Fig. 9. Experimentally observed 2D interference patterns registered by CCD_1 in the scheme of Fig. 7: (a) with the original pseudo-thermal light source in Fig. 2, averaged over 10000 frames; (b) with a Na lamp of extended illumination area as the light source; (c) with a Na lamp followed by a pinhole as the light source. (Zhang et al., 2009a)

3.3 Scheme III

In this subsection we propose an interesting interferometric scheme which displays unusual behavior in incoherent interferometry. The setup is a simple modification of Fig. 2 whereby a transparent glass rod is inserted into one arm (Zhang et al. 2009b). The length of the rod is selected in such a way that the two arms of the interferometer have equal optical path lengths although their physical lengths are different. Thus the interferometer does not require the driving beam to have better temporal coherence.

The rod can also upset the balanced diffraction between the two arms. When a wave travels over a distance z in a homogeneous material of refractive index n, the impulse response function is given by

$$h(x, x_0) = \sqrt{\frac{k_0}{i2\pi(z/n)}} \exp[ik_0(nz)] \exp\left[\frac{ik_0(x-x_0)^2}{2(z/n)}\right], \tag{23}$$

where k_0 is the wave number in vacuum. It should be noted that in the glass rod medium the optical path length nz is different from the diffraction length z/n. We rewrite Eq. (23) in the form of

$$H(x, x_0; nz, z/n) = \sqrt{\frac{k_0}{i2\pi(z/n)}} \exp(ik_0nz) G(x - x_0, z/n), \tag{24}$$

where nz and z/n are the optical path and diffraction lengths of the rod, respectively.
In successive free propagation through two regions of optical path length z_j and diffraction length $\bar{z}_j$ ($j = 1,2$), we obtain the impulse response function of the combined diffraction to be

$$H(x, x_0; z_1 + z_2, \bar{z}_1 + \bar{z}_2) = \int H(x, x'; z_1, \bar{z}_1) H(x', x_0; z_2, \bar{z}_2) dx'. \quad (25)$$

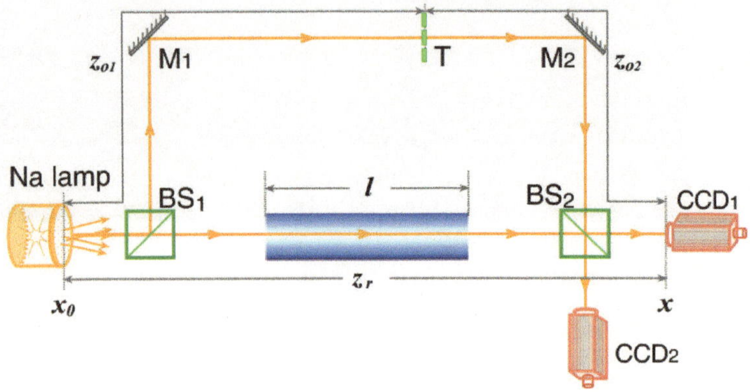

Fig. 10. Scheme III of incoherent interferometry, which is similar to the scheme of Fig. 2 but a glass rod is placed in one arm. The two ends of the rod are plane. (Zhang et al., 2009b)

In general, for a series of cascaded media, the impulse response function $H(x, x_1; Z, \bar{Z})$ corresponds to an optical path length $Z = n_1 l_1 + n_2 l_2 + \cdots$, and a diffraction length $\bar{Z} = \frac{l_1}{n_1} + \frac{l_2}{n_2} + \cdots$. As an example, we calculate the impulse response function of the reference arm in the present interferometer. The physical length of the reference arm is z_r while the glass rod is of length l and refractive index n. We thus require the impulse response function $H(x, x_0; Z, \bar{Z})$ for an optical path length of $Z = z_r - l + nl$ and diffraction length $\bar{Z} = z_r - l + l/n$.
In the other arm, object T(x) is placed at the position which is at a distance z_{o1} from the source and z_{o2} from either of the CCD cameras. The impulse response function is written as

$$h_o(x, x_0) = \int H(x, x'; z_{o2}, z_{o2}) T(x') H(x', x_0; z_{o1}, z_{o1}) dx'. \quad (26)$$

According to Eq. (7), the interference term can be calculated to be

$$\begin{aligned} \langle E_r^*(x) E_o(x) \rangle &\propto \int H^*(x, x_0; Z, \bar{Z}) H(x, x'; z_{o2}, z_{o2}) T(x') H(x', x_0; z_{o1}, z_{o1}) dx' dx_0 \\ &= \int H(x, x'; z_{o1} - Z, z_{o1} - \bar{Z}) H(x, x'; z_{o2}, z_{o2}) T(x') dx' \\ &\propto \exp[ik_0(z_o - Z)] \int T(x') G(x - x', Z_{eff}) dx', \end{aligned} \quad (27)$$

where the Gaussian function G has been defined by Eq. (8b), $z_o = z_{o1} + z_{o2}$ is the length of the object arm, and the effective diffraction length Z_{eff} is defined by

$$\frac{1}{Z_{eff}} = \frac{1}{z_{o2}} + \frac{1}{z_{o1} - \bar{Z}}. \tag{28}$$

As a result, Eq. (27) represents Fresnel diffraction with the effective diffraction length Z_{eff}. The equal-optical-path condition of the interferometer is thus

$$z_o = Z = z_r + (n-1)l. \tag{29}$$

Under this condition, the effective diffraction length becomes

$$Z_{eff} = z_{o2}\left[1 - \frac{z_{o2}}{l(n-\frac{1}{n})}\right]. \tag{30}$$

As long as a certain length l of a medium with index $n \neq 1$ satisfies Eq. (29), the interferometer is capable of realizing incoherent interferometry with a true incoherent light source. The effective diffraction length can be positive or negative, corresponding to normal or phase reversal diffraction, respectively.

In particular, when an object is located at the position $z_{o2}|_{img} = l(n - \frac{1}{n})$, the effective diffraction length is zero, i.e., $Z_{eff} = 0$. It turns out that $G(x - x', Z_{eff} = 0) \propto \delta(x - x')$ in Eq. (27). This means that the wave-front of a beam can be exactly recovered in the propagation. In other words, the image of the object is reproduced with an equal size on the CCD screen. Equation (30) is rewritten as

$$Z_{eff} = z_{o2}\left[1 - \frac{z_{o2}}{z_{o2}|_{img}}\right]. \tag{31}$$

Moreover, the distance of the object from the source for imaging to be obtainable is given by

$$z_{o1}|_{img} = z_o - z_{o2}|_{img} = z_r - l + \frac{l}{n} = \bar{Z}. \tag{32}$$

Thus this distance can be evaluated from the diffraction length of the reference arm.

In the experimental scheme of Fig. 10, the object and reference arms have lengths of $z_o = 41.8$ cm and $z_r = 33.8$ cm, respectively. When the length of the glass rod of refractive index n=1.5163 is $l = 15.5$ cm, the optical path of the reference arm is equal to that for the object arm, i.e. $Z = z_o = 41.8$ cm, although their physical lengths are different. Under this equal-optical-path condition, the two fields to be interfered at the detection plane come from the same wave-front of the source, hence we may use illumination from an incoherent thermal light source, such as the Na lamp mentioned above. Moreover, the diffraction length of the reference arm can be calculated to be $\bar{Z} = z_r - l + \frac{l}{n} = 28.5$ cm, so the imaging position in the object arm is $z_{o1}|_{img} = 28.5$ cm.

Figures 11(a)-(e) show the interference patterns for the double-slit placed, respectively, at the positions $z_{o1} = 31.0, 28.5, 24.2, 20.0$, and 10.6 cm, corresponding to the effective diffraction lengths $Z_{eff} = 2.0, 0, -5.7, -13.9$, and -42.0 cm. The experiment has verified that the position for the object to be imaged [Fig. 11(b)] is at $z_{o1} = 28.5$ cm, which is equal to the diffraction length of the reference arm $\bar{Z}$. The regions left and right of this position correspond to positive and negative diffraction behavior, respectively. However, for an amplitude-modulated object, the same pattern is obtained for both positive and negative diffraction with equal magnitude of diffraction length.

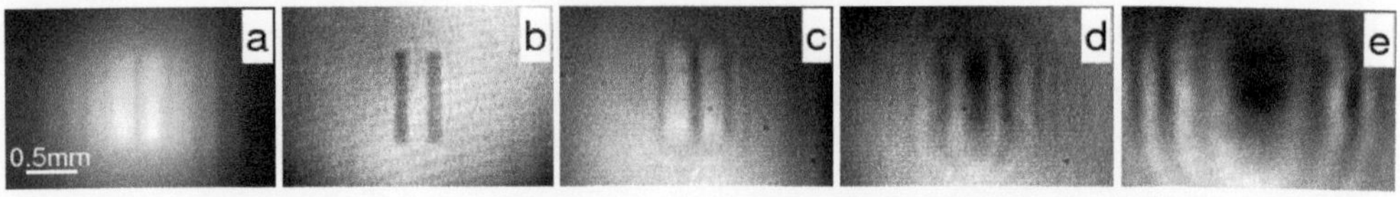

Fig. 11. Experimentally observed 2D interference patterns in the scheme of Fig. 10. [(a)-(e)]: the double slit is placed at the positions of $z_{o1} = 31.0, 28.5, 24.2, 20.0$, and 10.6 cm, respectively, where (b) is the image of the double slit (Zhang et al., 2009b).

We have learned in Section 2 that the effective diffraction length in the interferometric scheme of Fig. 2 can be either positive or negative depending on whether the reference arm is longer or shorter than the object arm. We see again the two kinds of diffraction occurring in the present setup. In particular, this scheme may be realized with a null effective diffraction length $Z_{eff} = 0$, which implies that phase sensitive imaging may be accomplished without using a conventional lens. This effect cannot occur in the coherent regime. Figure 12 compares the output patterns of the interferometer in the same configuration using spatially incoherent and coherent light, when the double-slit is placed at the position $z_{o1} = 28.5$ cm. The images of the double-slit when the interferometer is illuminated by the extended Na lamp are shown in the left column. After we insert a pinhole in front of the lamp to improve the spatial coherence, as expected, instead of obtaining an image we observe interference fringes, as shown in the right column of Fig. 12. In this case, lensless imaging cannot occur no matter where the object is placed.

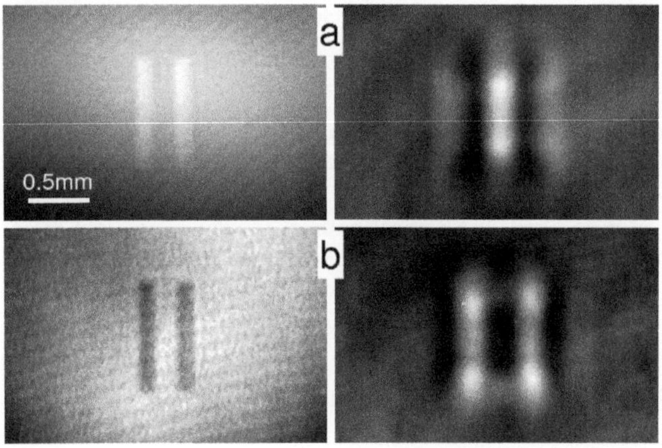

Fig. 12. Experimentally observed 2D patterns when a double-slit object is placed at the position $z_{o1} = \bar{Z} = 28.5$ cm. Left column: image patterns when the source is spatially incoherent; right column: fringe patterns when the source is spatially coherent (with a pinhole aperture in front of the source). Figure (a) was recorded by CCD_1 and (b) by CCD_2 (Zhang et al., 2009b).

A negative diffraction length in wave propagation means phase-reversal diffraction of the wave-front. Recent research on electromagnetic metamaterials shows that negative refractive index materials may be engineered. When an electromagnetic wave travels in a negatively refracting medium, its wave-front undergoes phase-reversal diffraction (Pendry, 2008), thus a slab of negative refraction material can play the same role as a curved lens to perform imaging. Interestingly, the two different physical systems, incoherent interferometry and negative refraction material, reveal some similarity in phase-reversal diffraction (Zhang et al., 2009b).

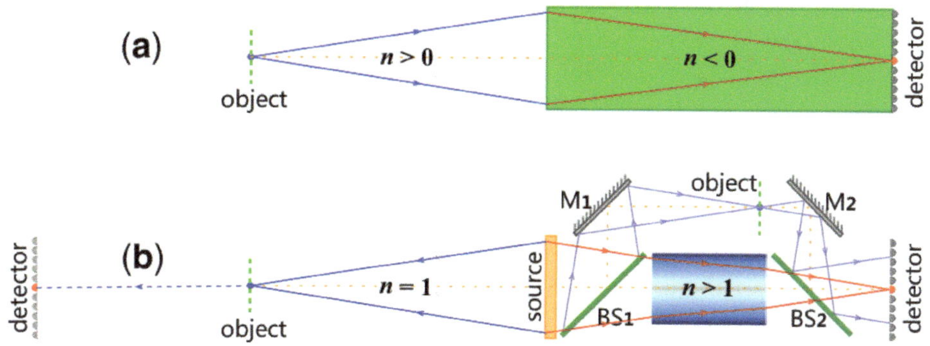

Fig. 13. Illustration of lensless imaging by two schemes. (a) Via successive diffraction through two media, where one is a negative refractive index medium. (b) Using an incoherent light interferometer; when the interferometer is opened out and the two arms are set along a line, the joint diffraction through them is comparable with that in (a) (Zhang et al., 2009b).

Figure 13 explains the similarity of the two models. In Fig. 13(a), we consider the successive diffraction of an object through two media, a positive refraction medium of length l_1 and index $n_1 > 0$ followed by a negative one of length l_2 and index $n_2 < 0$. Let $T(x_0)$ describe a transmittance object, illuminated by a plane wave E_0, then the output field after diffraction is written as

$$E(x) = E_0 \int T(x_0) H(x, x_0; n_1 l_1 + n_2 l_2, l_1/n_1 + l_2/n_2) dx_0. \quad (33)$$

When the net diffraction length in the successive propagation is null, i.e. $l_1/n_1 + l_2/n_2 = 0$, then $H(x, x_0; n_1 l_1 + n_2 l_2, 0) = \exp[ik_0(n_1 l_1 + n_2 l_2)]\,\delta(x - x_0)$ and we have

$$E(x) = E_0 \exp[ik_0(n_1 l_1 + n_2 l_2)]\, T(x). \quad (34)$$

Otherwise, Eq. (33) represents Fresnel diffraction with a diffraction length of $l_1/n_1 + l_2/n_2$

$$E(x) \propto E_0 \exp[ik_0(n_1 l_1 + n_2 l_2)] \int T(x_0) G(x - x_0, l_1/n_1 + l_2/n_2) dx_0, \quad (35)$$

which has the same form as Eq. (27)

To better compare the two schemes, in Fig. 13(b), the interferometer is opened out and the two arms are set along a line. We can see that the joint diffraction through the two arms is comparable with the geometry in Fig. 13(a).

Recently, various approaches for invisibility cloaking and transformation optics in complementary media with positive and negative refraction materials have been proposed, which can in theory accomplish exact optical cancellation between the object and its counterpart (Lai et al., 2009a, 2009b). In light of the similarity of time-reversal diffraction between incoherent interferometry and negatively refracting media, using the present scheme, we have conducted proof-of-principle experimental demonstrations of the theoretical proposals (Zhang et. al., 2010; Gan et al., 2011). The physical analogue between the two different systems may provide a convenient research platform. Moreover, a form of nonlocal imaging as well as interference effects that were previously regarded as the signature of two-photon entanglement or intensity correlation of thermal light can now be realized in incoherent interferometry, which is associated with the first-order field correlation (Gao et al., 2009; Gan et al. 2009).

4. Conclusion

Interference effects in incoherent interferometry show different physics from that in coherent interferometry. In the latter case, two interfering fields have well-defined spatial distributions, whereas in the former case, these fields fluctuate randomly in space and the interference pattern appears only in the statistical average. Furthermore, incoherent interference relies entirely on the first-order spatial correlation of the two fields, so the object information is contained in the joint diffraction of the two fields. For certain optical geometries, phase reversal diffraction can occur through the first-order spatial correlation of incoherent light, thus providing a method of wave-front recovery without using a lens, and also a means for optical transformation.

Incoherent interferometry as a novel interference mechanism exhibits richer phenomena than the coherent type. Since it is not dependent on spatial coherence, the present method may find potential application in holography and other interference technologies, especially in those areas where a coherent source is unavailable. Incoherent holographic interferometry is fundamentally different from the previously known incoherent holography, while it can play the same role as coherent holography.

5. Acknowledgment

The authors thank L. A. Wu for helpful discussions. This work was supported by the National High Technology Research and Development Program of China, Project No. 2011AA120102 and the National Natural Science Foundation of China, Project No. 10874019.

6. References

Cochran, G. (1966). New method of making Fresnel transforms with incoherent light. *J. Opt. Soc. Am.,* Vol.56, No.11, (November 1966) pp.1513-1517.

Gabor, D. (1972). Holography, 1948-1971, *Science,* New Series, Vol.177, No.4046, (July 1972) pp. 299-313.

Gan, Sh., Zhang, S. H., Xiong, J., & Wang, K., (2009). Interference from a nonlocal double-slit through one-photon process, *Optics Express* Vol.17, No.26, (December 2009) pp.23672-23677.

Gan, Sh., Zhang, S. H., Zhao, T., Xiong, J., Zhang, X, & Wang K., (2011). Cloaking of a phase object in ghost imaging, *Appl. Phys. Letts.* Vol.98, No.11, (March 2011) 111102, ISSN 0003-6951.

Gao, L., Zhang, S. H., Xiong, J., Gan, Sh., Feng, L. J., Cao, D. Zh., & Wang, K., (2009). Correlated imaging with one-photon interference, *Phys. Rev. A* Vol.80, No.2, (August 2009) 021806(R), ISSN 1050-2947.

Goodman, J. W. (1996). *Introduction to Fourier Optics* (Second Edition), p.371, The McGraw-Hill Companies, Inc. New York.

Lai, Y., Chen, H. Y., Zhang, Zh. Q., & Chan, C. T., (2009a). Complementary media invisibility cloak that cloaks objects at a distance outside the cloaking shell, *Phys. Rev. Lett.* Vol.102, No.9, (March 2009) 093901, ISSN 0031-9007.

Lai, Y., Ng, J., Chen, H. Y., Han, D. Zh., Xiao, J. J., Zhang, Zh. Q., & Chan, C. T., (2009b). Illusion optics: the optical transformation of an object into another object, *Phys. Rev. Lett.* Vol.102, No.25, (June 2009) 253902, ISSN 0031-9007.

Lohmann, A. W. (1965). Wavefront reconstruction for incoherent objects. *J. Opt. Soc. Am.,* Vol.55, Issue 11, (November 1965) pp. 1555-1556.

Mertz, L. & Young, N. O. (1963). Fresnel transformations of images, In: *Proceedings of the Conference on Optical Instruments and Techniques,* K.J. Habell, (Ed.), p. 305, John Wiley and Sons, New York, NY.

Pendry, J. B. (2008). Time reversal and negative refraction. *Science* Vol.322, (October 2008) pp. 71-73, ISSN 1095-9203.

Stroke, G. W. & Restrick, R. C. (1965). Holography with spatially noncoherent light. *Appl. Phys. Lett.,* Vol.7, Issue 9, (November 1965) pp.229-231, ISSN 0003-6951.

Zhang, S. H., (July 2010). The first-order field correlation effects of spatially incoherent light, In: *PhD thesis, Beijing Normal University,* Wang, K. (Dir), Available from: <http://d.wanfangdata.com.cn/Thesis_Y1776851.aspx>.

Zhang, S. H., Gao L., Xiong, J., Feng, L. J., Cao, D. Zh., & Wang, K., (2009a). Spatial Interference: From Coherent to Incoherent, *Phys. Rev. Lett.* Vol.102, No.7 (February 2009) 073904, ISSN 0031-9007.

Zhang, S. H., Gan, Sh., Cao, D. Zh., Xiong, J., Zhang X., & Wang, K., (2009b). Phase-reversal diffraction in incoherent light, *Phys. Rev. A* Vol.80, No.3, (September 2009) 031895(R), ISSN 1050-2947.

Zhang, S. H., Gan, Sh., Xiong, J., Zhang, X., & Wang, K., (2010). Illusion optics in chaotic light, *Phys. Rev. A* Vol.82, No.2, (August 2010) 021804(R), ISSN 1050-2947.

Part 2

Digital Holographic Microscopy

9

Alternative Reconstruction Method and Object Analysis in Digital Holographic Microscopy

Francisco Palacios et al *
University of Oriente
Cuba

1. Introduction

Holography is a method for storing and reconstructing both amplitude and phase information of a wave front. In digital holography the reconstruction process is accomplished by means of a computer (Yaroslavsky & Merzyalov, 1980) obtaining directly the phase distribution of the object wave front. Particularly with the improvement of the spatial resolution of CCD cameras and the increasing computational performance of personal computers digital holography has been widely applied in many fields such as deformation analysis (Schedin et al., 2001), object contouring (Wagner et al., 2000), microscopy (Takaki & Ohzu, 1999) and particle measurement (Murata & Yasuda, 2000). The technique of digital holography has been implemented in a configuration of an optical microscope (Schilling et al., 1997); the objective lens produces a magnified image of the object and the interference between this image and the reference beam is achieved by the integration of the microscope into one of the arms of a Mach-Zender interferometer. This configuration is called Digital Holographic Microscopy (DHM).
DHM is a powerful technique for real-time quantitative phase contrast imaging, since a single intensity image, called a hologram, allows the reconstruction of the phase shift induced by a specimen. This property of holograms offers phase-contrast techniques, which can then be used for quantitative 3D imaging (Palacios et al., 2005). Quantitative phase imaging is important because it allows the determination of the optical thickness profile of a transparent object with sub-wavelength accuracy (Yu et al., 2009). Through numerical processing of the hologram one can filter out parasitic interferences and the components of the image reconstruction: zero-order and twin image terms (Cuche et al., 2000) or to compensate for curvature introduced by the microscope objective (MO) (Pedrini et al., 2001),

* Oneida Font[1], Jorge Ricardo[1], Guillermo Palacios[1], Mikiya Muramatsu[2], Diogo Soga[2], Daniel Palacios[3], José Valin[4] and Freddy Monroy[5]
[1]*University of Oriente, Cuba*
[2]*University of Sao Paulo, Brasil*
[3]*University of Simon Bolivar, Venezuela*
[4]*Polytechnic Institute "José A. Echeverría", Cuba*
[5]*National University of Colombia, Colombia*

spherical aberration (Stadelmaier & Massig, 2000), astigmatism (Grilli et al., 2001) and anamorphism (De Nicola et al., 2005).

Compensation of aberrations is fundamental when quantitative phase determination is used in microscopic metrological applications. Several approaches have been proposed to remove the aberrations. A method proposed by Cuche (Cuche et al., 1999) used a single hologram, which involved the computation of a digital replica of the reference wave depending on two reconstruction parameters. A double-exposure technique (Ferraro et al., 2003) can compensate completely the inherent wave front curvature in quantitative phase contrast imaging, although it needs two hologram recordings with and without the sample and a subtraction procedure between the two holograms. In the paper (Colomb et al., 2006) the authors developed a method to compensate the tilt aberration by recording a hologram corresponding to a blank image to compute the first-order parameters directly from the hologram. However, the method is limited to selecting profiles or areas known to be flat in the hologram plane. A method was proposed by Miccio (Miccio et al., 2007) who performed a two-dimensional fitting with the Zernike polynomials of the reconstructed unwrapped phase, although the application of the method is limited to be the special case of thin objects.

For image reconstruction different algorithms have been developed. Among them, the Single Fourier Transform Formulation (SFTF) (Schnars, 1994), the convolution-based algorithm (CV) (Demetrakopoulos & Mittra, 1974) and the angular spectrum-based algorithm (ASA) (Yu & Kim, 2005) are most commonly used. The SFTF algorithm is fast and can be used with objects larger than a CCD. However, variation of the size of the reconstructed image as result of a change in the reconstruction depth poses problems in applications such as reconstruction of color holograms (Yamaguchi et al., 2002) and particle sizing (Pan & Meng, 2003). In contrast, the CV algorithm keeps the size of the reconstructed image the same as that of the CCD. However, it is applicable only to objects that are smaller than the CCD. In addition, when the CV algorithm is used for objects much smaller than a CCD, it degrades the image quality, since the image is represented by only a small number of pixels. The CV algorithm was extended to large objects by zero padding the holograms before reconstruction (Kreis et al., 1997). This approach, however, led to an increase in computational load.

To avoid a number of limitations of the previous algorithms, the double Fresnel-transform algorithm (DBFT) that allows the reconstruction of digital holograms with adjustable magnification was developed (Zhang et al., 2004). This algorithm involves two reconstruction steps implemented by a conventional single Fourier-transform algorithm. Through the adjustment of the distance parameter in the first stages, it is possible to control the size of the reconstructed images, independent of distance and wavelength, even for objects larger than a CCD without any computational penalty.

The method of image reconstruction proposed here is similar to the DBFT algorithm, since we also use two steps for image reconstruction, but they differ in the objectives and meaning of the two steps. Our formulation is specific for digital holographic microscopy; this way the objective of the first stage is the calculation of the objects' Fourier transform plane, where the complex wavefield contains all the information about the phase and intensity of object wavefield. Then, the second step is to reconstruct the complex amplitudes of the image wavefield starting from the objects' Fourier transform plane. We will describe and show advantages of the proposed method.

2. Experimental set-up

Figure 1 shows the experimental set-up used in this work. It is a Digital Holographic Microscope designed for transmission imaging with transparent sample. The basic architecture is that of a Mach-Zehnder interferometer. A linearly polarized He-Ne laser (15 mW) is used as light source. The expanded beam from the laser is divided by the beam splitter BS1 into reference and object beams. The microscope produces a magnified image of the object and the hologram plane is located between the microscope objective MO and the image plane *(x'-y')* which is at a distance *d'* from the recording hologram plane *(ξ-η)*. In digital holographic microscopy we can consider the object wave emerging from the magnified image and not from the object itself (VanLigten & Osterberg, 1966).

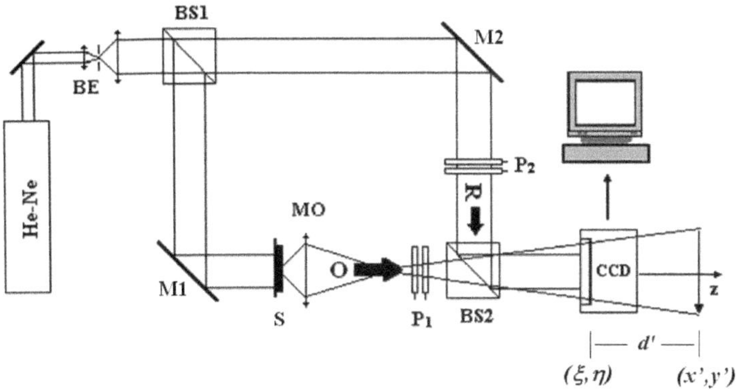

Fig. 1. Experimental set-up: BE, bean expander; BS, bean splitter (the splitting ratio of BS1 and BS2 are 90/10 and 50/50 respectively); M, mirror; MO, microscope objective; S, sample; P, dual polarizer; CCD, charge coupled device.

With the combinations of the dual polarizer P_1 and P_2 the intensities are adjusted in the reference arm and the object arm of the interferometer and the same polarization state is also guaranteed for both arms improving their interference. The specimen S is illuminated by a plane wave and a microscope objective, that produces a wave front called object wave *O*, collects the transmitted light. A condenser, not shown, is used to concentrate the light or focus the light in order that the entire beam passes into the MO. At the exit of the interferometer the two beams are combined by beam splitter BS2 being formed at the CCD plane the interference pattern between the object wave *O* and the reference wave *R*, which is recorded as the hologram of intensity $I_H(\xi,\eta)$,

$$I_H(\xi,\eta) = |O|^2 + |R|^2 + R^*O + RO^* \tag{1}$$

where R^* and O^* are the complex conjugates of the reference and object waves, respectively. The two first terms form the zero-order, the third and fourth terms are respectively the virtual (or conjugate image) and real image, which correspond to the interference terms. The off-axis geometry is considered; for this reason the mirror M2, which reflects the reference wave, is oriented so that the reference wave reaches the CCD camera with a small incidence angle with respect to the propagation direction of the object wave. A digital hologram is recorded by the CCD camera HDCE-10 with 1024x768 square pixels of size 4.65 μm, and transmitted to the computer by means of the IEEE 1394 interface. The digital hologram $I_H(j,l)$ is an array of $M \times N = 1024 \times 768$ 8-bit-encoded numbers that results from the two-dimensional sampling of $I_H(\xi,\eta)$ by the CCD camera,

$$I_H(j,l) = I_H(\xi,\eta)\,\mathrm{rect}\left[\frac{\xi}{L_x},\frac{\eta}{L_y}\right]\sum_{j=-M/2}^{M/2}\sum_{l=-N/2}^{N/2}\delta(\xi - j\Delta\xi, \eta - l\Delta\eta) \tag{2}$$

where j, l are integers defining the positions of the hologram pixels and $\Delta\xi = \Delta\eta = 4.65\ \mu m$ defines the sampling intervals in the hologram plane.

3. Basic principles of the alternative reconstruction method

In Digital Holographic Microscopy the field produced by the objective lens can be reconstructed in any plane along the field propagation direction. Traditionally the optical field $\psi(x',y')$ on the image plane $(x'$-$y')$ is calculated by propagation of the wavefront $\psi(\xi,\eta)$, a distance $z = d'$ from the hologram plane $(\xi$-$\eta)$, figure 2.

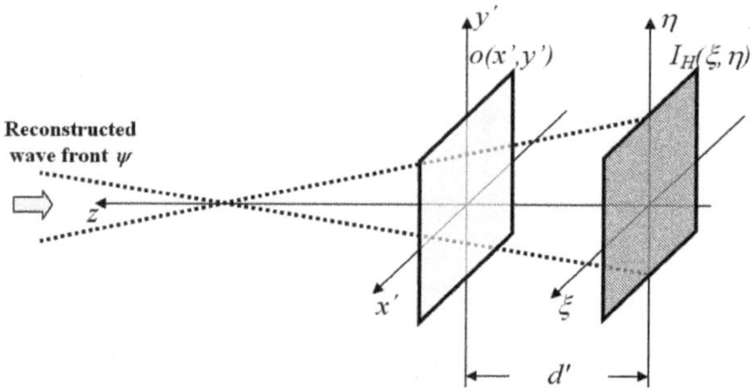

Fig. 2. Schematic diagram of traditional image reconstruction methods in DHM.

In our approach the reconstruction of the complex wave distribution $o(x',y') \equiv \psi(x',y';z=d')$ consists basically of two stages that involve two wavefield propagations. In the first stage, figure 3, we reconstruct the wave distribution $\tilde{o}(u,v)$ on the $(u$-$v)$ plane at reconstruction distance $z = D$ (first propagation).

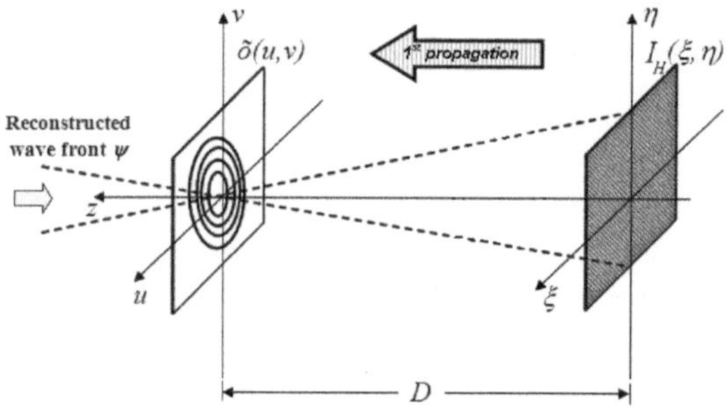

Fig. 3. Reconstruction of the wave distribution $\tilde{o}(u,v)$ on the $(u$-$v)$ plane at reconstruction distance $z = D$ (first propagation).

Applying the (ASA) algorithm the distance D is calculated, as shown in section 3.1.1. After the distance D has been calculated, the first propagation is carried out by means of the Fresnel approximation method, specifically the *Single Fourier Transform Formulation* (SFTF),

$$\psi_{SFTF}(u,v;z) = A\exp\left[\frac{i\pi}{\lambda z}\left(u^2+v^2\right)\right]\Im\left\{I_H(\xi,\eta)\exp\left[\frac{i\pi}{\lambda z}\left(\xi^2+\eta^2\right)\right]\right\} \qquad (3)$$

where z is the reconstruction distance, $\Im$ is an operator that denotes the Fourier transform, $A=exp(i2\pi z/\lambda)/(i\pi\lambda)$ and considering a plane reference wave with unit strength perpendicular to the recording plane.

At z = D, the reconstructed wavefield, consists of a zero order and the twin images (real and conjugate). The filtered complex wavefield $\psi^f_{SFTF}(u,v;z=D)$ can be expressed by Eq. (4) by replacing the specimen hologram $I_H(\xi,\eta)$ with a filtered hologram $I^f_H(\xi,\eta)=RO^*$ containing only spatial components of the real image (Cuche et al., 2000),

$$\psi^f_{SFTF}(u,v;z=D) = A\exp\left[\frac{i\pi}{\lambda D}\left(u^2+v^2\right)\right]\Im\left\{I^f_H(\xi,\eta)\exp\left[\frac{i\pi}{\lambda D}\left(\xi^2+\eta^2\right)\right]\right\} \qquad (4)$$

As has been proven (Palacios et al., 2008), the complex field $\psi^f_{SFTF}(u,v;z=D)$ is equivalent to the complex field distribution $\tilde{o}(u,v)$ on the back focal plane of the objective lens. From Abbe's theory of image formation (Lipson S. & Lipson H., 1981), the field on the back focal plane can be represented by the expression,

$$\psi^f_{SFTF}(u,v;z=D) \cong \tilde{o}(u,v) = S_\phi(u,v)\Im\left[o(x_o,y_o,\lambda f)\right] \qquad (5)$$

where $\Im\{o(x_o, y_o, \lambda f)\}$ is the Fourier transform of the object wave distribution at the plane $(u\text{-}v)$, with $o(x_o,y_o)$, the amplitude transmittance of the object,

$$\Im\left[o(x_o,y_o,\lambda f)\right] = \iint o(x_o,y_o)\exp\left[-i\frac{2\pi}{\lambda f}\left(x_o u + y_o v\right)\right]dx_o dy_o \qquad (6)$$

and $S_\phi(u,v)$ is a quadratic phase factor,

$$S_\phi(u,v) = \frac{i}{\lambda f}\exp\left[-ik\left(S_o+f\right)\right]\exp\left[\frac{ik}{2f}\left(u^2+v^2\right)\left(1-\frac{S_o}{f}\right)\right] \qquad (7)$$

where S_o is the distance from the object to the lens, f the focal distance of the lens, λ the wavelength of the incident plane wave.

From the wave theory of image formation, after the objective lens producing the diffraction pattern of the object in its back focal plane, a second Fourier transformation performed on the diffraction pattern it is associated with the image of the object (Goodman, 1968). Consequently, all the information about image wavefield at the hologram plane is contained in the complex wavefront $o(u,v)$ on the back focal plane, therefore the reconstruction of the optical wavefield $o(x',y';d')$ can be carried out from the $(u\text{-}v)$ plane, instead of traditionally the hologram plane $(\xi\text{-}\eta)$, figure 4.

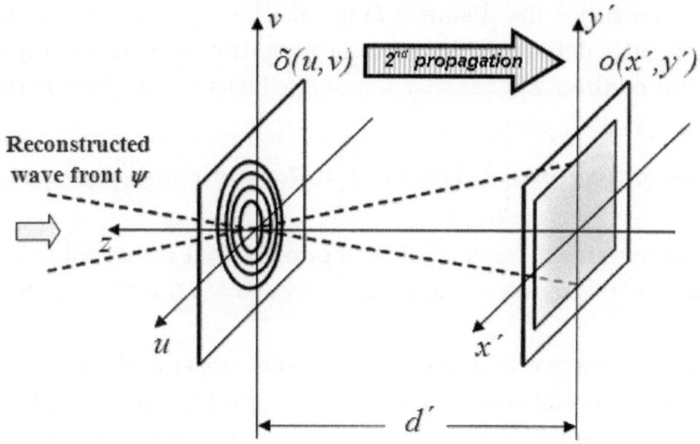

Fig. 4. Reconstruction of the optical wavefield $o(x',y';d')$ in the image plane from the *(u-v)* plane (second propagation).

In the second stage of the method, the complex wavefield $\psi(x',y';d')$ at an arbitrary distance d' can be obtained by propagation of the wavefield $\tilde{o}(u,v)$ through a distance d' and the result is inverse Fourier transformed,

$$\psi(x',y';d') = \Im^{-1}\left[\psi^{f}_{SFTF}(u,v;z=D)\exp\left(id'\sqrt{k^2+k_u^2+k_v^2}\right)\right] \tag{8}$$

where $\Im^{-1}$ symbolizes the inverse Fourier Transform, $k=2\pi/\lambda$, k_u and k_v are corresponding spatial frequencies of u and v respectively. The numerical implementation of Eq. (8), that we call the *Double-Propagation* algorithm (DPA), is given by,

$$\psi_{DPA}(m,n;d') = FFT^{-1}\Big\{\psi^{f}_{SFTF}(l,j;z=D) \times\exp\left[2\pi i d'\sqrt{(1/\lambda)^2+(l/N\Delta\xi)^2+(j/M\Delta\eta)^2}\right]\Big\} \tag{9}$$

where j, l, m, n are integers $(-M/2 < j, l < M/2)$, $(-N/2 < m, n < N/2)$ and $\psi^{f}_{SFTF}(l,j;z=D)$ is the discrete formulation of Eq. (4). From Eq. (9) we can obtain the intensity image $I_{DPA}(x',y';d')$ by calculating $|\psi_{DPA}(m,n;d')|^2$ and the phase image $\phi_{DPA}(x',y';d')$ by calculating $\arg[\psi_{DPA}(m,n;d')]$.

In short, the following steps describe the general procedure of the alternative reconstruction method:

1. Determining the distance D.
2. Calculate $\psi_{SFTF}(u,v;z=D)$ using Eq. (3), i.e., the complex amplitudes of the wavefield at $z=D$.
3. Obtaining $\psi^{f}_{SFTF}(u,v;z=D)$ by filtering the spatial components that correspond to the complex amplitudes of the Fourier transform of the objects.
4. Calculate $\psi_{DPA}(m,n;d')$, using Eq. (9), at an arbitrary distance d'.
5. Determining the intensity image or the phase image by calculating the square modulus or the argument of Eq. (9).

As we will demonstrate in next section, the formulation based on Eq. (9) guarantees that the reconstructed image maintains its size independently of depth d' and the phase curvature compensation can be done easily by techniques of image background subtraction.

3.1 Experimental validation of the method: advantages and limitations

Figure 5(a) shows the hologram of a USAF resolution target recorded by the experimental set-up. In figure 5(b) are represented the components of the reconstructed wavefields at $z = D$. The undiffracted reconstruction wave forms a zero-order image which is located in the center (already filtered), and corresponds to the first two terms on the right-hand side of Eq. (1).

Due to off-axis geometry the real image (delimited by the circle) and the conjugate image (delimited by the rectangle) are positioned at different locations on the reconstruction plane and they correspond to the interference terms on the right-hand side of Eq. (1). The separation of the interference terms depends on the angle between the reference and object wave. In figure 5b the expression $log[1+|(\psi_{SFTF}(x'y',d'= D)|^{2}]$ was used to compress the dynamic range of pixel values, which improves the visualization of the intensity distribution of the reconstructed wavefields (Lim, 1990).

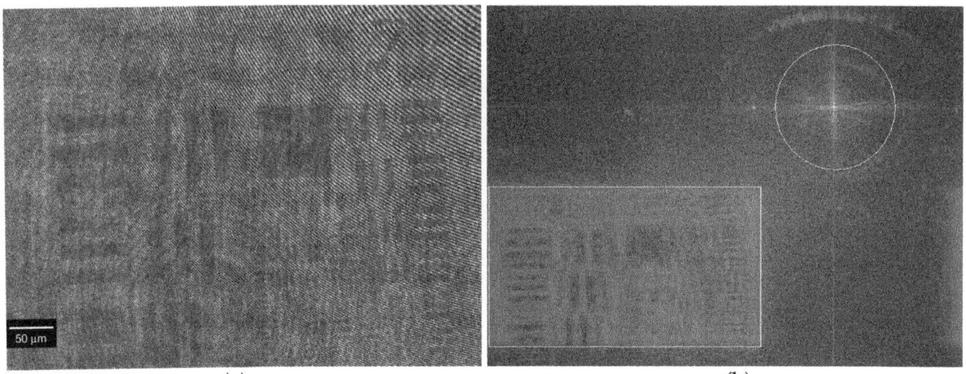

(a) (b)

Fig. 5. (a) Hologram of a USAF resolution target, (b) Components of the reconstructed wavefield at $z = D$. The circle delimits the real image and the rectangle the conjugate image.

At $z = D$ the reconstruction of real image is different of the genuine appearance of object image because at this distance the Fraunhofer diffraction pattern of the objects is reconstructed, figure 6, rather than the image of the objects.

Fig. 6. Intensity of field distribution on the focal plane.

Larger or lower spatial frequencies of object decomposition will be represented by intensity in the focal plane that is farther or closer from optical axis or equivalently farther or closer from the center of the pattern where is contained the undiffracted object wavefield.

3.1.1 Determination of the distance *D*

The distance D is determined only one time, and remains the same until there is some variation in the experimental set-up. Applying the (ASA) algorithm, the following steps were performed to calculate the distance D:

1. The angular spectrum $A(k_\xi k_\eta;z=0)$ of the hologram $I_H(\xi,\eta)$ at $z = 0$ is obtained by taking its Fourier transform. k_ξ and k_η are the corresponding spatial frequencies in the hologram plane ξ-η.
2. Filtering of the angular spectrum to suppress both the zero-order and the twin image. In this step a region of interest corresponding only to the object spectrum is selected and the modified angular spectrum $\tilde{A}_H\left(k_\xi,k_\eta;z=0\right)$ is obtained.
3. The new angular spectrum at plane z, $A(k_\xi k_\eta;z)$ is calculated from $\tilde{A}_H\left(k_\xi,k_\eta;0\right)$ as, $A\left(k_\xi,k_\eta;z\right)=\tilde{A}\left(k_\xi,k_\eta;0\right)\exp\left[ik_z z\right]$, where $k_z=\sqrt{k^2-k_\xi^2-k_\eta^2}$.
4. The reconstructed complex wave field at any plane $(x'$-$y')$ perpendicular to the propagating z axis is found by,
   $\psi_{ASA}\left(x',y';z\right)=\iint A\left(k_\xi,k_\eta;z\right)\exp\left[i\left(k_\xi x'+k_\eta x'\right)\right]dk_\xi\, dk_\eta$.
5. The reconstruction of the amplitude image $I_{ASA}(x',y';z)=|\Psi_{ASA}(x',y';z)|^2$, is carried out varying z from 0 to $300\ mm$ with $20\ mm$ as incremental step. The punctual maximum value $P(z) = [I_{ASA}(x',y';z)]_{max}$ is calculated and plotted for each z, figure 7.
6. Between the two points around the relative maximum the incremental step is reduced to $1\ mm$. The distance D is equal to the z value at which the absolute maximum of $I_{ASA}(x',y';z)_{max}$ is reached; the value of $z = D = 173\ mm$ is shows in figure 7.

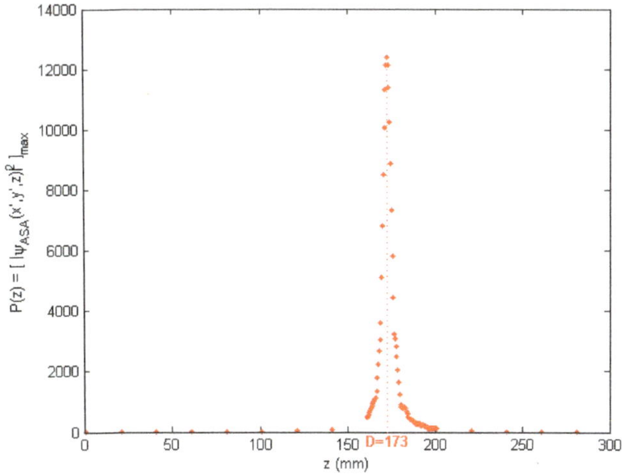

Fig. 7. Distance D determination using *ASA*.

Because the reference wave is plane, the distance D is the physical distance from the hologram to back focal point of the objective lens.

3.1.2 Behavior of the wavefield near to the back focal plane

Using the hologram of figure 5a, the behavior of the wavefield near the back focal plane is visualized. With z ranging from $141\ mm$ to $173\ mm$ and incremental step of $8\ mm$ a sequence of image reconstructions is presented in figure 8. The sequence shows the change of phase (A) and intensity (B) calculated from the expressions: $\arg(FFT^{-1}[\psi_{SFTF}(l,j;z)])$ and $\log[1+|(\psi_{SFTF}(l,j;z)|^2]$ respectively.

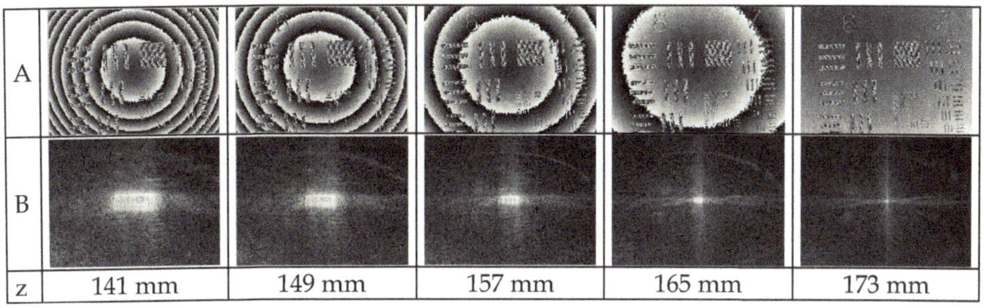

Fig. 8. Behavior of the wavefield in the region of the back focal plane.

From figure 8 it is corroborated that as the reconstruction plane approaches the focal plane, the phase jumps between the reference and object waves gradually disappear. These phase jumps totally disappear for $z = D = 173\ mm$, where the focal plane is reconstructed. This behaviour allows us to conclude that the curvature of the wavefront has a minimum on the focal plane and it is increased as the wavefield propagates away from this plane.

3.1.3 Intensity and phase image reconstruction

Calculating the square modulus and the argument of Eq. (9) in figure 9 provide the intensity (a) and phase (b) image reconstruction respectively at $z = D = 173\ mm$.

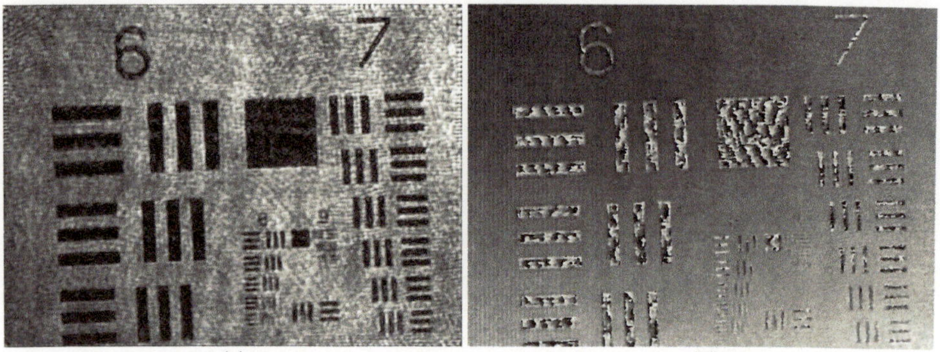

(a) (b)

Fig. 9. Reconstruction of the intensity image (a) and the phase image (b) at $z = D = 173\ mm$

Phase curvature compensation:

It should be noted from figure 9b that mixed phase images ϕ_o and ϕ_B appear, which are related with the phase of the objects and the quadratic constant phase factor S_ϕ respectively. Both phase images are shown more clearly in figure 10.

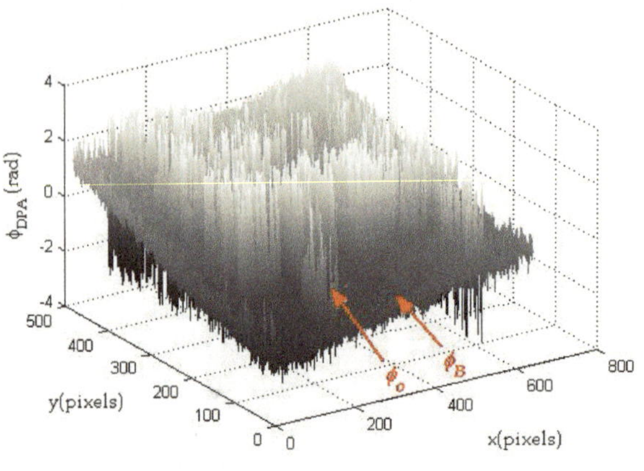

Fig. 10. Pseudo 3D rendering of the phase image ϕ_{DPA} presented in figure 9b.

Due to the slow variation of ϕ_B, it can be considered as a background contribution to the phase image of the objects ϕ_o, thus the problem of phase curvature compensation can be treated as a problem of phase image background subtraction. This way, alternative image background subtraction methods can be used. A procedure (Ankit & Rabinkin, 2007), that consists of applying a median filter with a large kernel size to the phase image represents a quick and simple way to obtain ϕ_B, figure 11.

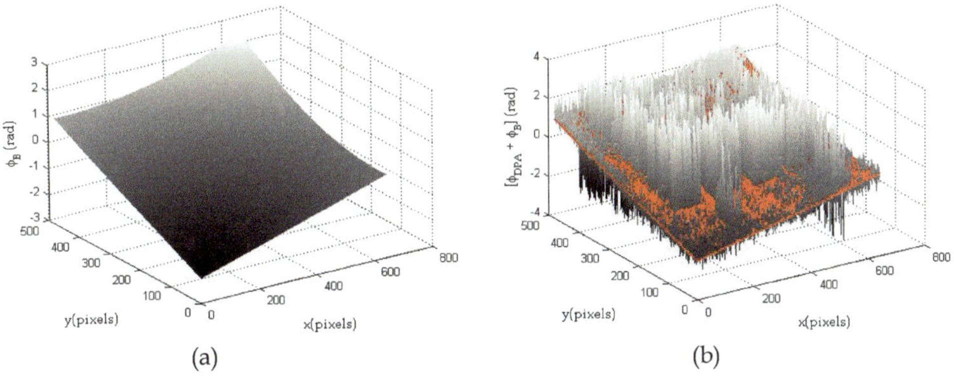

Fig. 11. Pseudo 3D rendering of the phase image ϕ_B (a) and simultaneous visualization of ϕ_{DPA} and ϕ_B *(color red)* (b)

As can be appreciated from figure 11b, a good fitting of ϕ_B is achieved when a median filter with a large kernel size is applied over ϕ_{DPA}.

Taking into consideration that the reconstructed complex wavefield ψ_{DPA} can be expressed as the superposition of two contributions corresponding to the object wave front $exp(i\phi_o)$ and the spherical wave front $exp(i\phi_B)$,

$$\psi_{DPA}(x',y';d') \propto \exp(i\phi_o)\exp(i\phi_B) \tag{10}$$

then, the phase curvature can be compensated numerically by the calculation of the corrected phase image $\phi_{DPA}^{corr}(x',y';d')$,

$$\phi_{DPA}^{corr}(x',y';d') \cong \phi_o(x',y';d') = \arg\left[\psi_{DPA}(x',y';d')\exp(-i\phi_B)\right] \tag{11}$$

In figure (12a) is presented the corrected phase image calculated by Eq. (11).

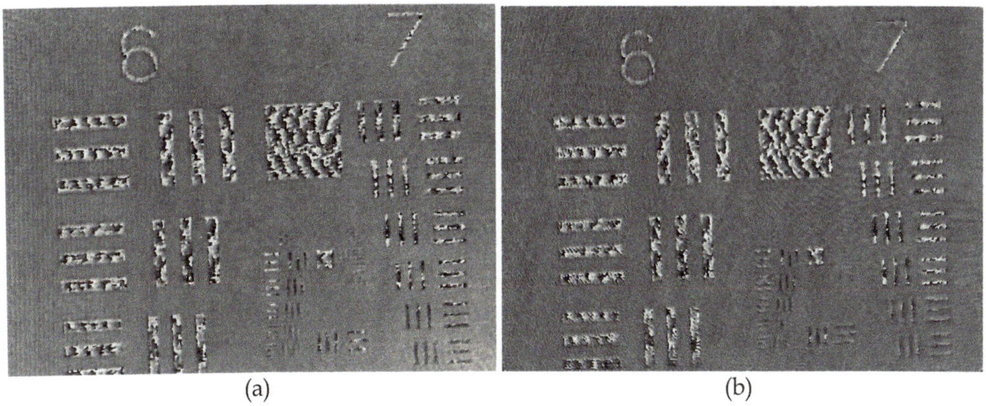

Fig. 12. Corrected phase image $\phi_{DPA}^{corr}(x',y';d')$ (a). Phase image calculated with ASA, but using a reference hologram (b).

For comparison figure 12b shows the phase image calculated from the same hologram and reconstructed with the ASA, but using a reference hologram (Colomb et al., 2006).

3.1.4 Refocusing at different reconstruction distances

With the analysis of the intensity image reconstruction we demonstrate the ability of DPA to refocus at different distances *d'*.

Figure 13 shows a comparison between DPA (row A) and ASA (row B) in the reconstruction of several holograms where the best focus of the image is: *d'=0 mm, d'=10 mm, d'=30 mm* and *d'=70 mm*.

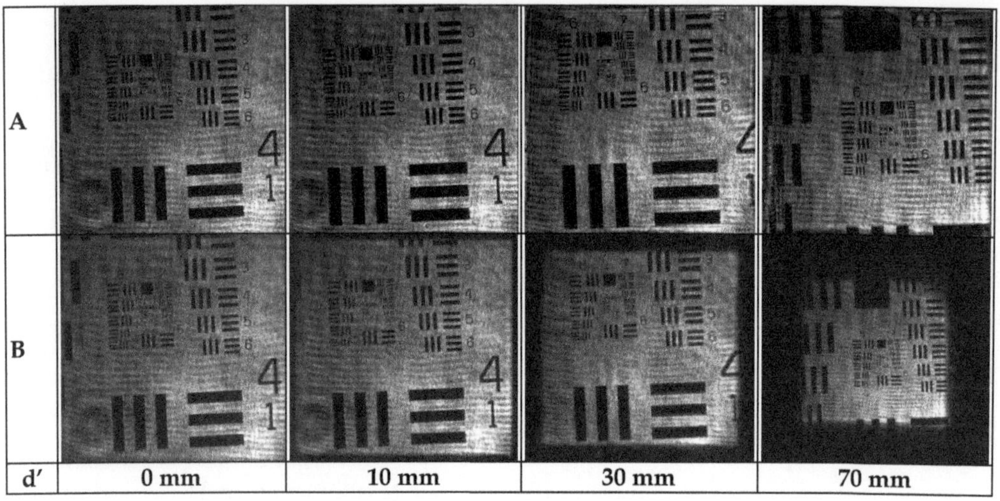

Fig. 13. Comparison between DPA (row A) and ASA (row B) in the intensity image reconstruction for different distances *d'*.

According to figure 13, as the reconstruction distance *d'* increases, the size of the reconstructed intensity image is constant for both algorithm until *d' = 10 mm*. For values of *d'* bigger than 20 mm, the size of the reconstructed intensity image is reduced when the ASA is used, whereas it remain constant when is used the DPA. The effect is most obvious for *d' = 70 mm*.

Summarizing, two main advantages can be attributed to the DPA algorithm: capability to maintain the size of a reconstructed image, independent of the reconstruction distance and wavelength for objects larger than a CCD and phase reconstruction with curvature compensation without the necessity of either a reference hologram or parameter adjustment. For a hologram of 1024 x 768 pixels using a standard PC computer (Pentium IV, 3.2 GHz) the required time for the calculation of the phase image, the intensity image and the distance *D* is 2.3, 1.2 and 2.6 seconds respectively. The limitations are related to the two manual filtering stages that exist in the reconstruction process. After the first propagation, the manual selection of Fourier transform components at the back focal plane can introduce others element that disturb the reconstructed plane.

4. Microscopic object analysis using DHM

In this section we study microscopic objects with regular forms starting from their Fraunhofer diffraction patterns obtained with DHM. Two types of analysis are considered: (i) analysis of objects according to their spatial distribution and (ii) analysis of individual objects.

4.1 Fourier transformation at the back focal plane

From Eq. (5) it can be seen that the wave field at the back focal plane $\tilde{o}(u,v)$ is proportional to the Fourier transform of the objects except for the spherical wave front S_{φ}, which represents a quadratic phase curvature factor that causes a phase error if the optical Fourier transformation is computed (Poon, 2007).
Using DHM it is possible to find the exact Fourier Transform of objects at the back focal plane. The complex conjugate of the constant phase factor $S_{\phi}(u,v)$, can be expressed through the well-known parameters of the experimental design presented in figure 1, i.e. d', D and f,

$$\tilde{S}_{\phi}(u,v)=\frac{i}{\lambda f}\exp\left[\frac{-ik}{2(D+d')}\left(u^2+v^2\right)-f(f+2)\right] \tag{12}$$

Multiplying Eq. (12) by Eq. (5), the constant phase factor is eliminated and the exact Fourier Transform at back focal plane $\Im_{BFP}$ is obtained,

$$\Im_{BFP}(u,v)=\psi^{f}_{SFTF}(u,v;z=D)\,\tilde{S}\phi(u,v)=\Im\left[o(x_o,y_o,\lambda f)\right] \tag{13}$$

Calculating the intensity distribution from Eq. (13), the object's Fraunhofer diffraction pattern $I_{FDP}(u,v)$ is obtained,

$$I_{FDP}(u,v)=\left|\Im_{BFP}(u,v)\right|^2 \tag{14}$$

Eq. (14) offers a powerful tool in microscopic analysis because the Fourier Transform plane can be manipulated and different techniques of Fourier optics can be applied digitally, such as pattern recognition, image processing and others.

4.2 System magnification

The knowledge of system magnification is important when quantitative relations between lineal dimensions of the enlarged image and the microscopic object have to be known. In DHM the total system magnification depends on where the camera CCD is placed.
The image and the back focal planes are related by a Fourier transformation, thus lineal distance in the image plane can be extracted by the reciprocal of the corresponding lineal distance in the focal plane. In the working conditions with the capture of one hologram, figure 14a, the Fraunhofer pattern is reconstructed, figure 14b. A micrometric scale *Mitutoyo®* with 100 lines per mm was used as object.
The magnification of the system $M_T = d_i/d_o$ can be determined by the relation between two distances d_i and d_o in the image and object plane respectively. We determine the distance $d_i = 0.89\ mm$ between two bar in the image plane as the reciprocal value of the measured distance $d_f = 1.12\ mm$ between two contiguous diffraction points on the Fraunhofer pattern.

This method for determining d_i is more accurate that their direct measure over the image plane. With the knowledge that the distance between the two bar in the object plane is d_o = $10\ \mu m$, $M_T = 0.89/0.010 = 89$.

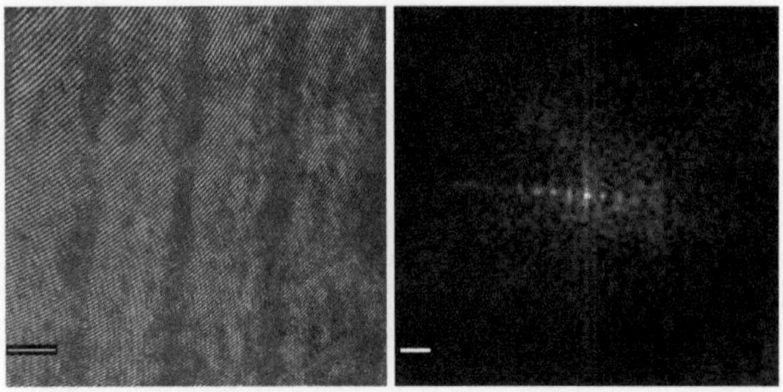

(a) (b)

Fig. 14. Digital hologram, scale bar 445 μm (a), Fraunhofer pattern, scale bar 2 mm^{-1} (b)

4.3 Analysis of objects according to their distribution

We consider objects with regular forms and two different forms of spatial distribution: randomly and periodically distributed. The objects' parameters can be determined by diffraction pattern manipulation in a simple and accurate way. This is an example of objects analysis that is important to biological and materials sciences.

4.3.1 Similar objects in random arrangement

We consider a sample of mouse blood cell as an example of random distribution of similar objects and use the proposed methodology to determine the diameter of cells. Figure 15a shows the hologram recorded with the experimental set-up.

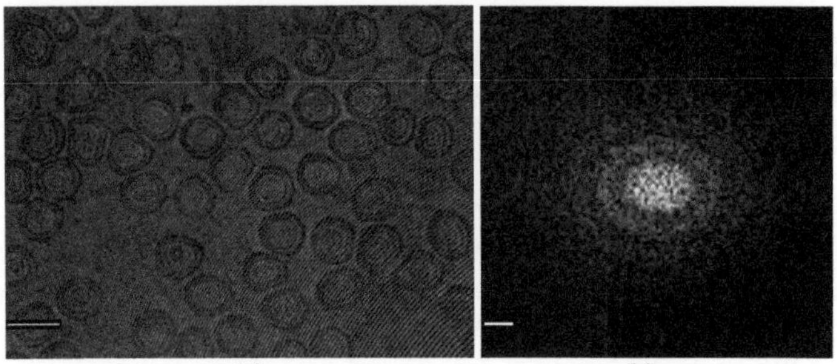

(a) (b)

Fig. 15. Digital hologram (a), Fraunhofer pattern, scale bar $2\ mm^{-1}$ (b).

Applying Eq. (14) the Fraunhofer pattern is obtained, figure 15b. As predicted by theory, the Fraunhofer diffraction pattern has a 'spotty' interference pattern, with a central peak and intensity in the diffraction plane that shows random fluctuations on a general background.

The radial intensity distribution $I_R(r)$, figure 16-upper, is measured by scanning the Fraunhofer diffraction pattern along radial lines (Palacios et al., 2001).

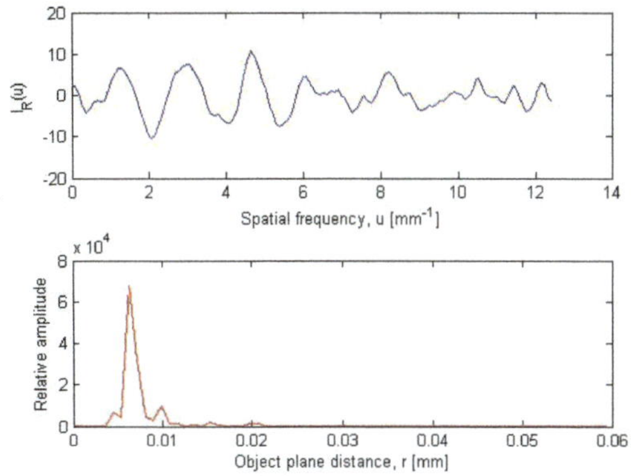

Fig. 16. Upper- the radial intensity curve $I_R(r)$, lower- spectrum of $I_R(r)$.

For each r value the intensity $I_R(r)$ is the result of averaging the intensity values $I_{FDP}(u,v)$ along the circumference from 0° to 360°, mathematically this operation can be represented by the expression,

$$I_R(r) = r^3 \frac{\sum_{\theta=1}^{360} I\left(x' = C_x(r,\theta), y' = C_y(r,\theta), d' = D\right)}{360} \tag{15}$$

where, $C_X(r,\theta) = N/2 + r\cos\theta$, $C_y(r,\theta) = N/2 - r\sin\theta$, with $0^o < \theta \leq 360^o$ and $0 < r \leq N/2$.
The spatial coordinates in the Fraunhofer diffraction pattern are defined on basis of the Fraunhofer diffraction pattern pixel resolution Δu, which is determined directly from the Fresnel diffraction formula at the reconstruction distance $z = D$. In this way, the radial distance $u=j\Delta u$, $j=0,1,\ldots,Np$, where Np is the points number of the radial intensity curve and $\Delta u= \lambda D/M\Delta\xi$. In the frequency spectrum, the spatial frequency is $f_u=j/Np\Delta u$, $j=0,1,..Np$.
The spectral analysis of the radial intensity curve $I_R(r)$ is carried out by the calculation of the square of the modulus of its 1D Fourier transform. In the resulting spectrum, figure 16-lower, the harmonic components are seen. As seen, only one fundamental harmonic that characterizes the diameter ($r_o = 6.3\ \mu m$) of the mouse blood cell appears.

4.3.2 Similar objects in periodical arrangement

In this section we consider a sample of periodically hexagonal structures inscribed on a plastic material with an ion beam as an example of regularly repeated identical objects.
In figure 17 is shown the hologram and intensity image reconstruction with the parameters for a hexagonal real space lattice.
In the reconstructed intensity image the parameters for the hexagonal real space lattice on the image plane are shown: diffraction angle for hexagonal crystal $\gamma = 120^o$ and the unit cell parameters a, b which for hexagonal crystal meet the condition, $a = b$.
The Fraunhofer diffraction pattern is calculated and shown in figure 18a. In figure 18b a section of the reciprocal lattice is specified and drawn through the diffraction points, the parameters a^*, b^* and γ^* are shown.

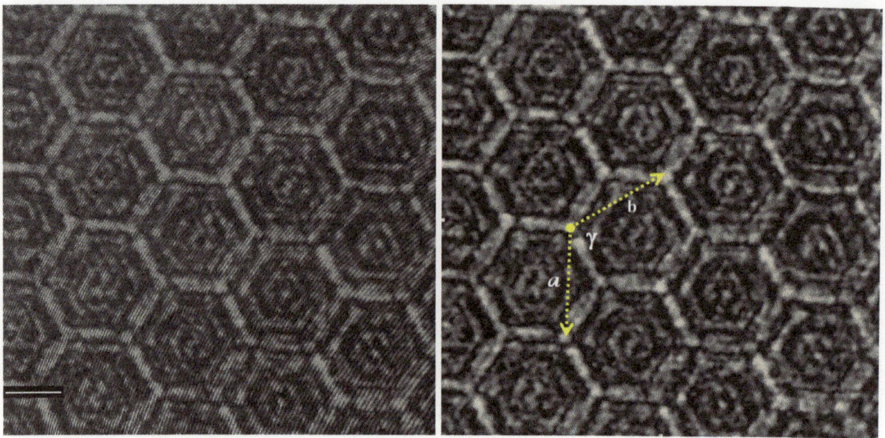

(a) (b)

Fig. 17. Digital hologram (a) and intensity image (b) of the sample. Scale bar $5\mu m$

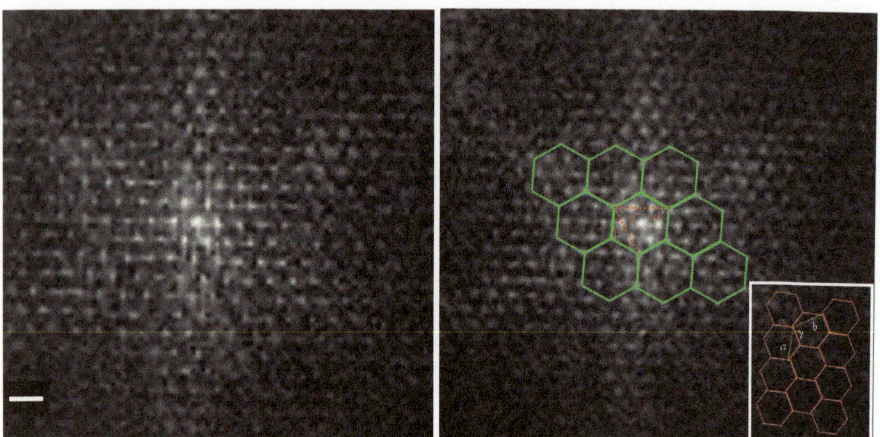

(a) (b)

Fig. 18. Fraunhofer diffraction pattern (a), scale bar 2 mm^{-1}. Over diffraction pattern it is drawn the reciprocal lattice (b). (*Inset: Representation of the real space*)

As shown in figure 18b, according to the theory of crystal diffraction, the reciprocal lattice edges of dimensions a^* and b^* are respectively perpendicular to the cell edges b and a, and quantities are related by the following expressions,

$$a^* = \frac{K}{a\sin\gamma^*} \; b^* = \frac{K}{b\sin\gamma^*} \text{ and } \gamma^* = 180^o - \gamma \tag{16}$$

where $K = \lambda L$ is the constant of diffraction, where λ the wavelength of monochromatic radiation and L is the camera length, i.e., the distance from the specimen to the diffraction plane. Applying the relations (16) the diffraction angle in the reciprocal lattice $\gamma^* = 60^o$ is determined and by measuring the parameters $a^* = b^* = 1.15\ mm$ in the reciprocal lattice the parameters $a' = b' = 863\ \mu m$ are calculated. It can be noticed that a' and b' are the real lattice parameters, but in the image plane. The parameters a, b of real lattice in the object plane are obtained dividing a' and b' by the total system magnification $M_T = 89$, i.e., $a = b = a'/M_T = b'/M_T = 9.69\ \mu m$. This value coincides with that obtained by AFM.

4.4 Calculation of nucleated cell dimensions

We demonstrate in this section the potentialities of Digital Holographic Microscopy in the determination of morphological parameters of nucleated cells. The spectral analysis of the radial behaviour of the Fraunhofer diffraction pattern allows the correlation between the peaks observed in the spectra and lineal dimensions of the cell.

As an example of application a sample of oral mucosa epithelial cell was selected. These cells have a nucleus inside a regular cytoplasm, figure 19.

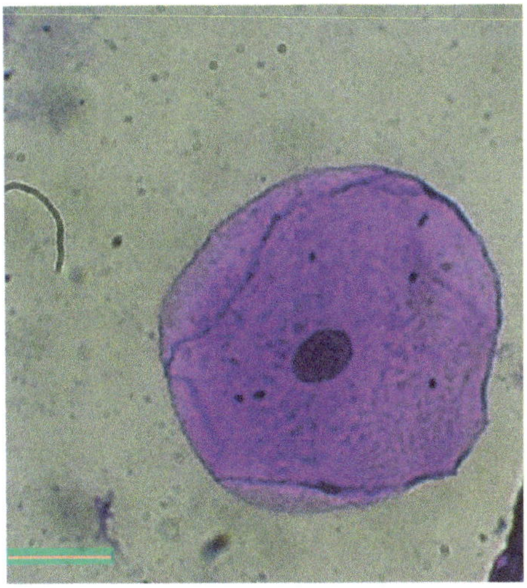

Fig. 19. Oral mucosa epithelial cell, optical microscopy image. Scale bar $20\ \mu m$

Two cases were analyzed, (i) the nucleus of the cell is located approximately at the centre of the cell and (ii) the nucleus is outside the centre.

Figure 20 shows the hologram (left) and the phase image reconstruction (right) of cell with the nucleus at the centre (A) and outside of centre (B). Both holograms were captured with the apparatus of figure 1 and the phase image was obtained by the calculation of the argument of Eq. (9) with $d' = 0$.

For both holograms the Fraunhofer pattern was calculated and is shown in figure 21 (left). Applying the spectral analysis of the Fraunhofer pattern radial intensity, the corresponding frequency spectra are obtained, figure 21 (right). In each spectrum a sequence of peaks are seen. In the case of cell with nucleus at the centre (A) four peaks appear in the spectrum, which is different for the case of the cell with nucleus outside the centre (B) where six peaks are observed in the spectrum.

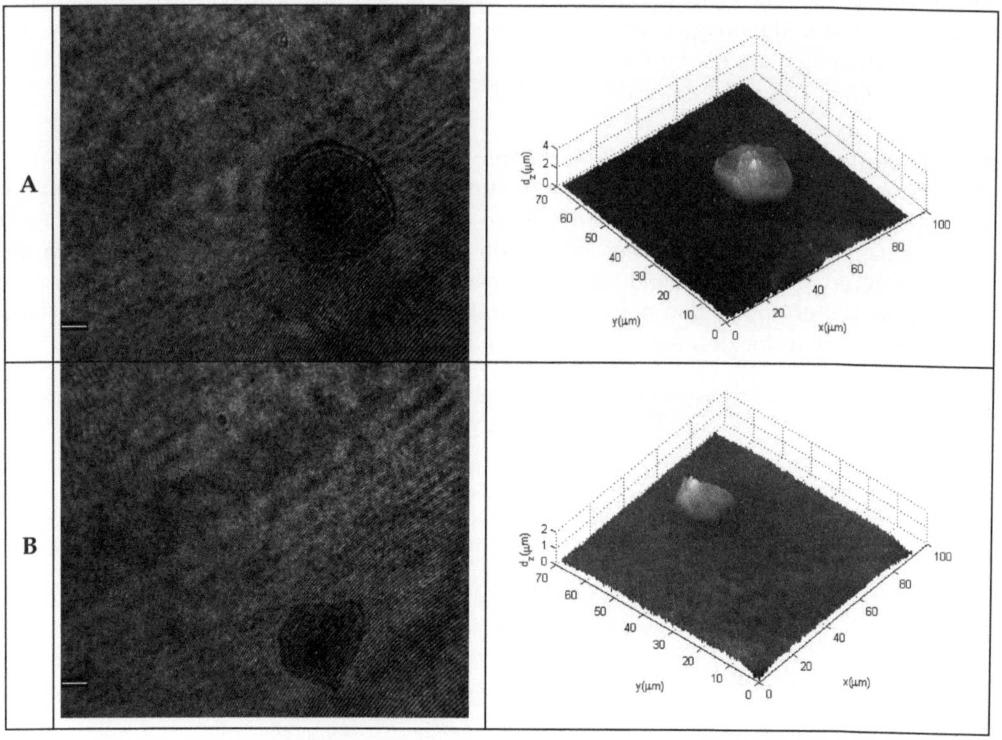

Fig. 20. Hologram (left) and phase image reconstruction (right) of cell with the nucleus at the centre (A) and outside of centre (B). Scale bar *20 μm*

The correlation between the peaks observed in the spectra and the diameter of cytoplasm and nucleus as well as other dimensions of the cell is definite, i.e., the peak position in the spectra is related with a lineal dimension in the cell.

In the case of the cell with the nucleus at the centre the peaks mean: peak (1) is characteristic of the nucleus diameter Dn, peak (2) characterizes half of the difference of both diameters, (Dc-Dn)/2, peak (3) characterizes half of the sum of both diameters, (Dc+Dn)/2 and peak (4) is characteristic of the cytoplasm diameter Dc.

If we consider the case of a cell with eccentric nucleus, which is shifted a distance *e* from the centre of cytoplasm two peaks correspond to the nuclear and cytoplasm diameters, (1) and (6) respectively, and the other four to half of the sums and differences of these two plus/minus the eccentricity 'e': peak (2) → (Dc-Dn)/2 – e, peak (3) → (Dc+Dn)/2 – e, peak (4) → (Dc-Dn)/2 + e, peak (5) → (Dc+Dn)/2 + e. The equivalent intervals in the cell are shown in figure 22.

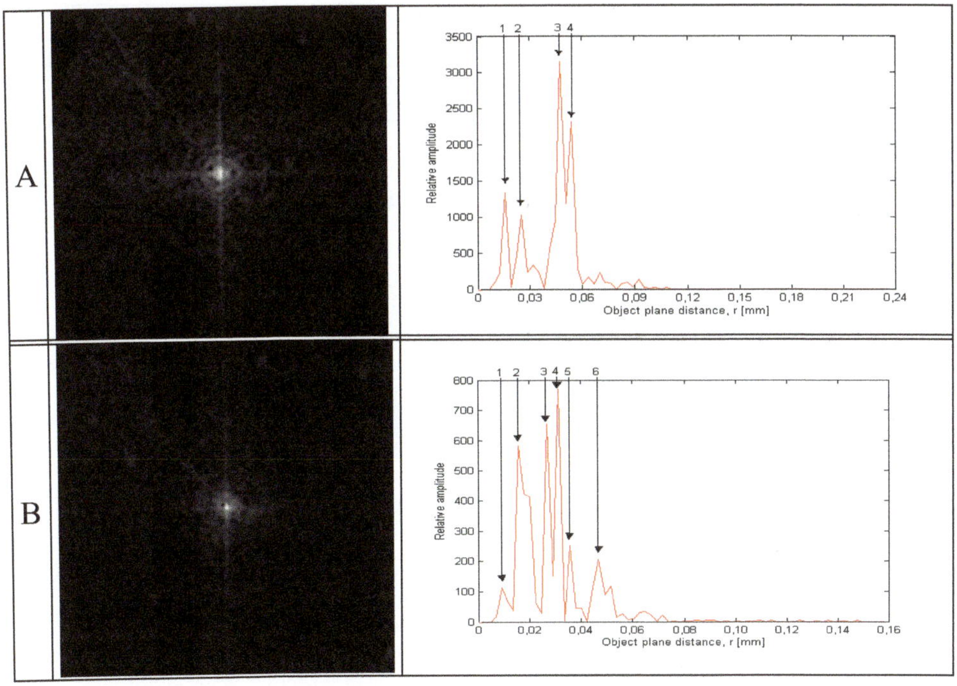

Fig. 21. Fraunhofer pattern (left) and the corresponding spectrum of their radial intensity (right) of cell with the nucleus at the centre (A) and outside of centre (B).

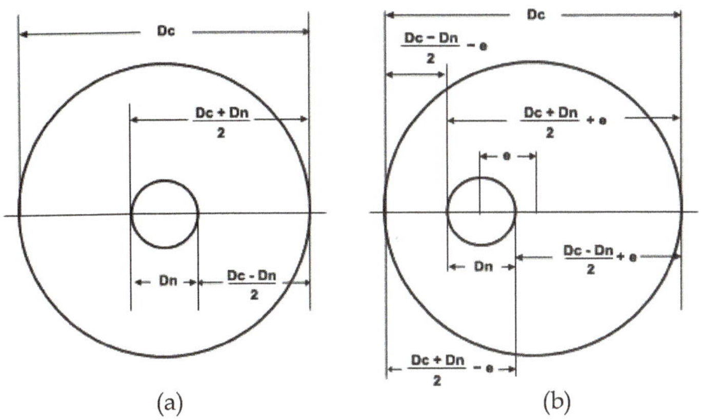

Fig. 22. Cellular dimensions reflected in the diffraction patterns: diagram (a) and (b) correspond to lineal dimensions of spectrums represented in figure 21 (right).

These results agree with theoretical predictions and experimental test presented in (Türke et al., 1978), although with the proposed method simpler the recording process and data processing are simpler.

4.5 Object detection

In this section we demonstrate that object detection is another example of objects analysis that can be realized by the examination of Fraunhofer diffraction patterns obtained by DHM.

The hologram of a microscopic character with the shape of a zero '0', figure 23a, is used to calculate the phase image shown in figure 23b.

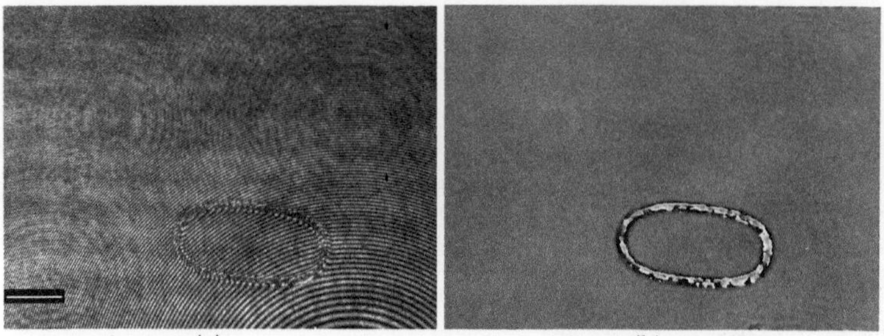

(a) (b)

Fig. 23. Hologram of a microscopic character with zero shape (a), scale bar: *50 μm*. Phase image reconstruction (b)

The microscopic character with zero shape will be detected when mixed with other microscopic characters. Figure 24 shows two holograms and the respective phase image reconstructions corresponding to microscopic characters with two different shapes: twenty '20' (A) and ten '10' (B). Here besides the microscopic character other elements exist to complicate the detection process.

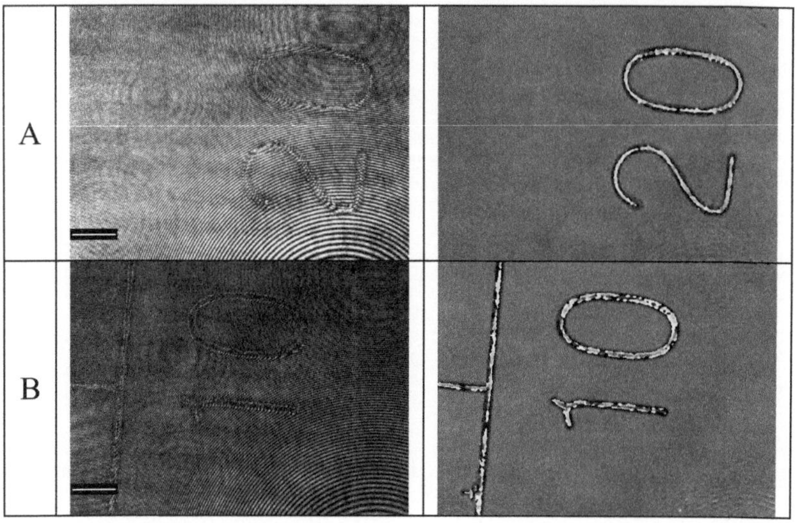

Fig. 24. Holograms, scale bar: *50 μm* (left) and phase image reconstruction (right) corresponding to microscopic characters with two different shapes: twenty '20' (A) and ten '10' (B).

The hologram of a microscopic character with zero shape, figure 23a, is used to calculate the Fraunhofer diffraction pattern shown in figure 25a. The radial intensity curve $I_R(u)$ and the corresponding frequency spectrum are shown in figure 25b.

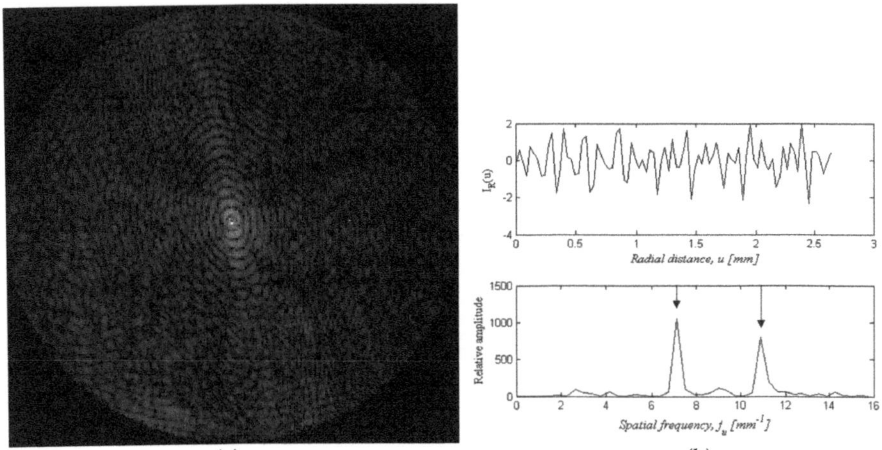

(a) (b)

Fig. 25. Fraunhofer diffraction pattern of the microscopic character with zero shape (a) and the radial intensity curve (b)-upper and the corresponding frequency spectrum (b)-lower.

According to figure 25a, the Fraunhofer diffraction pattern consists of elliptically shaped fringes with similar forms to that of the zero character. The fringes' width variation along their perimeter is detected by the radial intensity curve and with the subsequent spectral analysis the largest and smallest radius of the zero character can be obtained. The two peaks are signaled with arrows in the spectrum, figure 25b-down. They are positioned at the spatial frequencies $7.12\ mm^{-1}$ and $11.10\ mm^{-1}$, that represent the biggest and smallest radius of the zero character respectively. According to the peaks' spatial frequencies, the largest and the smallest radii are $140\ \mu m$ and $90\ \mu m$ respectively.

For object detection the holograms of a microscopic characters with '20' and '10' shapes, figure 24 (left), were used to calculate the Fraunhofer diffraction patterns shown in figure 26 (left). The radial intensity curve $I_R(u)$ and the corresponding frequency spectrum is shown in figure 26 (right).

In the frequency spectrum shown in figure 26 (right) appear, mixed with other peaks, two peaks at the same spatial frequency as those that are characteristic of the zero shape frequency spectrum. In this way, the object detection can be generalized for other objects with irregular forms, because the spectrum of the radial curve of the object's diffraction pattern presents a sequence of peaks that characterize the form of the object. A unique spectrum is associated with each form.

As has been shown, this method of object detection is similar to the qualitative analysis in x-ray diffraction, i.e. the presence of an object is characterized by the presence of peaks in appropriate positions in the spectrum. This analogy is very important because all the developed tools for the qualitative analysis in x-rays can be used for object detection.

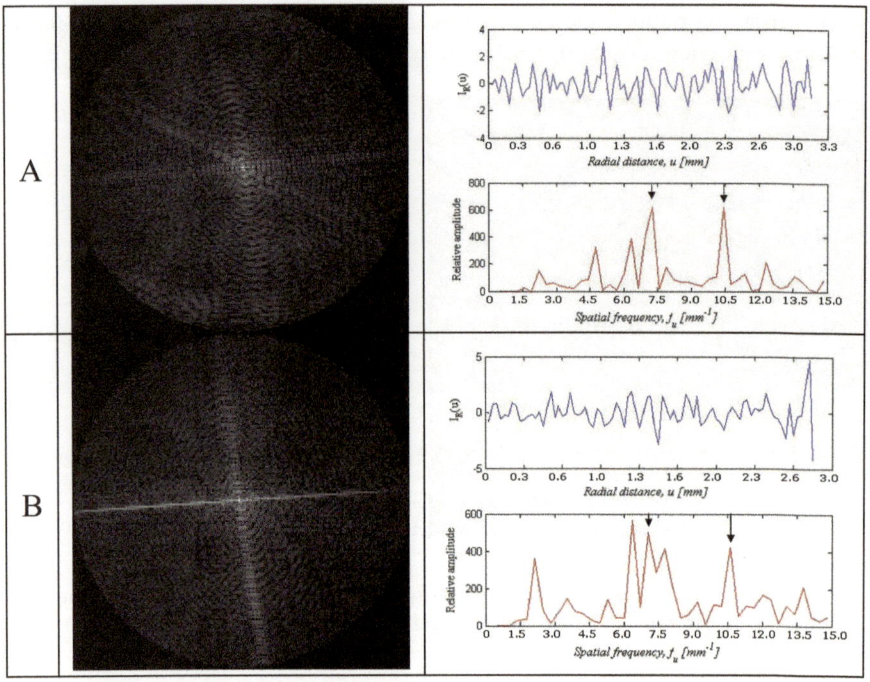

Fig. 26. Fraunhofer diffraction patterns of corresponding hologram of figure 24 (left) and the radial intensity curve with the corresponding frequency spectrum (right).

5. Conclusion

In this chapter, an alternative methodology for image reconstruction and object analysis in digital holographic microscopy is discussed. The model we use for image reconstruction is specific for digital holographic microscopy; it includes a first propagation from the hologram plane to obtain the Fourier Transform plane of the object at the back focal plane of the microscope objective lens and then a second propagation is applied to find out the image wave field at the reconstruction plane. Using digital hologram reconstruction as a method for calculating the amplitude and intensities distributions of the optical field on the Fourier Transform plane microscopic objects with regular forms are studied. Random and periodical distributions of objects are considered. The lineal dimensions of the objects are determined by manipulating the diffraction pattern in a simple and accurate way. The advantages of this methodology are summarized as follows: (a) using a single hologram the phase image is calculated with simple operation for phase curvature correction, (b) there are no limitations in the minimum reconstruction distance, (c) capability to maintain the size of a reconstructed image, independent of the reconstruction distance and wavelength for objects larger than a CCD, and (d) the spectral analysis of the radial intensity curve of the Fraunhofer diffraction pattern allows the determination of the lineal dimensions of the objects.

6. Acknowledgment

This work is supported by the Brazilian research agencies FAPESP and CAPES, University of Oriente, Cuba and the National University of Colombia.

7. References

Ankit, K. & Rabinkin, D. (2007). Edited by Luk, Franklin T. Proceedings of the SPIE, Volume 6697, pp. 66970E-66970E-17.

Colomb, T.; Kühn, J.; Charrière, F.; Depeursinge, C.; Marquet, P. & Aspert, N.(2006). Opt. Express 14, 4300.

Cuche, E.; Marquet, P. & Depeursinge, C. (1999). Appl. Opt. 38, 6994–7001.

Cuche, E.; Marquet, P. & Depeursinge, C. (2000). Appl. Opt. 39, 4070–4075.

De Nicola, S.; Finizio, A.; Pierattini, G.; Ferraro, P. & Alfieri, D. (2005). Opt. Express 13, 9935–9940.

Demetrakopoulos, TH.; Mittra, R. (1974). Appl Opt., 13:665-670.

Ferraro, P.; Nicola, S. D.; Finizio, A.; Coppola, G.; Grilli, S.; Magro, C. & Pierattini, G. (2003). Appl. Opt. 42, 1938–1946.

Goodman, J. W. (1968). McGraw-Hill, San Francisco, Calif.

Grilli, S.; Ferraro, P.; Nicola, S. D.; Finizio, A.; Pierattini, G. & Meucci, R. (2001). Opt. Express 9, 294–302.

Kreis, T.; Adams, M. & Jüptner, W. (1997). Proc. SPIE 3098, 224.

Lim, J. S. (1990). Englewood Cliffs, NJ, Prentice Hall, pp. 218-237.

Lipson, S. G. & Lipson, H. (1981) Second edition. Cambridge University Press.

Miccio, L.; Alfieri, D.; Grilli, S.; Ferraro, P.; Finizio, A.; Petrocellis, L. D. & Nicola S. D. (2007). Appl Phys. Lett. 90, 041104-1.

Murata, S. & Yasuda, N. (2000). Opt. Laser Technol. 32, pp. 567.

Palacios, D.; Palacios, F.; Sajo-Bohus, L. & Pálfalvi, J. (2001). J. Rad. Meas. 34, pp. 119-122.

Palacios, F.; Palacios, D.; Palacios, G.; Gonçalves, E.; Valin, J.; Sajo-Bohus, L. & Ricardo, J. (2008). J. Opt. Comm., Vol. 281, Issue 4, 15, pp 550-558.

Palacios, F.; Ricardo, J.; Palacios, D.; Gonçalves, E.; Valin, J. & De Souza, R. (2005) Opt. Comm. Vol. 248/1-3, pp 41-50.

Pan, G. & Meng, H. (2003). Appl. Opt. 42, 827.

Pedrini, G.; Schedin, S. & Tiziani, H. J. (2001). J. Mod. Opt. 48, 1035–1041.

Poon, T.C. (2007). *Optical Scanning Holography with Matlab®*, Springer Science+Business Media, LLC.

Schedin, S.; Pedrini, G.; Tiziani, H.; Aggarwal, A. & Gusev, M. (2001), Appl. Opt. 40, pp. 100.

Schilling, B.W.; Poon, T.C.; Indebetouw G.; Storrie B.; Shinoda K.; Suzuki Y. & Wu M.H. (1997). Opt. Lett. 22, pp. 1506-1508.

Schnars, U. (1994). J. Opt. Soc. Am. A 11, 2011.

Stadelmaier, A. & Massig, J. H. (2000). Opt. Lett. 25, 1630–1632.

Takaki, U. & Ohzu, H. (1999). Appl. Opt. 38, pp. 2204.

Türke, B.; Seger, G.; Achatz M. & Seelen W. (1978). App. Opt. 17:2754-2761.

VanLigten, R. F. & Osterberg, H. (1966). Nature, 211, pp. 282-283.

Wagner, C.; Osten, W. & Seebacher, S. (2000). Opt. Eng. 39, pp. 79.

Yamaguchi, I.; Matsumura, T. & Kato, J. (2002) Opt. Lett. 27, 1108.

Yaroslavsky, L. & Merzyalov, N. (1980), Consultants Bureau.

Yu, L. & Kim, M.K. (2005). Opt Lett 2005, 30:2092.
Yu, L.; Mohanty, S.; Zhang, J.; Genc, S.; Kim, M.K.; Berns, M.W. & Chen, Z. (2009) Opt. Express, Vol 17, No. 14.
Zhang, F.; Yamaguchi, I. & Yaroslavsky, L. P. (2004). Opt. Letters, Vol. 29, No. 14.

Part 3

Imaging

10

Study of Liquid Crystal on Silicon Displays for Their Application in Digital Holography

Angel Lizana[1], Laura Lobato[1], Andrés Márquez[2], Claudio Iemmi[3], Ignacio Moreno[4], Juan Campos[1] and María J. Yzuel[1]
[1]*Universitat Autònoma de Barcelona*
[2]*Universidad de Alicante*
[3]*Universidad de Buenos Aires*
[4]*Universidad Miguel Hernández*
[1,2,4]*Spain*
[3]*Argentina*

1. Introduction

Due to the significant capability of Liquid Crystal Displays (LCDs) to spatially manipulate the phase information of an incident light beam, this technology is been widely applied in a large number of optical applications. Nowadays, they are employed as Spatial Light Modulators (SLMs) in many areas, as for instance, in Optical Image Processing (Liu et al., 1985), in Holography Data Storage (Coufal et al., 2000), in Programmable Adaptive Optics (Dou & Giles, 1995), in Medical Optics (Twietmeyer et al., 2008), or in Diffractive Optics (Márquez et al., 2005), among others.

Recently, a new type of reflective LCDs, the Liquid Crystals on Silicon (LCoS) displays, have awakened a great interest due to their specific technical characteristics, which in general, are superior in many aspects to the ones provided by transmissive LCDs (Lee et al., 2004; Wu & Yang, 2005). For instance, as LCoS displays work in reflection, the light impinging such devices performs a double pass through the LC cell, leading to a larger phase modulation than that related to transmissive LCDs with the same thickness. This greater phase modulation capability allows LCoS displays to become very suitable devices for digital holography applications, as for instance, for laser beam shaping (Dickey et al., 2005; Rodrigo et al., 2011) or for optical micro-particle manipulation (Ashkin, 2006).

To maximize the efficiency of digital holograms generated by using LCoS displays, it is required to apply a suitable methodology for optimizing the performance of these devices when generating the specific phase and amplitude distributions. Nowadays, there exist different theoretical models proposed to improve the performance of LCDs (Azzam & Bashara, 1972; Gagnon, 1981; Márquez et. al., 2001). In general, most of these models are based on mathematical formalisms describing fully polarized light, as the Jones formalism (Jones, 1941) or the Berreman formalism (Berreman, 1972). However, some authors have discovered that LCoS displays may introduce non-negligible values of effective depolarization at the reflected beam (Lizana et al., 2008a; Márquez et al., 2008; Wolfe & Chipman, 2006), which are originated by the electrical addressing schemes applied in these devices (Hermerschmidt et al., 2007). This effective depolarization depends on the incident

angle (Lizana et al., 2009; Verma et al., 2010), the wavelength (Lizana et al., 2008c) and the state of the polarization of the incident beam (Márquez et al., 2008). In this situation, the application of an optimizing method that includes the unpolarized light contribution observed when working with LCoS display becomes mandatory.
In Ref. (Lizana et al., 2008b), another undesired phenomenon originated by the electrical addressing schemes applied in LCoS displays is reported. We refer to the time-fluctuations of the phase phenomenon, which may notably degrade the efficiency of the digital holograms generated with LCoS displays (Lizana et al., 2008b). Thus, to maximize the efficiency of digital holograms addressed to these devices, different ways to reduce the undesired influence of this damaging phenomenon must be applied.
This Chapter presents a study based on LCoS displays which can be useful as a guideline to optimize the performance of these devices for the generation of digital holograms. In section 2, a characterization and optimization methodology, based on Mueller-Stokes (M-S) formalism is described. This methodology considers the effective depolarization values observed in LCoS displays because the M-S formalism is able to describe fully polarized light, partial polarized light and unpolarized light contributions. In section 3, experimental evidences of the time-fluctuations of the phase phenomenon are presented and its effects on the generation of digital holograms are reviewed. In section 4, a method based on the minimum Euclidean distance principle, devised to reduce the undesired influence of this phenomenon is proposed and experimentally tested by analyzing the efficiency of different optimized digital holograms addressed to an LCoS display. Finally, conclusions are given in section 5.

2. Liquid Crystal on Silicon displays characterization and optimization methodology

In this section, we propose a characterization-optimization method, based on a combination of the Mueller-Stokes (M-S) formalism (Parke, 1949) and the Jones formalism (Jones, 1941), which is able to predict and optimize the intensity and the phase response of LCoS displays. The characterization of the LCoS display is conducted as a function of the addressed gray level (which is related to the applied voltages), by experimentally calibrating its corresponding Mueller matrices, which are able to describe depolarizing samples. In this situation, the intensity response of the LCoS display can be predicted but since Mueller matrices are calculated from intensity measurements, no information about the phase modulation response is obtained. As Jones matrices contain the required phase information, a conversion from Mueller matrices to Jones matrices is applied. To perform a proper conversion, a polar decomposition (Lu & Chipman, 1996; Foldyna et al., 2009) of the experimentally obtained Mueller matrices is carried on.
Once the intensity and phase response of the LCoS display can be predicted, an optimization procedure is applied, in order to get the desired LCoS display behaviour, i.e. to work as amplitude-only modulator or as phase-only modulator. In next subsections, the characterization and optimization methodology is described in detail and some results are provided.

2.1 LCoS display Mueller matrix characterization: Intensity prediction

The LCoS display under analysis is a commercial Philips LCoS model X97c3A0, sold as the kit LC-R2500 by HoloEye Systems. This prototype presents a monochrome reflective

Twisted Nematic (TN) LCoS display of 2.46 cm diagonal, in which the twist angle of LC is 45 degrees. The resolution of the display is XGA (1024x768 pixels), with a fill factor of 93%. The pixels are square and they are separated by a distance (center to center) of 19 μm. In addition, the device provides digitally controlled gray scales with 256 gray levels (8 bits).
To determine the Mueller matrix of the TNLCoS display, the set-up sketched in Fig. 1 is used.

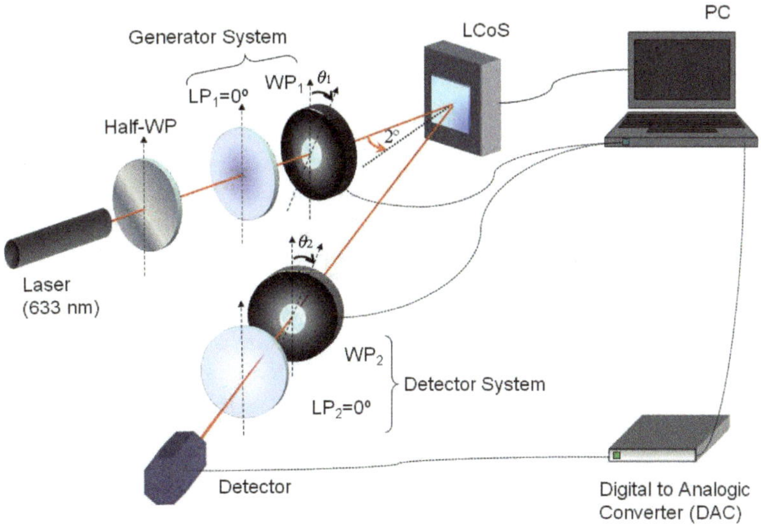

Fig. 1. Mueller matrix characterization set-up.

The TNLCoS display is illuminated by a linear polarized He-Ne laser source with a wavelength equal to 633 nm. In the incident beam, just following the laser source, is set a Half-Waveplate (HWP) that allows us to control the intensity of light transmitted by the linear polarizer LP_1. After the HWP element, a Polarization State Generator (PSG), which is formed by a linear polarized (LP_1) fixed at 0 degrees of the laboratory vertical and a Quarter-Waveplate (QWP_1), is placed. The QWP_1 is inserted into a rotating platform that allows us to electronically control the orientation of its fast axis, generating at each different orientation a different incident State of Polarization (SoP).
The angle between the incident and the reflected beams is equal to 4 degrees and so, it can be assumed that the TNLCoS display is operating under quasi-normal incidence. In the light beam reflected by the TNLCoS display is set a Polarization State Detector (PSD), which is formed by Quarter-Waveplate (QWP_2) followed by a linear analyzer (LP_2). Different orientations of the QWP_2 allow us to analyze different SoPs.
Finally, the radiometric measurements are performed by means of a commercial radiometer (Newport 1830-C) placed at the exit of the PSD. The combination of the PSD and the radiometer results in a complete Stokes polarimeter configuration (Chipman, 1995), which is capable of measuring any state of polarization exiting from the TNLCoS display.
The Mueller matrix M of a polarizing sample relates the incident and exiting (reflected, transmitted or scattered) states of polarization, described by the Stokes vectors S_{in} and S_{ex} respectively:

$$S_{ex}=M\,S_{ex}=\begin{pmatrix} m_{00} & m_{01} & m_{02} & m_{03} \\ m_{10} & m_{11} & m_{12} & m_{13} \\ m_{20} & m_{21} & m_{22} & m_{23} \\ m_{03} & m_{31} & m_{32} & m_{33} \end{pmatrix} S_{in} \tag{1}$$

where $m_{j,k}$ ($j = 0, 1, 2, 3$ and $k = 0, 1, 2, 3$) are the coefficients of the Mueller matrix.

Due to the linear relation shown in Eq. (1), the Mueller matrix M of a polarizing sample can be completely characterized by implementing a convenient independent linear equation system. In this way, the experimental LCoS display Mueller matrix is obtained by generating different incident SoPs (by means of the PSG in Fig. 1) and measuring their respective reflected ones by using different SoPs analyzers (implemented by means of the PSD in Fig. 1).

To completely characterize the 4x4 coefficients of the Mueller matrix describing the LCoS display, at least 4 independent incident SoPs and 4 independent analyzer SoPs must be used. Different sets of incident and analyzer SoPs are valid to this aim but the quality of the measurements will depend on different factors, such as the condition number (Taylor, 1974) of the matrices composed by the incident SoPs and the analyzer SoPs (De Martino et al., 2007; Peinado et al., 2010) or the number of SoPs selected (Peinado et al., 2010). Here, the calculation of the Mueller matrix is conducted as follows:

The intensity detected behind the PSD in Fig. 1 depends on the Stokes parameters of the light reflected by the LCoS display and on the orientation θ_2 of the quarter-waveplate QWP_2. It can be written as follows (Lizana et al., 2009):

$$I(\theta_2)=\frac{1}{2}\left[S_0+S_1\cos^2(2\theta_2)+S_2\sin(2\theta_2)\cos(2\theta_2)-S_3\sin(2\theta_2)\right] \tag{2}$$

where S_0, S_1, S_2 and S_3 are the Stokes parameters of the light reflected from the LCoS display. As the intensity in Eq. (2) is a periodical signal with respect to the angle θ_2, because it consists of different sinusoidal functions whose arguments are entire multiples of θ_2, the method of synchronous detection (Goldstein, 2003) is applied. The synchronous detection represents an estimation of the coefficients of the Fourier series of this function.

By performing a summation of intensities corresponding to N different equally spaced values of θ_2, completing a rotation of 360 degrees, some terms of Eq.(2) vanish due to the orthogonal properties of the sinusoidal sampled functions (Lizana et al., 2009). Besides, by performing another summation of different intensities, but now, multiplying the intensities by the sine or by the cosine of the corresponding angle θ_2, different terms in Eq. (2) vanish. In this way, the following mathematical expression for the Stokes parameters is obtained (Lizana et al., 2009):

$$\begin{pmatrix} S_0 \\ S_1 \\ S_2 \\ S_3 \end{pmatrix}=\begin{pmatrix} 2\cdot\sum_{r=1}^{N} I\left(\frac{\pi}{2},\theta_{2,r}\right)-4\cdot\sum_{r=1}^{N} I\left(\frac{\pi}{2},\theta_{2,r}\right)\cos(4\theta_{2,r}) \\ 8\cdot\sum_{r=1}^{N} I\left(\frac{\pi}{2},\theta_{2,r}\right)\cos(4\theta_{2,r}) \\ 8\cdot\sum_{r=1}^{N} I\left(\frac{\pi}{2},\theta_{2,r}\right)\sin(4\theta_{2,r}) \\ -4\cdot\sum_{r=1}^{N} I\left(\frac{\pi}{2},\theta_{2,r}\right)\sin(2\theta_{2,r}) \end{pmatrix} \tag{3}$$

where N is the number of selected angles θ_2 and $\theta_{2,r}=\frac{2\pi(r-1)}{N}$ with $r = 1, 2,\ldots, N$.

Once the SoP reflected by the LCoS display can be measured as described above, the next step is to fully characterize the LCoS display Mueller matrix. Next, the applied technique is described. The PSG in Fig. 1 is formed by the linear polarizer LP_1, fixed at 0 degrees to the laboratory vertical, followed by the quarter-waveplate QWP_1. The general expression for the states of polarization exiting from the PSG can be obtained by multiplying the Stokes vector of linearly polarized light in the horizontal direction (i.e. $S_{P1} = (1, 1, 0, 0)^T$) by the Mueller matrix of a linear retarder with a retardance equal to 90 degrees and whose orientation depends on the angle θ_1 (Goldstein, 2003). We obtain the following Stokes vector:

$$S_{PSG}=\begin{pmatrix}1 & 0 & 0 & 0\\ 0 & \cos^2 2\theta_1 & \sin 2\theta_1 \cos 2\theta_1 & -\sin 2\theta_1\\ 0 & \sin 2\theta_1 \cos 2\theta_1 & \sin^2 2\theta_1 & \cos 2\theta_1\\ 0 & \sin 2\theta_1 & -\cos 2\theta_1 & 0\end{pmatrix}\cdot\begin{pmatrix}1\\1\\0\\0\end{pmatrix}=\begin{pmatrix}1\\ \cos^2 2\theta_1\\ \sin 2\theta_1 \cos 2\theta_1\\ \sin 2\theta_1\end{pmatrix} \tag{4}$$

where different incident States of Polarization (SoPs) can be generated by varying the orientation angle θ_1 of the QWP_1. By replacing the Stokes vector given in Eq. (4) into Eq. (1), a general expression for the SoP exiting from the LCoS display is obtained:

$$S_k(\theta_1)=\left(m_{k0}+\frac{m_{k1}}{2}\right)+\frac{m_{k1}}{2}\cos 4\theta_1+\frac{m_{k2}}{2}\sin 4\theta_1+m_{k3}\sin 2\theta_1 \tag{5}$$

where k (k = 0, 1, 2, 3) indicates the different Stokes parameter and $m_{k,j}$ (j = 0, 1, 2, 3) are the different coefficients of the Mueller matrix.

Equation (5) comprises different sinusoidal functions whose arguments are entire multiples of θ_1, and thus, the method of synchronous detection can be applied again. In particular, by performing a summation of different reflected SoPs corresponding to N different equally spaced values of θ_1 we obtain the following expression:

$$\sum_{r=1}^{N} S_k(\theta_{1,r})=\sum_{r=1}^{N}\left(m_{k0}+\frac{m_{k1}}{2}\right)+\sum_{r=1}^{N}\frac{m_{k1}}{2}\cos 4\theta_{1,r}+\sum_{r=1}^{N}\frac{m_{k2}}{2}\sin 4\theta_{1,r}+\sum_{r=1}^{N} m_{k3}\sin 2\theta_{1,r} \tag{6}$$

where N is the number of SoPs, r (r = 1,2,…,N) indicates the specific term in the summation and with $\theta_{1,r}=\frac{2\pi(r-1)}{N}$.

By performing a summation of different reflected SoPs related to N different equally spaced values of θ_1 on a complete rotation of 360 degrees, some terms of Eq. (6) vanish because of the orthogonal properties of the sinusoidal sampled functions (Lizana et al., 2009). Besides, by performing a summation of different reflected SoPs related to N different equally spaced values of θ_1, but now, multiplying these SoPs by the sine or by the cosine of the corresponding angle θ_1, different terms in Eq. (6) vanish. By doing this computation, an expression for the Mueller matrix of the LCoS display is obtained:

$$M=\frac{1}{N}\begin{bmatrix}\sum_{r=1}^{N}S_{0,r}-2\sum_{r=1}^{N}S_{0,r}\cos 4\theta_{1,r} & 4\sum_{r=1}^{N}S_{0,r}\cos 4\theta_{1,r} & 4\sum_{r=1}^{N}S_{0,r}\sin 4\theta_{1,r} & 2\sum_{r=1}^{N}S_{0,r}\sin 2\theta_{1,r}\\ \sum_{r=1}^{N}S_{1,r}-2\sum_{r=1}^{N}S_{1,r}\cos 4\theta_{1,r} & 4\sum_{r=1}^{N}S_{1,r}\cos 4\theta_{1,r} & 4\sum_{r=1}^{N}S_{1,r}\sin 4\theta_{1,r} & 2\sum_{r=1}^{N}S_{1,r}\sin 2\theta_{1,r}\\ \sum_{r=1}^{N}S_{2,r}-2\sum_{r=1}^{N}S_{2,r}\cos 4\theta_{1,r} & 4\sum_{r=1}^{N}S_{2,r}\cos 4\theta_{1,r} & 4\sum_{r=1}^{N}S_{2,r}\sin 4\theta_{1,r} & 2\sum_{r=1}^{N}S_{2,r}\sin 2\theta_{1,r}\\ \sum_{r=1}^{N}S_{3,r}-2\sum_{r=1}^{N}S_{3,r}\cos 4\theta_{1,r} & 4\sum_{r=1}^{N}S_{3,r}\cos 4\theta_{1,r} & 4\sum_{r=1}^{N}S_{3,r}\sin 4\theta_{1,r} & 2\sum_{r=1}^{N}S_{3,r}\sin 2\theta_{1,r}\end{bmatrix} \tag{7}$$

where N is the number of SoPs present in the summation, r (r = 1, 2,..., N) indicate the specific term of the summation and with $\theta_{1,r}=\frac{2\pi(r-1)}{N}$.
By experimentally measuring the quantities given in Eq. (7), the Mueller matrices of the Twisted Nematic LCoS display for different gray levels (from 0 to 255 in steps of 20 gray levels) have been obtained. The validity of the obtained results is tested by analyzing their capability to predict the SoP of the light reflected by the LCoS display.
As an example, in Fig. 2 we have plotted the Stokes parameters of the reflected SoP as a function of the gray level, when using an incident linear SoP at 135 degrees to the laboratory vertical. The spots represent the experimental data, which is measured by applying the SoP measuring method described above (see Eq. (2)). The continuous lines represent the theoretical values, which are calculated by multiplying the Stokes vector of a linear polarized light at 135 degrees (i.e. S_{135} = (1, 0, -1, 0)^T) by the measured Mueller matrices of the TNLCoS display (see Eq. (1)). In all the cases, there is an excellent agreement between the predicted and the experimentally obtained Stokes parameters. Thus, the results given in Fig. 2 become an evidence of the effectiveness of the measured TNLCoS display Mueller matrices and validates the characterization method proposed.
It should be pointed out that depending on the specific application of the LCoS display, different physical parameters, such as the incident angle or the wavelength, can be required. In such situations, the LCoS display response can be notably different. However, the characterizing method described in this section can be extended to different experimental conditions, such as the incident angle (Lizana et al., 2009) or the wavelength (Lizana et al., 2008c).

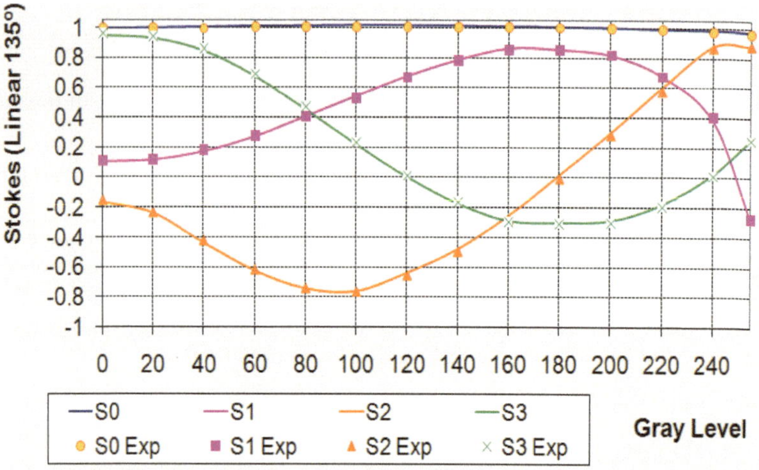

Fig. 2. Stokes parameters as a function of the gray level for an incident SoP linear at 135 degrees of the laboratory vertical. Whereas the spots represent the experimental data, the continuous lines represent the theoretical values.

2.2 Mueller to Jones conversion: Phase prediction

In this subsection, a methodology useful to predict the phase response of LCoS displays is proposed. To this aim, since Jones matrices contains the required phase information, a conversion of the experimentally obtained LCoS display Mueller matrices to Jones matrices is applied.

As stated above, LCoS displays introduce a certain amount of unpolarized light in the reflected beam (Lizana et al., 2008a; Márquez et al., 2008; Wolfe & Chipman, 2006). Since Jones matrices do not describe unpolarized samples, a direct conversion from the experimental Mueller matrices to the Jones matrices can not be directly performed. In this situation, a decomposition of the LCoS display Mueller matrices into different basic Mueller matrices becomes very helpful. There exist different ways to decompose depolarizing Mueller matrices (Ossikovski et al., 2008). All these different decompositions provide significant polarimetric information and they become more or less practical as a function of the specific application for which they are required.

Here, the Lu-Chipman decomposition (Goldstein, 2003), which is a natural generalization of the polar decomposition (Lu & Chipman, 1996) to the depolarizing case, is applied. The Lu-Chipman decomposition states that a general Mueller matrix can be decomposed as the product of three basic Mueller matrices: the Mueller matrix of a pure diattenuator M_D (an optical element that changes the orthogonal amplitudes unequally), the Mueller matrix of a pure retarder M_R (an optical element that introduces a phase-shift between the orthogonal components), and the Mueller matrix of a depolarizer M_Δ (an optical element that introduces certain amount of unpolarized light). In the case of LCoS displays, diverse polarimetric studies have proved that they are non-diattenuating devices (Lizana et. al, 2008c; Lizana et al., 2009; Márquez et al., 2008) and the corresponding M_D matrices can be approximated to the identity matrix. Therefore, when applying the Lu-Chipman polar decomposition to the LCoS display case, the following relation holds (Moreno et al., 2008):

$$M_{LCoS}=M_\Delta M_R=\begin{pmatrix}1 & \vec{0}^T\\ \vec{P}_\Delta & m_\Delta\end{pmatrix}\cdot\begin{pmatrix}1 & \vec{0}^T\\ \vec{0} & m_R\end{pmatrix} \quad (8)$$

where $\vec{0}^T=(0,0{,}0{,}0)$, $\vec{P}_\Delta$ refers to the polarizance vector (Goldstein, 2003), and the matrices denoted as m_Δ and m_R are 3x3 submatrices whose coefficients can be calculated by following the procedure described in (Goldstein, 2003).

The retarder component M_R of the Mueller matrix given in Eq. (8) is a non-absorbing, unitary and fully polarized Mueller matrix. Under these conditions, the equivalent Jones matrix of the retarder can be obtained from the M_R component (Moreno et al., 2008). It can be done by following the relationships between the Mueller matrix coefficients and the Jones matrix coefficients given in Ref. (Goldstein, 2003).

Once the equivalent Jones matrices of the TNLCoS are calculated, a methodology for the evaluation of the complex phase modulation is applied. We follow the technique proposed in Ref. (Moreno et al., 2003), being suitable to predict the phase response in LCDs. This technique considers that any non-absorbing reciprocal polarization device can be described by an unimodular unitary Jones matrix (Fernández-Pousa et al., 2000), the TNLCoS displays being a particular case. Then, such devices can be described by the following relation:

$$J_{LCoS}=e^{-i\beta}\begin{pmatrix}A & B\\ -B^* & A^*\end{pmatrix} \quad (9)$$

where $A=A_R-iA_I$ and $B=B_R-iB_I$ are complex parameters (subscripts R and I indicate the real and imaginary parts) which depend on the applied voltage V and fulfill the condition $|A|^2+|B|^2=A_R^2+A_I^2+B_R^2+B_I^2=1$.

From the calculated matrices M_R and the transformations given in (Goldstein, 2003), the complex parameters A and B are fully calibrated as a function of the gray level. However, no information about the external phase β in Eq. (9) is obtained. The knowledge of the external phase β is critical to accurately predict the phase modulation of LCoS displays and so, the application of a technique to determine its value becomes mandatory.

A general expression for the total phase δ of a LCD inserted between two polarizers (whose orientations are θ_1 and θ_2 respectively) is provided in Ref. (Moreno et al., 2003):

$$\delta=\beta+\delta_M=\beta+\operatorname{atan}\left(\frac{A_I\cos(\theta_1+\theta_2)+B_I\sin(\theta_1+\theta_2)}{A_R\cos(\theta_1-\theta_2)+B_R\sin(\theta_1-\theta_2)}\right) \tag{10}$$

where δ_M is a phase contribution that depends on the polarizer orientations θ_1 and θ_2.

As the coefficients A_R, A_I, B_R and B_I are known, the phase δ_M in Eq. (10) can be calculated for a given pair of orientations θ_1 and θ_2. Besides, the total phase δ as a function of the gray level can be experimentally determined, for instance, by using the interferometric based set-up given in Ref. (Lizana et al., 2008b). In this situation, the external phase β is readily obtained by isolating it from Eq. (10).

Since the external phase β does not depend on the incident and analyzer SoPs, once it is measured, the phase response of LCoS display as a function of the gray level can be completely predicted.

2.3 LCoS display optimizing procedure

In digital holography applications, the diffraction efficiency is maximized when operating in the phase-only modulation regime (Moreno et al., 2004). Therefore, in such applications is very desirable to operate with an LCoS display response achieving a linear phase modulation up to 360 degrees, without coupled depolarization or coupled amplitude modulation.

By following the methods given in subsections 2.1 and 2.2, we are able to predict the intensity and the phase response of LCoS display. By taking advantage of this prediction capability, an optimization procedure can be devised to maximize the efficiency of digital holograms generated with LCoS displays.

Here, we apply an optimizing method, based on a numerical search algorithm, which allows us to achieve pairs of incident and analyzer SoPs that provide an LCoS display working in the phase-only regime. By starting from an initial pair of incident and analyzer SoPs, the numerical search algorithm varies these parameters in order to maximize the figure of merit Q_{PO} shown below. The SoP searching process enables to use any type of fully polarized light (i.e. linearly polarized light, circularly polarized light and elliptically polarized light). In this way, better results than those obtained when using only linear polarized light are achieved (Márquez et al., 2008).

The figure of merit Q_{PO} is based on a specific criterion to evaluate the quality of the response in the phase-only regime. In particular, it is based on three conditions: minimum transmittance value as high as possible, maximum phase modulation and minimum transmittance contrast.

$$Q_{PO}=\frac{1}{\lambda_1+\lambda_2+\lambda_3}\left[\lambda_1\frac{\Delta\tau}{2\pi}+\lambda_2(1-\Delta T)+\lambda_3 T_m\right] \tag{11}$$

where $\Delta\tau$ is the maximum phase difference, ΔT is the maximum transmittance difference and T_m denotes the minimum transmission value. Moreover, the parameters λ_1, λ_2 and λ_3 are coefficients enabling us to vary the specific weight of the three parameters involved into the optimization process.

By maximizing the figure of merit given in Eq. (11), different TNLCoS display configurations have been obtained, providing excellent results for phase-only response (Márquez et al., 2008), as for instance, those shown in Fig. 3.

In Fig. 3, the intensity values (left axis) and the phase values (right axis) are represented as a function of the gray level, the former being plotted in pink and the latter plotted in brown. Whereas the predicted values are represented by curves, the measured experimental data are represented by spots. For the measurement of the phase (brown squares), the interferometer based set-up given in Ref. (Lizana et al., 2008b) has been applied. The experimental orientations of external waveplates (WP) and linear polarizers (LP) used to achieve the incident and analyzer SoPs for this optimized configuration are: incident (LP_1=45 degrees, WP_1=160 degrees); analyzer (LP_2= -41 degrees, WP_2=18 degrees).

We see an excellent agreement between experimental and theoretical values, validating the suitability of the employed optimization methodology. Besides, good results for phase-only modulation are achieved. In fact, an almost constant intensity curve as a function of the gray level, with a phase-shift approximately of 300 degrees is obtained.

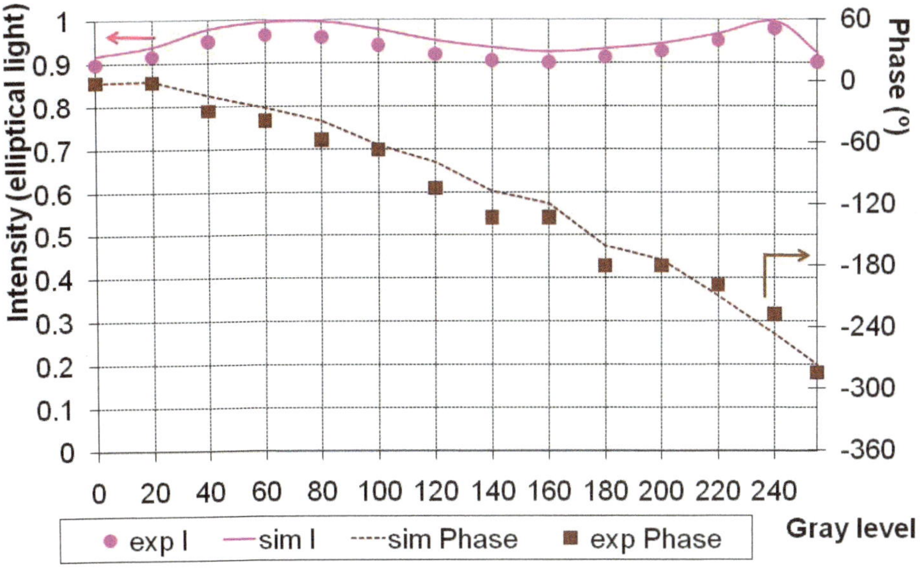

Fig. 3. TNLCoS display phase only response as a function of the gray level.

Finally, we have allowed the numerical search procedure to decrease the minimum value of transmittance T_m, in order to reach a higher phase-shift. The obtained result is shown in Fig. 4. The experimental orientations of external waveplates (WP) and linear polarizers (LP) are the following: incident beam (LP_1=88 degrees, WP_1=7 degrees); analyzer (LP_2= 90 degrees, WP_2= -15 degrees).

In Fig. 4, we see an excellent result in terms of phase-only regime. In particular, a continuous intensity response as a function of the gray level, which keeps almost constant around the value 0.65, is achieved. Besides, it is accompanied by a very large phase-shift, close to 360 degrees.

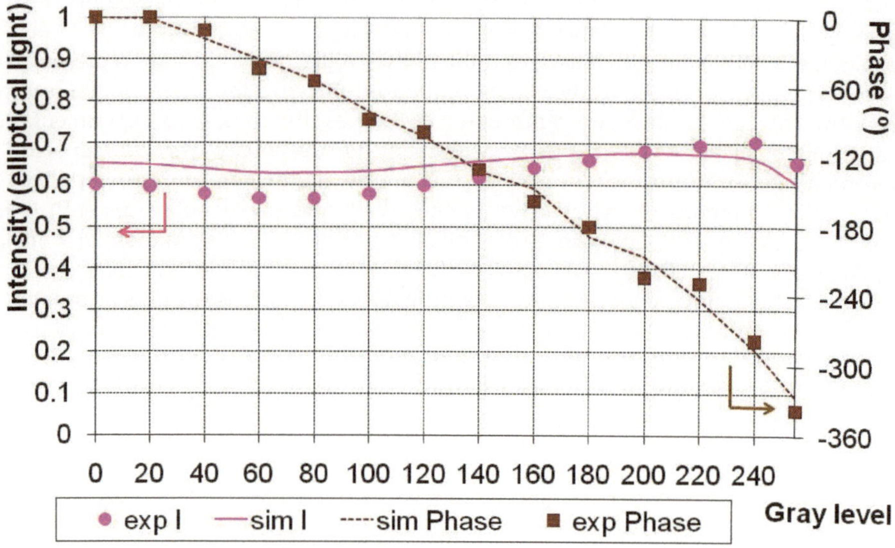

Fig. 4. TNLCoS display phase only response as a function of the gray level.

3. Time-fluctuations of the phase: Effects on the digital hologram efficiency

Experimental evidence of the time-fluctuations of the phase phenomenon, observed when working with LCoS displays (Lizana et al., 2008b; Lizana et al., 2010), are provided in this section. This phenomenon originates from the electrical addressing schemes (Hermerschmidt et al., 2007) applied in LCoS displays, and it may notably degrade the performance of LCoS displays when generating digital holograms (Lizana et al., 2008b). To provide greater insight into this topic, this section also includes a study of the time-fluctuations of the phase effects on the efficiency of diverse basic digital holograms generated with the TNLCoS display.

3.1 Time-fluctuations of the phase: Experimental evidence

To experimentally measure the time-fluctuations of the phase phenomenon, we have applied the diffraction based set-up sketched in Fig. 5.

We employ an unexpanded beam of a He-Ne laser (633 nm) to illuminate the TNLCoS display, with an angle of incidence equal to 2 degrees. A Polarization State Generator (PSG) is placed in the incident beam, composed of a Linear Polarizer (LP_1) and a Quarter Wave Plate (QWP_1), and a Half Wave Plate (HWP) is introduced in front of the PSG, allowing us to control the intensity of light incoming to the PSG. In the reflected beam, we have placed a Polarization State Detector (PSD), composed of a Quarter Wave Plate (QWP_2) followed by a Linear Polarizer (LP_2). We selected the PSG and PSD configuration in Fig. 5, which yields phase-only modulation with constant transmittance (Márquez et al., 2008). Then, we address a binary grating to the TNLCoS display, and by means of two radiometers (Newport 1830-C) placed in the far diffraction plane, the intensities of the zero and first diffracted orders are captured. The signals detected by the radiometers are synchronized and displayed on an oscilloscope (Tektronix TDS3012B), allowing us to perform intensity measurements of the diffracted orders as a function of the time.

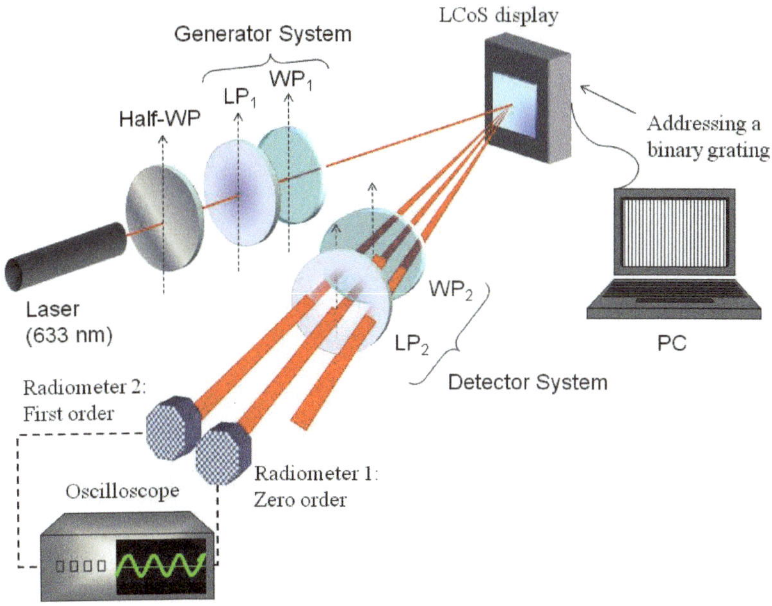

Fig. 5. Diffraction based set-up to measure the phase modulation as a function of the time.

Using the set-up sketched in Fig. 5, the intensity at the zero and at the first diffracted orders was measured as a function of the time, when addressing three different binary gratings. In particular, three binary gratings with a different pair of gray levels are used: (0,120), (0,211) and (0,255). The obtained results are given in Fig. 6. Whereas the intensity at the zero order is plotted in black, the intensity at the first order is plotted in red. The intensity values in Fig. 6 are normalized to the mean value of the intensity measured at the zero order, when a constant image is addressed to the LCoS display with the reference gray level (i.e. zero gray level).

Whereas in Fig. 6 (a) and in Fig. 6 (c), the zero order is in general more intense than the first diffraction order, in Fig. 6(b), the first order intensity is greater than the zero order intensity. This occurs because by changing the addressed gray level pair, the phase difference between the two parts of the binary grating varies. Thus, the quantity of light in each diffracted order in the far diffraction plane varies as well. However, the intensity of the diffraction orders is periodically fluctuating as a function of the time in all cases. This fact points out the existence of time-fluctuations of the phase phenomenon.

A well-known digital hologram is the binary grating with a phase difference of 180 degrees. Theoretically, when addressing such a binary grating, a null zero order has to be obtained. In addition, it has to be accompanied by the ±1 diffractive orders, whose intensity values should posses approximately the 40% of the input light. Therefore, the results obtained in Fig. 6(b) correspond in some instants of time to a binary grating with a phase difference equal to 180 degrees. In fact, in some intervals of time, the maximum value of the intensity at the first diffraction order reaches 0.4 and the zero order vanishes.

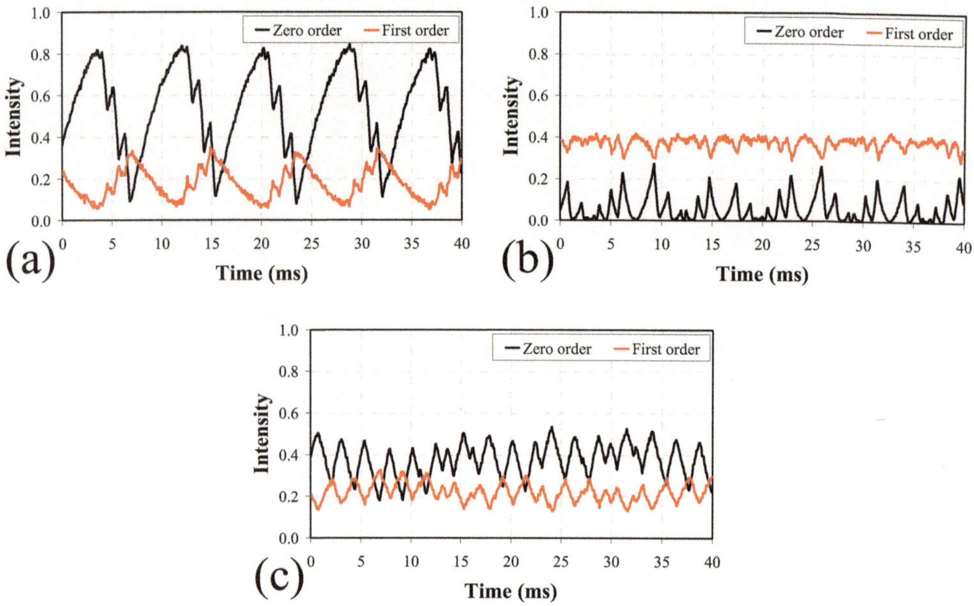

Fig. 6. Intensity at the zero (in black) and at the first (in red) diffractive orders, when addressing a binary grating containing the gray levels: (a) 0-120, (b) 0-211 and (c) 0-255.

As the TNLCoS display is working in the phase-only regime, the two gray levels applied for the binary gratings addressed to the TNLCoS display present the same values for the amplitude but a phase difference $\Phi(t)$ which depends on the time. At this point, an analytical expression for the intensities at the zero and the first diffracted orders can be derived as a function of Φ (Zhang et al., 1994). In particular, when the periodic structure of the grating is formed by two levels of the same size (i.e. a duty cycle of 50%), the normalized intensities as a function of the time, in the zero order and in the ±1 orders, can be described as follows:

$$I_0(t)=\frac{1}{2}\left(1+\cos\Phi(t)\right) \tag{12}$$

$$I_{\pm1}(t)=\frac{2}{\pi^2}\left(1-\cos\Phi(t)\right) \tag{13}$$

From the relations given in Eq. (12) and (13), an expression for the phase modulation as a function of the time can be deduced:

$$\cos\left(\Phi(t)\right)=\frac{4I_0(t)-\pi^2 I_1(t)}{4I_0(t)+\pi^2 I_1(t)} \tag{14}$$

From the obtained intensity values in Fig. 6, the phase $\Phi(t)$ as a function of the time can be calculated by applying Eq. (14). The results corresponding to the three binary gratings applied above (i.e. gray levels (0,120), (0,211) and (0,255)) are given in Fig. 7.

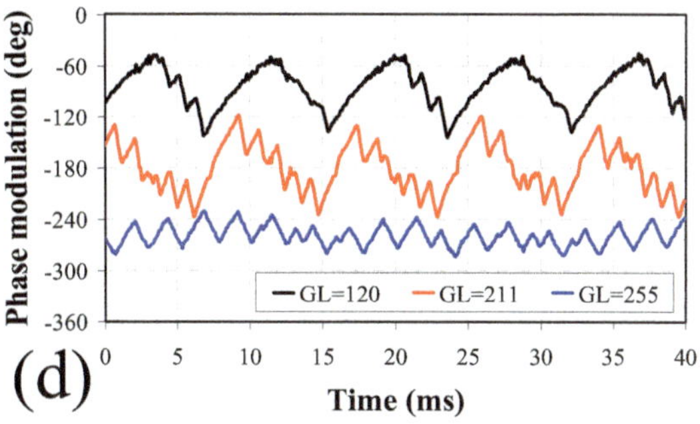

Fig. 7. Time fluctuations of the phase when addressing a binary grating containing the gray levels: 0-120 (in black), 0-211 (in red) and 0-255 (in blue).

The results shown in Fig. 7 constitute an experimental evidence for the time-fluctuations of the phase phenomenon, whose fluctuation amplitude depends on the gray levels applied to the binary gratings addressed. Note that the phase fluctuations can reach very high values, as in the case of the grating (0,211), where the mean phase difference is equal to 180 degrees, but it is accompanied with a great fluctuation amplitude of almost 120 degrees (Fig. 7, in red). As proved in this subsection, the phase-shift measured at the TNLCoS reflected beam is far from being constant along the frame period, leading to undesired effects that should be evaluated depending on the application.

3.2 Time-fluctuations of the phase: Effects on the digital holograms efficiency

In this section, we analyze the effect of time-fluctuations of the phase on the efficiency of two basic digital holograms, generated with the TNLCoS display. Both digital holograms have an average phase shift of 180 degrees.

The TNLCoS display has been configured to work in the phase-only regime by using the configuration related to Fig. 4. Then, by means of the combination of a spatial filter and a convergent lens, the whole TNLCoS display area has been illuminated with a collimated beam. The reflected image is captured with a CCD camera placed in the back focal plane of a second convergent lens set in the reflected beam. In this way, the CCD camera is able to capture the Fourier transform spectrum obtained by diffraction.

First, we address a binary phase-only grating with an average phase shift of 180 degrees to the TNLCoS display. The image captured with the CCD camera is shown in Fig. 8, where the typical diffraction orders generated by the binary grating are observed.

Although the intensity related to the ± 1 diffracted orders is stronger (brighter spots) than the one captured at the zero diffracted order, the later does not completely vanish, as would be expected in a binary grating with a phase difference of 180 degrees.

Fig. 8. Diffraction pattern captured with a CCD camera when addressing a binary grating with an averaged phase difference of 180 degrees.

This fact can be understood by comparing the results shown in Fig. 8 with those given in Fig. 6(b). As seen in Fig. 6(b), in those instants of time where the zero diffraction order is exactly equal to zero, the intensity of the first diffracted order is about 0.4. In such instants of time, the results are consistent with a binary phase grating with an instantaneous phase difference equal to 180 degrees. However, the intensity of the diffraction orders detected with the CCD camera (Fig. 8) corresponds to the mean values of the intensity measurements as a function of the time given in Fig. 6 (b). Therefore, as the intensity at the zero and at the ±1 diffracted orders varies as a function of the time (because the phase is fluctuating), the mean zero order is higher than 0 and the mean ±1 diffracted orders are lower than 0.4 (in particular, 0.059 for the zero order and 0.377 for the ±1 orders).

Next, we have displayed a binary phase computer generated hologram designed to reconstruct a butterfly. Again, the averaged phase-shift between gray levels is equal to 180 degrees. The results are shown in Fig. 9. We have added a linear phase along the diagonal direction with the aim of spatially separating the reconstructions of the different orders (i.e. the zero and ±1 orders are separated in this way). The butterfly is reconstructed at the ±1 orders, with an efficiency equivalent to those obtained in the first diffraction order in Fig. 8. The reconstructed butterflies are accompanied by a zero order diffracted with non-zero intensity that we can see as a bright peak located at the optical axis. As in the binary grating case, the intensity value measured at the zero diffracted order originates from time-fluctuations in the phase modulation.

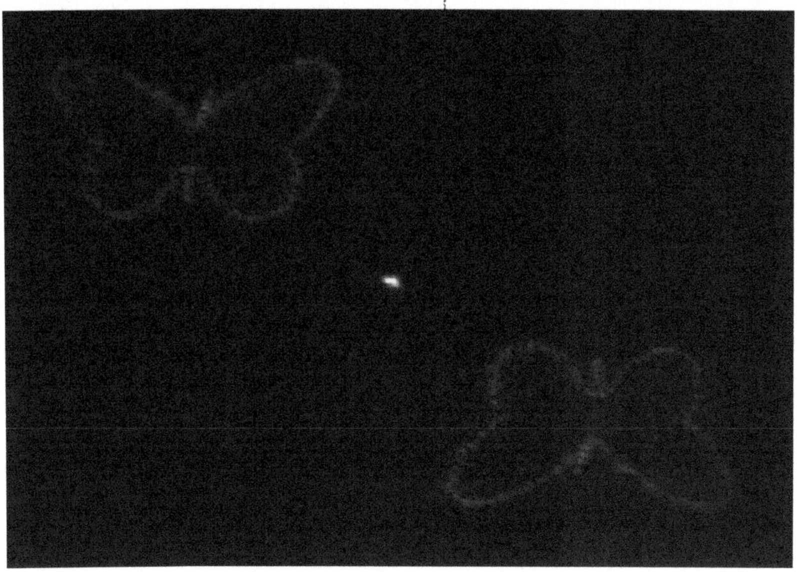

Fig. 9. Digital hologram designed to reconstruct a butterfly.

4. Enhancement of the digital hologram efficiency in the presence of time-fluctuations of the phase by applying the Minimum Euclidean principle

In this section, another type of LCoS display, the Parallel Aligned (PA) LCoS display model, is studied for use in the generation of digital holograms. The PALCoS display prototype under analysis is a commercial electrically controlled birefringence LCoS display distributed by HoloEye systems. This prototype is a PLUTO Spatial Light Modulator (SLM) with a diagonal display of 1.8 cm. The pixel pitch is of 8 μm with a fill factor of 87% and the display has a resolution of 1920x1080. Besides, this LCoS display model allows us to upload different electrical sequences formats for the digital electrical addressing, which are based on different pulsed-width modulation schemes. The use of different electrical sequences may result in different responses and efficiencies of the PALCoS display (Hermerschmidt et al., 2007; Moore et al., 2008).

A characteristic feature in PALCoS displays is that the LC molecules in such devices are all parallel aligned. Then, by illuminating the display with an incident linear State of Polarization (SoP) parallel to the LC extraordinary axis, the exiting SoP is constant as a function of the time, even in the presence of LC fluctuations. In this situation, the PALCoS display is working in the phase-only regime and the effective depolarization phenomenon can be avoided. However, the damaging effect of the time-fluctuations of the phase phenomenon described in section 3 is still present.

Here, the significance of the time-fluctuations phenomenon in our PALCoS display is experimentally analyzed for the available electrical sequences. Moreover, two different possible mapping schemes for the ideal phase-only function implementation of digital holograms are considered: a linear phase mismatching scheme and a mapping scheme based on the minimum Euclidean distance principle (Juday, 2001; Moreno et al., 1995). They are experimentally tested in the presence of time-fluctuations of the phase, in order to find the best configuration to maximize the efficiency of digital holograms generated with the PALCoS display (Lizana et al., 2010).

4.1 Time-fluctuations of the phase in the PALCoS display

To operate in the phase-only regime, the PALCoS display under analysis has been sandwiched between two polarizers oriented in the direction of the LC extraordinary axis. By using the diffraction based set-up shown in Fig. 5, we have measured the phase response as a function of the time for three different digital addressing sequences provided by HoloEye (since now labelled as the sequences #1, #2 and #3). For each sequence employed, the phase as a function of the time is measured for different gray levels. In all the cases, the measurements are made by using an unexpanded He-Ne laser beam (633 nm) and at quasi-normal incidence (i.e. an incident angle of 2 degrees).
Figure 10 shows the PALCoS phase response as a function of the time for the addressing sequences #1 (Fig. 10(a)), #2 (Fig. 10(b)) and #3 (Fig. 10(c)).

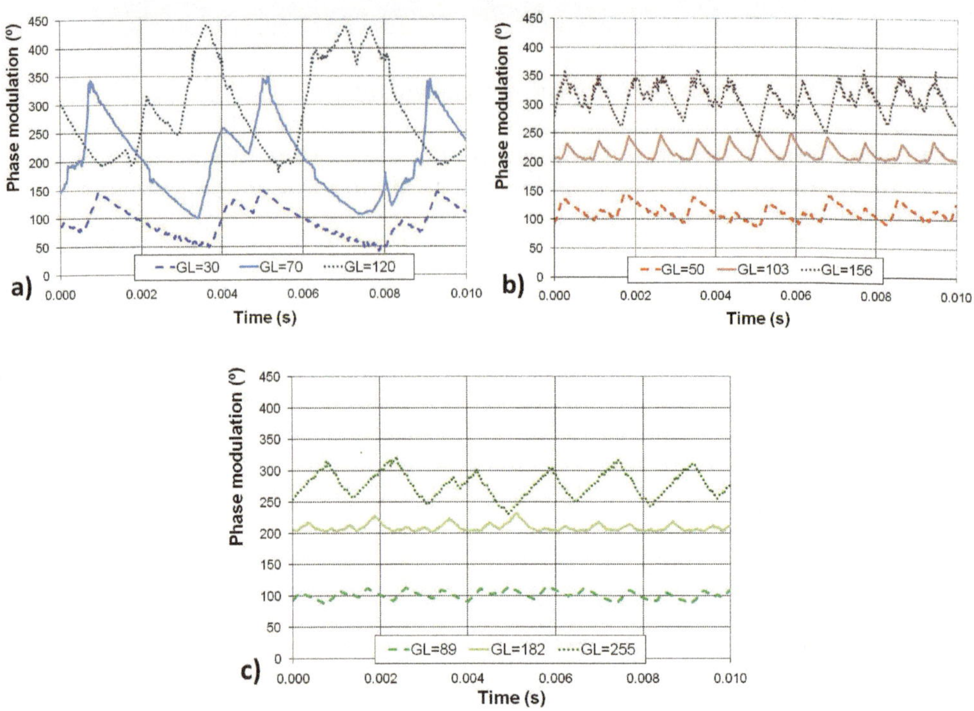

Fig. 10. Phase modulation as a function of time for different gray levels, for the addressing sequences: (a) #1, (b) #2 and (c) #3.

In every case, the measurements are done for three different gray levels, which lead to average phase values of 100 degrees, 200 degrees and 300 degrees respectively. In this way, we can assume that an estimation of the whole phase domain is conducted.

Figure 10 shows that for all the addressing sequences, periodic time-fluctuations of the phase are obtained. Besides, the amplitude of the phase fluctuations depends on the addressed sequence. In particular, whereas the fluctuations for the sequence #1 (Fig. 10(a)) are very large, they become much smaller for the sequences #2 and #3 (Fig. 10(b) and Fig. 10(c) respectively).

To generate digital holograms with LCoS displays, the required phase distribution has to be implemented. To this aim, the significant parameter is the mean phase. Thus, the mean phase modulation curves for the addressed sequences #1, #2 and #3 are measured as well. The mean phase values have been obtained by calculating the difference between the eigenvalues of the PALCoS display equivalent Jones matrices, which are obtained as described in subsection 2.2.

The results are shown in Fig. 11, where the phase modulation for the sequences #1 (in blue) and #2 (in red) are larger than 360 degrees (i.e. 520 degrees and 480 degrees). On the other hand, the phase depth for the sequence #3 is clearly smaller (about 280 degrees).

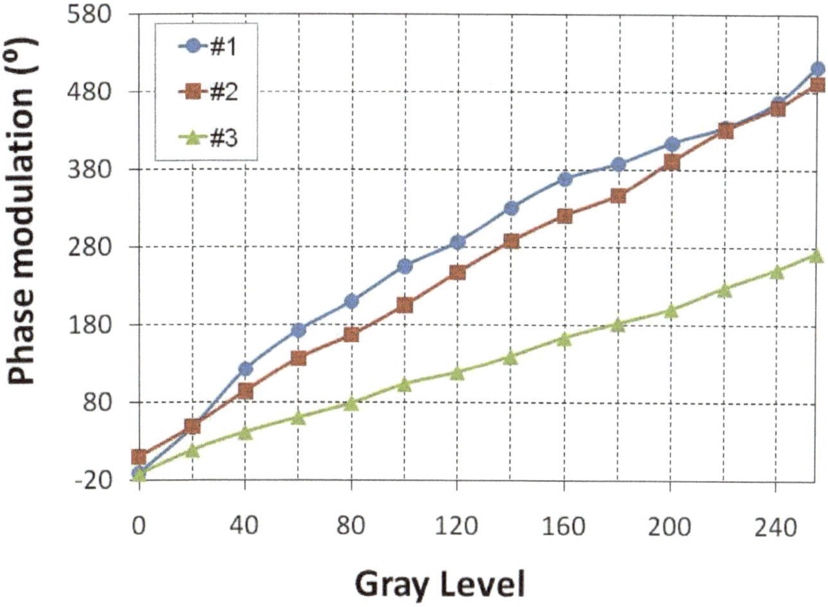

Fig. 11. Phase modulation as a function of the gray level for the sequence #1 (in blue), #2 (in red) and #3 (in green).

4.2 Digital hologram efficiency evaluation: Application of the minimum Euclidean distance principle

Figure 11 shows how the different electrical sequences available with the PALCoS display provide different phase modulation depths. In general, phase modulations larger than 360 degrees are desired to maximize the efficiency of holograms generated with LCDs. However, as shown in Fig. 10, the electrical sequences applied in LCoS display produce time-fluctuations of the phase, which degrade the efficiency of the generated holograms. Thus, small amplitude of the fluctuations is desired as well.

In this situation, to maximize the efficiency of digital holograms generated with LCoS displays, a trade-off between phase modulation depth (as large as possible) and amplitude of the time-fluctuations of the phase (as small as possible) has to be found.

Next, the suitability of the different electrical sequences to maximize the efficiency of the implemented holograms is experimentally analyzed. Besides, different mapping schemes for phase-only distribution implementation are also reviewed.

4.2.1 Linear phase mismatching and saturated mismatching encoding schemes

To accurately implement digital holograms with LCDs, it is very important to experimentally generate a real phase distribution as close as possible to the designed one. However, as a consequence of diverse non-linearities related to the experimental implementation, the ideal phase distribution is never achieved. To reduce different LCDs degradation sources, some authors have proposed diverse strategies (Davis et al., 1998; Márquez et al., 2001). For instance, if the available dynamic range is less than 360 degrees, the minimum Euclidean distance projection principle (Juday, 2001; Moreno et al., 1995) can be applied. In this way, the modulation diffraction efficiency may be greatly enhanced by selecting an appropriate mapping scheme for the implementation of the phase function onto the restricted phase-only domain (Moreno et al., 1995).

Next, two possible mapping schemes, for the ideal phase-only distribution implementation of holograms, are reviewed. These schemes are represented in Fig. 12, where the truly

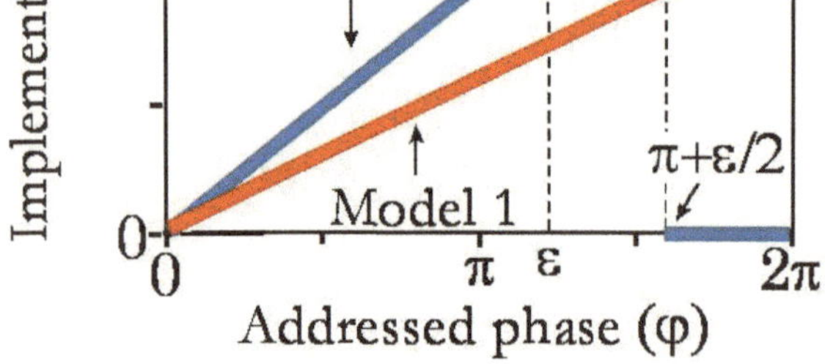

Fig. 12. Mapping scheme implementation: In red, the linear mismatching encoding (model 1) and in blue, the saturation mismatching encoding (model 2).

implemented phase (denoted as p) is represented as a function of the addressed phase (denoted as φ). Note that the diagonal dotted line represents the correct matching between the designed phase φ and the displayed phase p.

We assume that the phase values available with the applied LCD are in the range $(0, \varepsilon)$, with $\varepsilon < 2\pi$. First, the red line (model scheme 1) represents a linear phase mismatching. Second, the blue line (model scheme 2) represents a more efficient encoding scheme, which we denote as saturated mismatching encoding. Model 2 represents the perfect phase matching up to the maximum modulation depth $\varphi = \varepsilon$, while there is a saturation for values $\varphi > \varepsilon$. Then, for each phase $\varphi > \varepsilon$, the closest available phase value in the modulation domain is taken by following the minimum Euclidean distance principle (see Fig. 12, in blue).

4.2.2 Experimental results

In this subsection, we compare the efficiency of a basic continuous digital hologram, the blazed grating (Fujita et al., 1982), generated with our PALCoS display when uploading the different electrical sequences (i.e. sequences #1, #2 and #3). For all the sequences, the implementation is conducted by using the linear phase mismatching scheme. In this way, we have limited the phase range between 0 and 360 degrees and we have applied a look-up-table to produce a linear increment for the average phase values. Besides, the phase modulation provided by sequence #3 is lower than 360 degrees (see Fig. 11), and so, the saturated mismatching encoding is applied for this sequence as well.

In all the cases, the blazed grating is written to the modulator and the corresponding intensity of the zero and of the first diffracted orders is measured as a function of the time by using the diffraction based set-up given in Fig. 5 and by illuminating the PALCoS display with a He-Ne laser (633 nm). The period of the grating is fixed to 16 pixels, being the sufficient number of pixels to neglect the effect of the quantification of the phase levels.

The measurements obtained for the intensity diffracted to the zero (in blue) and to the first (in red) orders are plotted in Fig. 13. In Fig. 13(a) and Fig. 13(b), we have plotted the results obtained when using the sequence #1 and #2 respectively. Next, Fig. 13(c) and Fig. 13(d) show the results obtained when using the sequence #3, the former when applying the linear mismatching and the latter when applying the saturation mismatching encoding.

The largest intensity fluctuations are measured when addressing the sequence #1 (Fig. 13(a)) and the best diffraction efficiency is obtained for the sequence #3 with the saturated encoding (Fig. 13(d)). In this way, the sequence #3, even providing a phase modulation lower than 360 degrees (see Fig.11), is the most stable and efficient addressing sequence.

The results here provided evidences that to maximize the efficiency of digital holograms generated with LCoS displays, it is important to find a trade-off between the phase modulation depth and the amplitude of the time-fluctuations phenomena in LCoS displays. In this framework, a mathematical model suitable to evaluate the LCoS display response in presence of time-fluctuations on the phase becomes helpful (Lizana et al., 2010).

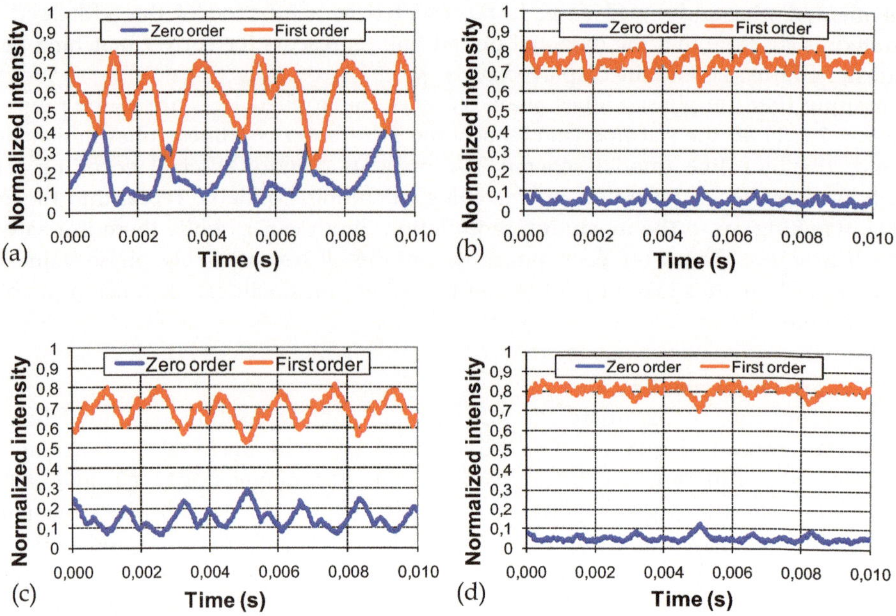

Fig. 13. Normalized intensity for the zero and first orders obtained when addressing a blazed grating and the addressing sequence: (a) #1, (b) #2, (c) #3 with linear mismatch, and (d) #3 with saturated mismatch.

5. Conclusions

This Chapter provides a study of LCoS displays for their application in the generation of digital holograms. In particular, the analysis presented in this Chapter can be used as a guideline to maximize the efficiency of LCoS display in digital holographic applications.
In particular, this Chapter provides a characterization and optimization method, based on a combination of the Mueller formalism and the Jones formalism, suitable to maximize the efficiency of the addressed holograms. Besides, experimental evidence is provided for the time-fluctuations of the phase phenomenon, which degrades the performance of LCoS displays. In this way, a discussion of the damaging effect of such phenomenon, when generating digital holograms, is included. Finally, we experimentally prove that, to maximize the efficiency of digital holograms generated with LCoS displays in presence of phase fluctuations, is important to find a trade-off between phase modulation depth and amplitude of the time-fluctuations. To this aim, different mapping schemes for phase-only distribution implementation are reviewed.

6. Acknowledgment

We acknowledge financial support from the Spanish Ministerio de Educación y Ciencia (grants FIS209-13955-C02-01 and FIS209-13955-C02-02). C. Iemmi gratefully acknowledges the support of the Universidad de Buenos Aires and CONICET (Argentina).

7. References

Ashkin, A. (2006). *Optical Trapping and Manipulation of Neutral Particles Using Lasers: A Reprint Volume With Commentaries,* World Scientific Publishing Company, ISBN 9810240589, Hackensack, USA.

Azzam, R.M.A. & Bashara, N.D.N. (1972). Ellipsometric Measurement of the Polarization Transfer Function of an Optical System. *Journal of the Optical Society of America,* Vol. 62, No. 3, pp. 336-340, ISSN 0030-3941.

Berreman, D.W. (1972). Optics in Stratified and Anisotropic Media: 4x4-Matrix Formulation, *Journal of the Optical Society of the America,* Vol. 62, No. 4, pp. 502-510, ISSN 0030-3941.

Coufal, H.J.; Psaltis, D. & Sincerbox, B.T. (2000). *Holographic Data Storage,* Springer-Verlag, ISBN 0470749628, Berlin, Germany.

Chipman, R. A. (1995); *Polarimetry,* in Handbook of Optics, McGraw-Hill, New York, USA.

Davis, J.A.; Moreno, I. & Tsai, P. (1998), Polarization eigenstates for twisted-nematic liquid-crystal displays, *Applied Optics,* Vol. 37, No. 5, pp. 937-945, ISSN 0003-6935.

De Martino, A.; Kim, Y.; Garcia-Caurel, E.; Laude, B. & Drévillon, B. (2003). Optimized Mueller polarimeter with liquid crystals, *Optics Letters,* Vol. 28, No. 8, pp. 616-618, ISSN 0146-9592.

Dickey, F.M.; Holswade, S.C. & Shealdy, D.L. (2005). *Laser Beam Shaping Applications,* CRC Press, ISBN 0824759419, Albuquerque, USA.

Dou, R. & Giles, M.K. (1995). Closed-loop adaptive optics system with a liquid crystal television as a phase retarder, *Optics Letters,* Vol. 20, pp. 1583-1585, ISSN 0146-9592.

Fernández-Pousa, C.R.; Moreno, I.; Bennis, N. & Gómez-Reino, C. (2000). Generalized formulation and symmetry properties of reciprocal nonabsorbing polarization devices: application to liquid-crystal displays, *Journal of the Optical Society of America A,* Vol. 17, No. 11, pp. 2074-2080, ISSN 0740-3232.

Foldyna, M.; Garcia-Caurel, E.; Ossikovski, R.; De Martino, A. & Gil, J.J. (2009). Retrieval of a non-depolarizing component of experimentally determined depolarizing Mueller matrices, *Optics Express,* Vol. 17, No. 15, pp. 12794-12806, ISSN 1094-4087.

Fujita, T.; Nishihara, H. & Koyama, J. (1982). Blazed gratings and Fresnel lenses fabricated by electron-beam lithography, *Optics Letters,* Vol. 7, No. 12, pp. 578-580, ISSN 0146-9592.

Gagnon, R.J. (1981). Liquid-crystal twist-cell optics, *Journal of the Optical Society of America,* Vol. 71, No. 3, pp. 348-353, ISSN 0030-3941.

Goldstein, D. (2003). *Polarized Light, Mercel Dekker,* New York, USA, ISBN 0-8247-4053-X.

Hermerschmidt, A.; Osten, S.; Krüger, S. & Blümel, T. (2007). Wave front generation using a phase-only modulating liquid-crystal based micro-display with HDTV resolution, *Proceedings of SPIE,* Vol. 6584, pp. 65840E, ISBN. 978-0-8194-6712-6.

Jones, R. C. (1941). A new calculus for the treatment of optical systems, *Journal of the Optical Society of America,* Vol. 31, No. 7, pp. 488-493, ISSN 0030-3941.

Juday, R.D. (2001). Generality of matched filtering and minimum Euclidean distance projection for optical pattern recognition, *Journal of the Optical Society of America A*, Vol. 18, No. 8, pp. 1882-1896, ISSN 0740-3232.

Lee, Y.; Gourlay, J.; Hossack, W.J.; Underwood, I. & Walton, A.J. (2004). Multi-phase modulation for nematic liquid crystal on silicon backplane spatial light modulators using pulse-width modulation driving scheme, *Optics Communications*, Vol. 236, pp. 313-322, ISSN 0030-4018.

Liu, H.K.; Davis ,J.A. & Lilly, R.A. (1985). Optical-data-processing properties of a liquid-crystal television spatial light modulator, *Optics Letters*, Vol. 10, pp. 635-637, ISSN 0146-9592.

Lizana, A.; Moreno, I.; Iemmi, C.; Márquez, A.; Campos, J. & Yzuel, M.J. (2008a). Time-resolved Mueller matrix analysis of a liquid crystal on silicon display, *Applied Optics*, Vol. 47, No. 23, pp. 4267-4274, ISSN 0003-6935.

Lizana, A.; Moreno, I.; Márquez, A.; Iemmi, C.; Fernández, E.; Campos, J. & Yzuel, M.J. (2008b). Time fluctuations of the phase modulation in a liquid crystal on silicon display: characterization and effects in diffractive optics, *Optics Express*, Vol. 16, No. 21, pp. 16711-16722, ISSN 1094-4087.

Lizana, A.; Márquez, A.; Moreno, I.; Iemmi, C.; Campos, J. & Yzuel, M.J. (2008c). Wavelength dependence of polarimetric and phase-shift characterization of a liquid crystal on silicon display, *Journal of the European Optical Society - Rapid Publications*, Vol. 3, No. 08011, pp. 1-6, ISSN 1990-2573.

Lizana, A.; Martin, N.; Estapé, M.; Fernández, E.; Moreno, I.; Márquez, A.; Iemmi, C.; Campos, J. & Yzuel, M.J. (2009). Influence of the incident angle in the performance of Liquid Crystal on Silicon displays, *Optics Express*, Vol. 17, No. 10, pp. 8491-8505, ISSN 1094-4087.

Lizana, A.; Márquez, A.; Lobato, L.; Rodange, Y.; Moreno, I.; Iemmi, C. & Campos, J. (2010). The minimum Euclidean distance principle applied to improve the modulation diffraction efficiency in digitally controlled spatial light modulators, *Optics Express*, Vol. 18, No. 10, pp. 10581-10593, ISSN 1094-4087.

Lu, S.Y. & Chipman, R.A. (1996), Interpretation of Mueller matrices based on polar decomposition, *Journal of the Optical Society of America A*, Vol. 13, No. 5, pp. 1106-1113, ISSN 0740-3232.

Márquez, A.; Iemmi, C.; Moreno, I.; Jeffrey, J.A.; Campos, J. & Yzuel, M.J. (2001), Quantitative prediction of the modulation behavior of twisted nematic liquid crystal displays based on a simple physical model, *Optical Engineering*, Vol. 40, No. 11, pp. 2558-2564, ISSN 0091-3286.

Márquez, A.; Iemmi, C.; Campos, J.; Escalera, J.C. & Yzuel, M.J. (2005). Programmable apodizer to compensate chromatic aberration effects using a liquid crystal light modulator, *Optics Express*, Vol. 13, No. 3, pp. 716-730, ISSN 1094-4087.

Márquez, A.; Moreno, I.; Iemmi, C.; Lizana, A.; Campos, J. & Yzuel, M.J. (2008). Mueller-Stokes characterization and optimization of a liquid crystal on silicon display showing depolarization, *Optics Express*, Vol. 16, No. 3, pp. 1669-1685, ISSN 1094-4087.

Moore, J.R.; Collings, N.; Crossland, W.A.; Davey, A.B.; Evans, M.; Jeziorska, A.M.; Komarcevic, M.; Parker, J.R.; Wilkinson, T.D. & Xu, H. (2008). The silicon backplane design for an LCOS polarization-insensitive phase hologram SLM, *IEEE Photonic Technology*, Vol. 20, No. (1/4), pp. 60 - 62.

Moreno, I.; Campos, J.; Gorecki, C. & Yzuel, M.J. (1995). Effects of amplitude and phase mismatching errors in the generation of a kinoform for pattern recognition, *Japanese Journal of Applied Physics*, Vol. 34, No. 12 A, pp. 6423-6432, ISSN 0021-4922.

Moreno, I.; Velásquez, P.; Fernández-Pousa, C.R.; Sánchez-López, M.M. & Mateos, F. (2003). Jones matrix method for predicting and optimizing the optical modulation properties of a liquid-crystal display, *Journal of the Applied Physics*, Vol. 94, No. 6, pp. 3697-3702, ISSN 0021-8979.

Moreno, I.; Iemmi, C.; Márquez, A.; Campos, J. & Yzuel, M.J. (2004). Modulation light efficiency of diffractive lenses displayed in a restricted phase-mostly modulation display, *Applied Optics*, Vol. 43, No. 34, pp. 6278-6284, ISSN 0003-6935.

Moreno, I.; Lizana, A.; Campos, J.; Márquez, A.; Iemmi, C. & Yzuel, M.J. (2008). Combined Mueller and Jones matrix method for the evaluation of the complex modulation in a liquid-crystal-on-silicon display, *Optics Letters*, Vol. 33, No. 6, pp. 627-629, ISSN 0146-9592.

Ossikovski, R.; Anastasiadou, M.; Hatit, S.B.; Garcia-Caurel, E. & De Martino, A. (2008). Depolarizing Mueller matrices: how to decompose them?, *Physica Status Solidi A*, Vol. 205, No. 4, pp. 720-727, ISSN 0031-8965.

Parke, N.G. (1949), Optical Algebra, *Journal of Mathematics and Physics*, Vol. 28, No. 2, pp. 131-139, IDS XT812.

Peinado, A.; Lizana, A.; Vidal, J.; Iemmi, C. & Campos, J. (2010). Optimization and performance criteria of a Stokes polarimeter based on two variable retarders, *Optics Express*, Vol. 18, No. 10, pp. 9815-9830, ISSN 1094-4087.

Rodrigo, J.A.; Alieva, T.; Cámara, A.; Marínez-Matos, O.; Cheben, P. & Calvo, M.L. (2011). Characterization of holographically generated beams via phase retrieval based on Wigner distribution projections, *Optics Express*, Vol. 19, No 7, pp. 6064-6077.

Taylor, P. (1974). *Theory and Applications of Numerical Analysis*, Academic Press, ISBN 0-12-553560-0, London, United Kingdom.

Twietmeyer, K.M.; Chipman, R.A.; Elsner, A.E.; Zhao, Y. & VanNasdale, D. (2008). Mueller matrix retinal imager with optimized polarization conditions, *Optics Express*, Vol. 16, No. 26, pp. 21339-21354, ISSN 1094-4087.

Verma, R.S.; Swami, M.K.; Manhas, S.S. & Gupta, P.K. (2010). Mueller matrix-based optimization of reflective type twisted nematic liquid crystal SLM at oblique incidences, *Optics Communications*, Vol. 283, No. 12, pp. 2580-2587, ISSN 0030-4018.

Wolfe, J.E. & Chipman, R.A. (2006). Polarimetric characterization of liquid-crystal-on-silicon panels, *Applied Optics*, Vol. 45, No. 8, pp. 1688-1703, ISSN 0003-6935.

Wu, S.T. & Yang, D.K. (2005). Reflective Liquid Crystal Displays, John Wiley & Sons Inc., Chichester, United Kingdom.

Zhang, Z.; Lu, G. & Yu, F.T.S. (1994). Simple method for measuring phase modulation in liquid crystal television, *Optical Engineering*, Vol. 33, No. 9, pp. 3018-3022.

11

Synthetic Image Holograms

Jakub Svoboda, Marek Škereň and Pavel Fiala
Czech Technical University in Prague
Czech Republic

1. Introduction

This chapter is dedicated to synthetic image holograms - the elements which can create a reconstruction of a 3D object for observation with the human eye. Holography as a technique of image recording and reconstruction has been extensively developed from sixties of the twentieth century. During this time there have been various attempts to synthesize holograms artificially without the presence of the real object in the classical recording setup. Different approaches have been used, several trying to synthesize the three-dimensional object from two-dimensional views using the classical recording setup, the others trying to calculate the microstructure of the hologram completely in a computer. Today, we can divide synthetic holography into two major streams, the first containing the methods for creating the image for observation by human eye and the second consisting of approaches for designing the synthetic diffractive structures for general wavefront generation. The former techniques can exploit various imperfections of human vision and omit several parameters of the optical wave. The latter techniques are usually based on the direct calculation of the microstructure and they try to create the reconstruction in its full complexity. Only the first group of synthetic image holograms will be analyzed in this chapter.

The synthetic approach to hologram creation can have several advantages, but also noticeable disadvantages. The most important advantages are connected with flexibility in modifying the recorded object. First, the object need not to exist in reality in a form of a physical model. For most synthetic approaches, it is fully sufficient to have a 3D computer model for preparing the recording data. Also for real physically existing objects it could be tricky to perform the recording process in a classical setup. For example, various outdoor scenes such as buildings and others could not be included in the laboratory setups. Generally, the scaling possibility is very limited in classical holography, so the recorded object (or its model) must be of final size. On the other hand, it is easy to scale the computer model of an object. The next problem is in various corrections of color properties, surface textures, and general fine tuning of the recorded object. While such operations are very simple in the case of computer models, they could bring insoluble problems for real physical models. The stability of the object is also very important. It is crucial to highly stabilize the object for recording in classical holographic recording setup (when exposing with a continuous-wave laser), whereas in a computer stability is not a problem. This can apply also for holograms of living objects or dynamic scenes, where it is easy to take snaphots using photographic techniques, but holographic exposure is almost impossible. Finally, according to the recording technology chosen, other parameters of the synthetic hologram can be highly superior to those of classical holograms (e.g. fidelity of color mixing, contrast of the image, etc.).

Unfortunately, the synthetic approach brings also a whole range of specific problems and disadvantages. Generally, the recording setup is much more complicated in synthetic holography than in classical holography (although it can be much easier to operate some synthetic writers than to tune the classical laboratory recording setup). Furthermore, most synthetic techniques are limited in final size of the hologram (e.g. for directly written synthetic elements the typical area is in cm^2, while for classical holograms it is usually in dm^2). The recording process itself takes usually much longer in synthetic holography (although the preparation for exposure could be longer when using the classical recording). It is also important to realize, that in most synthetic approaches, the reconstructed optical wavefront does not exactly correspond to the one from a similar classical hologram. Various parameters are often omitted due to the limited performance of human eye, which can not evaluate the optical signal in full complexity.

The most important features of the synthetic hologram can be summarized as follows: the object data could be obtained without necessity to work with a physical model, various modifications of the object could be possible; the object could be potentially dynamic, spacious, etc.; the hologram itself could be as large as possible, it could enable general color mixing, the reconstruction could be possible in white light.

In the following sections the key steps of the design and recording processes are described together with the common technology used for recording the synthetic holograms. Section 1 contains analysis of the human vision with focus on 3D perception by human eyes. Section 2 describes the approaches to the hologram synthesis using two most common techniques - synthesis at the hologram plane and synthesis at the eye-pupil plane. Various details concerning the color mixing, special 3D properties, and kinetic behavior of the holograms are also discussed. Finally, in Section 3 the most common devices and recording materials are briefly mentioned.

1.1 Human vision

The visual sense of a 3D scene is caused by the brain interpreting the image information from the eye nerves from both eyes. A significant advantage is that the eyes are quadratic detectors (they do not feel the phase); therefore, 3D vision can be "faked" by only transmitting proper intensity images into the eyes as it will be shown.

Due to the relative positions of our eyes, the image from the left eye slightly differs from the image from the right eye. This is the crucial aspect causing the 3D sense and that is why it is used in hologram synthesis.

The image of the scene in front of our eyes is displayed by the eye's optical system on the photosensitive layer of the eye - the retina. This image information is processed by the photoreceptors (rods and cones) and transmitted to the brain through the optic nerve for further processing. What happens in the eye is not the only relevant factor for vision. There are also factors supporting depth perception, which relate to the shape of the object. Vision cues can be divided into physiological and psychological ones [Najdek, 2008].

The physiological cues relate to the eye physiology and the principles of vision - the necessary information from the viewed scene is either obtained by tracking the position of the image on the eye's retina (*stereopsis, movement parallax*) or by sensing the tension in the eye's muscles (*accommodation, convergence*). The physiological cues express the depth of the viewed scene by the means of 2D images. These cues are for example *perspective, relative size, occlusion, shading, depth of field*, etc.

The most important cues in 3D imaging are the following physiological cues - stereopsis and movement parallax. *Stereopsis* is based on the difference between the image viewed by the left eye and the image viewed by the right eye. Each eye has slightly different angle of viewing due to the different spatial position of the eyes (common distance between the eyes is about 65 mm). The effect is called binocular disparity or binocular parallax. The image information from the two different images is then processed in the brain's visual center and the relative spatial position of the viewed objects is conceived. Stereopsis is probably the most important and strongest cue, from close up to medium viewing distances. The principle of stereopsis is illustrated in Fig. 1.

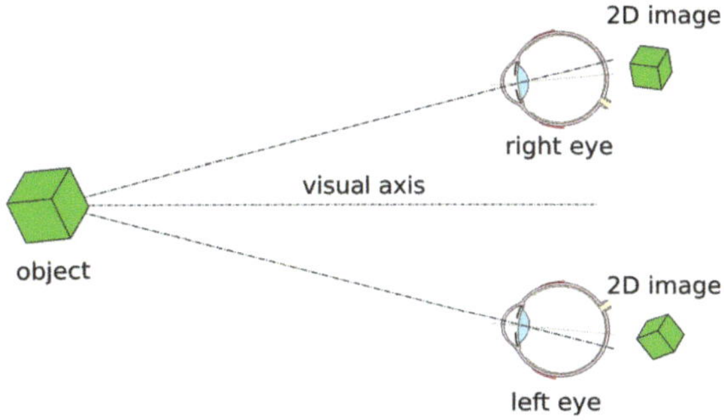

Fig. 1. Stereopsis: When viewing a 3D object, the images that are incident on each eye's retina slightly differ in angle. Brain interprets this change in angle as the depth of the object. Besides movement parallax, this is a crucial cue, that contributes to 3D vision.

Movement parallax is also based on processing an image from retina, but it is a monocular cue. The difference in the position of the image is caused by the observer moving his head. This technique can be found in animals whose eyes are placed on the opposite sides of their head - e.g. some types of birds that swing their heads to achieve motion parallax. There are also surveys showing that people with monocular vision use the movement parallax to obtain the depth perception [Faubert, 2002; Ferris, 1972].

1.2 Stereoscopic methods and holography

There is a variety of methods of providing 3D perception for the observer. Until the "invention" of binocular vision, 3D space was represented mainly by the monocular means of psychological cues as the relative size, occlusion, and shading. By using perspective, the images started to be more real. The first 3D perception as we know it from real life came with Sir Charles Wheatstone's invention in 1838 [Wheatstone, 1838; Wheatstone, 1852]. He described the binocular vision and proposed the device called the *stereoscope*. This device used stereopsis for imaging depth - it created two separated viewing zones with two different images, one for each eye. Since the stereoscope was invented before photography, the first 3D images were paintings.

From that time on, many stereoscopic methods have been developed. Modern 3D display methods can be classified according to the number of viewing zones and the way they are separated. The methods with two viewing zones only are denoted *binocular methods* or *methods of selective observation*. The observer has to wear some kind of glasses to split the particular views into the left and right eye. On the other hand, there are methods of spatial display, which create more viewing zones. In such a case, these zones need to be spatially displaced by the displaying device itself. These methods are therefore denoted *the methods of selective display* or *autostereoscopic methods*.

As shown in Fig. 2a, in the binocular methods, there are two 2D images (binocular pair) overlapping each other, which are transmitted in the direction of the observer [Takanori, 1976]. To view the correct image by the corresponding eye, the observer must wear glasses with some sort of filters that will block the image for the other eye and let only the correct one to pass. The first device ever, that was used for this kind of display was the already mentioned *stereoscope*. Typical representatives of these display methods nowadays are e.g. anaglyphs or the 3D cinema. In *anaglyphs* [Zone, 2003], the two observation channels are superimposed, but each one of them is encoded by a different color filter. The observer wears glasses with colored filters that separate properly the two images into the eyes. Using a pair of pure colors from R-G-B for filtering enables only monochromatic observation. Therefore, to achieve a color result, mixed color filters are used for the glasses. The most common is red-cyan (50 % blue and 50 % green). Other color filters have been invented to improve color rendering. An example is Infitec - narrow band interference filters.

The modern 3D *cinemas* use either polarization filters or shutter systems [Turner, 1986]. The polarization principle is the same as in anaglyphs - the two viewing channels are superimposed, each encoded into a different polarization state. The glasses are equipped with proper polarization analyzers to split the two channels. The shutter systems split the binocular pair in the time domain. The movie frames are projected with about twice the frequency of a 2D movie. The shutter glasses are synchronized with the projector and they block periodically the left and the right eye, so that the observer sees only one frame with the correct eye at a time.

The main disadvantages of all the binocular methods are the necessity of wearing the filtering glasses, the lack of motion parallax, and the distortion of the image. The image can be viewed undistorted only from the same position from which it was taken. That means, for example, that in the cinema, most spectators see the scene in a somewhat distorted way. Nevertheless, these methods are satisfying for 3D imaging and are widely used in the entertainment industry.

Unlike separating the channels at the observer by using the selective observation methods, autostereoscopic methods create the separated viewing zones using the displaying device itself [Halle, 1997]. These methods are capable of creating more than two spatially separated viewing zones and motion parallax can be obtained without the necessity of wearing any glasses. Another advantage over binocular imaging is the fact, that when multiple viewing zones are involved, there is no constraint on the observer's position. Each zone contains the image information from the particular direction and wherever the observer is located relative to the displaying device, he sees the 3D image in the correct perspective.

A typical representative of this imaging method is integral photography. The object is captured through an array of tiny convex lenses. Each of the lenses creates an image of the object with a slightly different perspective. These images are recorded on a photographic plate. After being developed, the plate is illuminated from behind and the light travels again through the array of lenses. Since the setup is exactly the same as it was during the recording, the light rays from various lenses meet again in the original position, where the surface of the object was located (Fig. 2b). For viewing such an image, the observer does not need to wear any glasses.

Another improvement in autostereoscopic methods comes with using diffractive structures. The tiny lenses are replaced with a diffractive structure, that controls the distribution of the

image information into proper directions. The diffractive methods are discussed in detail in Section 2. An illustration of a synthetic diffractive structure is shown in Fig. 2c.
The most perfect autostereoscopic method is "true" holography. The hologram is capable of reproducing a wavefront of an object in its full complexity and therefore it works in fact as an autostereoscopic display with infinite number of viewing zones. An illustration of reconstruction from a hologram is shown in Fig. 2d.

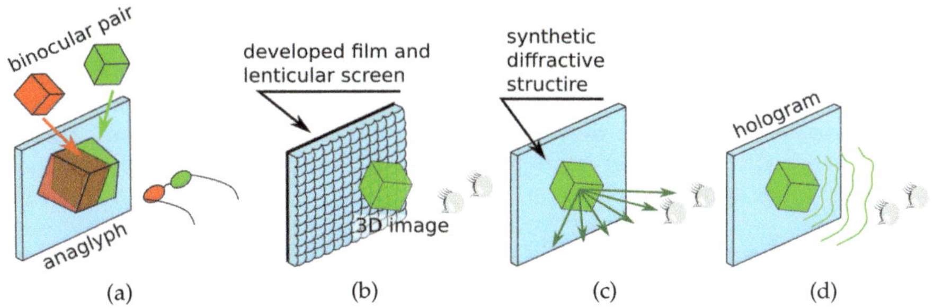

Fig. 2. 3D displaying methods: **(a)** In binocular methods, only two channels are employed which overlap each other in space. In this case, an anaglyph is shown. **(b)** Integral photography is an example of non-diffractive autostereoscopic methods, in which multiple viewing zones are spatially separated. **(c)** Diffractive autostereoscopic methods create the spatially separated viewing zones as in case (b). **(d)** Hologram is capable of reconstructing the image wavefront in its full complexity. It acts as if there is infinite number of viewing zones.

2. Synthetic holography

In this section, the basics of synthetic image holography are discussed. According to the above described principles of autostereoscopy, it is desirable to exploit the directionality of the diffraction of light on the hologram (or more generally on the diffractive structure) for creating the spatially separated image channels in space. The problem consists of two main tasks, first the acquisition of image data for recording of particular channels and the second dealing with the recording process itself. Thus in Section 2.1 the general ideas of decomposition of the three-dimensional signal to the set of two-dimensional views are described. In Section 2.2 two particular approaches to the multiplexing of the 2D information within a single holographic element are presented. In the last part (Section 2.3), more detailed aspects of the recording and reconstruction processes are analyzed such as the true-color mixing based on holographic elements, various spatial properties of the reconstructed image, and possible synthesis of the holograms with kinetic behavior.

2.1 3D image as a set of 2D views

The analysis of human vision (Section 1.1) showed that three-dimensional image perception can be artificially generated by creating the spatially separated discrete image channels in space. It has been also mentioned, that due to the inability of the human eye to register the phase properties of the incident optical wave (except for the directionality of the wave), the real image signal can be substituted by two-dimensional intensity views. The question is, how the discretization of the spatial channels should be done to create a satisfactory undisturbed 3D view. The method is obviously based on imperfections of the detecting system (in this case the human eye) in terms of a pupil size and limited resolution. It has also been denoted, that classical holography is the most perfect autostereoscopic method, as it reconstructs the signal wavefront in its full complexity. The interesting question arises if limiting the viewing channel size by the synthesis of the hologram from 2D views will converge to the "analogue" optical hologram. From the point of view of intensity distribution the classical hologram is apparently the limiting case of the synthetic construction, but there is a huge difference in phase properties of the reconstructed optical field. Although the phase behavior of the reconstruction must possess certain resemblance to that of the "ideal" hologram as the signal directional distribution is similar, particular spatial views are not mutually phase-synchronized as they would be from the classical hologram. This effect in principle does not affect the observation by human eye, but can cause problems when using the element for other applications.

Acquiring the source data

Both approaches presented in this chapter need to somehow capture the set of 2D images before the synthesis can be processed. These images can be obtained using a real camera taking pictures of a real object or using a virtual camera capturing a virtual 3D object in a PC. There is a variety of methods that can be used for this sampling of an object. The methods can be classified according to the shape of the trajectory of the camera and according to the direction of the optical axis of the camera. The way of capturing also depends on the demands on the 3D behavior of the reconstruction - if the final hologram should be composed of both parallaxes or if it should be single parallax only (Section 2.3.2). Fig. 3 shows one of the possible capturing setups for both the single-parallax and the full parallax case.
Assuming the single parallax case, the above figure shows the method in which the camera is moved through equal length steps on a straight line, while the optical axis of the camera

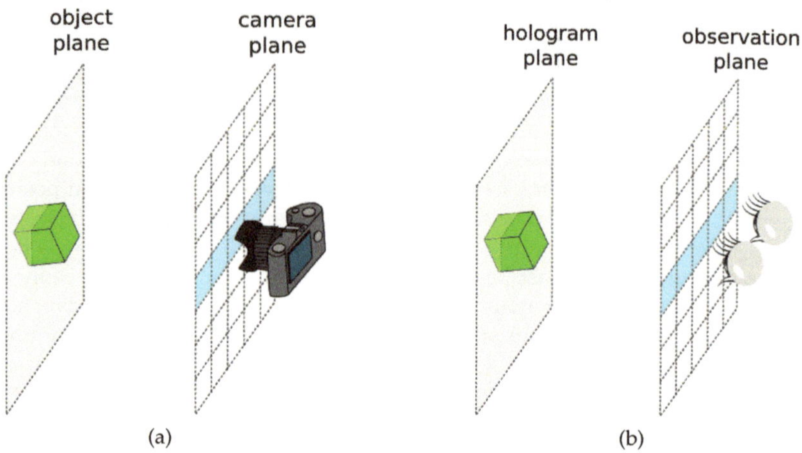

Fig. 3. **(a)** Capturing of the 2D images. The figure shows the equidistant capturing with linear movement of the camera. The blue line represents the single parallax case - in each "square" one snapshot is taken. The whole array represents the case of capturing both parallaxes when the camera moves in two mutually perpendicular directions. **(b)** The positions of the viewing zones are coincident with the positions of a camera during the capturing process.

remains perpendicular to the direction of movement. In this case, taking the snapshots with a real camera would require a special objective, so-called shifted-lens camera. However, this method is more often used in the case of capturing the virtual 3D object in a PC. Therefore, the need for a special objective can be bypassed by using "wide angle" capturing and cropping the recorded image. A big advantage of this method is a relatively simple recording setup for the master-hologram - with only linear translations. This is the method used in the experiments described in this chapter.

The other possible setup for a single parallax case is the equiangular method with the circular movement of the camera. This method is the simplest from the capturing point of view, since the camera can be static while the object rotates around a vertical axis. It is also the closest approach to the viewing setup when the observer holds the hologram and tilts it around the vertical axis. However, if the recording medium is not coated on a cylindrical surface, this method demands sophisticated image preprocessing of the 2D images to match the flat surface of the hologram plate.

The closest method to the viewing setup is the one when the camera makes equidistant steps on a straight line and the camera axis tilts towards the object. It resembles the case when the hologram is hanging on a wall and the observer moves around, turning his head towards the hologram. Nevertheless, it is much more complicated from the capturing point of view and, similarly to the equiangular method, the images would have to be processed in a computer before recording the master-hologram. The above mentioned methods are the ones that are the most widely used for the 2D data capturing.

If the final hologram should include both the horizontal parallax and vertical parallax, the process of camera movement should also include the vertical parallax. There are again several ways how to capture the spatial information. Similarly to the equiangular method with the circular movement of the camera, there is a "spherical" method for capturing the two parallaxes. In such a case the object turns around its vertical axis and the snapshots for the horizontal parallax are taken. The next horizontal "row" of exposures is taken after the object is turned about a small angle around its horizontal axis. In this manner the required range of angles in both parallaxes is taken. In Fig. 3, the planar method of capturing the two parallaxes is also provided - the "single parallax" line of capturing is repeated in the vertical direction. The axis of the camera is again kept perpendicular to the plane of movement.

This is how the continuous 3D information is sampled into a set of discrete 2D views. The following sections describe how various synthetic holograms are created from such a set of views.

2.2 Multiplexing views within a single hologram

In this section the basic approaches to the multiplexing of angular views obtained in the previous section are presented. There are many different ways how to realize different spatial channels but only those based on holographic or, more generally, on diffractive principles will be analyzed. As has been already mentioned, the decomposition of the 3D information into the two-dimensional channels is conditioned by the limited performance of the human eye. The first of the presented methods, named "synthesis at the hologram plane" is based on the limited resolution of the human eye in the image plane. The next approach, called commonly "synthesis at the eye-pupil plane" is based on the limited (finite) pupil size of the human eye which limits the monocular 3D perception. Both methods exploit holographic principles for image creation.

2.2.1 Data synthesis at the hologram plane

In optical holography, the relative positions of the object and the final hologram (the physical plate on which the recording has been made) can be arbitrary. However, during the copying process the object is usually placed in the close neighborhood of the final hologram for several reasons. The most important are dispersion and sensitivity to the imperfections of the reconstructing light source. When the image is formed close to the hologram, the blurring caused by the dispersion effects is negligible as is the blurring caused by the spatial incoherence of the light source. If we suppose such a geometry, we can assume that the image points and their projections to the hologram plane are approximately of the same size.

When the hologram is observed from the conventional distance $\sim$ 25 cm, the approximate resolution limit of the human eye is $\sim$ 0.1 mm. On the other hand, from the point of view of diffractive structures such a dimension is large enough for carrying much more than just the information about a single point. As a rough estimate, if we assume an elementary diffraction grating with period $\sim$ 800 nm and area $4 \times 4\mu$ m (which contains 5 periods of such a grating), we can place more than 600 different gratings within the area of $0.1 \times 0.1\,\text{mm}^2$. The main idea of image multiplexing at the hologram plane is based exactly on this calculation.

To understand the method one has to realize how the principle of stereoscopy applies for the given geometry. Let us assume the situation according to Fig. 3. If the observer sees the 3D image floating in the neighborhood of the hologram plane, it means, that each point on the hologram should send different information to the different directions (meaning that the visibility of the particular point on the hologram surface in terms of color and intensity is dependent on the observation angle). It is impossible to achieve such an angular functionality from conventional photography, where the angular luminosity is a smooth function. On the other hand, the idea of realizing an angularly selective image on the base of diffraction is pretty straightforward. Let us assume the diffraction of white light by the elementary regular diffraction grating according to Fig. 4a.

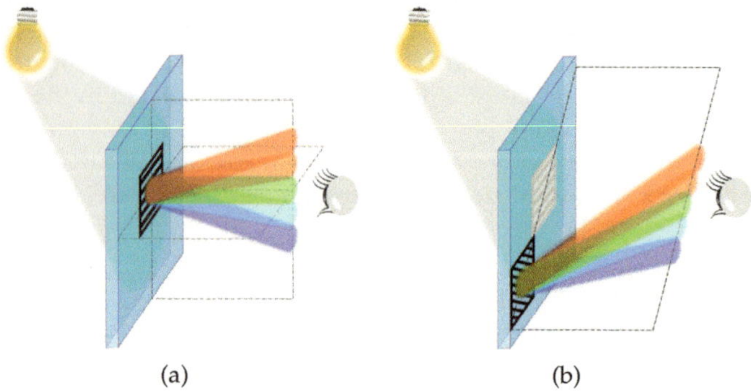

Fig. 4. Diffraction of white light on the thin diffraction grating under the common geometry - for horizontal grating **(a)** and rotated grating **(b)**.

Let the diffraction grating have the period 750 nm, the incidence angle is 45°, and the grating lines are perpendicular to the plane of incidence. If the light is monochromatic (with wavelength in the middle of the visible region, e.g. $\lambda = 530\,\text{nm}$), there will be only two diffracted beams according to the grating equation

$$\sin \alpha_m = \sin \alpha_i + m \frac{\lambda}{\Lambda}, \tag{1}$$

namely the 0-th order with $\alpha_0 = \alpha_i$ and the first diffraction order with $\alpha_1 \approx 0°$. When the light is polychromatic, there will be one "white" 0-th order and many 1st orders, one for each spectral color. They will slightly differ in direction according to the equation (1). If we specify the direction of observation to be the normal to the hologram at its midpoint, one will observe only a narrow spectral band of colors and, in our example, he will see the grating as green ($\sim$ 530nm). The described principle is used in synthetic holography for generation of the different colors from a single white light source. Namely if the grating period changes within the range from 600nm to 950nm, an observer looking in the direction perpendicular to the hologram will perceive the wavelengths from 430nm to 670nm, which is approximately the range of the visible light. Of course, if he changes his viewing direction, the perceived color will change according to the grating equation. Each image point can in principle change its color from blue to red through all rainbow colors; this is why such holograms are also called rainbow holograms [Benton, 1969]. The rainbow hologram can be observed in correct color composition only under the defined geometry (relative position of the light source, hologram, and observer). Otherwise the colors are shifted according to the equation (1). If there is more than one color in the image, particular colors will maintain their relative relations, but they will be all shifted in the same manner when the geometry is changed (e.g. when the hologram is vertically tilted).

It has been shown that the period of the elementary grating on the hologram surface determines the color observed from the particular direction if the hologram is illuminated with white light in the defined geometry. The next important parameter of the grating is the orientation of the grating lines. In the above example they were perpendicular to the plane of incidence of the reconstructing light beam. In general, the grating orientation can be rotated within the range of 360°. The effect of such a change is demonstrated in Fig. 4b.

Based on this analysis we can suggest the method for generating a synthetic 3D hologram from 2D views. As has been mentioned, in the given geometry 3D imaging requires the ability of the particular point on the hologram surface to send different signal in the different directions. Such a functionality can be accomplished using the elementary point segmentation according to Fig. 5.

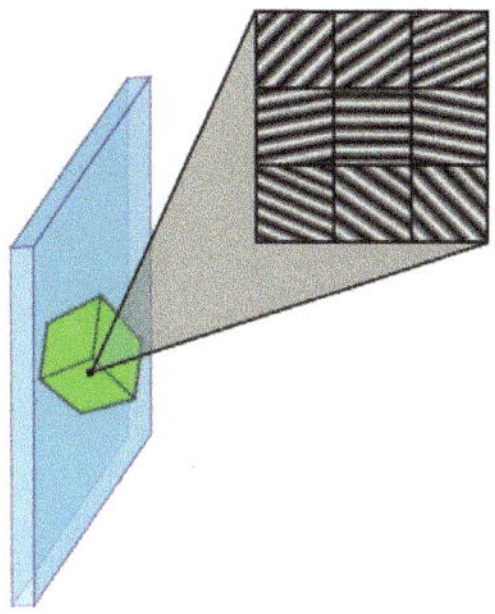

Fig. 5. Basic idea of generating the spatial image channels using diffraction gratings.

Each distinguishable point on the hologram surface (further called a macro-dot) is divided into a set of smaller units (further called micro-dots). Each micro-dot is filled with a regular grating with appropriate period and orientation. As shown earlier, the number of micro-dots with dimensions sufficient to contain a functional grating, which could fit within a single macro-dot, is several hundreds, so there is plenty of space for representing a high number of angular views of an object. In fact, it is not necessary to use such a high number of micro-dots for two reasons. First, most holograms synthesized using this method are of rainbow type, thus the vertical parallax is omitted and angular views are captured only in the horizontal direction. This significantly reduces the number of necessary micro-dots. Second, the finite angular resolution of the human eye and the finite eye-pupil size limit the number of horizontal views (of course this number is also influenced by the desired angular range). In practice, only a relatively small number of horizontal views is needed to create a satisfactory 3D perception. Usually 9-16 angular channels are used.

At this point it is necessary to mention that the spatial separation of the micro-dots within the elementary resolvable macro-dot is not the only method of multiplexing the elementary gratings. Depending on the holographic recording techniques, particular gratings can also partially or fully overlap within the macro-dot area. The multiple exposure of a single macro-dot area can record all gratings while maintaining their particular properties. However, it can be shown, that the potential drawbacks of this method have more serious consequences than those of the spatial multiplexing. Moreover, the exposure parameters are dependent on the particular grating parameters and the number of multiplexed gratings when exposing the same area multiple times, whereas they are completely independent when the gratings are separated spatially.

Unfortunately, the spatial separation also brings several unwanted effects to the image reconstruction, the fragmentation of the microstructure to the micro-dots being the worst of them. In practice, the often used dimension of the micro-dot is below 10μm. Thus the diffraction by the grating is affected with the diffraction by the aperture of the grating. Mathematically, the diffraction pattern consists of the convolution of the diffraction by the grating and by the rectangular aperture. This leads to the spreading of the diffraction order, introduces noise, and decreases the observed luminosity of the point. As a consequence, the micro-dot size must be chosen as a trade off between resolution and luminosity of the image. In Fig. 6 there are examples of the desired and real microstructure which has been synthesized

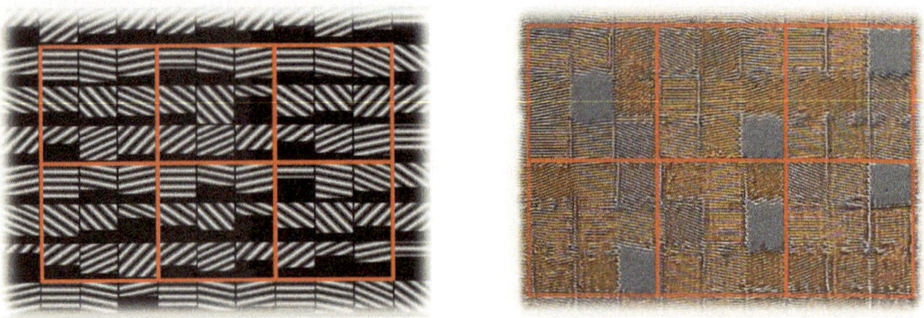

(a) (b)

Fig. 6. Example of the macro-dot segmentation **(a)** and the corresponding real micro-structure **(b)**.

using the described technique. The micro-dots are clearly visible together with the gratings inside them which have a periodicity of $\sim 600 - 950$nm depending on particular micro-dot and the dimension of micro-dots is $\sim 13\mu$m. The macro-dot (or the observable image point) has a dimension of $\sim 39\mu$m and is segmented into 9 micro-dots (there are 9 angular views distributed in horizontal direction, the hologram is of rainbow type).

The described synthesis at the hologram plane has many advantages but also several disadvantages in comparison with other hologram synthesis techniques. Because of the direct calculation of properties of each image point, the approach is very flexible. It enables us to calculate full 3D holograms with true-color RGB color mixing (see Section 2.3.1), kinetic behavior (see Section 2.3.3), and high contrast with low noise. There is one more aspect which should be mentioned, namely the possibility to fully adjust the luminosity of each image point. The overall luminosity of the micro-grating is given by its area and diffraction efficiency. To manage the luminosity continuously, either of these two parameters can be used. If the area of micro-dot is maintained, the diffraction efficiency can be changed by tuning the profile shape or the depth of the modulation (usually the depth of the modulation is tuned as it can be done easily by changing the exposure dose for the particular micro-grating). Unfortunately, it is relatively tricky to maintain the proper luminosity relations as the dependency of the depth of the modulation on the exposure dose can be highly nonlinear, apart from the potential problems with changing the dose independently for each grating within the chosen recording setup. Thus, the luminosity is usually driven by changing the micro-dot area and simultaneous maintaining of other parameters of the grating (shape and depth of the modulation). The micro-dot is divided into two parts, one filled with grating and the second unexposed (see Fig. 6a). This allows one to perfectly (linearly) tune the luminosity and set the desired luminosity relations over the hologram area. The main drawback is the further segmentation of the micro-dots, which enhances the diffraction effects on the grating aperture (which is now even smaller) and can lead to higher noise levels in reconstruction. Thus the proper choice of the dot size as a kind of trade off between the mentioned effects crucially influences the final quality of the reconstructed image.

2.2.2 Data synthesis at the eye-pupil plane

In the previous case, the area of the hologram is divided into the macro- and micro-dots, where the primary gratings are finally directly recorded using one of the approaches described in Section 3.1. In the case of synthesis at the eye-pupil plane, the microstructure of the final synthetic hologram is continuous without any segmentation. The discretization of the spatial information is achieved using holographic principles - an interference pattern of one reference beam and multiple signal beams is recorded to the synthetic hologram. Each of the signal waves would represent one particular viewing zone (see Fig. 3). To achieve this effect, the data synthesis at the eye-pupil plane consists of two steps. First, the synthetic master-hologram is created, where the particular 2D views are recorded and then the master-hologram is transferred into the final hologram. The name of the method is derived from the fact that the position of the former master-hologram coincides with the eye-pupil plane when viewing the final hologram.

The most instructive one and probably the most known method is the one shown in Fig. 7. There are four main steps included in this type of synthesis: **1.** sampling of the 3D object (obtaining the set of 2D views - Fig. 3a), **2.** recording of these 2D views (spatially separated) into the synthetic master-hologram (Fig. 7a), **3.** transferring these recorded views into the final hologram by a single shot (Fig. 7b), and **4.** viewing the final hologram (Fig. 3b).

The 2D images are sequentially displayed on the ground glass screen shown in Fig. 7a. The holographic plate is placed in front of the screen approximately within the plane, where the snapshots were taken from. The plate is covered with a shade with a vertical slit, so that only a narrow stripe of the same width (or smaller than the width) of eye pupil is exposed at each shot. The reference wave is incident from the same side as the signal wave. When the stripe

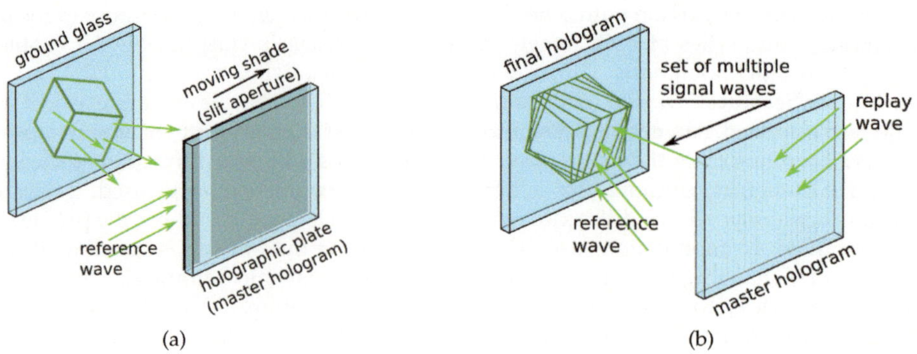

Fig. 7. Hologram synthesis in the eye pupil: **(a)** The 2D views are sequentially displayed on a ground glass screen and recorded as elementary holograms in the form of narrow stripes. This means that the master-hologram is composed of a number of these elementary holograms placed next to each other. **(b)** The transfer scheme. All the stripes from the master-hologram are replayed at once using a conjugate "reference wave" from (a). All the reconstructed images overlap in the former ground glass plane and the final hologram is recorded.

is exposed, the image on the ground glass is changed to the next view, the shade is shifted to the next position and the next exposure is taken. In this manner, all the views of the object are processed.

After developing the plate, this synthetic master is placed into the transfer scheme (Fig. 7b). All the recorded stripes are reconstructed at once and all the 2D views overlap in the former plane of the ground glass screen. The final hologram plate is placed exactly in this plane and the hologram is exposed. When viewing the final hologram, the stripes of the synthetic master-hologram are reconstructed in the exact position where the master was placed during the transfer process. The plane of the observer's eyes coincide with these stripes and each eye perceives only the view of the object from the correct direction (Fig. 3).

The main advantage of this method over the method of synthesis at the hologram plane is in the achievable size of the final hologram. Unlike the directly written synthetic holograms, there is no high demand on accuracy of the positioning of the recorded stripes (let us say about 0.1 mm). Therefore there are no big constraints on the recording device. Furthermore, the size of the final image is not limited by the size of the master-hologram, but by the size of the image on the screen. There are, of course, also several disadvantages depending on the particular setup. For example the main disadvantage of the setup in Fig. 7 is wasting the energy on the slit aperture. Although the energy loss can be lowered by shaping the reference beam, the reference beam must be shifted along with the shade, which makes the exposure complicated. There are, however, ways how to make the synthesis more efficient in a simpler setup (Section 3.2). The general disadvantage of these methods is in lower contrast compared to the directly written synthetic holograms.

The hologram described in the above section lacks vertical parallax. Therefore, it is used for creating a single-parallax rainbow hologram. The setup can be modified to create various types of final holograms (full parallax, color, etc.). The following section shows the possibilities of creating such types of holograms using the two discussed synthetic approaches.

2.3 Advanced properties of synthetic holograms

The following paragraphs provide information about the most common types of synthetic holograms and their properties, that can be created using the discussed methods. It is concerned with color holograms, full parallax holograms and kinetic holograms.

2.3.1 Color mixing in synthetic holography

Since the reconstruction of the hologram is based on the diffraction of the light on the hologram structure, the observed colors are spectral ones. If a non-spectral color is required (e.g. white), the additive color mixing must be introduced.
Additive color mixing is based on the colorimetric theory. It can be described by the means of CIE 1931 color space. The CIE defines a so called standard colorimetric observer and the color sensation experienced by the observer is determined by the spectral sensitivity of his eye cells (the cone cells). There are three peaks of their sensitivity. Hence, the sensation can be described using three parameters (X, Y, and Z), so called tristimulus values. These values can be used for computing the coordinates of a color in the well known CIE xy chromaticity diagram. More details of colorimetry can be found e.g. in [Wyszecki and Stiles, 2000].
The most common way of achieving a mixed color is to use two or three spectral colors, usually red, green, and blue (RGB), although the color mix can be achieved from another colors. As was stated the hologram is a diffractive structure, thus these three colors are obtained by the means of diffraction gratings.
Considering the data synthesis at the hologram plane, these gratings are created directly in each particular dot. The true-color dot is divided into three smaller areas, each filled with one grating. The relative areas of these three gratings control the contributions of each of the colors to the true-color mix. Such an array of "mixed" dots can be seen in Fig. 8.

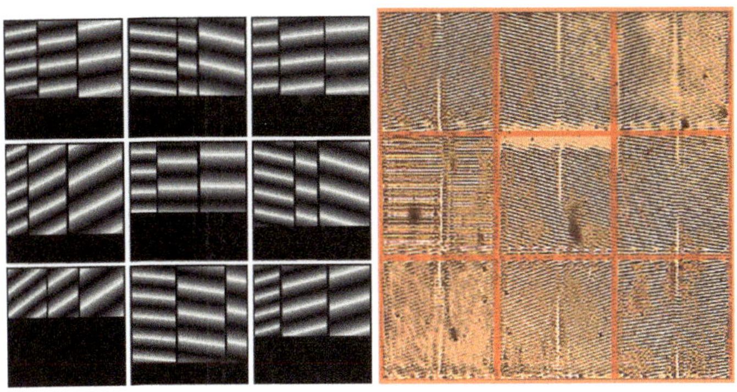

(a) (b)

Fig. 8. Macro-dot segmentation for RGB color mixing **(a)** theoretical idea and **(b)** image of a real microstructure.

The area of the dot is divided into two major parts - one filled with grating and the second empty. The ratio of these areas defines the overall luminosity of the image point. The grating part is further segmented into three parts each of them filled with one basic grating for color mixing. In principle, the color can be mixed from two, three or more gratings, but the most common is the three-color mix. The particular basic gratings are chosen individually for each mixed color as the closest components in the colorimetric triangle. In a case of a true-color 3D synthetic hologram, each dot is divided into micro-dots corresponding to the spatial channels (see Section 2.2.1), where each such micro-dot is further segmented into the color components. As already mentioned, such complicated fragmentation can negatively affect the efficiency of the hologram and overall noise level, so the dot size must be carefully chosen.
In the case of the eye-pupil synthesis, the area of the final hologram is not composed of any elementary dots. Thus the color mixing must be synthesized during the recording of the master hologram. If three different lasers are used for synthesis (with proper wavelengths corresponding to the desired RGB mix), the synthesis is relatively simple, all angular channels are exposed three times (once for each color) within the same area of the recording medium. For each exposure, the recorded 2D view contains only particular color channel of the image. However, three-laser synthesis is relatively rare, usually the complete recording process is performed using a single laser (including the copying of the master to the final hologram). Thus the master hologram for color mixing (in a case of color rainbow hologram synthesis) consists of three independent vertically shifted rows of exposures. The shift between the three rows (color channels) can be calculated from analysis of the dispersion effects accompanying the diffraction by the final hologram [Saxby, 1994]. Each of the signal waves coming from the three channels carries the same spatial information about the object. The correct true-color mix is achieved by controlling the relative intensities of these waves. Fig. 9a shows copying of such an RGB synthetic master into the final hologram. Fig. 9b shows the diffraction of white light by the final rainbow hologram.

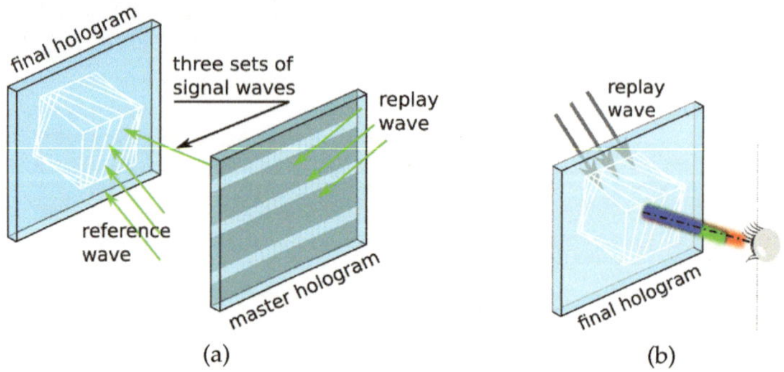

Fig. 9. **(a)** Transfering three color masters into final RGB color rainbow hologram.
(b) Reconstruction of the final hologram from the step (a) with white light replay beam. Only the orders that are diffracted in the direction of the observer are indicated in the figure.

In Fig. 10a there is an example of the RGB color master for a rainbow hologram synthesized using the method described in Section 3.2. The final hologram (copy of the master) is displayed in Fig. 10b.
Color mixing as shown in Fig. 9a is exactly correct for the plane of the final hologram. If the color point is outside of the hologram plane, it gets distorted due to the different axial

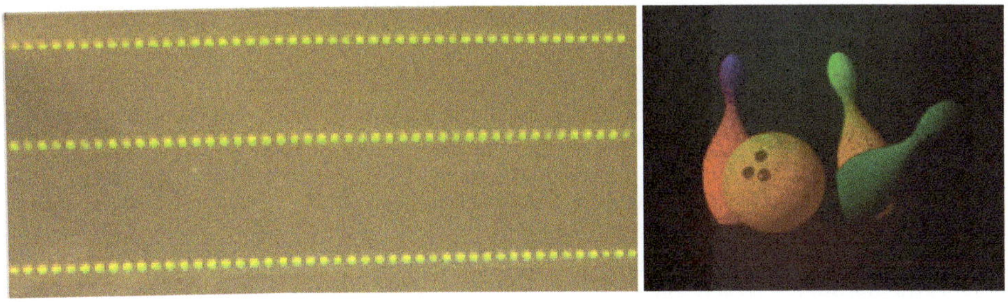

(a) (b)

Fig. 10. **(a)** Photograph of the "RGB color" master-hologram. The three color channels can be clearly seen. **(b)** Photograph of the final color synthetic rainbow hologram, that was obtained using the master from step (a).

magnification for the three wavelengths that actually compose this particular point. This issue can be solved by a proper "pre-distortion" of the image data before the recording process.[1]
The described techniques for color mixing utilized the rainbow effect (diffraction of white light by the volume transmission or thin grating). However, the color can be synthesized also in volume reflection holograms, where all steps are complicated by the Bragg condition (the dispersion cannot be used for generation of whole spectral range of colors as the grating recorded with the given geometry will not work for other than the design color). Nevertheless, a reflection color hologram can be synthesized using special techniques [Hariharan, 1996].

2.3.2 Single and full parallax synthetic holograms

In general, the classical hologram can record the complete 3D properties of the object - including both horizontal and vertical parallaxes. However, the parallax perpendicular to the fringes on the hologram surface can potentially suffer from strong dispersion effects. If the reconstructing light source is monochromatic, the dispersion will not apply and full three-dimensionality can be easily achieved. In a case of polychromatic reconstruction (which is the usual demand), the dispersion effects can lead to unacceptable rainbow blurring of the image.
Let us assume the common geometry of reconstruction according to Fig. 4a. The reconstruction beam is coming from the top and the observer is viewing the image from the front of the holographic plate. Thus the main direction of the fringes is horizontal. Besides

[1] The position of the non-compensated reconstructed color point differs from the original object point due to the different axial magnification for the particular reconstruction wavelengths. In the case of green recording wavelength ($\lambda_G = 532\,\text{nm}$), assuming the reference and reconstruction waves are planar, the axial magnifications for the particular color channels are $\alpha_R = 0.84$, $\alpha_G = 1$, and $\alpha_B = 1.13$ (for the red wavelength $\lambda_R = 633\,\text{nm}$ and blue wavelength $\lambda_B = 470\,\text{nm}$). Since the hologram is mostly synthesized from the 2D views obtained from the computer, this distortion can be compensated by pre-distorting the 3D model on the computer in the opposite manner. To overlap in the plane of the hologram the red point has to be located on the "beam" coming from the red channel and the blue point on the beam coming from the blue channel. To compensate for the distortion in the z direction, the red "object" has to be axially magnified by the ratio $1/\alpha_R$ and the blue one shrank by the ratio $1/\alpha_B$. During the 2D images processing, the proper pre-distortion can be introduced to each color channel. When reconstructed, the spectral deformation causes all the three channels to overlap in the position of the green channel.

the reconstruction by the monochromatic source (which is also possible today thanks to the wide availability of semiconductor laser sources) there are two principal solutions of this problem. The first is to use volume reflection holograms which can be highly selective to the reconstruction wavelength - it means the holographic reconstruction will appear only for a narrow band of wavelengths also in a case when the light source is polychromatic. The second approach exploits the idea of rainbow holography - the vertical parallax is omitted and the object maintains its 3D properties only in horizontal plane. When observing such a hologram from different directions in vertical plane, the object is seen in different "rainbow" colors. However, such a single parallax hologram can offer a very truthful 3D perception thanks to the geometry described in Fig. 4a and the horizontal position of human eyes. The missing vertical parallax disrupts the observation less than one might expect.
In Fig. 11, there is an example of the recorded master for the rainbow hologram created using

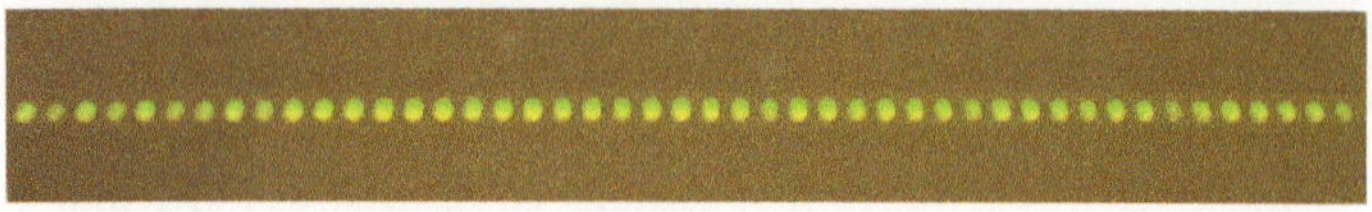

Fig. 11. Photograph of single parallax master-hologram. Each "shining" holopixel represents one view of the object.

the synthesizing method described in Section 2.2.2. Each of the "shining" spots in the figure represents recorded interference pattern of signal and reference wave. These spots are further denoted "holopixels". As it is seen, the process is very simple as only a small number of exposures is needed. Also the data acquisition from the object or computer model is much simpler than in the general full parallax case. The "single row" master is copied to the final plate in the transfer scheme and the rainbow hologram is created. The final hologram can be thin or thick, white light reconstruction is possible and the dispersion effects will only create the rainbow reconstruction under different directions of observation in the vertical plane. By contrast the monochromatic reconstruction is undesirable, in such case the reconstruction is visible only in a narrow stripe.
The full parallax hologram can be also synthesized at the eye-pupil plane. In such a case, the views must be acquired in both vertical and horizontal directions and recorded in a two-dimensional field. In Fig. 12a, there is a typical full parallax master recorded using the method described in Section 2.2.1. The exposed area corresponds to the future region where the observer's eyes will move. Finally, the recorded master must be copied to the reflection copy in the classical holographic transfer setup. In Fig. 12b there is an example of the image reconstructed from a full parallax reflection hologram fabricated using the described method. When using the synthesis at the hologram plane, usually rainbow holograms (single parallax) are synthesized. The rainbow effect is used for achieving proper color composition and white light diffraction is essential for proper image synthesis. The common techniques for recording such holograms are also usually capable of recording thin gratings only, so the volume reflection hologram cannot be easily recorded. However, the volume recording is theoretically possible using one of the following methods.
If the hologram is designed for monochromatic reconstruction, it can maintain both parallaxes and, based on the thin gratings of various period and orientation, the full 3D images can be constructed. Such a "master" can be copied in laser light to a different plane and then again back to the original plane using volume recording materials and a reflection holographic recording setup. This method gives the desired result, but the complexity of the approach is pretty high and most advantages of the direct recording at the hologram plane will vanish.

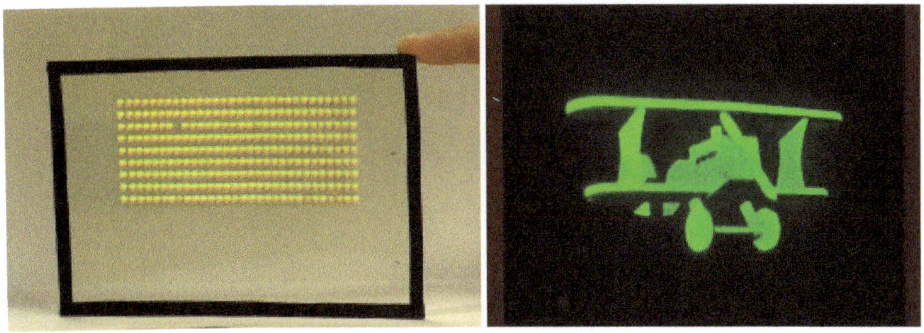

(a) (b)

Fig. 12. **(a)** Photograph of full parallax master-hologram. The holopixels are recorded in both the horizontal and the vertical direction. **(b)** Photograph of reconstructed image from reflection hologram, that was created using the synthetic master from (a).

The second possible approach to synthesizing a volume reflection hologram directly at the hologram plane is based on using a volume material already for the primary recording. However, in such a case the elementary micro-gratings must be recorded as volume reflection elements. The method requires full control over the direction of the recording beam and in the case of color holograms also the possibility to change the reference beam direction. Recording of this kind is very complicated and most of the advantages of the synthetic approach will be suppressed. In Fig. 13, there is an example of the rainbow 3D hologram with true-color RGB mix recorded at the hologram plane using a dot-matrix writer.

2.3.3 Kinetic behavior and holographic animations

The above described methods demonstrated the ability of the diffractive structures to create a system of spatially separated image channels in observation space. Such a system can be used not only for simulating the three-dimensional behavior of the object, but also for introducing dynamic effects into the viewed scene. The main idea of the synthesis of holographic animation is straightforward. Instead of the different angular views of the object, particular channels can contain the views of the object at the different time instants. However, it is necessary to analyze how the dynamic effect will influence another parameters of the reconstructed image, particularly how can the dynamic behavior be combined with the three-dimensional effects. There are in principle two main groups of dynamic effect which can be synthesized. The first is the continuous animation when the same scene is observed from all viewing directions but with some continuous incremental changes between neighboring channels. The second group consists of effects when the observed scene completely changes if the observer moves his head or tilts the hologram - such effects are usually called flip-flop.
When the continuous holographic animation is to be synthesized, the crucial problem can arise from interference between the animation and the 3D effects. If the time development is introduced in the horizontal plane, each eye of the observer will perceive a different object (taken at the different time instant). Such disproportion can rapidly disturb the observation. To remove the problem, the reconstruction can be observed with one eye only, which means

Fig. 13. Example of the real reconstruction of the true-color RGB synthetic 3D rainbow hologram designed using the synthesis at the hologram plane and fabricated using dot-matrix laser lithography (in scale).

the three-dimensionality will be perceived only through the movement parallax. Another solution could be based on introducing the time development in the vertical direction. Such an approach automatically suspends application of rainbow holograms as the rainbow effect occupies the vertical parallax. However, for full parallax holograms the approach is possible. The recording process is very similar to the one from Section 2.3.2 with the only difference in image contents for particular rows of exposures (while there the rows corresponded to different vertical views of the object, here each row corresponds to different time states of the scene). As known from rainbow holography, the vertical parallax is not crucial for 3D perception so this approach can offer satisfactory 3D reconstruction with dynamic behavior when moving the observer's head up and down.

Practically, because of the complicated process of recording of volume reflection holograms, many holographic animations are based on the combination of rainbow hologram with dynamic behavior. Both 3D effect and the time development are included in the horizontal direction and the observer eyes are forced to perceive a kind of unnatural view. However, if the time development of the scene is slow and smooth (there is no complete change of the objects but only small parts are moved), the reconstruction can be satisfactorily observed including both the 3D and the dynamic effects.

When the dynamic flip-flop behavior is desired, the above described problems will not apply. For example, if two different objects should be observed, one from the left and second from the right, all channels left of the normal will contain only angular views of the first object and all channels right from the normal will contain the views of the second one. The transition zone, when the left eye of the observer sees the left object and the right eye sees the right one, is usually ignored. In principle, several different objects can be recorded in a single hologram. Both approaches - full parallax (volume reflection holograms) and single parallax (rainbow holograms) can be used when appropriate requirements are met.

In principle, the analysis above is independent of the chosen recording technology. Both synthesis at the hologram plane and eye-pupil plane can be used. Of course, limitations similar to those discussed in Section 2.3.2 can arise when synthesizing the volume reflection hologram at the hologram plane.

3. Recording technology and materials

There are various techniques which can be used for recording of synthetic holograms. In principle, they can be divided into two major groups. The first group consists of approaches where the diffractive microstructure is recorded point-by-point with high resolution. The devices used are usually the commonly available direct writers such as electron beam or laser beam lithography writers. The second group comprises approaches based on classical holography where the microstructure is created "automatically" using exposure with an interference field of two or more large laser waves.

3.1 Direct-write lithography techniques

When synthetic image holograms are calculated point-by-point, a recording device with very high resolution is needed. The typical periodicity of the hologram microstructure is for most holograms within the range 500 – 1000nm. In this case, direct-write lithography is usually used for recording. In principle, three different approaches can be chosen according to the particular application, namely the e-beam lithography, focused laser lithography, and matrix laser lithography.

The finest microstructure can be recorded using a focused electron beam. Very high resolution can be achieved using the e-beam writing technology, where feature size can be on the nanometric scale. However, for image holograms such extreme resolution is not necessary. As has been shown in Section 2.2.1, the typical synthetic image hologram consist of a set of regular microgratings with periodicity just below one micron and dimensions of several microns. The e-beam lithography can easily achieve these parameters, but with very high costs and several restrictions following from the principles of the technique. The strongest limitations are usually the overall size of the hologram which can hardly exceed several cm^2 and complications connected with the fact, that the recording must take place in vacuum and on conductive recording material. In case of electron beam the interference effect cannot be exploited in order to simplify the exposure process. The exposure itself is usually extremely long (many hours to days according to the area to be exposed).

A very similar technique to e-beam writing is laser lithography with a focused laser beam. The recording process is practically identical to the e-beam recording except that instead of an e-beam, a focused laser beam is used. Usually visible lasers with wavelength in the blue part of the spectrum are used. The biggest advantage of the process is operation in normal conditions (no vacuum is needed) and overall costs of the process. The biggest disadvantage is limited resolution given by the focusability of the laser beam (which depends on the recording wavelength and numerical aperture of the focusing system). The grating periodicity of hundreds of nanometers is more or less on the edge of capabilities of the technology.

The most advanced technology for recording the optical microstructures is the matrix laser lithography. In these devices, the elementary grating is recorded as a pattern of interference between two focused laser beams. Thus the parameters of the grating (periodicity and orientation) are given by the angle between the beams and their plane of incidence (instead of the size of the focused spot). Moreover, a whole micro-grating with area of hundreds of μm^2 is exposed at once in a single exposure, which rapidly speeds up the recording process. The original so-called dot-matrix devices used two physical laser beams with mechanical alignment of their geometry. However, such an approach could not ensure sufficient resolution for shaping the micro-gratings, precise alignment of the gratings within larger areas, or massive segmentation needed by multi-channel RGB stereograms. Recently, the advanced devices have been developed based on the projection of a microstructure from the computer driven micro-display with large demagnification. The basic idea of such a device is depicted in Fig. 14.

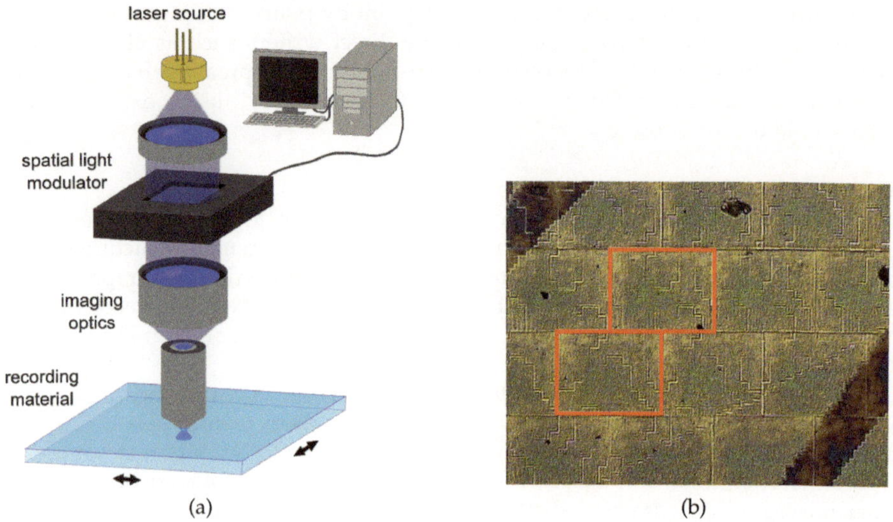

Fig. 14. **(a)** Basic setup of the matrix laser writer. **(b)** Photograph of exposed photoresist.

The system of micro-gratings (with general parameters) is projected on a micro-display (usually a liquid crystal based spatial light modulator - SLM). The display is imaged on the recording material in laser light with large demagnification (several hundred times). In a single exposure a relatively large area is exposed at once, which can contain different gratings with arbitrary parameters. Micro-gratings within this area (typically $\sim 0.01 - 0.05\,\text{mm}^2$ are perfectly phase-synchronized and also a completely aperiodic microstructure can be recorded. The exposure is performed multiple times and between subsequent exposures the recording material is moved. The device can operate in typical room environment and can easily reach a recording speed of several cm^2 per hour. In Fig. 14b there is an example of the microstructure obtained using the described technology. Particular exposures in the form of rectangular tiles are clearly visible. Each tile contains a system of micro-gratings.
The advantages of the direct writing are high flexibility of the recorded microstructure and excellent properties of the reconstructed image (high contrast, low noise, true-color RGB, etc.). The main disadvantages are high costs of the technology and the area of the hologram, which rarely exceeds several tens of cm^2.

3.2 Wide-beam interference based methods

As has been stated in Section 2.2.2, the synthesis in eye-pupil plane requires two steps to create the final synthetic hologram. The second step is more or less the standard holographic transfer copying. Concerning the first step - recording the synthetic master-hologram, there are various approaches [DeBitetto, 1969; Huff, 1981; Ratcliffe, 2003]. The one that we are going to present in more detail omits the ground glass with the focused image (Fig. 7a) and it also saves energy by shaping the reference beam and omitting the slit aperture. This particular setup differs from the one shown in Fig. 7 - the image wavefront is recorded before it actually creates the image. The principle of the method is explained in Fig. 15.

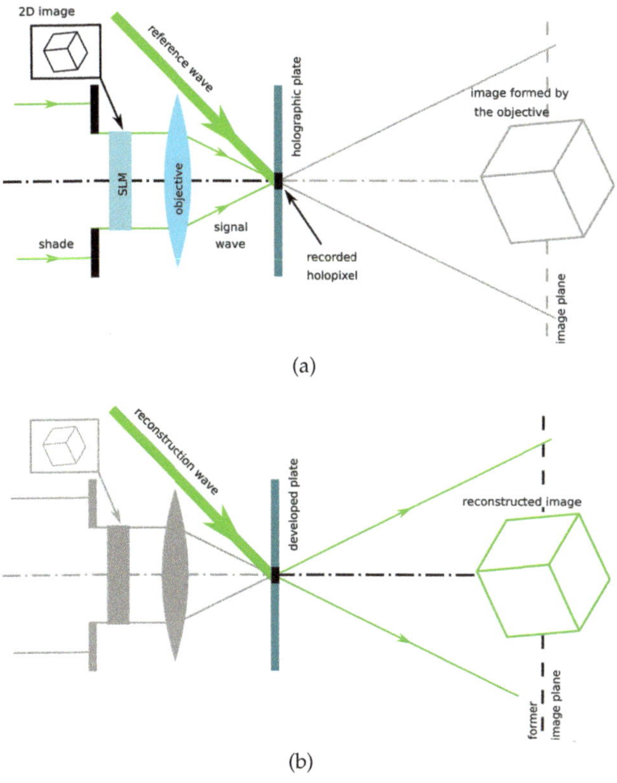

Fig. 15. **(a)** Recording a single 2D image near the output pupil of the objective. **(b)** While replaying with the "reference" wave, the magnified image is focused in the former image plane. The magnified image in the image plane is two-dimensional, lying in the plane perpendicular to the plane of paper (the correct projection of the "big cube" in the image should be a line segment).

The recording process itself is in the following sequence. The 2D image, representing a particular view, is uploaded to a transparent SLM. The signal wave passes through the SLM and the 2D image is focused by the "writing" objective onto its image plane with some magnification. Near the output pupil of the objective, the pattern of interference between the signal and reference beam (the *holopixel*) is recorded onto a holographic plate. Then the plate is moved, the image on the SLM is switched for the next view and the next holopixel is recorded, and so on. The process is more complicated, since the movement of the holographic plate needs to be compensated by moving also the SLM.[2] The scheme of the particular setup can be seen in Fig. 16. All the views obtained by sampling the object are recorded in this manner and the plate is developed and bleached. Since the reference wave is collimated, the replay beam can be a wide collimated wave that reconstructs all the recorded holopixels at once. The reconstructed images from all the holopixels overlap at the place where the object was situated during capturing. The final hologram plate is placed into this place and the final hologram is recorded (similar to Fig. 7b). The device is driven by a PC - the images are sequentially uploaded to the SLM and the two $x - y$ stages are operated as shown in the figure.

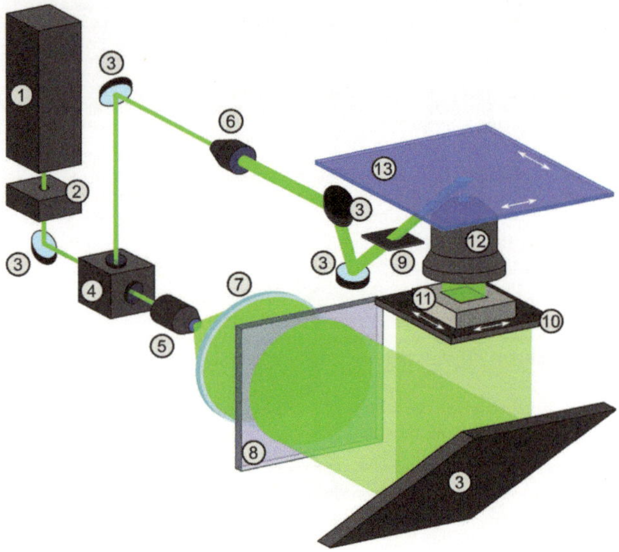

Fig. 16. Scheme of the eye pupil synthesis device: **1** laser, **2** shutter, **3** mirror, **4** beam splitter, **5** microscope objective with spatial filter, **6** beam expander, **7** collimator, **8** holographic diffuser, **9** square aperture, **10** signal shade, **11** SLM, **12** special objective, **13** holographic plate.

3.3 Commonly used recording materials, light sources, and mass replication techniques

A very important issue in the image holography is the replication process. For many applications it is desirable to produce the final hologram in large quantities (e.g. holograms

---

[2] Since the master-hologram plate is moved during the recording process, the relative position of the recording device and the already recorded images is changed. But in the final hologram, all the recorded views have to "emerge" from one particular place to form the 3D impression. That is the reason why all the possible recording setups either have to include tilting the objective or have to implement some kind of movement compensation. There are several possibilities for the mechanical movement that can be considered. However, the simplest and the most robust way from the mechanical point of view is one where the "writing" objective is static. In such a case, the signal spot is always in the same position and the reference beam also does not have to be moved. The only moving elements would be the holographic plate and the SLM (as the compensating element). The step of the holographic plate is equal to the size of the holopixel. The step of the SLM is in the opposite direction and it is smaller simply in the ratio of the objective magnification.

for document security applications are often made in series of a million). Thus it is extremely important to look for a proper technology which could enable cost effective production of large series. In classical image holography the holograms are usually copied optically in an optical setup, which is still a relatively expensive process. Volume gratings could not be copied in any other way. However, if the grating is of the relief type, it can be also copied using some of the imprint techniques (like mechanical embossing, etc.). To make this possible, the hologram must be exposed in proper relief recording material. The light source used usually depends on the requirements of the particular recording medium.

For exposure of relief gratings photoresists are often used (if exposing with laser beam). For e-beam exposure the electron beam sensitive resists are used. Unfortunately, the gratings in photoresists are usually thin, so this material can be used only when the volume properties are not needed (theoretically, the relief gratings in resist can be "thick" and possess volume effects, but they can not be copied using an embossing technology). Photoresist can be used for all exposure steps or the hologram can be finally transfered to the resist material from a different recording medium. The spectral sensitivity is maximum in UV part of the spectrum so photoresists are usually exposed using short wavelength visible sources (such as semiconductor lasers with wavelength $\sim$ 400nm or gas lasers with wavelengths within the range 400 − 460nm). Because of their overall low sensitivity the photoresists are not perfectly suitable for primary exposure of large areas.

After the hologram is recorded in the relief material such as photoresist it is usually metalized and copied to a hard metal (usually Nickel) relief copy using the electro-forming process. Such a matrix is then used for mechanical embossing.

For recording of the volume gratings whole range of holographic materials can be used according to particular needs. Silver halide gelatin, dichromated gelatin, and some photopolymers can satisfy the requirements. In contrast to the resist materials, they can be easily sensitized for laser sources within the whole visible range. Recently, diode pumped solid state lasers are widely used in holography because of their high output power, very good coherence properties, high efficiency, and relatively low operational costs. Very common are 532nm sources based on the second harmonic from Neodymium doped active material. For further details concerning the recording materials and other components necessary for the recording process see [Bjelkhagen, 1993; Collier et al., 1971].

The synthetic image holograms can be constructed in various ways according to the particular application and other demands on the hologram properties. Within this chapter only the basic ideas have been presented. Within the Optical Physics Group at the Faculty of Nuclear Sciences and Physical Engineering of the Czech Technical University in Prague, various approaches to image synthesis are researched. Besides research in the field of holographic techniques, also recording materials are developed and automated recording devices are constructed. All samples presented in this text have been fabricated using the technology available at the Optical Physics Group.

4. References

Benton, S. A. (1969). A method of reducing the information content of holograms, *Journal of the Optical Society of America*: -59–1545

Bjelkhagen, H. I. (1993). *Silver-Halide Recording Materials for Holography and their Processing*, Springer-Verlag.

Collier, R. J., Burckhardt, C. B. & Lin, L. H. (1971). *Optical Holography*, ACADEMIC PRESS, INC.

DeBitetto, D. J. (1969). Holographic panoramic stereograms synthesized from white light recordings. *Applied Optics*: -8–1740.

Faubert, J. (2002). Influence of optical distortions and transverse chromatic aberration on motion parallax and stereopsis in natural and artificial environments, *in* Javidi, B. & Okano, F. *Three-dimensional Television, Video, and Display Technologies*, Springer-Verlag New York Inc, pp 359-396.

Ferris, S. H. (1972). Motion parallax and absolute distance. *Journal of experimental psychology*: -5–258.

Halle, M. (1997). Autostereoscopic displays and computer graphics. *Computer Graphics*: -31–58.

Hariharan, P (1996). *Optical Holography*, Cambridge University Press, 2nd edition.

Huff, L. & Fusek, R. L. (1981). Cylindrical holographic stereograms. *International Symposium on Display Holography*: -1–149.

Imax (2011). URL: http://www.imax.com.

Infitec (2011). URL: http://www.agentur-best.de/grobritannien-uk/index.php.

Najdek, D. (2008). *Synthesis of three-dimensional full-color images using the optical diffractive structures*, PhD thesis, CTU in Prague.

Ratcliffe D., Vergens, F., Rodin, A. & Grichine, M. (2003). *Holographic printer*, U.S. Patent 2003/0156308 A1.

RealD (2011). URL: http://www.reald.com.

Saxby. G. (1994). *Practical Holography*, Prentice Hall International.

Škereň, M. (2006). *Computer Generated Optical Diffractive Structures*, PhD thesis, CTU in Prague.

Takanori, O. (1978). *Three-Dimensional Imaging Techniques*, Academic Press.

Turner, T. L. & Hellbaum, L. F (1986). Lc shutter glasses provide 3-d display for simulated flight. *Information Display Magazine*: -2–22.

Wheatstone, C. (1838). On some remarkable, and hitherto unobserved, phenomena of binocular vision. *Philosophical Transactions*: -128–371.

Wheatstone, C. (1852). On some remarkable, and hitherto unobserved, phenomena of binocular vision (continued). *Philosophical Transactions*: -142–1.

Wyszecki G. & Stiles W. S. *Color Science - Concepts and Methods, Quantitative Data and Formulae*, John Willey & Sons, Inc.

Zone, R. (2003) Good old-fashion anaglyph: High tech tools revive a classic format in spy kids 3-d. *Stereo World*: -29–11.

12

Holoimages on Diffraction Screens

José J. Lunazzi
Campinas State University
Brazil

1. Introduction

What is a hologram? Even if holography was described in a single paper by its creator, there are many descriptions for such a widely divulged phenomenon, known all around the world. Many techniques and elements are entitled "holographic", but they can be classified in two main groups, the "academic" and the "popular" ones. I realized this in July 1989 in a Bulgarian holography meeting when showing my white light holographic screen to Yuri Denisyuk, whom I consider to be the second inventor of holography. He asked if the image I was showing came from a hologram, and my answer was the question "What is a hologram?" His answer was: "Does it employ a reference beam?". My answer was no and then I learned how to introduce the holographic screen techniques in science, not as holography, which is a combination of interference, recording and diffraction, but as a combination of interference, recording, projection of images acquired by any other technique, and diffraction. The projection is made on a fine diffracting structure of about 1,500 lines/mm in such a way that each eye receives a different image which corresponds to the parallax of a 3D scene. But when I showed my projections to people they mostly believed they saw holograms. For them, a hologram is an element which shows 3D in an at least apparent parallax without needing any complementary goggles for the eyes. I call this a popular definition of holography and it can be applied to holographic screens and to auto-stereoscopic systems, provided they reach at least apparent continuity. Non-diffracting auto-stereoscopic techniques are hardly trying to reach this.

A holographic screen, which from now on I will name commonly as a diffractive screen, consists basically of the hologram of a diffuser whose format is designed to create an observation space for the image projected on the screen. This observer's position field is obtained using reverted illumination, i.e., illuminating the screen in the opposite direction to the reference beam. In this way we can generate the more directional screen which is possible nowadays, in large format and employing lightweight and unbreakable materials. Gabor himself tried some ways to make stereoscopic screens without the need of additional goggles or filters (1). The screen obtained by recording an interference pattern, in a holographic manner, is a way for doing that.

2. The hologram as a diffuser

The construction of a surface that generates a luminous distribution at will is not a simple task. Even assuming that, as the light is going to reach a long distance, its distribution in a plane may correspond to the Fourier transformation of the transmission or reflection surface properties. It is not completely correct to suppose that, because the Fourier transformation can only be applied in a paraxial approximation, and the angles we need to achieve do not always have this limitation. But, assuming the hypothesis, we could obtain the desired surface profile by using the inverse Fourier Transformation of a light distribution that we define as the desired one on the lighted plane.

How can one construct such a microstructure on the surface? Which machining technique would be useful? Could we replicate it in a rapid and cheap way?

We understand that if we make the hologram of a surface of any profile the surface of the resulting hologram is effectively a diffuser with the diffracted light intensity profile of the original surface. It is a fact that diffractive elements made by holography begin to be employed for designing illuminating systems, working with monochromatic or nearly monochromatic light (2). Although the efficiency of an easily replicated hologram is less than 25%, the concentration of the light strictly at a desired region may compensate and even overcome the problem in many cases, and holographic diffusers are on sale by many companies, one of the applications being the internal illumination of computer displays (3).

3. The diffractive screen helping to fulfill the observation field in multiple projections

The surface which, like a plane mirror or a hologram, should generate a light ray distribution which is equivalent to the originally generated by a three-dimensional object must have the capability to emit in a large angular field with directional intensity and color values corresponding to the scene which is being reconstructed. Each ray is the component of a luminous point to be constructed outside the surface, generally, in front of it. The three-dimensional image which floats in front of a screen is the one that most impresses the public, even more when the observer can pass his hand through.

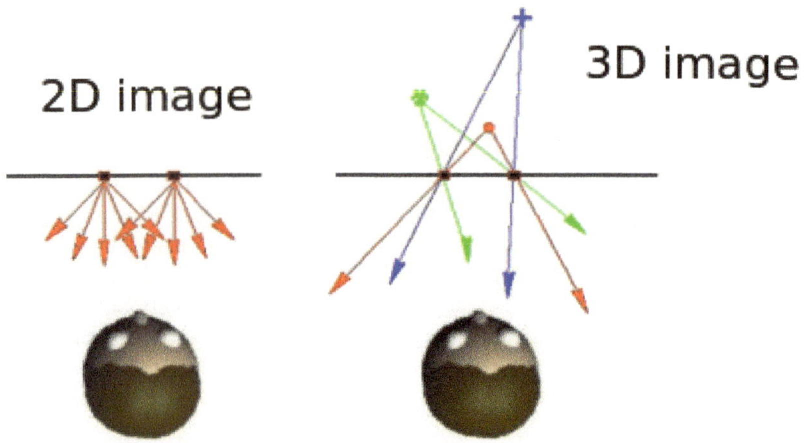

Fig. 1. Left: Ordinary diffusion at a screen. Right: Rays in an ideal three-dimensional display.

The capability to make such an element within a so-called pixel for a TV or computer display does not yet exist. A theoretical way to achieve it was patented (4) based on the location of a 2D micro-display at the focus of a small lens, each micro-pixel element being a ray generator. The number of pixels of a modern display should correspond to the number of volumetric pixels, also called "voxels". Each voxel is the origin of many rays and each ray comes from a sub-pixel of the ideal complete screen, so that we may think these as "ray pixels". An image in a volumetric space with the present pixel display capability for each voxel involves the need for much more pixelation at the emitting surface. The required resolution can be considered as the square of the present capability of 2D screens.

The addressing of those micro-pixels could be facilitated by coding techniques because in each micro-display one pixel's position would always have a corresponding pixel position in its neighbor micro-display, both beginning to a unique three-dimensional display point so that its color and intensity would be very close. Usually, colour does not change with the observation perspective, and the intensity angular change is a relationship which corresponds to a diffusion surface law that could be predetermined and does not need to be received as object information, as it happens in 3D computer drawing rendering.

A fixed relationship interconnects then each family of micro-pixels, each one belonging to one specific micro-pixel display element, and its implementation could be automatic to reduce the need for transmitted information. As such a system is not yet possible, the Holografika company (5) produces one in which a certain number of projectors is located behind a diffracting screen whose function is to produce a diffuse lobe so that as the observer moves, the transition of the light coming from one projector to the light coming from the neighbor projector does not have dark regions. The space region for the observer is then continued and no dark regions are present. The number of projectors must be about 18 and until now only images made by the computer have been shown and animated through simple movements. We can understand that the name "holographic display" given by the company to the system can only be accepted within the popular meaning described above; no light interference is present during the process, neither in practice nor in concept. It is the only commercial system claiming to have the appearance of a continuous parallax. It also claims that the light is directed so that the observer's eye focalizes effectively at the point where the image is represented, eliminating the difference between convergence and accommodation, one important element of visual discomfort.

To support such an assertion, it is necessary to prove that more than one ray exiting from different positions on the screen, those rays converging at the image, are seen by one eye. The commercial system has only a horizontal parallax, so that a certain degree of vertical astigmatism should be present in direct proportion to the eye's aperture. A similar system employing LEDs instead of projectors was proposed (6)

4. Non-diffractive screens for stereo imaging

Dennis Gabor asserted (1) that he and Semioj Ivanov simultaneously studied the possibilities of achieving a screen which could eliminate the need of special goggles. The original idea was developed almost fifty years before by Gabriel Lippmann (1), who employed thin cylindrical transparent elements assembled side by side. Gabor's description is very complete, but it was Ivanov who succeeded in installing the first (and still the only) 3D auto-stereoscopic cinema (7) in 1945.

Based on two cinema projectors placed side by side, the 3 m x 4 m screen was made of a set of thin vertical cylindrical lenses whose surfaces were as shown in the upper part of Figure

2, where the light that is focused at a short distance of the back surface is reflected with different angular orientation due to the different transversal position of each projector and in a multiple way because each lenticular element receives the reflected light with a different angle. Figure 3 shows the conical distribution of the elements necessary to converge the whole scene at the observer's positions. It is to be noted that much accuracy is needed to keep the distance corresponding to the right and left eyes positions so that there are limitations to the positions where observers can be, something common to every stereo system, if mainly auto-stereoscopic.

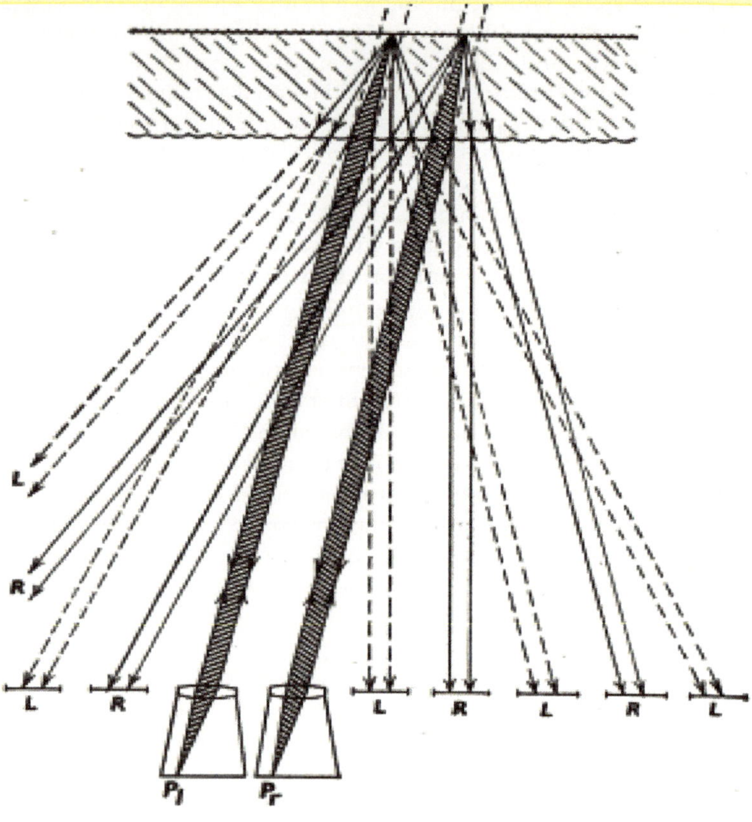

Fig. 2. Surface structure of Ivanov's screen (from Ref.1).

It is reported (1, 7) that Ivanov's theater could receive 180-200 spectators and that some movies were made for it, but it was discontinued because of the fact that the spectator needs to keep his or her head in an almost fixed position.

After Ivanov died Gabor tried to improve the technique by experiencing with mirrored elements with only vertically diffusing properties, eliminating the need for the conical assembly of elements, but he finally concluded that the production of such a screen would be too expensive and abandoned the idea.

After having performed some tests he thought that, if the depth of the scene could be limited, the observer could see the scene normally when one eye is receiving any view or even when the eyes are at positions when inverted views were received. That is, the monocular or pseudoscopic image could be seen as an orthoscopic one.

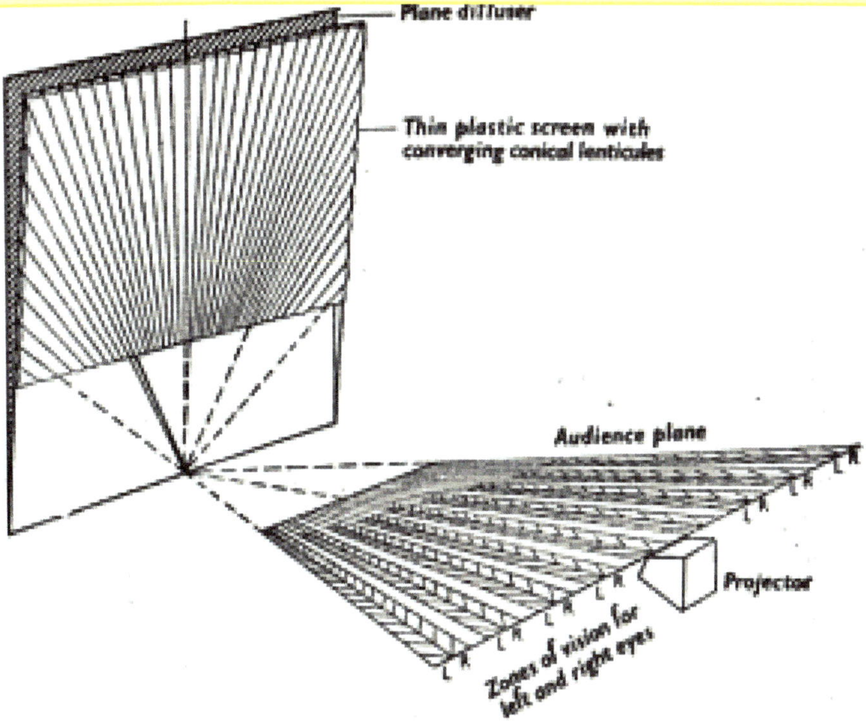

Fig. 3. Elements and vision zones at Ivanov's autoestereopic system (from Ref.1).

In my test of similar situations employing diffractive screens, I could mention a few people who could tolerate those cases. When there is lateral movement of the camera or some relative movement between objects, some 3D perception can occur to be explained by a sort of Pulfrich effect. For people who are very keen on 3D observation, as Gabor would be, it is certainly possible to perceive the depth in such situations.

5. Diffractive screens for stereo imaging

Once holography was discovered and its first practical applications performed, the possibility of employing a hologram not for creating images but for generating vision zones of projected images appeared. The roughness properties that Gabor wanted for the screen to

diffuse light vertically for the observer to see the whole projected image can then be obtained by making the hologram of a rectangular shape diffuser. The converging image which results from illuminating in a direction which is opposite to that corresponding to the reference beam occupies the zone where one observer's eye must be located. And the focusing property of the screen generates another observation zone for the other eye, provided that the angular separation of the two projectors to one screen point matches the binocular separation for the observer, obtaining by diffraction the effect of Ivanov's screen by refraction and reflection.
The idea was patented (8), but I do not know if any prototype was made.
It was conceived as a thick (Lippmann-Bragg-Denisyuk) hologram to avoid the simultaneous viewing of more than one image due to the spectral dispersion, so that the proposed solution was purely monochromatic. It still presents two of Ivanov's screen problems: different angular separation for binocular vision at different distances from the screen and the observer having to keep his head at an almost fixed position. While adding the lack of poly-chromaticity, it dispenses with a three-chromatic procedure to be applied for color reproduction.

6. Diffractive screens applied to holographic cinematography

To obtain holographic-like images employing incoherent light it is common to mount a set of discrete images in sequence giving the illusion of a continuous parallax system. The first example was maybe from Lloyd Cross with his integral hologram of 1972 (9), made of more than one hundred pictures acquired by means of a laterally translating movie camera. Many views are necessary to cover a wide parallax angle. We may calculate that by considering that the pupil's eye is about 3 mm wide and that it must pass from one observation region to its neighbor region without perception of a discontinuous jump when following any point in the image.
Holography is the only recording system which provides continuous parallax. A holographic image can be enlarged by means of a concave mirror, for example, but the viewing zone is restricted to only one observer and the longitudinal magnification is always different from the transverse one. The way to cinematography was paved by Komar (10), who succeeded in projecting a large size holographic image by means of a diffractive screen. He recorded a large size scene of about 1 m x 1 m, capable of including a person, through a large aperture objective (200 mm) having also a large angular aperture on a 70 mm format film. Reconstructing by inverting the direction of the reconstruction beam puts the image at the precise position occupied by the object, enabling the correction of distortions or aberrations. But this image cannot be seen by an observer because he must receive the rays in a position from which the image has a reversed depth and can only receive rays coming from the aperture of the lens, much smaller than the whole scene. A conventional diffusing screen would only show a plane image. A diffracting screen made as the hologram of a concave mirror (11) may direct each of the viewpoints on the scene to a continuous viewpoint sequence and, through the proper managing of the diffraction order in a horizontal direction, invert the depth, showing an orthoscopic image. Multiple exposures with changing reference beam angles give the possibility to provide full parallax to more than one observer.

In that way a Lippman-Bragg-Denisyuk hologram was made for the screen and its evolution made it capable of rendering color images by means of the three-chromatic procedure. It was discontinued, but constituted the only cinema system that can receive the academic term of "holographic", because recording was made by interference of the object light (a sequence of laser pulses) with a reference beam, recording and diffraction followed, and the concept of a holographic screen was introduced, giving continuous parallax and allowing the observer to move his head. The only visual shortcoming maybe came from the strict monochromaticity of the laser light beams. Living subjects were reproduced. The system cost was, of course, very high. While in need of more precise information, it was certain that at least eight people could watch a twenty minutes movie on the 1 m x 1 m square screen at a few meters distance.

7. On the parallax of a lens image

The very common use of photography has always rendered plane images and that is why it is normal to think that a lens image is plane. But we know that the images are three-dimensional and that it is the detecting system which makes the result to be in a plane. By closing or opening the aperture of a lens, one always captures the same plane scene but the sharpness of those elements not precisely focused diminishes at large apertures. When opening the diaphragm from his smaller diameter, the out-of-focus situation of the additional rays, which do not fit in the image generated by the center of the lens, happens because of the new perspectives added by the lens area being unobstructed. It is important to notice that they correspond to the viewpoint of each area part being unobstructed on the lens. Continuous parallax is then allowed.

One of the first applications of this property was a three-dimensional photography system employing only one camera, opening to light only one area on each side of the lens, one transmitting through one color filter and the other through another, photographing in only one shot a direct anaglyph.

The horizontal parallax transmitted by a slit and placed on a lens may be employed for generating multiple views and even continuous parallax.

Lunazzi (12) projected a scene directly from objects on a diffractive screen of 15 cm x 30 cm employing an ordinary slide projector objective. The 6x enlarged image gives the precise impression of a holographic one, but has more focusing depth limitations. "Direct holoprojection of objects" was a name given to this technique in which the horizontal extension of a lens is the fundamental property (Figure 4).

Son (13) employed this parallax property to project sequentially a set of views from a multimedia projector. Each slit position on the projector corresponded to a vision zone for the screen and the observer could have a different view within a discrete sequence of lateral positions. The image persistence on the retina gives the illusion of simultaneous viewing, but it is necessary for the system to put all views in the time of one ordinary movie frame (1/24s), so that a set of views may be projected. At 24 views per second, for example, the frame capability of the electronic multimedia system needed is about the square value of that of an ordinary projector. In the present state of the art it seems not possible to achieve this at a high definition resolution. Employing many projectors at close lateral positions is a possibility to reduce the frame rate needs and to obtain a brighter image, but it is only possible if the screen has a low scattering level in order to avoid the simultaneous view of all projections.

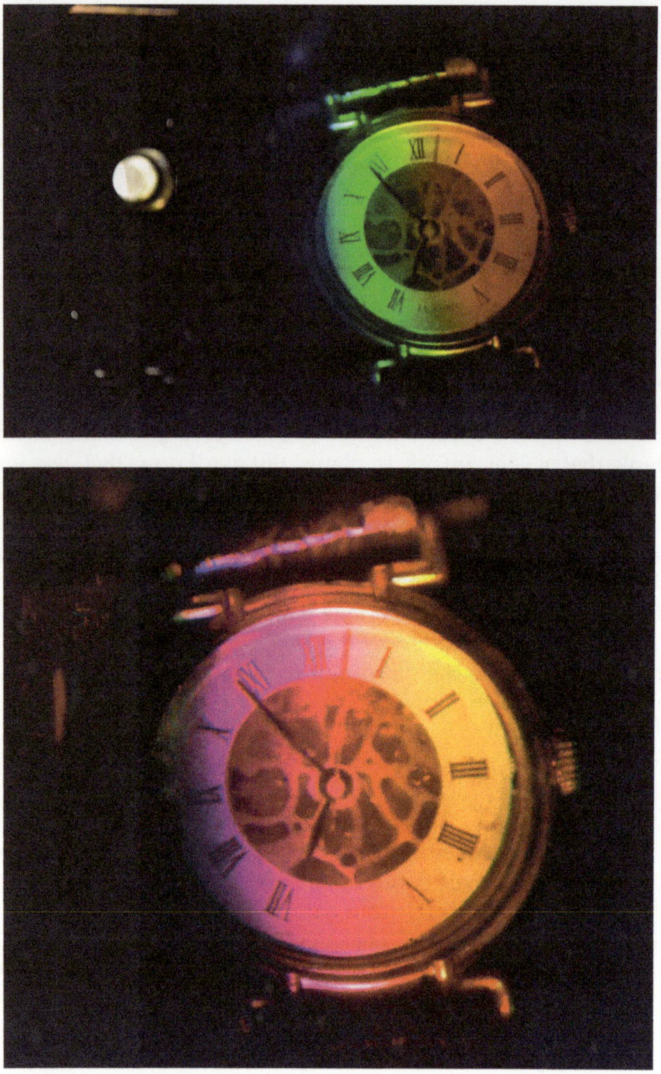

Fig. 4. Up: Photograph of an object directly projected on a holographic screen. Above: anaglyphic stereo representation of the same scene, with color channel.

8. The continuous parallax

We see the need to approach the perception of a continuous view sequence to have a three-dimensional image with good quality. There is a white-light imaging process whose parallax is inherently continuous, as in holography. It matches depth coding by diffraction naturelly happening when the diffracted light is collected in a small region after a diffraction grating, with its also natural decoding happening after projecting that light on a second diffractive element (14, 15). Each wavelength represents a viewpoint based on a small area of the first diffracting element and as the spectrum is continuous the parallax also is. We can better understand the basic process of a diffractive screen considering it as a primary element, a diffraction grating. If we further approximate the projecting lens to a simple pinhole camera corresponding to its central part, we can see in Figure 5 how the ray tracing based on an object point explains the resulting image by central symmetry (16).

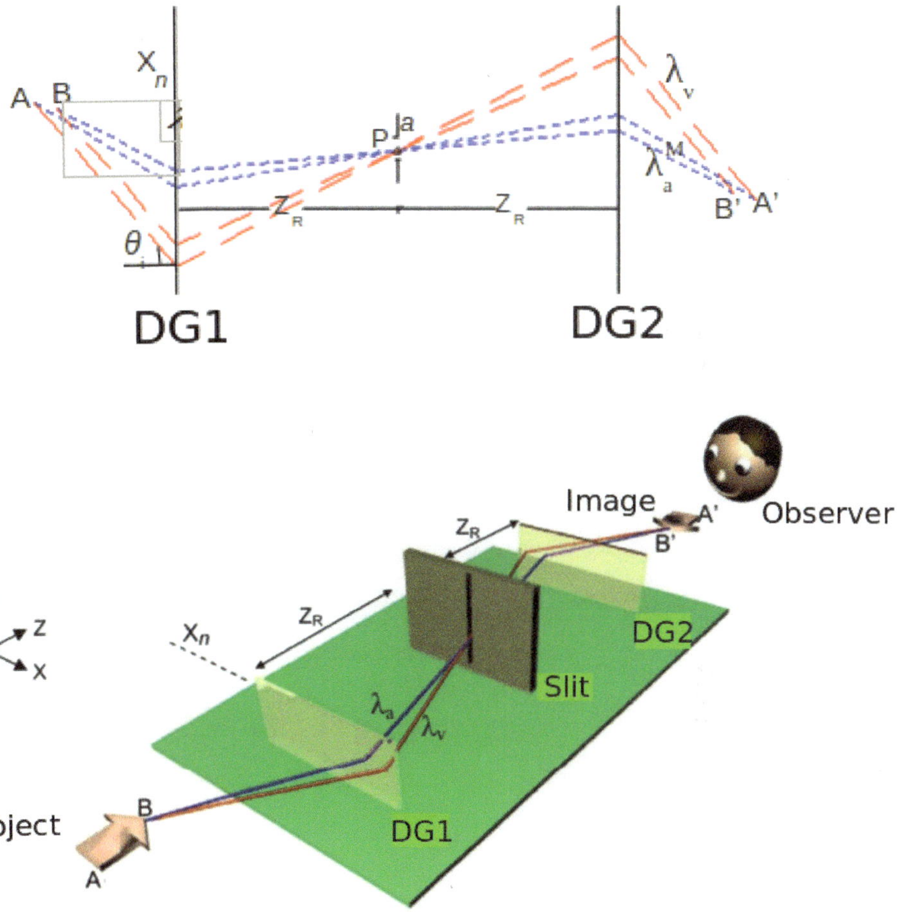

Fig. 5. Symmetry in double diffraction imaging intermediated by a slit.

DG1 and DG2 are two identical diffraction gratings intermediated by a slit. Each object point has a corresponding image point symmetrically to a central point. Because the observer looks to the image from behind, he sees an inverted depth. In this case of perfect matching between two diffractions, it is interesting to notice a property which resembles a holographic one: the diffraction at a symmetric order, on the second grating for example, generates an image with inverted depth. An image in normal depth is so obtained (17).

In a second approximation, we can consider the diffractive screen as a diffracting lens, that is, a bi-dimensional grating which puts the light it receives converging to a unique position, as a convergent ordinary optical element. A diffracting lens is obtained directly by a hologram made with two point sources. If we project in monochromatic light, the screen acts as the one of Komar, but, projecting in white light and making the screen with the point sources from the same side of the film, the diffracted transmitted images are affected by a horizontal dispersion. The same basic property that gives orthoscopic and pseudoscopic images with two gratings corresponds now to the same images but seen all over the screen extension. When the observer moves laterally, he receives continuous view sequences of the object. In this way it has been possible to observe the enlarged image of objects on a one square meter screen but an intense reduced size projection lamp and a dark ambience is necessary. To avoid the need of having the observer watching at a very precise height, one point source in the the interference process process is substituted by a thin vertical diffuser. It gives the vertical size of the observation region but with a reduced image brightness. Besides the limited diffraction efficiency, another brightness limitation results from the need of a thin slit on the projecting lens to get maximum focal depth.

9. White light holographic cinematography

The spectral depth coding by diffraction was first discovered in holograms (18) and matches perfectly the projection on diffracting screens generating the image through the decoding

Fig. 6. Hologram made in 35 mm film enlarged x40 by using white light

property a second diffracting element may have. Enlargement is the same for all three dimensions of the object. A small transmission hologram made with a lateral reference beam on 35 mm film can be enlarged on the screen by illuminating it with a halogen lamp (19). To obtain a better luminous efficiency the scene was recorded employing a photographic objective covered with a horizontal slit and it was projected by reversing the light path as in Komar's technique but enlarged and in white light. It was possible then to have an image on a 0,84 m x 1,10 m screen (20) at x40 enlargement (Figure 6).

The observer's space depends vertically on the height of the diffuser and laterally on its width and on the screen dispersion. The angle between the object and the reference beams being of 45 degrees, no more than four observers seated in two rows can see the scene simultaneously. A similar system not enlarging holograms but projecting pictures of classic movies was presented to the public (21). There, the images appeared from six meters behind the screen coming closer little by little until traversing it to one meter from the observer, located two meters away from the screen.

The recording and white light projection of holographic movies was not accomplished due to the lack of resources and concentration of efforts in the application of electronic images. The recording of holograms in white light which is based on an interesting proposal (22) was not accomplished up to now.

10. Continuous parallax in electronic images projected on diffractive screens

The depth coding-decoding diffractive principle allows projection of a point source image at any position with respect to a diffractive screen. A computer-controlled figure generator was developed having a thin white light beam being focused at some millimeters from a diffraction grating whose lines were vertical. Because the beam was reflected on a mirror which was made to rotate through a vertical axis, the distance from the focus of the beam to the grating changed. This distance made the necessary degree of coding, which constitutes the horizontal size of a white-light spectrum. The movement was complemented with another two computer-controlled rotating mirrors to generate a luminous point located in any three dimensional position floating in front of the screen (23). A software was in charge of generating animations in the format of line figures. Figure 7 shows that a cube could be drawn without the need to correct any distortion. The image volume is about 100 l, no more is possible due to its reduced brightness. The possibility of white-light laser beams now

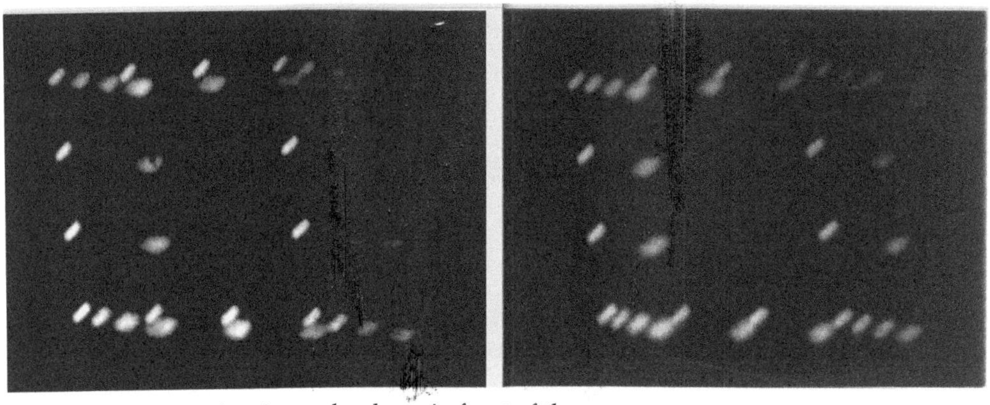

Fig. 7. Luminous points in a cube shape in front of the screen.

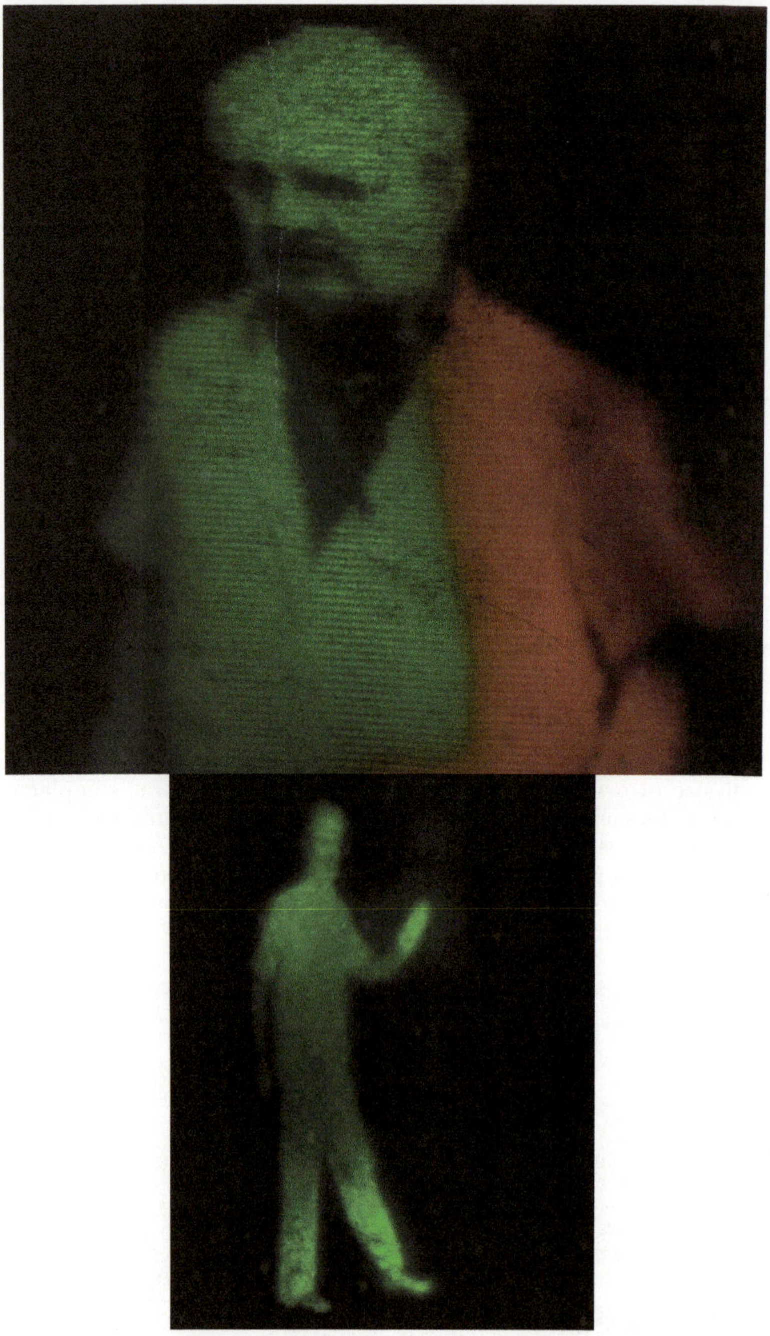

Fig. 8. Vídeo scene appearing in an oblique plane in front of the screen. a) detail. b) complete image, 80 cm high, made of two hundred video lines.

obtained by TiSa lasers whose spectrum is broadened by means of Bragg optical fibers may give to the application of this technique new possibilities.

The depth coding-decoding diffractive principle also allows us to project an electronic image at any plane in respect to a diffractive screen. By choosing an oblique plane in which to position a video image, an interesting approach to a holographic TV was obtained: The scene has depth as well as transversal characteristics, giving the illusion of a perfect 3D to most spectators (24). Although the images we show (Fig.8) shows limited resolution, this is due to electronic equipment limitations. The diffractive imaging process has in principle no resolution limitations but those due to diffusion of the screen or speckle noise.

After some time observing, some people notice the lack of a perfect relief on the image. The scene can generate a volumetric image by projecting many parallel planes in a rapid sequence, each plane having a the image of a corresponding slice in which the whole image was divided. Animated scenes were made by three-dimensional computer drawing from which four slices were made. They were projected through a computer-controlled rotating mirror to make that at each mirror position the corresponding slice was projected (25, 26). Figure 9 shows a vertical sequence of stereo pairs corresponding to how the scene is viewed.

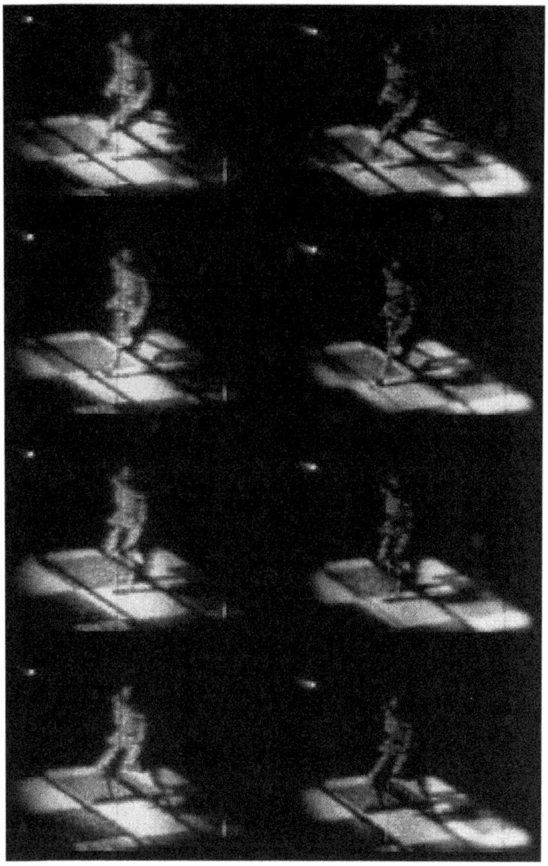

Fig. 9. Four stereo pair frames of a volumetric scene distributed in four assembled slices.

In this proposal discontinuity exists on the slicing, the more slices dividing the scene, the more perfect it appears. Slicing on video scenes should be made by a vertical white-light strip shaped beam sweeping on the scene, but has not been implemented yet.

11. Lensless projection on diffractive screens

The projection by means of lenses or mirrors put limitations in focalization because it is usual to deal with oblique projections. A proposal to solve this problem by elliminating focusing was made by Lunazzi (27, 28) based on employing a linear luminous source, a long filament of a halogen lamp horizontally located behind the screen. The image is a consequence of the shadow projected by each part of the filament. The shadow on the screen can only be seen from a position which is precisely opposite to the filament part. A continuous sequence of shadows is generated rendering horizontal parallax. The angular extension of the filament in respect to the screen center defines the angular field for observation. The image is very peculiar (Fig. 10) because although it does not have inverted depth it shows an object whose closer parts are smaller than those which are farther away from the observer, as in a common shadow. The objects can only be transmission and not diffusing objects, which is not a problem because we can envision its application for elements like liquid crystal displays (Fig.11).

Fig. 10. Continuous parallax image projected with a linear source. Left: original view. Right: anaglyphic stereo representation with no color channel.

Fig. 11. A LCD transmission watch as seen in the linear source projection

An improvement of this technique that can be expected would be a way to make the image appear in front of the screen and a way to illuminate from inside of its transparent support, in a similar way to the so-called "edge lit" holograms (29).

12. Not holographic diffractive screens

If the term "holographic" corresponds to Gabor's idea of wave reconstruction, it should be applied to cases in which the interest is precisely the reproduction of waves, like in the case of imaging or holographic interferometry techniques. In that sense, the construction of a diffractive element by interferential means does not give to it the holographic characteristics. If the name "holographic" is given because a three-dimensional continuous parallax image results on the element, it is because the popular sense of the term is being employed. That is why the term "diffractive screen" was widely employed in this text. To reinforce the idea

the fact can be mentioned that similar elements can be constructed by direct recording techniques like lithography, for example. A 5 m diameter diffractive lens was made for an astronomic space project (27), a technique that could be similar to one for the construction of diffractive screens.
It was made in the form of a mosaic, already explored with holographic techniques (10).

Fig. 12. Picture of a 5 m diameter diffractive lens made by litography (by R. Hyde- S. Dixit).

13. Comparation of electronic image systems employing diffractive screens with other volumetric systems

Non-glasses autostereo systems are not practical for its wide-spread use for commercial purposes yet (30)(31), so that the many systems which has been proposed for it are to be kept under discussion.
Let us consider the following volumetric systems: lenticular, a parallax barrier, a curvature controlled mirror, a rotating screen under projection, a rotating LED display, an electro-optic hologram generator, a fast recording of integral hologram, a laser focalization on crystal, a laser focalization on air, and flying luminous sources.
Among the lenticular systems let us consider only those in which small semi-cylindrical vertical lenses are in contact with a conventional screen. To reach the number of views necessary to produce an apparent continuous view sequence many views must be displayed simultaneously, making a large bandwidth necessary or much processing by very special digital techniques. This is the same as happens for the diffractive screen systems to add several slices or views. Maybe an advantage can be proved in the case of the slicing oblique plane system, because the continuous parallax does not demand electronic processing power.

The parallax barrier system can be disregarded for its long duration presentation applications because of the necessity to maintain the head position during the observation. Relief is inverted when deviating from the right position. Although more than two views can be provided, the adding of views reduces the brightness of the image.
To make large diffractive screens seems to be as easy as making large holograms, while examples of lenticular or parallax barrier screens which are larger than one square meter are not known yet.
The curvature controlled mirror employs fast moving parts, being limited in size to a few decimeters by air resistance and noise generation. The rotating screen systems presents the same problem; its advantage is the ability to show 360 degrees images. The images of the rotating screen systems can not fill more space than the screen does, appearing in fact within a transparent cylinder, never in front or in the back of it, the effect that more impresses the public.
The electro-optic system, already named as "holographic video", developed by Benton, never recorded live video scenes but computer made ones. It has size limitations and consumes much computer processing power.
The fast recording of holograms has been possible in telecommunications with new photosensitive materials. A scanning frame was recorded every two seconds, which is still ten times more than required for video, and not at large size yet. To reach the video velocity much bandwidth and processing capability will also be needed and, if reaching the 1/20s frame speed, the persistence of the images on the retina makes the photosensitive material unnecessary; a diffractive screen can accomplish the task equally.
There are systems creating luminous points within crystals and images can be seen from almost any position, having more than 360 degrees viewing capability, but the images remains within the supporting crystal and it seems neither practical nor possible to have large size images. A new interesting possibility appears to make the same in a liquid, while not employing visible light for that (32).
Other systems focussing intense laser beams on air may generate 360 degrees views, but presents a low resolution and high cost, as well as noise or dangerous luminous or temperature levels. One can think that this systems could be useful for working at large sizes, filling large volumes to be seen from a large distance.
Finally, the same can be said of the recent system made of small flying sources, mini-helicopters whose position can be remotely computer controlled: large sizes could be achieved at a considerably energy cost, low speed and short duration.

14. How to evaluate parallax with ordinary images of a system

It is common to say that holography is the technique allowing to reconstruct the wave amplitude and phase, but if we remain with this sole idea we could not properly analyze the holographic image. Considering that the phase of light cannot be seen, not even a detector with enough capability exists. What our eyes see are rays, directions of propagation. My personal vision of holography comes from my initiation as a 14-year-old photographer making stereo pictures.
This allows me to concentrate on the images and not on a mental visualization of Maxwell equations and their application, in which even complex-conjugated light fields are present. The third dimension can only be clearly perceived through binocular vision, which most popular imaging systems do not permit at present time. To demonstrate the existence of

perspective selective viewing with ordinary cameras it is necessary to show the parallax, something that the photograph of any three-dimensional image, or the video made with the camera at a fixed position, cannot prove. In the case of photographs, a sequence, during which the camera changes laterally its position, allows to observe the relative change of positions of corresponding points at the scene, and also to mount a stereo pair for binocular observation. It improves the knowledge which the photographs offer, but it is not yet enough. A video registering gives more close frames than a series of photographs makes possible, and should detect the presence of dark regions or jumps on the transitions. But this is still not enough: the aperture of the camera lens must be as close as possible to the eye's pupil, about 3 mm to avoid mixing a dark and a luminous region at the same time giving the appearence of not existing dark regions.
In the internet, there are many sites claiming to show three-dimensional or holographic imaging systems, but which are shown from a single viewpoint.
A properly made system must be justified with images that can prove it. The worst case is certainly when simulated pictures are shown without declaring the simulation. It is usual in this cases to see part of the figure exiting the limits of the frame of the screen, what is impossible because light does not travels in curved path.

15. Conclusions

After describing many systems that project goggle-less three-dimensional images on diffractive screens made by recording light interference on photosensitive material, it can be understood that, although these techniques are not the ones which are chosen for the industry at the moment, interesting properties have yet to be explored. Advances in the diffraction and dispersion efficiency of the elements may recall the attention to this subject in the future.

16. References

[1] Gabor, D. Three-Dimensional Cinema, The New Scientist, July 14th, 1960, p141-145, http://books.google.com/ or precisely http://books.google.com/books?id=vPmlBbHCOTIC&pg=PA141&lpg=PA141&dq=gabor+screen+three-dimensional+cinema&source=bl&ots=V5lK36onTg&sig=JyWoWWe2uq76kZBq4AgH-kAoq4s&hl=es&ei=p0tgTYn8DYL-8Aa5np3SCw&sa=X&oi=book_result&ct=result&resnum=4&ved=0CD4Q6AEwAw#v=onepage&q=gabor%20screen%20three-dimensional%20cinema&f=falseK or, simply: http://tinyurl.com/gaborstereoscreen

[2] *"Holographic non-isotropic diffusing screen"*, Donald H. Mc Mahon US patent 3,708,217 (1973)

[3] *Holographic diffuser for back-lit display* United States Patent 5471327 http://www.freepatentsonline.com/5471327.html

[4] *"Pixel element for a three-dimensional screen"*, Tibor Balogh, USPTO 6736512

[5] www.holografika.com

[6] *"3D displays using light from an undiffused point-source array"*, Jung-Young Son and Vladimir V. Saveljev, SPIE Newsroom 10.1117/2.1200902.1414

[7] Ref.1 p.142

[8] *"Projektionsschirm fur Projektion mit dreidimensionalem Eindruk"*, Bestenreiner, Friedrich, 2011417, Offenlegungstag 23 September 1971. AGFA is Patent's owner.

[9] *"Springer handbook of lasers and optics"*, Frank Trägger ed. 2007, Table 20.1 p.1206

[10] Description of V. Komar's development ot holographic cinematography at: http://wiki.answers.com/Q/Who_invented_3D_movies

[11] *"Holographic Screens and Their Applications Applications"*, Son et al, SPIE 3358 p337-345 1998

[12] *"Processo para projeção de imagens com paralaxe horizontal sobre tela holográfica"*, J.J. Lunazzi, INPI (BR), No 9302553-0

[13] Jung-Young Son, et al., "A Multiview 3-D Imaging System With Full Color Capabilities", SPIE Proc., Stereoscopic Displays and Virtual Reality IV in Electronic Imaging, 1998.

[14] *"Holophotography with a diffraction grating"*, Lunazzi,J.J., Opt.Eng.V29 N1 (1990) pp.15-18.

[15] "New possibilities in the utilisation of holographic screens", J.J. Lunazzi, Proc.of the SPIE meeting "Electronic Imaging", conference "Practical Holography VI", San Jose-CA-USA, 9-14.02.92, p.289-293, SPIE 1667.

[16] "Pseudoscopic imaging in a double diffraction process with a slit: critical point properties," J. J. Lunazzi and N. I. Rivera J. Opt. Soc. Am. A, JOSA A, Vol. 23, Issue 5, pp. 1021-1026 (May 2006).

[17] *Orthoscopic white-light imaging by means of two bi-dimensional diffracting elements and a pinhole"* J. J. Lunazzi , N.I.R. Rivera, 5th Iberoamerican Meeting on Optics and 8th Latin American Meeting on Optics, Lasers, and Their Applications; Aristides Marcano O., Jose Luis Paz; Eds. 3-7/10/04. Proc. SPIE Vol. 5622, p. 1469-1473.

[18] *"3D Photography by holography"*, Lunazzi, J.J., Opt.Eng. V29 N1 (1990) p.9-14 http://arxiv.org/pdf/physics/0703209

[19] *"Enlarging Holograms Under White Light"*, J.J. Lunazzi, Proc. of the 17th Gral. Meeting of the International Commission for Optics, Taejeon, Korea, 19-23.08.96, SPIE V 2778 p.469-470

[20] Lunazzi SPIE News, "Enlarging Holograms Under White Light, a Way to Save Holographic Material" http://arxiv.org/pdf/0905.1157

[21] Lunazzi, J.J. e Benito, R., KARAS Holografia, Madrid, Espanha, 1995, at "Tres Rios" cinematography Theaters

[22] *"White Light Colour Photography for Rendering Holoimages in a Diffractive Screen"*, J.J. Lunazzi, Proceedings of the Fourth International Conference on Holographic Systems, Components and Applications, Neuchatel, Switzerland, 13-15 September 1993. http://arxiv.org/pdf/0904.2598 (Conf. Publ. No.379). IEE, London, UK. pp. 1536, 1993

[23] J.J. Lunazzi and M. Diamand, *"Volume Images Vector Display Based on a Diffractive Screen"*, Optical Review (Japan) Vol. 6, No. 6 (1999), p.513-517 http://arxiv.org/pdf/1101.3340

[24] *"Holo-Television System with a Single Plane"*, J.J. Lunazzi, D.S.F. Magalhaes, N.I.R. Rivera, R.L. Serra, Opt. Lett. 34, 533-535 (2009) , http://arxiv.org/pdf/0902.4705

[25] "A Holographic Visualisation System: A Sequel", E.G. daFonseca, P.L. de Geus, C.F.X. de Mendonça, J.J. Lunazzi, E. Bertini, Proc. of the X " International Symposium on Computer Graphics, Image Processing and Vision", SIBGRAPI'98, pages 135-141, October 1998. Biblioteca digital do SIBGRAPI: http://fenix.sid.inpe.br:1906/col/sid.inpe.br/banon/1998/06.04.18.20/doc/tag.html

[26] J.J. Lunazzi, *"The use of diffractive screens for electronic imaging"*, Proc. of the Holographic Display Artists and Engineers Club-HODIC meeting, Japan Association of Display Holography, Tokyo, Japan, 1996.08.30 http://arxiv.org/pdf/physics/0609243

[27] "*3D shadowgram projection using a simple diffractive screen*", José J. Lunazzi, N I. R. Rodríguez, Proc. of the XXIX Enc. Nac. de Fis. da Mat. Cond., SBF, São Lourenço-MG, 9-13 de maio de 2006. http://www.sbf1.sbfisica.org.br/procs/2006/pdfs%20optics/Interferometry,%20 Holography%20and%20Applied%20Optics/1427.pdf

[28] "*Imagens Por Difração com Luz Branca Sem Elementos Intermediários*", N. I. R. Rodriguez, PhD Thesis, Campinas State University, 2007 http://webbif.ifi.unicamp.br/teses/apresentacao.php?filename=IF284

[29] *"Edgelit holography: Extending Size and Color"*, Nessbit, R.S., MIT Thesis September 1999 http://citeseerx.ist.psu.edu/viewdoc/download?doi=10.1.1.38.5280&rep=rep1&type=pdf

[30] CEOS 2011 video demonstration: "*Sony Autostereoscopic 3D TV without glasses at CES 2011 "Which first look review""*, http://www.youtube.com/watchv=8a66ENGh16Q&feature=relmfu

[31] CEOS 2011 video demonstration: "*LG Autostereoscopic 3D TV without glasses at CES 2011 - Which first look review""*, http://www.youtube.com/watch?v=MdLXeXfoiJk

[32] Ivan T. Lima, Jr., Val R. Marinov *"Volumetric Display Based on Two-Photon Absorption in Quantum Dot Dispersions"* Journal of Display Technology, V6, N6, JUNE 2010, 221

13

Two and Three Dimensional Extreme Ultraviolet Holographic Imaging with a Nanometer Spatial Resolution

P. W. Wachulak[1] and M. C. Marconi[2]

[1]*Military University of Technology, Institute of Optoelectronics*

[2]*NSF ERC for Extreme Ultraviolet Science & Technology and Department of Electrical and Computer Engineering, Colorado State University*

[1]*Poland*

[2]*USA*

1. Introduction

The word "hologram" (from the greek *"holos"*: whole, complete and *"graphos"*: writing, drawing) means "total recording". Holography is a well known technique originally proposed in 1948 by Gabor, who also coined the name, as a new microscopy alternative. He realized that the interference of two mutually coherent waves, one called the reference wave and the second one - the object wave, allows for recording of information consisting of both amplitude and phase of diffracted or scattered beam from an object (Gabor, 1948). This coding of the amplitude and phase of the object beam into an interference pattern allowed him to demonstrate that from this complicated holographic pattern, ultimately the image of the original object can be obtained. Several years after the appearance of Gabor's paper, Baez (Baez, 1952) suggested extension of this idea to the X-ray region, but it remained as an interesting proposal till the early 1960s, when holography started to be widely applied. It was after the paper by Leith and Upatnieks, who proposed the off-axis holography - scheme which overcomes many of Gabor configuration drawbacks (Leith & Upatnieks, 1962). Since that time holography was widely used in numerous applications, some of them requiring increased spatial resolution. On this path, reducing the illumination wavelength is a direct way to improve spatial resolution both in nanopatterning (Solak et al., 1999; Wachulak et al., 2008a) and holographic imaging, described herein. This is the reason why short wavelength sources such as synchrotrons, extreme ultraviolet (EUV) and soft X-ray (SXR) lasers, high harmonics generation sources (HHG), etc., became an interesting alternative for high resolution imaging.

This chapter is devoted to 2-D and 3-D holographic imaging using a capillary discharge EUV laser. The chapter is organized as follows. In section 2 recent developments in high resolution holographic imaging will be briefly presented including different imaging techniques and short wavelength sources. In section 3 some general information about Gabor in-line EUV holography will be presented with detailed analysis of the resolution limitations due to coherence of the EUV source and digitization process. Starting from section 4 through 6 recent developments in holographic 2-D and 3-D imaging will be presented (Wachulak et al., 2010). In section 4 a 2-D holographic imaging using a compact EUV laser, the experimental details, results and resolution estimation using a wavelet decomposition and correlation method will be discussed. The resolution of the 2-D holographic imaging was further improved by increasing the recording/reconstruction numerical aperture, leading to spatial resolution comparable to the illumination wavelength, approximately 46nm (section 5). A novel method of resolution and feature size assessment, based on a Gaussian filtering and correlation was applied and the results were compared with well established, knife-edge resolution test. Finally in section 6 a 3-D holographic recording and reconstruction, that allowed for successful 3-D information retrieval from a single high numerical aperture EUV hologram, will be presented. Section 7 concludes the chapter.

2. Developments in high resolution EUV and soft X-ray holographic imaging

The potential for holography in the EUV region was recognized very early. However realization of this goal become very difficult, thus only in the early 1970s the first images of very simple objects were obtained for the first time (Giles, 1969; Aoki & Kikuta, 1974). The main obstacle to record and reconstruct good quality holograms in this region of e-m spectrum was lack of sufficiently bright and coherent sources. It was not until 1987 that the high resolution x-ray imaging was realized by use of 2.5–3.2 nm SXR radiation from the National Synchrotron Light Source (NSLS), where spatial resolution of 40 nm was demonstrated. Fourier transform holography at the NSLS achieved spatial resolution of 60 nm (McNulty et al., 1992). Gabor holography with an early X-ray laser pumped by two beams of the fusion-class NOVA laser at Lawrence Livermore National Laboratory demonstrated a spatial resolution of 5 μm (Trebes et al., 1987).
Other experiments utilized synchrotron light to image biological samples, nano structures, and magnetic domains (Jacobsen et al., 1990; Lindas et al., 1996; McNulty et al., 1992). Lens-less diffractive imaging, based on iterative phase retrieval, following the proposal by Sayre, (Sayre et al., 1998) have demonstrated SXR imaging with 50 nm spatial resolution utilizing $\lambda = 1.5$ nm source (Elsebitt et al., 2004). The first experimental demonstration of lens-less diffractive imaging using coherent soft X-rays generated by a tabletop SXR source allowed for image acquisition with spatial resolution of 214 nm (Sandberg et al., 2007) later improved to 72 nm (Sandberg et al., 2008). High resolution scanning X-ray diffraction microscopy proved to be a useful imaging technique employing coherent, short wavelength radiation, reaching spatial resolution better than 70 nm (Thibault et al., 2008).
The practical demonstration of EUV and SXR holography proved to be difficult in particular because the lack of sufficiently bright and coherent sources at short wavelengths, and to the fact that coherent EUV and SXR laser sources have historically been restricted to large user facilities. The first demonstration of a coherent table-top holographic imaging achieved 7 μm spatial resolution with a spatially-coherent HHG source (Bartels et al., 2002); this resolution has been extended to a resolution of 0.8 μm in later experiments (Morlens et al., 2006). Time resolved holographic imaging was also implemented with HHG sources to study the ultrafast dynamics of surface deformation with a longitudinal resolution of less than 100 nm and a lateral resolution of less than 80 μm (Tobey et al., 2007). Holography was also used to demonstrate 100 nm-resolution holographic aerial image monitoring based on lens-less Fourier transform holography at EUV wavelengths, using synchrotron-based illumination (Hun Lee et al., 2001). Femtosecond EUV radiation provided by the free-electron laser

FLASH was used for digital in-line holographic microscopy to image particles, diatoms and critical point dried fibroblast cells with 620 nm spatial resolution at 8 nm wavelength (Rosenhahn et al., 2009). Digital in-line SXR holography (DIXH) was used to image immobilized polystyrene and iron oxide particles with spatial resolution of 850 nm at wavelength range of 3.7-5.6 nm to take advantage of selective contrast in this wavelength range (Rosenhahn et al., 2008). Holographic measurement scheme to monitor the X-ray-induced explosion of microscopic objects was performed by a femtosecond time-delay X-ray holography, inspired by Newton's "dusty mirror" experiment, allowed to see the changes in EUV induced explosion of 140 nm diameter polystyrene beads (Chapman et al., 2007). By combining HHG holography with iterative phase retrieval algorithm, usually employed in diffractive lens-less imaging, reconstructed hologram spatial resolution was improved to ~53 nm (Sandberg et al., 2009). Holograms can be also obtained in very short exposure times. Using uniformly redundant arrays (URA) instead of a single or multiple reference pinholes in Fourier type holography the throughput of the imaging system might be sufficiently large to acquire a hologram with a single 15 fs EUV pulse and reconstruct with spatial resolution approaching 50 nm (Marchesini et al., 2008). Naturally, the body of knowledge related to this topic is so immense, that we are not able to mention all the work done in the field, only some aspects of it.

3. Gabor in-line EUV holography

The acquisition of holographic images is a two step process consisting of recording and reconstruction phase. The holographic recording in Gabor's in-line configuration is depicted in Fig. 1a. During the recording step the interference pattern between two mutually, collinear and coherent beams is stored in the recording medium. The two interfering beams are the reference beam (black dashed lines) and the object beam (green solid lines). The recording medium is a material used to record the interference pattern that can provide a linear mapping between the incident intensity and some kind of change in the medium such as the reflection, transmission or height modulation. If the object and reference wavefronts are both expressed as a two complex fields having amplitudes and phases:

$$o(x,y) = |o(x,y)| e^{j\phi(x,y)} \tag{1}$$

$$r(x,y) = |r(x,y)| e^{j\psi(x,y)} \tag{2}$$

then the interference between these two complex fields occurring at the location of the recording medium can be expressed as the intensity distribution of sum of two fields:

$$\begin{aligned} I(x,y) &= |r(x,y)+o(x,y)|^2 = [r(x,y)+o(x,y)][r(x,y)+o(x,y)]^* = \\ &= |r(x,y)|^2 + |o(x,y)|^2 + r^*(x,y)\cdot o(x,y) + o^*(x,y)\cdot r(x,y) \end{aligned} \tag{3}$$

The first two terms are the intensities of both interfering beams, while the last terms depend also on their phases. That is why the recording medium, sensitive only to the intensity, is in fact capable of storing the intensity and the phase information simultaneously.
The linear mapping of the recording medium can be described as transmission change of the recording medium (for example photographic film) as a function of the incident intensity:

$$t(x,y) = t_0 + \kappa \cdot I(x,y) \tag{4}$$

where t_0 is the uniform transmittance of the film introduced by the constant exposure and κ is a linear factor relating the transmission to the incident intensity $I(x,y)$. By substituting equation Equation (3) into Equation (4) the transmission of the recording medium can be thus expressed as:

$$t(x,y) = t_0 + \kappa \cdot \left[|r(x,y)|^2 + |o(x,y)|^2 + r^*(x,y) \cdot o(x,y) + o^*(x,y) \cdot r(x,y) \right] \tag{5}$$

This interference pattern, stored in the recording medium during the reconstruction step, acts as a complicated diffraction grating. If the intensity of the interference pattern during the recording step is linearly translated into a recording medium transmission then the reconstruction step is similar to the one depicted in Fig. 1b. The hologram is placed in the same geometry and illuminated by the same wavefront as in the recording step, now called reconstruction wavefront. This can be expressed as:

$$\begin{aligned} u(x,y) = r(x,y) \cdot t(x,y) = r(x,y)\left(t_0 + \kappa \cdot |r(x,y)|^2\right) + \kappa \cdot r(x,y) \cdot |o(x,y)|^2 + \\ \kappa \cdot o(x,y) \cdot |r(x,y)|^2 + \kappa \cdot o^*(x,y) \cdot |r(x,y)|^2 \end{aligned} \tag{6}$$

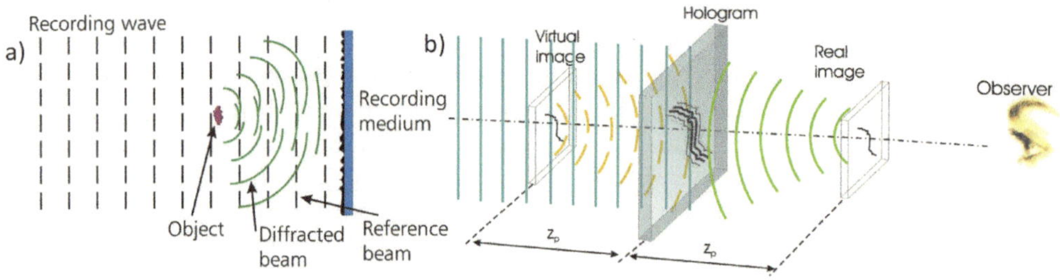

Fig. 1. Holographic recording scheme a) using Gabor's in-line configuration. The interference pattern between flat wavefront (reference beam) and beam diffracted from the object (object beam) is stored in the recording medium. Optical reconstruction b): the transmission hologram is illuminated by the reconstruction beam (the same as reference beam) and the image of the object appears where the object was initially located during the recording phase.

The first term is a background, the second is very small due to Gabor holography requirement for the object with high transmittance, thus the $|o(x,y)|^2 \to 0$. The wavefront diffracted from complicated transmission diffraction grating (hologram) converges behind the hologram generating the real image (fourth term $\sim o^*(x,y)$) while divergent wavefront generates the virtual image in front of the hologram where the object was placed during the recording step (third term $\sim o(x,y)$). The reconstructed intensity image is then $U(x,y) = u(x,y) \cdot u^*(x,y)$.

In recent experiments a high spatial resolution photoresist, usually utilized in electron beam lithography, was used for the recording of large numerical aperture (NA) holograms (Wachulak et al., 2006, 2007, 2008b). After developing, the holographic interference pattern was translated into a relief pattern in the surface of the photoresist and digitized using an atomic force microscope (AFM). Digitized hologram was used as an input for a numerical code that reconstructs the image using a Fresnel propagator. This calculation back-propagates the object beam wavefront to obtain the amplitude and phase distribution of the field in the image plane (Goodman, 1996; Schnars & Juptner, 1994, 2002). To evaluate Fresnel-Kirchhoff integral, the product of spatial frequency representation of the hologram, obtained through a two dimensional fast Fourier transformation, and a quadratic phase free space Fresnel propagator in the spatial frequency domain was computed. Using this method the reconstruction of the hologram was performed numerically.

3.1 Coherence limitation to the spatial resolution in EUV holography

A photoresist was used to record the hologram. The interference pattern was converted into a height modulation after the developing process. To reconstruct the hologram a digitization is necessary. The digitization converts the photoresist height modulation into a gray-scale image that can be processed numerically in order to reconstruct the object.

The spatial resolution of the holographic recording is dictated by a numerical aperture $NA = n \cdot \sin(\Theta)$, where Θ is a maximum half-angle of cone of light that can enter the imaging system and n is an index of refraction of a medium. Also, the spatial resolution depends on the resolution of the recording medium. Consequently, the highest spatial frequency that can be recorded in the recording medium sets a limit to the NA. To avoid this limitation the holograms were recorded in a high resolution photoresist, PMMA, that has the spatial resolution of ~ 10nm for e-beam exposure (Hoole et al., 1997, Yamazaki et al., 2004).

The spatial resolution of the hologram is also limited by spatial and temporal coherence of illumination source and by digitization process. The coherence limitations to the hologram NA manifest themselves when a path difference between radiation diffracted by the object and reference beam exceeds either the longitudinal or transverse coherence lengths. The spatial resolution is often given by:

$$\Delta = \frac{a \cdot \lambda}{NA} \tag{7}$$

where $a \in < 0.3, 1 >$ depending on the method used to measure the resolution and coherence of the source (Heck et al., 1998), λ is the wavelength of the illumination.

The limitation to the resolution set by the spatial coherence can be understood by the scheme depicted in Fig. 2. The reference beam and the beam diffracted from the point object will interfere only within coherence area depicted as a circle, with coherence radius R_c. Beyond that region one can assume that the interference will not occur. If the angle between two beams is Θ_{sc} then the recording numerical aperture is equal to:

$$NA_{sc} = \frac{R_c}{\sqrt{R_c^2 + z_p^2}} \tag{8}$$

where z_p defines a distance from the object to the recording medium. The spatial resolution is thus limited to:

$$\Delta_{sc} = \frac{a\lambda\sqrt{R_c^2 + z_p^2}}{R_c} \tag{9}$$

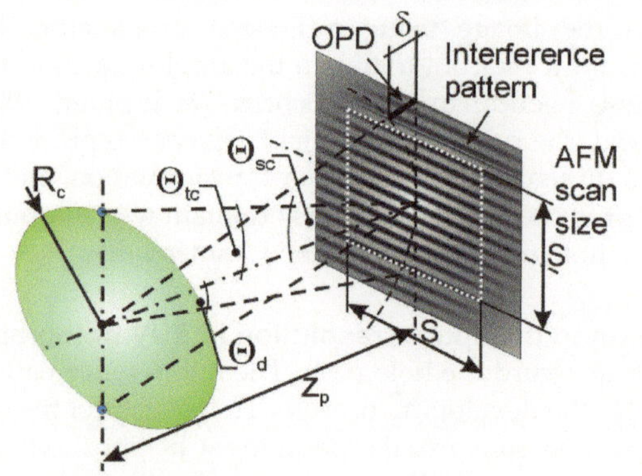

Fig. 2. Schematic description of spatial coherence, temporal coherence and scanning size limitations to the hologram spatial resolution.

The temporal coherence limits the recording numerical aperture as well. This can be seen in the scheme in Fig. 2 as well. The optical path difference δ between the reference and diffracted beams has to be smaller than coherence length of the illumination source $\delta \leq l_c$ in order to observe the interference. This limits region where the interference takes place, that can be defined by Θ_{tc} angle. The numerical aperture is thus restricted to:

$$NA_{tc} = \sqrt{1-\left(\frac{z_p}{z_p + l_c}\right)^2} \tag{10}$$

and similarly the resolution:

$$\Delta_{tc} = \frac{a\lambda}{\sqrt{1-\left(\frac{z_p}{z_p + l_c}\right)^2}} \tag{11}$$

Another limitation to the resolution is digitization process of the hologram. The resolution of reconstructed image is affected by scanning size s because for the object located at the center of scanning area, the finest interference fringes, carrying information about the highest spatial frequencies, are located at the edges of the scanned area. This can be related to the numerical aperture of the digitization as:

$$NA_d = \frac{s}{\sqrt{4z_p^2 + s^2}} \tag{12}$$

and similarly the spatial resolution:

$$\Delta_d = \frac{a\lambda\sqrt{4z_p^2 + s^2}}{s} \tag{13}$$

It is important to notice, that according to Equation (7) improvements to the resolution can be done either by decreasing the wavelength of illumination using an output of a short wavelength EUV source, or by increasing the numerical aperture in the recording and reconstruction steps. Decreasing distance between the object and recording medium allows for storing in the photoresist surface finer interference fringes, thus improving the resolution. During reconstruction, the AFM scan has to be large enough and with sufficiently fine sampling to read these fringes, to be able to reconstruct the information about corresponding spatial frequency components in the final, reconstructed image. The experiments, described in this chapter, will show, that increasing the recording and reconstruction NA allows to reach the resolution in EUV holographic imaging comparable to the wavelength of illumination.

3.2 Sampling considerations during the digitization process

The information about the highest spatial frequencies in the hologram has to be properly preserved in the recording medium and subsequently retrieved to reconstruct the image of the object with the highest possible resolution. For the simplest case of point objects corresponding Gabor holograms are Fresnel Zones (FZ). This simple hologram has a maximum spatial frequency defined by the outermost zone width Δr.

The resolution of the recording medium has to be better than the highest spatial frequency component in the hologram $\Delta_{rec} \leq \Delta r$, in order to faithfully reconstruct the object. Similarly pixel size in the AFM scans has to be equal or smaller than the highest spatial frequency component in the hologram $\Delta_{AFM} \leq \Delta r$. Given the relationship between the NA and Δr (Attwood, 1999), the minimum number of samples required for a given spatial resolution is:

$$N_{samples/line} = \frac{s}{\Delta r} \tag{14}$$

where s is the size of digitized hologram. The numerical aperture defined by the outermost zone width can be expressed as:

$$NA = \frac{\lambda}{2\Delta r} \tag{15}$$

Substituting Equation (7) yields to a resolution:

$$\Delta = 2a\Delta r \tag{16}$$

Moreover, using Equation (7) and digitization NA, expressed in terms of trigonometric functions $NA_d = \sin\left[\arctan\left(\frac{s}{2z_p}\right)\right]$, hologram scan size can be expressed as:

$$s = 2z_p \tan\left[\arcsin\left(\frac{a\lambda}{\Delta}\right)\right] \tag{17}$$

if $0 \leq \frac{a\lambda}{\Delta} \leq 1$. Finally the number of sample points obtained with the AFM in single scan line has to be equal to:

$$N_{samples/line} = \frac{2a \cdot s}{\Delta} = \frac{4az_p}{\Delta} \tan\left[\arcsin\left(\frac{a\lambda}{\Delta}\right)\right] \tag{18}$$

In case of small angle approximation it will be equal to:

$$N_{samples/line} = \frac{4a^2 z_p \lambda}{\Delta^2} \tag{19}$$

The total number of sampling points for two dimensional interference pattern can be expressed as:

$$N_{total} = \left(N_{sample/line}\right)^2 = \frac{16a^4 z_p^2 \lambda^2}{\Delta^4} \tag{20}$$

Equation (20) shows that the number of points in the digitalization, necessary to attain a given image spatial resolution Δ, scales as $N_{total} \sim \Delta^{-4}$. This imposes a practical limitation in Gabor's scheme if the distance z_p is not kept small. Due to the fact that the number of points per scan line in the AFM is often limited to 1024, 2048 etc., the NA is practically limited and the best option to increase the NA is to decrease the distance z_p.

4. Holographic 2-D imaging using EUV lasers

Two holograms of an atomic force microscope tip were obtained using a compact table top EUV laser in a Gabor's in-line configuration, as described in detail in (Wachulak et al., 2006). This configuration is very easy to set up and robust, moreover, it requires neither optics nor critical beam alignment.

4.1 Experimental details

In this experiment a table top discharge pumped capillary Ne-like Ar laser, radiating at 46.9 nm wavelength, was used for hologram recording. The laser was configured to produce 0.1 mJ pulses at repetition rate of 1 Hz. With a ratio $\lambda/\Delta\lambda \approx 10^4$, the capillary discharge laser has a longitudinal coherence length of $l_c \approx 470$ μm. Using Gabor's geometry, shown in Fig. 3, two holograms with two different NA were obtained by changing the distance between the object and the recording medium z_p. Two selected distances were $z_p \approx 4$ mm (NA = 0.038) and $z_p \approx 120$ μm (NA = 0.172). The temporal coherence limits the image spatial resolution to $\Delta \approx 64$ nm for $z_p = 4$ mm and $\Delta \approx 30$ nm for $z_p = 120$ μm. However, spatial coherence imposes a more severe limitation. In both cases this is the limiting factor to the maximum attainable resolution, $\Delta \approx 340$ nm for the smaller NA at $z_p \approx 4$ mm and $\Delta \approx 30$ nm in the second case.

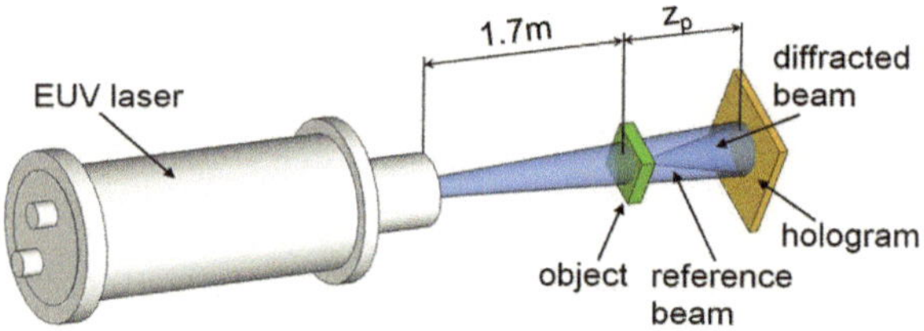

Fig. 3. Experimental set up used to record the EUV holograms in Gabor's configuration.

The recording was done in a 120 nm thick layer of PMMA (MicroChem 950,000 molecular weight) spin-coated on top of a silicon wafer. Exposures of approximately 100 seconds were necessary with the experimental set up utilized in this work. After the exposure, the photoresist was developed using a standard procedure: the sample was immersed in a solution of MIBK - methyl isobutyl ketone (4-Methyl-2-Pentanone) with IPA (isopropyl alcohol) 1:3 for 30 seconds and rinsed with IPA for 30 seconds. Finally the sample was dried using compressed nitrogen.

4.2 Results of 2-D holographic imaging with EUV laser

The exposure was adjusted to be in region of linear response of the photoresist and consequently a relief pattern height printed in its surface was equivalent to the interference intensity pattern of the hologram. The developed photoresist surface was mapped with the AFM to generate digitized holograms. Fig. 4a shows the hologram with low numerical aperture (NA = 0.038) recording, digitized area ~300x300 μm^2, and pixel size corresponding to 270 nm, while Fig. 4c depicts digitized hologram with higher numerical aperture (NA = 0.172). In this case the area scanned is 42x42 μm^2 with pixel size equivalent to 41 nm. The holograms were reconstructed by numerically simulating illumination with a short wavelength EUV readout wave. The amplitude and phase distributions of the field in the image plane were obtained calculating the field emerging from the hologram illuminated by a plane reference wave and numerically back-propagating the fields with a Fresnel propagator (Schnars & Juptner, 2002). The reconstructed image was found by taking two dimensional inverse fast Fourier transform (2D-IFFT) of the product of spatial frequency Fresnel propagator and the 2D-FFT of the hologram. This calculation allowed to obtain the amplitude and phase distribution of the field in the image plane. Fig. 4b and Fig. 4d are, respectively, the reconstructed images of corresponding holograms shown in Fig. 4a and Fig. 4c. The inset in Fig. 4b is a magnified region showing the end of the tip. In both reconstructed images a triangular profile of the AFM tip is clearly revealed. In case of Fig. 4d it was evident after the reconstruction that the tip was broken and partially contaminated, as can be observed in protrusion, marked by an arrow, in the upper part of the tip.

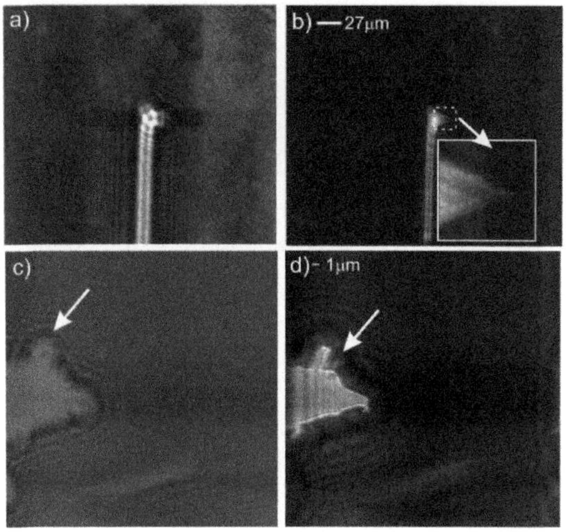

Fig. 4. Hologram recorded in surface of photoresist a) and reconstruction b) with low NA= 0.038 and, consequently, hologram c) and reconstruction d) with higher NA= 0.172.

4.3 Resolution estimation using wavelet decomposition and image correlation method

An optimum resolution in the reconstructed image was assessed by correlation with a synthesized image - template, used as a reference image. This reference image was constructed as a binary template that has by definition 1 pixel resolution. From the reference image a set of lower resolution images (wavelet components) was generated by wavelet decomposition of the reference image, each one having a spatial resolution relative to the reference image given by $Y = 2^X$, where Y is the relative resolution between the images in the wavelet decomposition and X is the order of the wavelet. In this case "haar" wavelet was applied. The reconstructed images were obtained by running the Fresnel propagator code with slightly different z_p around 4 mm and 120 μm, respectively. Then all reconstructions were correlated with the set of decreasing resolution wavelet components, shown in Fig. 5. To perform the correlation, higher wavelet orders were resized to have the same image size. The correlation coefficients between wavelet components and the reconstructed images for slightly different z_p provided a quantitative resolution of the reconstructions, relative to the synthesized reference images. This procedure also allowed for selection of an optimum reconstruction distance z_p. Fig. 5 shows a binary template that constitutes the order zero (X=0) wavelet component and the first four wavelet components with decreased resolution given by a factor 2^X.

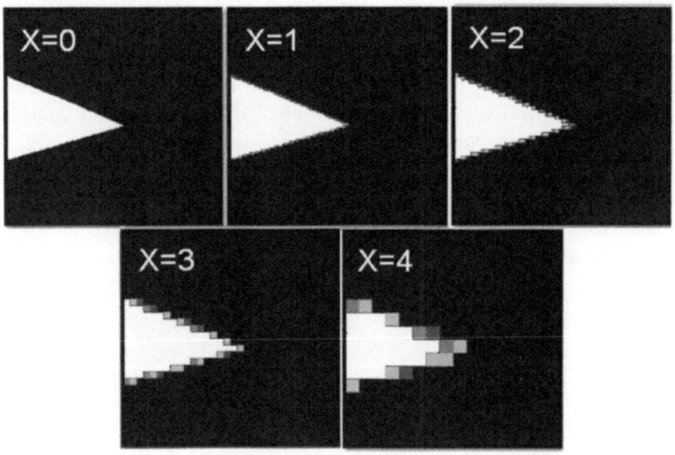

Fig. 5. Binary template (X=0) and wavelet components used to perform 2-D correlation with the reconstructed images to assess a relative resolution of the holograms.

Fig. 6 shows obtained correlation values for different images, reconstructed at different object-hologram distances z_p and different wavelet components, as a function of the wavelet scale X. By slightly changing the reconstruction distance the correlation coefficient changes but the shape of the curves remains unchanged.

Fig. 6a corresponds to smaller numerical aperture hologram (NA = 0.038). This plot shows that the largest value of the correlation function for all z_p corresponds to wavelet scale X = 0, and does not significantly change in the interval between X = 0 to X=1, decreasing faster for wavelet orders larger than 1. We conservatively assumed that the best correlation curve corresponds to X = 0.5, indicating that the resolution for this image is $2^{0.5} = 1.41$ relative to the reference image. The synthesized binary reference image has a pixel size of 270 nm and consequently this analysis indicates that the reconstructed image has the spatial resolution equivalent to Δ=380 nm. Fig. 6b corresponds to higher numerical aperture recording (NA = 0.172). This plot clearly shows the maximum in the correlation values for X = 2 at all reconstruction distances ranging from 118 to 132 μm.

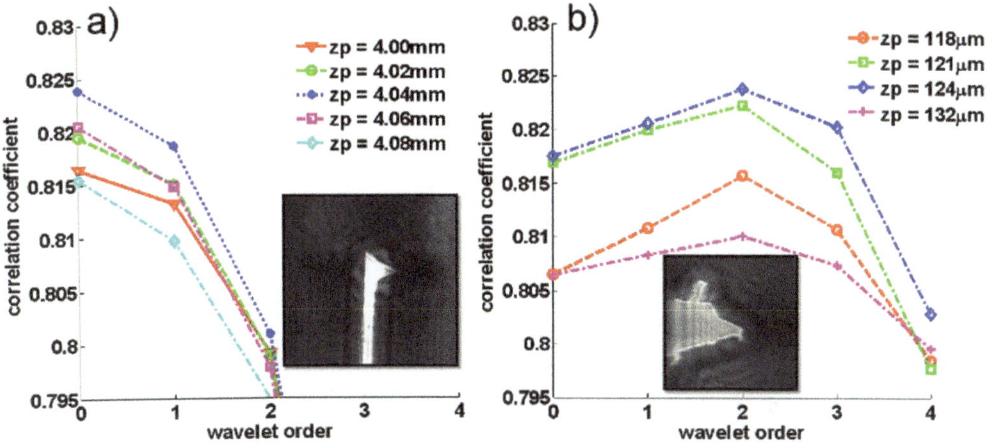

Fig. 6. Correlation values for different object-hologram distances z_p as a function of the wavelet scale X, a) recording with NA=0.038, pixel size 270 nm, b) recording with NA= 0.172, pixel size 41 nm.

This analysis indicates that the spatial resolution of the reconstruction is 4 pixels. In this case, the pixel size was 41 nm, setting the spatial resolution of the image to Δ=164 nm. Based on the recording NA, the spatial resolution limited by Rayleigh criterion Δ= (0.61 λ)/NA gives value of Δ_R=166 nm that compares very well with the results of wavelet analysis. The results are in good agreement with theoretical calculations and coherence limitations and were summarized in Table 1.

Rayleigh criterion based $a = 0.61$	Spatial coherence $\frac{a\lambda\sqrt{R_c^2 + z_p^2}}{R_c}$	Temporal coherence $\frac{a\lambda}{\sqrt{1-\left(\frac{z_p}{z_p+l_c}\right)^2}}$	AFM pixel size $\Delta_{AFM} = \Delta r$	AFM sampling limit $\Delta = 2a\Delta r$	Experimental resolution
z_p = 4 mm	337.8 nm	64.1 nm	270 nm	329.4 nm	380 nm
z_p = 120 μm	30.4 nm	29.2 nm	41 nm	50 nm	164 nm

Table 1. Coherence limitation to the resolution, AFM sampling limitations and experimentally obtained resolutions estimated based on the correlation analysis.

5. Wavelength resolution holography

5.1 Experimental setup

To reach near-ultimate spatial resolution z_p has to be reduced even further, according to Equations (7) and (20). A sample was composed of carbon nanotubes (CNT) with diameters between 50 and 80 nm and length 10-20 μm, placed on a 100 nm thick silicon membrane acting as a support. It was imaged using a table top EUV laser in Gabor's in-line configuration as depicted in Fig. 7 and described in more details in (Wachulak et al., 2008b).

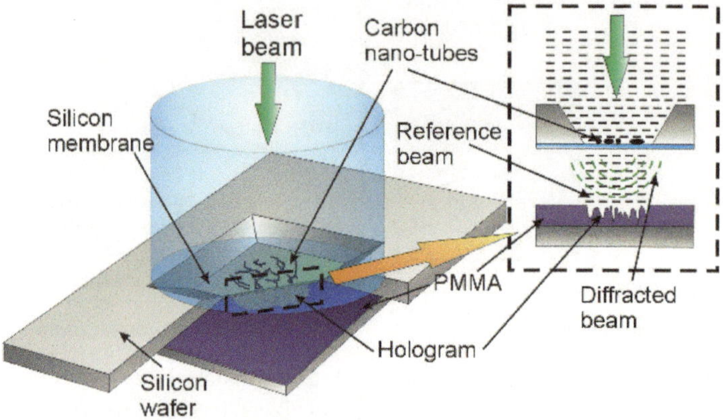

Fig. 7. Experimental setup in Gabor's hologram configuration. The EUV laser beam illuminates the object composed of carbon nano-tubes deposited on top of Si membrane.

The thin Si membrane has a transparency of approximately 60% at $\lambda = 46.9$ nm [1]. The sample was placed approximately $z_p = 2.6$ μm away from the Si wafer spin-coated with 120 nm thick layer of PMMA photoresist, resulting in recording NA practically equal to 1. However the limited area of the AFM scan that digitizes the hologram (9.9×9.9 μm^2) limits the NA to 0.88. The photoresist was later developed and the hologram stored in the resist was digitized with the AFM and numerically reconstructed using the same Fresnel propagator code. Fig. 8a,c show the holograms digitized with the AFM. The corresponding numerical reconstructions are shown in Fig. 8b,d. The CNTs are clearly visible as white lines in a black background.

The spatial resolution in the reconstructed image was found applying the knife-edge method. It is based on measuring 10-90% rise of the intensity in the line-cuts through the image. Two examples of such line-cuts depicted in Fig. 9a,b were taken in regions indicated in Fig. 8b,d. These cuts were realized in region where a "plateau" in the maximum and minimum intensities was clearly reached, to recreate the knife-edge resolution test. The measurements yielded the spatial resolution of ~46 nm. A mean value of resolution was found in similar cuts realized in different areas in the reconstructed images resulting in the statistical resolution of 45.8±1.9 nm, where error is assigned by a standard deviation spread in the measurements (Wachulak et al., 2008b).

---

[1] *from CXRO database, "http://www-cxro.lbl.gov/".*

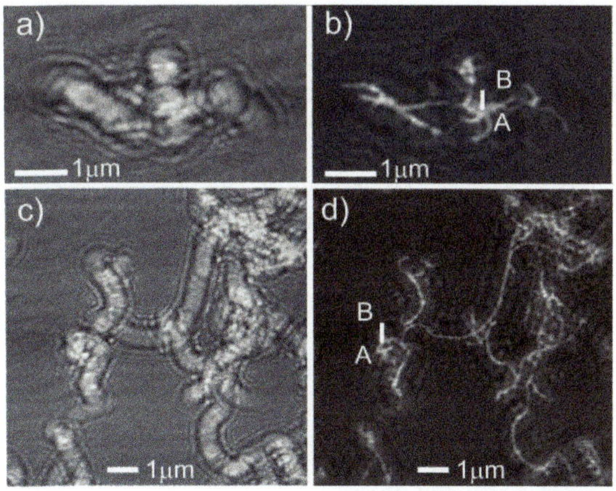

Fig. 8. a,c) Holograms and b,d) reconstructed images of 50-80 nm diameter carbon nanotubes. The holograms were obtained by scanning the photoresist surface with the AFM. Knife-edge test was applied to estimate the resolution of the reconstructed image in b,d).

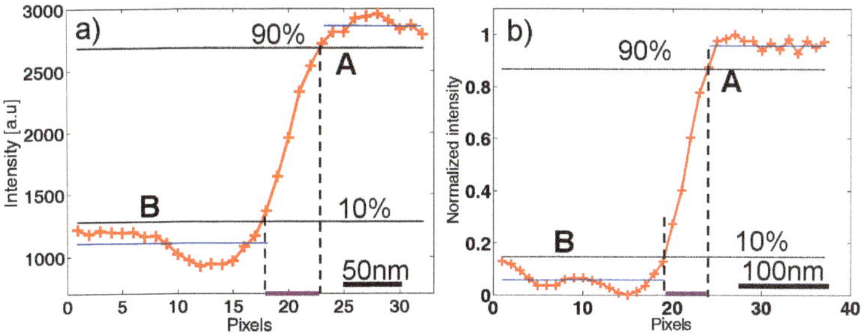

Fig. 9. The intensity lineouts through the reconstructions indicating a 10-90% intensity modulation over approximately 4.5 pixels. Figure a,b) corresponds to the reconstructed images in Fig. 8b,d) respectively. A mean value of multiple cuts obtained at different locations yielded the spatial resolution of 45.8±1.9 nm.

5.2 Gaussian filtering and correlation method for resolution and feature size assessment

To obtain a global assessment of the image spatial resolution, a correlation analysis on the reconstructed holograms was performed (Wachulak et al., 2008c). This method is based on the correlation between the reconstructed holographic image and a 2-D set of templates with calibrated resolution and diameter of the nanotubes, all generated from a master binary template. The set of master binary templates is derived from the original image, depicted in Fig. 10a, by skeletonizing the image, (Yatagai et al., 1982). The result of skeletonization is shown in Fig. 10b. It represents the shape of the nanotube, but has thickness of only 1 pixel. Then the skeleton is convolved with a set of circular templates of different diameters representing different diameters of the CNTs. One of those templates after the convolution is shown in Fig. 10c. From these master binary templates, a sub-set of templates with variable and calibrated resolutions was obtained by applying a Gaussian filter of variable width. Fig. 10d shows one example of these templates, where the edges of the nanotube in Fig. 10c were blurred by the filtering process. Then the 2-D correlation between all the templates with variable resolution and variable nanotube diameter has been computed resulting in a 2-D correlation map. The global maximum in this map points out to the template that is the most similar to the reconstructed hologram. Both, resolution and nanotube diameter in the reconstructed image are the same as in the template that maximizes the correlation coefficient in the correlation map. Fig. 10e depicts the 2-D correlation map that indicates the CNT diameter equal to 70.6±5 nm and the spatial resolution 47.5±5 nm, where error is assigned as a one step in calculation of the surface shown in Fig. 10e.

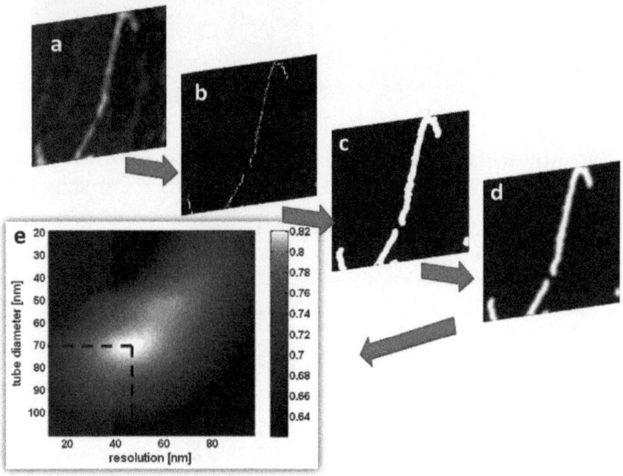

Fig. 10. Procedure to generate the templates with calibrated resolution. Filtered image a) with the low frequency background fluctuations removed. Skeletonizing b) of the image. Binary template c) obtained by convolution of the skeleton template (b) with a circular template. Template with degraded resolution d) obtained by convolution of the binary template (c) with a 2-D Gaussian filter. Correlation coefficients e) of the reconstructed image with different templates plotted as a function of nanotube diameter and template resolution. The nanotube diameter and resolution of the image are determined to be 70.6±5 nm and 47.5±5 nm by localizing a global maximum in the correlation plot.

The obtained resolution exceeds by ~19-21 nm the best possible obtainable resolution of 27 nm set by $\lambda/2NA$ (Jacobsen et al., 1990). Several factors degrade the resolution. It was verified that the spatial and temporal coherence of the laser source is not a limitation in this experiment, because $z_p << l_c$, R_c and NA~1. The longitudinal coherence length of the EUV laser is approximately l_c ~ 470 μm, $\Delta\lambda/\lambda \approx 10^{-4}$, and the spatial coherence radius is approximately 0.25 mm at the location of the recording, 0.75 m from the source (Liu et al., 2001). With these parameters, the spatial resolution set by the laser coherence properties is approximately 29 nm. Another limiting factor could arise from the digitalization because the AFM maps the hologram into a 1024×1024 matrix over a surface of approximately 9.9×9.9 μm^2, giving a pixel size of 9.7 nm. This pixel size sets the maximum spatial frequency that can be recorded, establishing a limit for the spatial resolution of ~12 nm that practically does not play a role in this experiment. However, the resolution is affected by photoresist resolution (estimated to be ~20 nm for photon exposure (Solak et al., 1999)) and by resolution of the AFM that is defined by radius of curvature of the tip equal to 10 nm. The latter is not a fundamental limitation, as it can be overcome using a high resolution AFM tips. Taking all these factors in the convolution, the best possible resolution in our experiment is estimated to be ~39 nm, slightly better than the observed resolution. An additional possible factor degrading the resolution is roughness of Si membrane. Variations in the membrane's thickness can introduce a random background noise that degrades the image resolution. Detailed modeling indicates that a surface roughness of ~20 nm, which is similar to a measured roughness of the wafer that contains the membrane, would degrade the resolution to 45-46 nm adding a noise background similar to the noise measured in the reconstructed images.

6. Holographic 3-D imaging using EUV lasers

The possibility of volume three dimensional imaging by numerical sectioning obtained from a single high numerical aperture hologram (Wachulak et al., 2007) will be discussed in this section. Three dimensional images were obtained from Gabor holograms recorded in the photoresist after exposure using a table-top EUV laser. Digitized holograms were numerically reconstructed over the range of image planes by numerically sweeping the reconstruction distance, resulting in numerical optical sectioning of image depths.

6.1 Experimental details

The experimental set up is schematically illustrated in Fig. 11. The same source was used in this experiment as in the previous ones. The test object used in holographic volume imaging experiment consisted of a tilted metallic surface covered with opaque spherical markers.

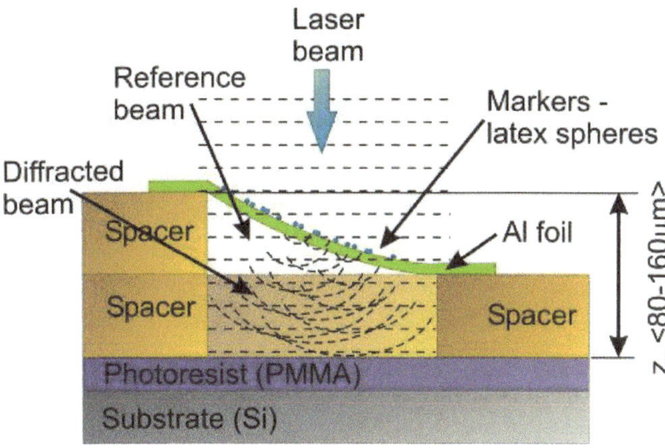

Fig. 11. Diagram of the 3-D EUV holography experimental set up showing details of the test object used.

This test object was fabricated by covering a semicircular hole 1.5 mm in diameter made in an 80 μm thick Mylar sheet with a 100 nm thick aluminum foil. The hole was partially covered

with a second Mylar sheet 80 μm thick, as schematically indicated in Fig. 11. The aluminum foil contours over the semicircular aperture to produce a variable height surface with desirable characteristics for this test and has a transmittance of approximately 35% at λ=46.9 nm limited mainly by a layer of native oxide [2]. The Al filter also suppresses lower photon energy plasma emission (i.e., long wavelength background) from the Ar laser source.
The object was prepared by placing a drop of water with heavily diluted latex spheres (2.62% solution in water) 465 nm in diameter, [3] on top of Al foil. Evaporation of the water left a random distribution of latex spheres (markers) deposited over the partially transparent tilted Al foil membrane. These spheres are completely opaque to 46.9 nm EUV laser radiation. With this deposition procedure, the markers are randomly distributed over the supporting Al foil and at predictable distances to the photoresist (heights) imposed by the foil profile. To activate the PMMA with 46.9 nm radiation requires exposures of 240 laser shots, 4 minutes exposure time at the repetition rate employed in this experiment. After exposure, the photoresist was developed using standard developing procedures.

6.2 Results of 3-D holographic imaging with EUV laser

Digitized hologram is shown in Fig. 12a. The digital reconstruction of the hologram digitized with the AFM is based on a numerical Fresnel propagator. Fig. 12a shows a small (42×42 μm²) section of the hologram. The numerical reconstruction of the hologram provided images of the object described above and shown in Fig. 12b.

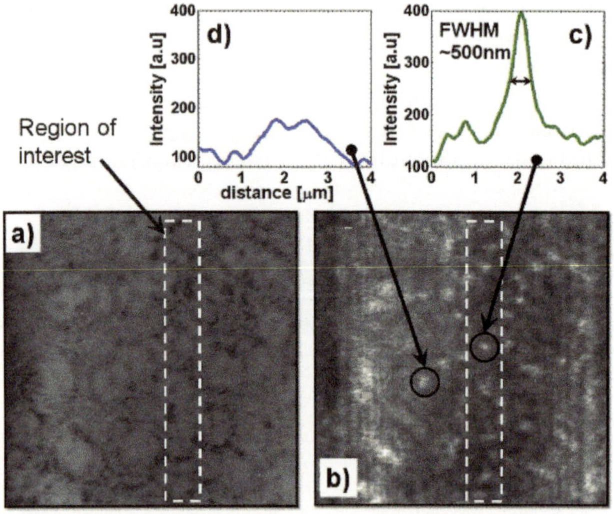

Fig. 12. Hologram of 3-D sample a) recorded in the photoresist surface and digitized with the AFM with a random distribution of 465nm diameter latex spheres - markers over tilted surface of Al foil. Numerical reconstruction b) of the hologram using Fresnel propagator algorithm. Intensity cuts in vertical direction of one marker that is "in focus" (c) and one "out of focus" (d). The white dotted lines indicate region in the reconstruction that is "in focus".

---

[2] *from CXRO database, "http://www-cxro.lbl.gov/".*
[3] *From Polysciences Inc.*

One of the critical parameters in the reconstruction code is a distance between the recording medium and the object, indicated in the diagram in Fig. 11 as z_p. Small changes in z_p reconstructs slightly different images. To determine the value of z_p corresponding to the optimum reconstruction, a 2-D image correlation was used and is described in details in references (Wachulak et al., 2006, 2007, 2008c).

To demonstrate retrieval of the depth information from the hologram the numerical reconstruction of the digitized hologram, shown in Fig. 12a, has to be performed for different values of distance z_p. The different runs produced different reconstructed images in which the latex spheres - markers located at the correct z_p generated a sharper image than those markers "out of focus". Fig. 12b shows one of these reconstructed images. In this case the reconstruction is optimum for z_p that matches height of a central part of the hologram, indicated in the figure by white dotted rectangle. In this region the height is such that the markers located there are reconstructed "in focus", while the latex spheres above and below this level are reconstructed blurred. This can be observed in Fig. 12c,d where the intensity profiles obtained in a vertical cut of one "in focus" marker (c) and one "out of focus" marker (d) are plotted. By changing z_p in the reconstruction code only the latex sphere markers located at the height equal to z_p produce sharper reconstruction images as compared to those markers out of focus. This is a similar effect to optical sectioning, however, performed on a digitally reconstructed image. The depth information in the hologram can finally be retrieved varying the reconstruction parameter z_p. To determine a value of z_p corresponding to the best reconstruction the reconstructed image was correlated with a template of the marker consisting of a circular mask with known size representing the latex sphere. Finding the maximum value of correlation between the reconstructed image of each marker and the mask determines corresponding optimum height. Combining this information with *x-y* coordinates of each marker allowed for placing each marker uniquely in a 3-D space and the reconstruction of surface of the test object with depth resolution.

Fig. 13a,c show the surface topography obtained from the reconstructed images in two different regions of the hologram. In case of Fig. 13a, the AFM scan was performed in region of the test object close to edge of Mylard spacer, where the slope of the Al foil is expected to be high. A similar scan performed at distance approximately 200 μm away from this edge, produced image with smaller slope, as shown in Fig. 13c. Fig. 13b,d show reconstructed heights for all markers as a function of transversal coordinate x in the same two regions of the object plotted in Fig. 13a,c. The surface plot from xyz space was projected into xz plane. These plots give a measure of spread of the calculated heights for all the markers and also show, as indicated by the best linear fit, different slopes in these two regions. The statistical dispersion of data points relative to the best linear fit are $\Delta z = 2.64$ μm for the region with high slope, Fig. 13b, and $\Delta z = 1.32$ μm for the region with lower slope Fig. 13d. This spread in the measured heights of markers compares well with expected accuracy in z direction determined by the NA of the hologram. As pointed out by Rogers, if one assumes the hologram as a superposition of Fresnel Zone Plates (FZPs) (Rogers, 1950), the resolution in z coordinate can be related to its depth of focus. For a FZP the depth of focus is given by $\delta z = \lambda / NA^2$ (Attwood, 1999). The NA corresponding to higher slope region, where $z_p = 160$ μm, is NA = 0.13, yielding a depth of focus $\delta z = 2.77$ μm. In region, where the Al foil has a lower slope, the latex markers were closer to the hologram, at a distance $z_p = 140$ μm. For this reconstruction the expected vertical resolution is $\delta z = 2.12$ μm.

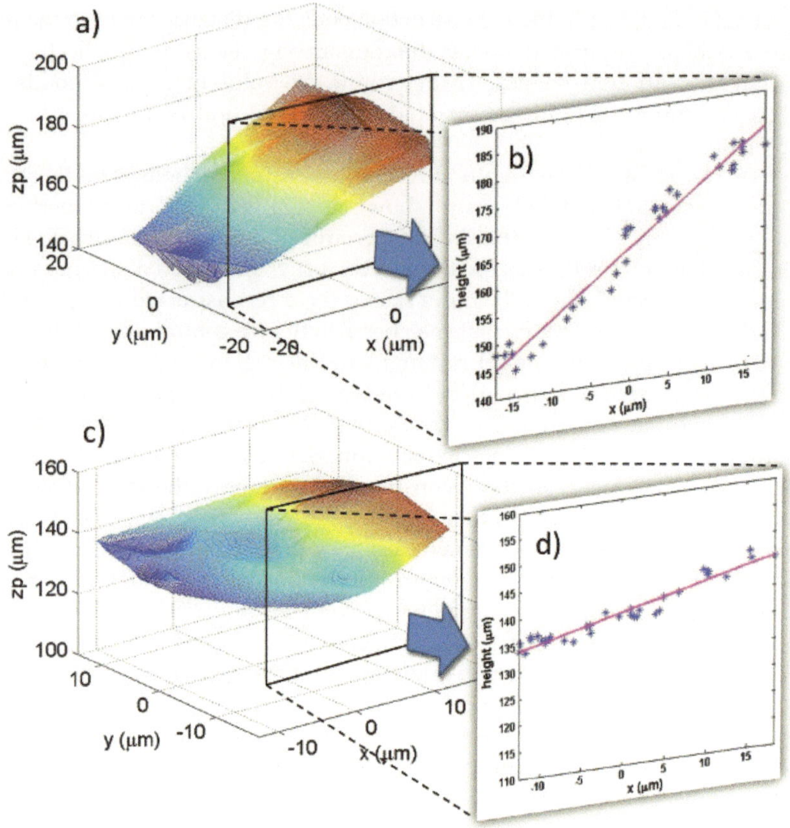

Fig. 13. Surface topographies obtained from reconstructed images and heights of the markers in different regions of the object. a), b) show the region close to the edge of the spacer where high slope is revealed in the Al foil while c), d) show region approximately 200 μm away from the edge with lower Al foil slope. b), d) are the projections of the 3-D plots on x-z plane to better visualize the slopes.

7. Conclusions

In this chapter the EUV table-top holographic imaging using a compact EUV laser as the illumination source was presented. The spatial resolution of the images of AFM tips, obtained by numerical reconstruction, was assessed utilizing a wavelet image decomposition and image correlation method leading to 164 nm. By increasing the numerical aperture of the recording and digitization wavelength-resolution EUV holograms were generated. The images were numerically reconstructed from the hologram recorded in surface of the photoresist and digitized with the AFM. Images of carbon nanotubes were obtained with 46±2 nm resolution determined by a knife edge test. Continuing development of highly coherent table-top EUV/SXR lasers in the vicinity of 10 nm (Wang et al., 2006) can be expected to enable future holographic imaging only limited by the photoresist resolution. Increasing flux of the EUV and SXR table top lasers opens a perspective in the future for single shot recording, permitting full field time resolved holographic imaging. This imaging method allows hologram recording without any previous object preparation, as required in electron microscopy, and free of any interaction with a probe that may occur in scanning microscopes. Photon based imaging systems also allow spectroscopic contrast, an important characteristic in imaging with shorter wavelength radiation. It also opens the possibility to study specimens in different environments, for example in the presence of external magnetic or electric fields. Moreover detailed processing of the reconstructed holographic images, performed by changing object-hologram distance in the reconstruction code was presented. It enables retrieving the depth information from a single high NA hologram. Using a specially fabricated 3-D object the numerical reconstruction and analysis of the hologram allowed to map the surface topography with a depth resolution close to 2 μm.

8. Acknowledgements

This work was supported by the National Science Foundation ERC for Extreme Ultraviolet Science and Technology, Award Number EEC-0310717. The authors thank to Prof. Randy Bartels, Prof. Carmen Menoni and Prof. Jorge Rocca for their constructive comments and helpful discussions during the experiments presented herein.

First author would also like to acknowledge the support from Foundation for Polish Science, Homing 2009 Program, award number HOM/2009/14B, related to novel, high resolution imaging techniques.

9. References

Aoki, S.; Kikuta, S.; (1974). X ray holographic microscopy. *Jpn.J Appl.Phys.*, Vol. 13, pp. 1385

Attwood, D.T.; (1999). *Soft X-ray and extreme ultraviolet radiation,* Cambridge University Press, ISBN 0-521-65214-6, Cambridge, UK

Baez, A. V., (1952). A study in diffraction microscopy with special reference to X-rays. *Journal of the Optical Society of America,* Vol. 42, pp. 756

Bartels, R.A.; Paul, A.; Green, H.; Kapteyn, H.C.; Murnane, M.M.; Backus, S.; Christov, I.P.; Liu, Y.W.; Attwood, D.; Jacobsen, C.; (2002). Generation of spatially coherent light at extreme ultraviolet wavelengths. *Science.* Vol. 297, No. 5580, pp. 376-378

Chapman, H. N.; Hau-Riege, S.P.; Bogan, M.J.; Bajt, S.; Barty, A.; Boutet, S.; Marchesini, S.; Frank, M.; Woods, B.W.; Benner, W.H.; London, R.A.; Rohner, U.; Szoke, A.; Spiller, E.; Moller, T.; Bostedt, Ch.; Shapiro, D.A.; Kuhlmann, M.; Treusch, R.; Plonjes, E.; Burmeister, F.; Bergh, M.; Caleman, C.; Huldt, G.; Seibert, M. M.; Hajdu, J.; (2007). Femtosecond time-delay X-ray holography. *Nature,* Vol. 448, No. 9, pp. 676

Elsebitt, .S.; Schlotter, W. F.; Lorgen, M.; Hellwig, O.; Eberhardt, W.; Stohr, J.; (2004). Lensless imaging of magnetic nanostructures by X-ray spectro holography. *Nature,* Vol. 432, pp. 885-888

Gabor, D., (1948). A new microscopic principle. *Nature,* Vol. 161, pp. 777

Giles, J. W.; (1969). Image reconstruction from a Fraunhofer X-ray hologram with visible light. *Journal of the Optical Society of America,* Vol. 59, pp. 1179

Goodman, J. W.; (1996). *Introduction to Fourier Optics.* (3rd edition), Roberts & Company, ISBN 0-9747077-2-4, Englewood, Colorado, USA

Heck, J.M.; Attwood, D.T.; Meyer-Ilse, W.; Anderson, E.H.; (1998). Resolution determination in X-ray microscopy: an analysis of the effects of partial coherence and illumination spectrum. *Journal of X-Ray Science and Technology*. Vol. 8, pp. 95-104

Hoole, A. C. F.; Welland, M. E.; Broers, A. N.; (1997). Negative PMMA as a high-resolution resist - the limits and possibilities. *Semicond. Sci. Technol.* Vol. 12, pp. 1166-1170

Hun Lee, S.; Naulleau, P.; Goldberg, K. A.; Hyun Cho, C.; Tae Jeong, S.; Bokor, J.; (2001). Extreme-ultraviolet lensless Fourier-transform holography. *Applied Optics*, Vol. 40, No. 16, pp. 2655

Jacobsen, C.; Howells,M.; Kirz, J.; Rothman, S.; (1990), X-Ray Holographic Microscopy Using Photoresists, *Journal Of The Optical Society Of America A*, Vol.7, No.10, pp. 1847-1861

Leith, E. N.; Upatnieks J., (1962). Reconstructed Wavefronts And Communication Theory. *Journal of the Optical Society of America*, Vol. 52, pp. 1123

Lindaas, S.; Howells, H.; Jacobsen, C.; Kalinovsky, A.; (1996). X-ray holographic microscopy by means of photoresist recording and atomic-force microscope readout. *Journal of the Optical Society of America A*, Vol. 13, No. 9, pp. 1788-1800

Liu, Y.; Seminario, M.; Tomasel, F. G.; Chang, C.; Rocca, J. J.; Attwood, D. T.; (2001). Achievement of essentially full spatial coherence in a high-average-power soft-x-ray laser. *Physical Review A*, Vol. 63, No. 3, pp. 033802

Marchesini, S.; Boutet, S.; Sakdinawat, A. E.; Bogan, M. J.; Bajt, S.; Barty, A.; Chapman, H. N.; Frank, M.; Hau-Riege, S. P.; Szoke, A.; Cui, C.; Shapiro, D. A.; Howells, M. R.; Spence, J. C. H.; Shaevitz, J. W.; Lee, J. Y.; Hajdu, J.; Seibert, M. M.; (2008). Massively parallel X-ray holography. *Nature Photonics*, Vol. 2, pp. 560

McNulty,I.; Kirz, J.; Jacobsen, Ch.; Anderson, E. H.; Howells, M. R.; Kern, D. P.; (1992). High-Resolution Imaging by Fourier Transform X-ray Holography. *Science*, Vol. 256, No. 5059, pp. 1009-1012

Morlens, A.S.; Gautier, J.; Rey, G.; Zeitoun, P.; Caumes, J.P.; Kos-Rosset, M.; Merdji, H.; Kazamias, S.; Casson, K.; Fajardo, M.; (2006). Submicrometer digital in-line holographic microscopy at 32 nm with high-order harmonics. *Optics Letters*. Vol. 31, No. 21, pp. 3095-3097

Rogers, G.L.; (1950). Gabor diffraction microscopy. The hologram as a generalized zone-plate. *Nature*. Vol. 166, pp. 236-237

Rosenhahn, A.; Barth, R.; Staier, F.; Simpson, T.; Mittler, S.; Eisebitt, S.; Grunze, M.; (2008). Digital in-line soft x-ray holography with element contrast. *J. Opt. Soc. Am. A*, Vol. 25, No. 2, pp. 416

Rosenhahn, A.; Staier, F.; Nisius, T.; Schäfer, D.; Barth, R.; Christophis, Ch.; Stadler, L.M.; Streit-Nierobisch, S.; Gutt, Ch.; Mancuso, A.; Schropp, A.; Gulden, J.; Reime, B.; Feldhaus, J.; Weckert, E.; Pfau, B.; Günther, Ch. M.; Könnecke, R.; Eisebitt, S.; Martins, M.; Faatz, B.; Guerassimova, N.; Honkavaara, K.; Treusch, R.; Saldin, E.; Schreiber,S.; Schneidmiller, E.A.; Yurkov, M.V.; Vartanyants, I.; Grübel, G.; Grunze, M.; Wilhein, T.; (2009). Digital In-line holography with femtosecond VUV radiation provided by the free-electron laser FLASH. *Optics Express*, Vol. 17, No. 10, pp. 8220

Sandberg, R. L.; Paul, A.; Raymondson, D. A.; Hadrich, S.; Gaudiosi, D. M.; Holtsnider, J.; Tobey, R. I.; Cohen, O.; Murnane, M. M.; Kapteyn, H. C.; (2007). Lensless Diffractive Imaging Using Tabletop Coherent High-Harmonic Soft-X-Ray Beams. *Physical Review Letters*, Vol. 99, No. 098103, pp. 098103-1

Sandberg, R.L.; Song, C.; Wachulak, P.W.; Raymondson, D.A.; Paul, A.; Amirbekian, B.; Lee, E.; Sakdinawat, A.; Vorakiat, C.L.; Marconi, M.C.; Menoni, C.S.; Murnane, M.M.; Rocca, J.J.; Kapteyn, H.C.; Miao, J.; (2008). High numerical aperture tabletop soft x-ray diffraction microscopy with 70 nm resolution. *Proceedings of the National Academy of Science*, Vol. 105, No. 1, pp. 24-27

Sandberg, R.L.; Raymondson,D.; La-o-vorakiat,C.; Paul, A.; Raines, K.S.; Miao, J.; Murnane, M.M.; Kapteyn, H.C.; Schlotter, W. F.; (2009). Tabletop soft-X-ray Fourier transform holography with 50 nm resolution. *Optics Letters*. Vol. 34, No. 11, pp. 1618

Sayre, D.; Chapman, H.N.; Miao, J.; (1998). On the extendibility of X-ray crystallography to noncrystals. *Acta Crystallographica Section A*, Vol. 54, pp. 232-239

Schnars, U.; Juptner, W.P.O.; (1994). Digital recording and reconstruction of holograms in hologram interferometry and shearography. *Applied Optics*. Vol. 33, No. 20, pp. 4373-4377

Schnars, U.; Juptner, W.P.O.; (2002). Digital recording and numerical reconstruction of holograms. *Measurement Science & Technology*. Vol. 13, No. 9, pp. R85-R101

Solak, H. H.; He, D.; Li, W.; Singh, S.; Cerrina, F.; Sohn, B. H.; Yang, X. M.; Nealey, P., (1999). Exposure of 38 nm period grating patterns with extreme ultraviolet interferometric lithography, *Applied Physics Letters*, Vol. 75, No. 15, pp. 2328-2330

Thibault, P.; Dierolf, M.; Menzel, A.; Bunk, O.; David, Ch.; Pfeiffer, F.; (2008). High-Resolution Scanning X-ray Diffraction Microscopy. *Science*, Vol. 321, pp. 379

Tobey, R.I.; Siemens, M.E.; Cohen, O.; Murnane, M.M.; Kapteyn, H.C.; Nelson, K.A.; (2007). Ultrafast extreme ultraviolet holography: dynamic monitoring of surface deformation. *Optics Letters*. Vol. 32, No. 3, pp. 286-288

Trebes, J. E.; Brown, S. B.; Campbell, E. M.; Matthews, D. L.; Nilson, D. G.; Stone, G. F.; Whelan, D. A.; (1987). Demonstration of X-Ray holography with an X-Ray laser. *Science*, Vol. 238, No. 4826, pp. 517-519

Yamazaki, K.; Yamaguchi, T.; Namatsu, H.; (2004). Three-Dimensional Nanofabrication with 10-nm Resolution. *Jpn. J. Appl. Phys*. Vol. 43, pp. L1111-L1113

Yatagai, T.; Nakadate, S.; Idesawa, M.; Saito, H.; (1982). Automatic fringe analysis using digital image processing techniques. *Optical Engineering*. Vol. 21, pp. 901

Wachulak, P.; Bartels, R.A.; Marconi, M.C.; Menoni, C.S.; Rocca, J.J.; Lu, Y.; Parkinson, B.; (2006). Sub 400 nm spatial resolution extreme ultraviolet holography with a table top laser. *Optics Express*. Vol. 14, No. 21, pp. 9636-9642

Wachulak, P.; Marconi, M.C.; Bartels, R.A.; Menoni, C.S.; Rocca, J.J.; (2007). Volume extreme ultraviolet holographic imaging with numerical optical sectioning. *Optics Express*. Vol. 15, pp. 10622-10628

Wachulak, P. W.; Capeluto, M.G.; Menoni, C. S.; Rocca, J. J.; Marconi, M. C.; (2008). Nanopatterning in a compact setup using table top extreme ultraviolet lasers, *Opto-Electronics Review*, Vol. 16, No. 4, pp. 444–450

Wachulak, P.W.; Marconi, M.C.; Bartels, R. A.; Menoni, C. S.; Rocca, J.J.; (2008). Soft x-ray laser holography with wavelength resolution. *J. Opt. Soc. Am. B*. Vol. 25, No. 11, pp. 1811-1814

Wachulak, P.W.; Brewer, C.A.; Brizuela, F.; Chao, W.; Anderson, E.H.; Bartels, R.A.; Menoni, C.S.; Rocca, J.J.; Marconi, M.C.; (2008). Simultaneous determination of feature size and resolution in soft x-ray microscopy images. *Journal of the Optical Society of America B*. Vol. 25, pp. B20-B26

Wachulak, P. W.; Marconi, M. C.; Bartels, R. A.; Menoni, C. S.; Rocca, J. J.; (2010). Holographic imaging with a nanometer resolution using compact table-top EUV laser. *Opto-Electronics Review*, Vol. 18, No. 1, pp. 28-38

Wang, Y.; Granados, E.; Larotonda, M.A.; Berrill, M.; Luther, B.M.; Patel, D.; Menoni, C.S.; Rocca, J.J.; (2006). High-brightness injection-seeded soft-x-ray-laser amplifier using a solid target. *Physical Review Letters*. Vol. 97, No. 12, pp. 123901-123904

14

Computer-Generated Phase-Only Holograms for Real-Time Image Display

Edward Buckley
Pixtronix, Inc.
USA

1. Introduction

Despite their esoteric sounding title, computer-generated holograms (CGHs) are now commonplace in a wide variety of applications and are a vital component in some surprisingly familiar consumer products. Such devices can be realized as fixed, etched structures - and are commonly called diffractive optical elements (DOEs) - or displayed on dynamically addressable liquid-crystal on silicon (LCOS) microdisplays. In either case, the principal attraction is the ability of these devices to generate arbitrary complex-valued optical fields from a small, thin device.

As discussed in Bernhardt et al. (1991), one CGH is able to perform the entire functionality associated with a multiple element glass lens design, leading to low-cost, lightweight optical assemblies. Furthermore, the process by which CGHs are made is simple, and lends itself to volume manufacturing through embossing and injection molding techniques; as demonstrated by Buckley & Wilkinson (2006), it is even possible to obtain adequate performance from CGHs patterned onto overhead transparencies from a standard office laser printer. Furthermore it is possible to fabricate phase-modulating DOEs which do not absorb incident optical illumination, leading to very high efficiencies.

Naturally, the flexibility and potential of CGH technology and its ability to implement multiple optical functions and exert control over optical fields - including very near-field evanescent waves as demonstrated by Brauer & Bryngdahl (1997); Elschner & Schmidt (1998); Gupta & Peng (1991); Kowarz (1995); Liu & Kowarz (1998; 1999); Madrazo & Nieto-Vesperinas (1997); Schmitz et al. (1996); Thompson et al. (1999) - has resulted in huge commercial utilization. For example, CD and DVD drives contain a diffractive optical element to appropriately condition and direct the laser beam onto the disc surface and, with the advent of the DVD disc, simultaneous optical pick-up from multiple disc layers can be achieved by employing an injection molded hybrid refractive-diffractive lens.

In addition to fixed holograms, there exist numerous methods for representing dynamic CGHs on reconfigurable microdisplay devices. There are a wealth of papers describing dynamic CGHs in applications as diverse as laser beam shapers in Dresel et al. (1996), fanouts and splitters for dynamic routing and multiplexing of laser beams into fibers in telecommunications applications Bengtsson et al. (1996); Gillet & Sheng (2003); Jean-Numa Gillet (1999); Keller & Gmitro (1993), optical traps for biophotonics Jesacher et al. (2004); Sinclair et al. (2004), performing transformations upon optical fields Case et al.

(1981); Gu et al. (1986); Roux (1991; 1993); Stuff & Cederquist (1990), self-adjusting CGHs Lofving (1997), aspheric testing Tang & Chang (1992), and wavelength discrimination for wavelength-division multiplexing (WDM) applications Dong et al. (1998; 1996); Layet et al. (1999); Yang et al. (1994).

Despite the obvious benefits of computer-generated holography for a wide range of applications, however, it is only recently that CGHs have been demonstrated for the projection and display of two dimensional video-style images. Indeed, such a method of image projection and display has long been desired, but was never previously realized, due to high computational complexity and poor quality of the resultant images.

1.1 CGHs for two-dimensional image display

Presenting visual information using a phase-only holographic approach provides a significant efficiency advantage compared to conventional video projection techniques. Unlike conventional projection displays, which utilize amplitude-modulating microdisplays to selectively block incident optical energy to form the desired image, a holographic display employing an ideal dynamic phase-modulating CGH has a transmission of near unity. Significant efficiency gains could therefore be realized compared to conventional LCOS or DLP®-based projectors, in which the illumination is set at a level sufficient to produce a peak white value regardless of the average pixel level (APL) of the scene. Furthermore, the use of an LCOS display as the dynamic modulating element in a laser-based holographic projector allows the removal of front polarizer which serves to waste an additional 50% of the available light in LED-illuminated systems.

The properties of diffraction potentially also allow for projection angles several times greater than currently possible in conventional LCOS-based systems. Conventional LCOS-based projection systems are limited by the necessity for a relatively large projection lens, since the function of the projection lens assembly is to enlarge an already sizeable image; to miniaturize the projection optics, then, the resultant image size must be shrunk concomitantly or be subject to severe aberrations, which can only be reduced through the use of highly complex and expensive lens systems. A phase-only holographic projector, on the other hand, is able to exert control over the entire optical field and consequently Buckley et al. (2009) was able to demonstrate that ultra-wide projection angles and novel projection geometries could be achieved without residual optical aberrations.

In addition to the compact, simple opto-mechanical assembly of a CGH-based projector, the use of solid-state light sources and LCOS-based light modulators would result in a system containing no moving parts. Fault tolerance of the optical system, which is achieved since the hologram pattern is decoupled from the desired image by a Fourier relationship, is also an attractive property in some applications where display integrity is required and "dead pixels" are unacceptable.

Although there have been plenty of examples of using fixed holograms for 2D image formation by Heggarty & Chevalier (1998); Kirk et al. (1992); Lesem & Hirsch (1969); Taghizadeh (1998; 2000); Takaki & Hojo (1999), previous attempts at real-time image projection and display using CGHs have been mainly limited to the 3D case and the demonstrations by Ito et al. (2005); Ito & Okano (2004); Ito & Shimobaba (2004); Sando et al. (2004) have required significant computational resources. The few attempts at an implementation of real-time 2D holographic projection by, for example, Mok et al. (1986); Papazoglou et al. (2002); Poon et al. (1993) have been affected by critical limitations imposed both by the computational complexity of the hologram generation algorithms required, and by the poor quality of images produced by the binary holograms they generate.

Recently, a great deal of progress has been made in using binary-phase CGHs for projection as detailed in Buckley (2008a;b; 2011a) and a new approach to hologram generation and display, based on a psychometrically-determined perceptual measure of image quality, has been shown to overcome both of these problems and has resulted in the commercialization of a real-time 2D holographic projector. This chapter will bring together, for the first time, the recent theoretical and practical advances in realizing 2D and 3D holographic projection systems based on binary phase CGHs.

2. Motivation

For video display applications, in which the APL is significantly less than the full-white maximum, a projection display based on phase-only computer generated holography could offer a significant efficiency advantage compared to amplitude-modulating LCOS displays, since light is not blocked from the desired image pixels. Quantifying this benefit has proven difficult, however, since there is widespread disagreement in the published literature from, for example, Bhatia et al. (2009; 2007); Buckley et al. (2008); Lee et al. (2009); Weber (2005) as to an acceptable value to use for the APL. The variation in reported values appears to result from the point at which the APL measurement is defined.

In a generalized display, the light intensity produced L_{out} is related to the video signal voltage V by $L_{out} \propto V^{\gamma}$, where γ is the display gamma. To obtain a display intensity response L_{out} which is linear with respect to the video image P, the transmit video signal V is encoded by an inverse gamma correction function so that $V \propto P^{1/\gamma}$. To ensure a uniform perceptual response, the display gamma is typically set to $\gamma = 2.2$ to match the approximate lightness sensitivity of a human viewer.

In a projection architecture in which the light sources can be modulated in response to average scene or per-pixel brightness, the resultant efficiency benefit is directly related to the mean value of L_{out}, $\mathrm{E}[L_{out}]$, which is clearly not equal to $\mathrm{E}[V]$ when $\gamma \neq 1$. In order to calculate this mean value, and since neither the form of L_{out} nor V are known a priori, we must derive a statistical model for the pixel distribution pre- and post-gamma.

Consider an image pixel P that can take a value in the range $[0, p)$, quantized into n bins of size b so that $b = p/n$. The number of occurrences of a pixel value within the bin $[p_{i-1}, p)$ is k_i, so that the total number of occurrences of that pixel value is

$$\sum_{i=1}^{\infty} k_i = k \tag{1}$$

and the total number of occurrences k is fixed, so that

$$\frac{1}{k} \sum_{i=1}^{\infty} k_i = \epsilon \tag{2}$$

where ϵ is some positive constant.

We define $\mathrm{Pr}_n(b)$ to be the probability that the pixel value will fall into the bth bin n times. Since each pixel has an equal probability of taking a value in the range $[0, p)$, the probability

that a pixel is addressed once with a value in the bth bin is $\Pr_1(b) = \lambda b$, where λ is a constant, and the probability that a pixel is not addressed is $\Pr_0(b) = 1 - \lambda b$.
We wish to find $\Pr(P > p)$, where P is the smallest pixel value, which is equivalent to finding the probability that a pixel is not addressed with any value in the range $(0, p)$. If we suppose further that the pixel value probabilities in any bin are independent of each other, then we obtain

$$\Pr(P > p) = \Pr_0^n(b) = \left(1 - \frac{\lambda p}{n}\right)^n \tag{3}$$

From elementary calculus,

$$\lim \left(1 - \frac{\lambda p}{n}\right)^n = \exp\left(-\lambda p\right) \tag{4}$$

so that

$$\Pr(P > p) = \lim_{b \to 0} \Pr_n(P > p) = \exp(-\lambda p) \tag{5}$$

and it follows that the corresponding probability density function (PDF) $f_P(p)$ is

$$f_P(p) = -\frac{d}{dp}\Pr(P > p) = \lambda \exp(-\lambda p) \tag{6}$$

where $p > 0$, thus completing the proof that the pixel values are exponentially distributed with mean λ.
Let the image pixels P be subject to a gamma encoding process with value γ such that $V \propto P^{1/\gamma}$. If P is exponentially-distributed with parameter λ, written as $P \sim \exp(\lambda)$, then Leemis & McQueston (2008) provides the standard result that the PDF of the random variable V, $f_V(v)$, is Weibull distributed $V \sim \text{Weibull}[\alpha, \beta]$, or

$$f_V(v; \alpha, \beta) = \frac{\alpha}{\beta}\left(\frac{v}{\beta}\right)^{\alpha - 1} \exp\left(-\frac{v}{\beta}\right)^{\alpha} \tag{7}$$

with mean value given by

$$\mathrm{E}\left[V\right] = \alpha\Gamma\left(1 + \frac{1}{\beta}\right) \tag{8}$$

where Γ is the Gamma function, $\alpha = \lambda^{\frac{1}{\gamma}}$ and $\beta = \gamma$. A number of measurements of V for typical TV content are provided by Jones & Harrison (2007); Lee et al. (2009); Stobbe et al. (2008); Weber (2005) and Jones & Harrison (2007) presents curves for experimentally-measured APL data by country, to which a Weibull-distributed variable $V \sim \text{Weibull}[\alpha = 0.43, \beta = 2.2]$ with mean of approximately 38% is an excellent fit to the average measured APL.

Since we know from experimentally-measured transmission data and equation 6 that the average pixel value in a video image is $\mathrm{E}\left[P\right] = \lambda = \alpha^{\gamma} = 16\%$, then we can reasonably state that, due to the nature of a typical video image, the average optical utilization efficiency of a holographic projector should be a factor of six greater than an LCOS-based system excluding all other inefficiencies. When comparing to LED-illuminated systems which require a front polarizer and careful étendue matching, the efficiency gain could approach one order of magnitude and clearly motivates the investigation of a projection system based on phase-only holography.

3. 2D Fourier holography

A holographic display employs a phase-modulating display element in combination with a coherent light source to form images by diffraction, rather than projection. A Fraunhofer (or far-field) holographic display is based on the result that, when a hologram $h(u,v)$ is illuminated by coherent collimated light of wavelength λ, the complex field $F(x,y)$ formed in the back focal plane of the lens of focal length f due to Fraunhofer diffraction from the pattern $h(u,v)$ is the two-dimensional spatial Fourier transform of the hologram pattern:

$$F(x,y) = \int_{-\infty}^{\infty} h(u,v) \exp\left\{\frac{2j\pi}{\lambda f}(ux+vy)\right\} dudv \tag{9}$$

The relationship of equation 9 is illustrated in Figure 1.

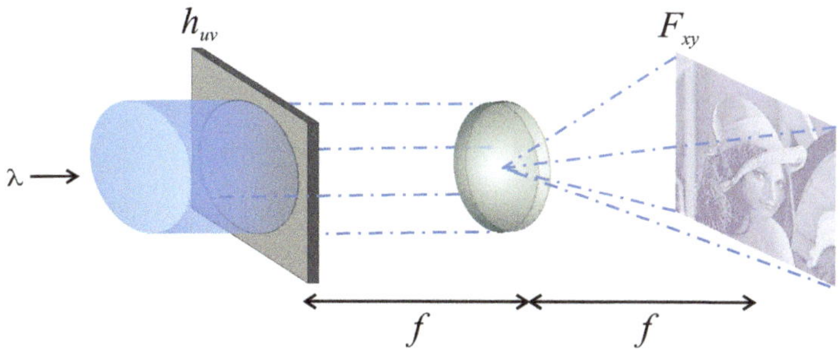

Fig. 1. The relationship between hologram $h(u,v)$ and image $F(x,y)$ present at the back focal plane of a lens of focal length f, when illuminated by coherent monochromatic light of wavelength λ.

If the continuous hologram pattern is then replaced by an element with pixel size Δ then the image F_{xy} formed (or replayed) in the focal plane of the lens is related to the pixellated hologram pattern h_{uv} by the discrete Fourier transform $\mathcal{F}[\cdot]$, and is written as

$$F_{xy} = \mathcal{F}[h_{uv}] = \sum_{u=0}^{P-1}\sum_{v=0}^{Q-1} \exp 2j\pi\left(\frac{ux}{P}+\frac{vy}{Q}\right) \tag{10}$$

Despite the potential advantages of a holographic display, previous attempts at constructing such a system as detailed by Georgiou et al. (2008); Heggarty & Chevalier (1998); Mok et al. (1986); Papazoglou et al. (2002); Poon et al. (1993) have been unable to overcome two fundamental technical problems.

The first difficulty is that of calculating a hologram h_{uv} such that, when illuminated by coherent light, a high quality image F_{xy} is formed. It is not possible to simply invert the Fourier transform relationship of equation 10 to obtain the desired hologram h_{uv}, since the result of this calculation would be fully complex and there is no material in existence that can independently modulate both amplitude A_{uv} and phase φ_{uv} where $h_{uv} = A_{uv} \exp j\varphi_{uv}$. Even if such a material became available, the result contains amplitude components which would absorb incident light and reduce system efficiency. A much better approach is to restrict the hologram h_{uv} to a set of phase only values $\exp j\varphi_{uv}$. Performing this operation on h_{uv} whilst maintaining high image quality in F_{xy} is absolutely non trivial, and requires computation to mitigate the effects of information lost in the quantization.

The second problem is one of computation. Until recently, there was no hologram-generation method in existence that could simultaneously produce images of sufficient quality for video style images, whilst calculating the holograms quickly enough to allow real-time image display. Figure 2 provides a good example; the 512 × 512-pixel hologram h_{uv} of Figure 2(b) took 10 hours to compute using the standard direct binary search (DBS) algorithm proposed in Dames et al. (1991); Seldowitz et al. (1987) and the resultant reconstruction F_{xy}, shown in Figure 2(c), is a very poor representation of the desired image T_{xy} of Figure 2(a).

In this section it is shown that the twin barriers to the realization of a real-time, high quality holographic display can be overcome by defining a new, psychometrically-determined, measure of image quality that is matched to human visual perception. A method of displaying phase holograms that is optimized with respect to this new measure is presented, and is shown to result in high-quality image reproduction.

(a) T_{xy} (b) $h_{uv} = \exp j\phi_{uv}$ (c) $|F_{xy}|^2 = |\mathcal{F}[h_{uv}]|^2$

Fig. 2. Image $|F_{xy}|^2$ resulting from the reconstruction of a desired image T_{xy} from a binary phase-only hologram h_{uv} calculated using the DBS algorithm.

3.1 An improved method for hologram generation

Conventional hologram generation algorithms such as DBS, and the Gerchberg-Saxton (GS) described in Gerchberg & Saxton (1972), attempt to exhaustively optimize a hologram to minimize some metric J, which is calculated by comparing the projected image F_{xy} with respect to a target image T_{xy} within some region Ω. Typically, such algorithms employ the mean-squared error (MSE) measure where

$$J = \sum_{\Omega} |F_{xy} - \gamma T_{xy}|^2 \tag{11}$$

and γ is a normalizing constant chosen to minimize equation 11 - which seems intuitively satisfying, since zero MSE implies a perfect reconstruction. Unfortunately, this metric is particularly insensitive for the low MSE values typically encountered in holographically-generated images.

An effective demonstration of the deficiency of the MSE measure is provided in the following example, in which three images F^1_{xy}, F^2_{xy} and F^3_{xy} are generated from a target image T_{xy}. Image F^1_{xy} is equivalent to T_{xy} except for a small contrast change, F^2_{xy} contains additive Gaussian noise of variance σ_n^2, and F^3_{xy} exhibits both a change in contrast and additive noise. If the change in contrast is given by c and the mean value of the image pixels is μ , then the MSE metric for each of the images can be shown to be

$$J = (\mu + c)^2 + \sigma_n^2 \tag{12}$$

The resultant images are shown in Figure 3, together with MSE figures calculated using equation 12. Although F^1_{xy} exhibits the highest perceptual image quality, and F^3_{xy} the lowest, the MSE metrics in fact indicate the opposite. It is clear from equation 12 and Figure 3 that MSE is in fact dominated by mean image errors caused by the contrast change, rather than the additive Gaussian noise which corresponds with poor perceptual image quality.

(a) F^1_{xy}: $J = 103$ (b) F^2_{xy}: $J = 81$ (c) F^3_{xy}: $J = 67$

Fig. 3. The poor correspondence of the MSE metric with image quality. Whilst F^1_{xy} exhibits the highest image quality, and F^3_{xy} the worst, the MSE metric J indicates the reverse.

In order to determine an improved optimization metric, it is necessary to derive the properties of noise in holographic replay and, in particular, the noise resulting from the approximation of the complex Fourier Transform information by a phase only hologram.

3.2 General properties of holographic replay

Without loss of generality, we consider the one-dimensional image F_x, which is the discrete Fourier transform (DFT) of the corresponding P-pixel hologram h_u, and is termed the replay field (RPF):

$$F_x \triangleq \sum_{u=0}^{P-1} h_u \exp\left(-\frac{2j\pi xu}{P}\right) \tag{13}$$

Since the form of h_u is not known a priori, because it is the result of some unspecified calculation, it can only be assumed that h_u is a random variable with some, as yet unknown, distribution. The quest to determine the properties of holographic replay therefore begins by considering the properties of samples of the DFT of a random sequence h_u, proceeding to determine the properties of the absolute value of the signal and noise components of the image as would be detected by the eye.

Consider a $P \times 1$ vector of independent identically distributed (i.i.d) random variables $h_1, h_2, \ldots, h_P$, each of which has the same arbitrary probability distribution function (PDF) $f_{h_i}(u)$, $i = 1 \cdots P$. The central limit theorem (CLT) states that the sum of these i.i.d random variables will tend to the Normal distribution which, remarkably, holds true even if the random variables are not themselves Normally distributed, provided that the sample size P is large enough.

Since equation 13 shows that the DFT of h_u is merely a weighted average of h_u, with the weights being complex exponential factors, the samples resulting from the DFT operation will therefore be governed by the CLT. Hence, regardless of the distribution of the samples h_u, the real and imaginary parts of the DFT will be Normally distributed provided that P is large enough. This is an important result in determining the properties of noise occurring in holographic replay.

We consider further a $P \times Q$ set of complex random samples h_{uv} which can be written as

$$h_{uv} = \Re[h_{uv}] + j\Im[h_{uv}] \tag{14}$$

where the real and imaginary parts of h_{uv} have mean and variance (μ_r, σ_r^2) and (μ_i, σ_i^2) respectively. The DFT of these samples, obtained from equation 10, is therefore Normally-distributed in real and imaginary parts

$$F_{xy} = \Re[F_{xy}] + j\Im[F_{xy}] \tag{15}$$

and, following some lengthy calculations, the samples of the DFT are found to be distributed as

$$\begin{aligned} \Re[F_{xy}(0)] &\sim \mathrm{N}[\mu_r P, \sigma_r^2 PQ] \\ \Im[F_{xy}(0)] &\sim \mathrm{N}[\mu_i P, \sigma_i^2 PQ] \\ \Re[F_{xy}] &\sim \mathrm{N}[0, (\sigma_r^2 + \sigma_i^2)PQ/2] \\ \Im[F_{xy}] &\sim \mathrm{N}[0, (\sigma_r^2 + \sigma_i^2)PQ/2]. \end{aligned} \tag{16}$$

where $F_{xy} \sim \mathrm{N}[\cdot]$ indicates that the samples F_{xy} are Normally distributed.

3.2.1 Effect of hologram quantization upon the image

In order to determine the properties of noise in holographic replay, it is necessary to determine the effects of quantizing the hologram h_u. Let the samples e_u represent the error introduced into the hologram by quantization, and $E_x = \mathcal{F}[e_u]$ be the resultant noise introduced into the image. It is clear from equations 16 that, regardless of the PDF of the error in the samples e_u, the image error samples E_x are always Normally distributed in real and imaginary parts and, hence, the amplitude of this error is Rayleigh distributed and is given by

$$f_{|\mathrm{E}|}(x) = \frac{x}{\sigma_n^2} \exp\left(-\frac{x^2}{2\sigma_n^2}\right) \tag{17}$$

where $\sigma_n^2 = (\sigma_r^2 + \sigma_i^2)\, PQ/2$ and depends upon the nature of the quantization performed. It follows that the noise amplitude in any holographically-formed image - regardless of the

algorithm used to generate the hologram - will always be Rayleigh distributed and dependent only upon the noise variance σ_n^2.
It follows that a holographically generated image will consist of a desired signal component of average value V plus additive noise E_{xy} due to the hologram quantization, and therefore the samples of the total complex image amplitude F_{xy} are distributed as

$$\begin{aligned} \Re[F_{xy}] &\sim \mathrm{N}\left[V, \sigma_n^2\right] \\ \Im[F_{xy}] &\sim \mathrm{N}\left[0, \sigma_n^2\right] \end{aligned} \tag{18}$$

so that the magnitude of the image $|F_{xy}|$ is Ricean distributed and described by

$$f_{|\mathrm{F}|}(x) = \frac{x}{\sigma_n^2} I_o \left(\frac{xV}{\sigma_n^2}\right) \exp\left(-\frac{x^2+V^2}{2\sigma_n^2}\right) \tag{19}$$

with energy

$$\mathrm{E}\left[|F_{xy}|^2\right] = V^2 + 2\sigma_n^2 \tag{20}$$

Equation 20 is the crucial result for deriving an improved hologram generation algorithm, because it describes the statistical properties of the images produced by any holographic display. Surprisingly, equation 20 shows that holographically-generated images can be completely characterized by just two parameters, V and σ_n^2, regardless of the algorithm used to create the hologram. By appropriate manipulation of these parameters, therefore, it is possible to control the noise properties of a holographic display.

3.2.2 Perceptual significance of noise in holographic replay

Although equation 20 characterizes the statistics of holographic replay with just two parameters, the relationship between the choice of values for each parameter and the resultant perceived image quality is not clear. Since it is not obvious what values a human viewing an image with Ricean distributed pixel values would assign to V and σ_n^2, the only logical way to proceed is to characterize the perceptual degradation of image quality with respect to these parameters by performing a suitably-designed psychometric test on a representative sample of the population as shown in Cable et al. (2004).
The general question of the comparative perceptual importance of artifacts in images is too broad to consider in this chapter. Instead, we deal with the more tractable problem of the relative perceptual significance of noise (that is, the deviation of the RPF from the target) that is inevitably present in any holographic reproduction, and how the statistical parameters of the noise affect perception.
The psychometric test was designed to present the subject with 300 sequential stimuli, examples of which are shown in Figure 4. Each stimuli comprises a pair of images, which have each been generated from a set of basis images, and the images are presented and random positions with random intensities. To simulate the effect of holographic replay, intensity noise $|E_{xy}|^2$ with mean μ and variance σ_n^2 was added to each image pair, according to equation 19.
A subject was placed in front of a monitor screen displaying such stimuli, which in combination are termed the 'veridical field'. To give the impression of a video image, the stimuli were updated 20 times per second. The subject was then asked to record their subjective interpretation of the most pleasing image or, if no distinction was possible, to

record no preference. This is known as the three-alternative forced choice (3AFC) paradigm, described in Greene & d'Oliveira (1999). To ensure that the subjective choice of image quality was made instinctively, as it would be for a typical video stream, a time limit of four seconds per image was imposed and, if the response time of the subject was longer, the result was discarded.

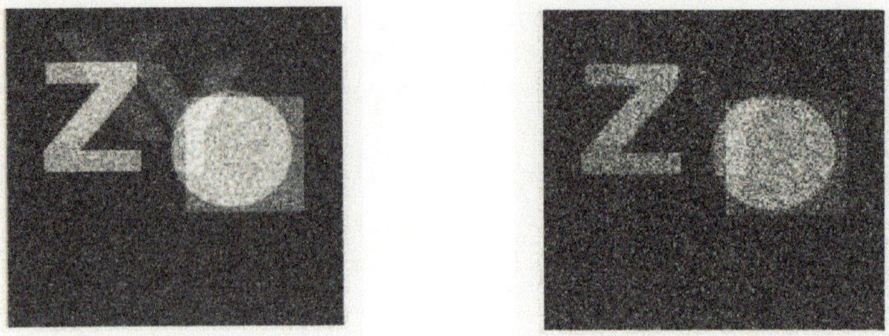

Fig. 4. Sample psychometric test stimuli. Each of the left and right images contains additive noise of mean μ and variance σ_n^2, updated 20 times per second as per a video image.

The results were analyzed by constructing a scatter plot of Figure 4 indicating, for each sample, the user's preference and demarcating the scatter plot into regions where the subject considers the left image to be superior ("left preferred"), reverse ("right preferred") or has no preference ("cannot tell"). Boundaries of best fit between these three regions were then constructed using a linear least-squares measure.

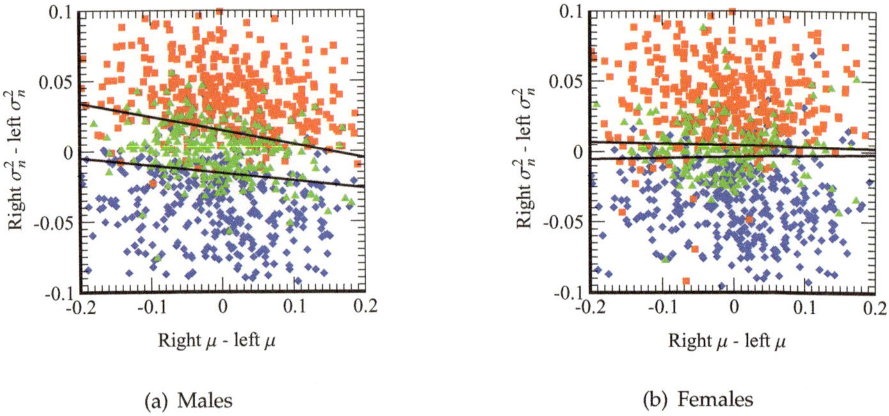

Fig. 5. Psychometric test results showing mean difference and variance difference between left and right stimuli, and the associated viewer preference. Key: ■ - Left preferred, ◆ - Right preferred, ▲ - No preference.

The results contained in Figures 5(a) and 5(b) clearly show, as indicated by the dominant horizontal component in the boundary lines, that noise variance in holographic replay is far more significant than the mean as a determinant of the perceptual significance of noise. This experiment suggests that a hologram generation algorithm which employs an error metric that minimizes noise variance σ_n^2 is likely to produce RPFs that are subjectively regarded as far higher in quality than the equivalent RPF obtained from other metrics, such as MSE minimization, which attempts to minimize noise energy $\mu^2 + \sigma_n^2$.

3.3 Reduction of noise variance

The conclusion that noise variance is an improved determinant of the perceptual significance of noise in a video image suggests a method for perceptual reduction of noise by exploiting temporal averaging. Consider a holographic display which generates N video subframes which are the result of some, as yet unspecified, hologram generation algorithm. The intensity of the ith displayed image is $I = \left|F_{xy}^{(i)}\right|^2$, and has mean μ and variance σ^2 and $i = 1, \cdots, N$. If the average of all such subframes is displayed, the time-averaged percept is

$$V_{xy} = \frac{1}{N} \sum_{i=1}^{N} |F_{xy}^{(i)}|^2 \tag{21}$$

and, from the CLT, it follows that the variance of this time-averaged field is given by

$$\text{Var}[V_{xy}] = \frac{\sigma^2}{N} \tag{22}$$

which is N times smaller than the variance of each individual subframe $\left|F_{xy}\right|^2$. Hence, a reduction in the noise variance of a video frame can be achieved by displaying the average of N noisy subframes. This property precisely fulfils the requirements suggested by the analytical and psychometric test results.

3.4 Practical implementation

A simple method for the creation of the time averaged percept of equation 21 relies upon the properties of the human visual system. The eye, because it responds to intensity, is a square-law detector and due to its composition has a finite response time. Kelly & van Norren (1977) performed a series of experiments using flickering veridical fields to deduce the temporal frequency characteristics of the eye, which resulted in the frequency response curves of Figure 6. Since the rod and cone structures respond slightly differently to flicker, there are disparities between pure luminous and chromatic (red-green) flicker responses - nevertheless, the frequency response of the human eye can be well approximated by a brick-wall filter function with a temporal bandwidth of approximately 25 Hz.

Using this approximation and accounting for the square-law response, the time-averaged intensity percept V_{xy} is approximately equal to the integral of the veridical field $\left|F_{xy}\right|^2$ within a 40 ms time window, and can be expressed as

$$V_{xy} = \int_{t-0.04}^{t} |F_{xy}(\tau)|^2 d\tau \tag{23}$$

If a suitable microdisplay is used to show N subframes within this 40 ms period, then the integral of equation 23 becomes the summation of equation 21. Hence, by displaying N frames quickly enough to exploit the limited temporal bandwidth of the eye, a human subject will

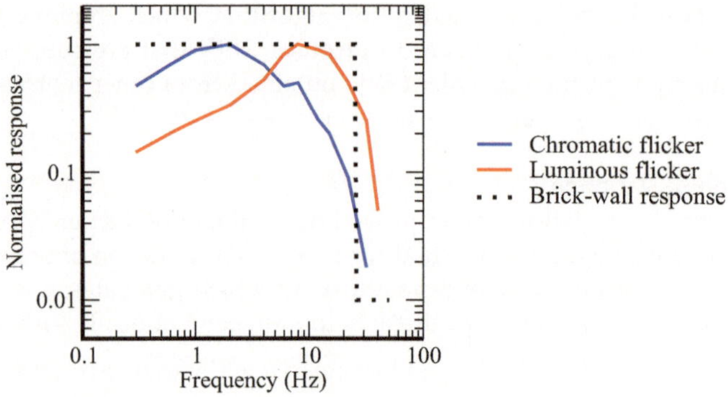

Fig. 6. Temporal frequency response of the eye to luminous and chromatic flicker (after Kelly & van Norren (1977)). The curves show that the eye can be modeled as a brick-wall filter function with temporal bandwidth of approximately 25 Hz.

perceive an image which is the average of N noisy subframes which is, from equation 22, substantially noise-free.

3.4.1 The one-step phase retrieval (OSPR) algorithm

What remains is to design a hologram-generation algorithm that has the capability to generate N sets of holograms both efficiently and in real time. A simple and computationally-simple method for generating $i = 1, \cdots, N$ holograms, each of which gives rise to a reconstruction $\left|F_{xy}^{(i)}\right|^2$ containing the desired image T_{xy} in addition to i.i.d noise $\left|E_{xy}^{(i)}\right|^2$, is provided in the OSPR Algorithm 1 below.

inputs: $P \times Q$ pixel target image T_{xy}, N
output: N binary phase holograms $h_{uv}^{(n)}, n = 1, \cdots, N$, of size $P \times Q$ pixels
for $n \leftarrow 1$ **to** $N/2$ **do**
Let $T_{xy}^{(n)} = \sqrt{I_{xy}} \exp j\varphi_{xy}^{(n)}$ where $\varphi_{xy}^{(n)}$ is uniformly distributed in the interval $[0; 2\pi]$
Let $g_{uv}^{(n)} = \mathcal{F}^{-1}\left[T_{xy}^{(n)}\right]$
Let $m_{uv}^{(n)} = \Re\left\{g_{uv}^{(n)}\right\}$ where $\Re\{\cdot\}$ represents the real part
Let $m_{uv}^{(n+N/2)} = \Im\left\{g_{uv}^{(n)}\right\}$ where $\Im\{\cdot\}$ represents the imaginary part
Let $h_{uv}^{(n)} = \begin{cases} -1 \text{ if } m_{uv}^{(n)} < 0 \\ 1 \text{ if } m_{uv}^{(n)} \geq 0 \end{cases}$
end

Algorithm 1: The OSPR algorithm for calculating N $P \times Q$ pixel binary phase holograms $h_{uv}^{(n)}, n = 1, \cdots, N$ from a $P \times Q$ pixel target image T_{xy}.

The results of the previous sections allow us to verify that Algorithm 1 generates holograms with the correct properties. Provided that mean quantization error (introduced into the hologram in the last step of Algorithm 1 is zero, which follows from thresholding about zero, the results of equation 16 can be applied to show that the real and imaginary parts of the reconstruction error E_{xy} - given by the Fourier transform of the quantization error e_{uv}, $\mathcal{F}[e_{uv}]$ - are independently distributed with zero mean and a variance which depends on the second moment of the reconstruction error only. It follows from equation 17 that the magnitude of the reconstruction error has a Raleigh distribution and that we can ensure that each of the N holograms generated will exhibit i.i.d. noise in its RPF if each φ_{xy} in step 1 is i.i.d.

3.4.2 Diffraction efficiency

A simple expression for maximum diffraction efficiency η of a phase-only quantized hologram is provided by Goodman & Silvestri (1970) and is

$$\eta = \text{sinc}^2\left(\frac{1}{M}\right) \tag{24}$$

where $\text{sinc}(x) \triangleq \frac{sin(\pi x)}{\pi x}$ and M is the number of phase levels uniformly distributed in the interval $[0, 2\pi]$. For binary phase devices with $M = 2$, the maximum achievable diffraction efficiency is just 41%. This figure can be further refined to account for the desired image pattern, per Wyrowski (1991), the computation algorithm - as shown in Mait (1990) - and spatial quantization effects as covered by Arrizón & Testorf (1997); Wyrowski (1992).

It is relatively straightforward to determine the maximum achievable diffraction efficiency when the OSPR algorithm is used to calculate the hologram patterns h_{uv}. We first consider an unquantized hologram pattern $m_{uv} \sim \text{N}\left[0, \sigma^2/PQ\right]$ resulting from Algorithm 1, which reconstructs to form an image F_{xy} with total energy σ^2 by equations 16 and Parseval's theorem. A quantization operation is applied to the random variable m_{uv} to obtain a quantized random variable h_{uv} such that

$$h_{uv} = \begin{cases} a \ m_{uv} < q \\ b \ m_{uv} \geq q \end{cases} \tag{25}$$

where q is the quantization threshold. Restricting the analysis to one dimension for a moment, then the noise e_u introduced into the hologram by quantization pixel values about a point q at reconstruction points a, b can be calculated using

$$e_u = \begin{cases} u - a \ u < q \\ u - b \ u \geq q \end{cases} \tag{26}$$

which results in a mean quantization noise $\text{E}\left[e_u\right]$ and quantization noise energy $\text{E}\left[e_u^2\right]$ of

$$\begin{aligned} \text{E}\left[e_u\right] &= \int_{-\infty}^{q} (u-b) f_m(u)\, du + \int_{q}^{\infty} (u-a) f_m(u)\, du \\ \text{E}\left[e_u^2\right] &= \int_{-\infty}^{q} (u-b)^2 f_m(u)\, du + \int_{q}^{\infty} (u-a)^2 f_m(u)\, du \end{aligned} \tag{27}$$

where $f_m(u)$ is the PDF of the random variable m_u. In the case of binary phase holography, the mean hologram quantization noise $\text{E}\left[e_u\right]$ is minimized for $q = \frac{a+b}{2}$; since $a = -b$ then $q \longrightarrow 0$ as and $\text{E}\left[e_u\right] \simeq 0$ previously shown. $E_{xy} = \mathcal{F}[e_{uv}]$ then represents an upper bound for the RPF noise resulting from hologram quantization, since

$$E_{xy} = \left|\mathcal{F}\left[e_{uv}\right]\right| = \left|\mathcal{F}[m_{uv}] - \mathcal{F}[h_{uv}]\right| \geq \left|\left|\mathcal{F}[m_{uv}]\right| - \left|\mathcal{F}[h_{uv}]\right|\right| \tag{28}$$

by the triangle inequality.
The noise in the RPF due to quantization can be determined by evaluation of equations 27 at the appropriate reconstruction points a and b. For binary phase quantization, the points lie on the centroids of the positive and negative real axes respectively so that

$$a = \int_{q \longrightarrow 0}^{\infty} u f_m(u) du = \sigma \sqrt{\frac{2}{\pi}} \tag{29}$$

Using equations 27 it can further be shown that the RPF noise due to binary phase quantization is

$$\mathrm{E}\left[e_{uv}^2\right] = \frac{\sigma^2}{PQ}\left(1 - \frac{2}{\pi}\right) = \frac{\sigma_n^2}{PQ} \tag{30}$$

so that the reconstruction $E_{xy} \sim \mathrm{N}\left[0, \sigma_n^2\right]$ from equations 16 and it follows that $\sigma_n^2/\sigma^2 \simeq 36\%$ of the reconstruction energy resulting from a binary phase hologram generated by the OSPR algorithm is noise. The diffraction efficiency η, defined as the proportion of usable energy directed into the first-order intensity samples in the presence of a RPF noise energy σ_n^2, is then

$$\eta = \begin{cases} \frac{1}{2}\left(1 - \frac{\sigma_n^2}{\sigma^2}\right) & M = 2 \\ 1 - \frac{\sigma_n^2}{\sigma^2} & \text{otherwise} \end{cases} \tag{31}$$

and is approximately 32% for binary phase holograms generated using OSPR. A similar calculation by Buckley & Wilkinson (2007) results in a figure of 88% for OSPR-generated continuous phase holograms.

3.4.3 Signal-to-noise ratio

Signal-to-noise ratio (SNR) is an important metric for image display applications since it defines the maximum achievable contrast ratio. In a holographically formed two-dimensional image, the SNR is defined as the ratio of the mean signal energy to the mean noise energy, where the RPF, $|F_{xy}|^2$, contains the desired target image T_{xy} with mean value V^2 in addition to additive noise $|E_{xy}|^2$ caused by hologram quantization.
If each subframe i contains noise components $E_{xy}^{(i)}$ with amplitudes $\left|E_{xy}^{(i)}\right|$ that are Rayleigh distributed as per equation 17, it is simple to show that the noise intensity $\left|E_{xy}^{(i)}\right|^2$ is distributed as the exponential distribution

$$f_{|E|^2}(x) = \frac{1}{\sigma_n^2} \exp\left(-\frac{x}{\sigma_n^2}\right) \tag{32}$$

In an OSPR-based holographic display system, the overall noise field $\left|E_{xy}\right|^2$ is the time average of N such contributions due to the square-law detection properties of the eye as shown by equation 21. Using a standard result it can be shown that the sum of N such independent, identically distributed exponential random variables is distributed according to the Gamma distribution

$$f_{|E|^2}(x) = \frac{x^{n-1}}{\Gamma(N)\,\sigma_n^{2N}} \exp\left(-\frac{x}{\sigma_n^2}\right) \tag{33}$$

where $\Gamma(\cdot)$ is the complete Gamma function. Since the mean of the Gamma distribution is $N\sigma_n$ then it follows from equation 21 that the mean noise energy in a veridical field V_{xy} composed of N frames is

$$\mathrm{E}\left[V_{xy}\right] = \sigma_n^2 \tag{34}$$

The noise energy present in the field V_{xy}, and hence the SNR, is clearly independent of the number of subframes N and, as for the diffraction efficiency, is defined by the number of hologram phase levels and choice of hologram computation algorithm.

If we further define a fractional coverage value η to be the ratio of the sum of the normalized pixel values to PQ, so that $\frac{1}{PQ} \leq \eta \leq 1$, then, since the quantization noise is determined by the number of phase levels and is constant, then the SNR S can be defined as

$$S \triangleq \frac{\mathrm{E}\left[T_{xy}\right]}{\mathrm{E}\left[\left|E_{xy}\right|^2\right]} = \frac{V^2/\eta}{\sigma_n^2} \tag{35}$$

and, since the total RPF energy $\sigma^2 = V^2/\eta + \sigma_n^2$, then equation 30 can be used to show that

$$S \simeq \frac{1.75}{\eta} \tag{36}$$

Since typical video images exhibit $\eta = 0.24$, giving $S \simeq 7$ independent of the number of subframes N, this immediately highlights an obvious limitation of binary phase holographic video projection. There are several algorithmic methods capable of improving the contrast ratio of a holographically-generated image, each of which depend upon quantizing the hologram in such a way that noise can be selectively placed in the RPF. The Gerchberg-Saxton (GS) and Direct Binary Search (DBS) hologram generation algorithms can both be modified so that each algorithm attempts to minimize the quantization noise energy within a predefined signal window within the RPF, thereby obtaining a local signal-to-noise ratio (SNR) improvement as Brauer et al. (1991); Meister & Winfield (2002); Wyrowksi et al. (1986); Wyrowski & Bryngdahl (1988) previously found. However, both algorithms generate RPFs of insufficient quality and impose computational burdens that are incompatible with a high-quality real-time holographic display.

The error diffusion (ED) algorithm, whilst not capable of generating holograms, was shown to be able to quantize holograms to generate RPFs with this useful characteristic by Kirk et al. (1992). As demonstrated in Buckley (2011b), it is possible to employ a multiple subframe approach, using OSPR to calculate holograms which are subsequently binarized using ED, to combine the benefits of image uniformity and high contrast. By implementing a parallel-processor design, the ED algorithm can be realized at the rate required by a multiple-subframe holographic projection system.

3.4.4 Choice of microdisplay

The requirements imposed upon the microdisplay used in the holographic projection system described previously are very different to those for the equivalent imaging system in terms of the liquid crystal material, backplane circuitry and pixel geometry. For a microdisplay employed in an imaging system, the choice of pixel size is usually chosen to represent a compromise between maintaining an adequate aperture ratio whilst minimizing diffractive effects - in a projection system which exploits diffraction, however, such a restriction does not apply. It is a standard result, given in Hecht (1998), that the diffraction angle θ from a hologram pattern of pixel size Δ placed behind a lens and illuminated with coherent collimated light of wavelength λ, is given by

$$\theta = \arctan\left(\frac{\lambda}{2\Delta}\right) \tag{37}$$

This inverse relationship between diffraction angle and feature size suggests that the pixel size in a microdisplay employed in a holographic projection system should be as small as possible, so that subsequent lens power to achieve the desired projection angle is minimized.

It is also of paramount importance to provide predictable phase modulation over a wide temperate range and, because multiple subframes are displayed per video frame for the purposes of noise reduction, a high frame rate is required. These requirements can be fulfilled by the use of a ferroelectric Liquid Crystal on Silicon (LCOS) device operating as a phase-only modulator, as shown by O'Brien et al. (1994).

In phase modulating mode, a ferroelectric LCOS device with a cell gap providing optical retardation Γ can act as a pixellated binary phase hologram in which each of the pixels can independently impose a phase shift of either 0 or π radians. To achieve phase modulation, the direction of polarization or the incident light (with components E_x and E_y) is aligned to bisect the switching angle 2θ of the two LC states. The resultant modulated light components E'_x and E'_y in switched and unswitched states can be written in Jones Matrix notation, and are given by

$$\begin{bmatrix} E'_x \\ E'_y \end{bmatrix} = \begin{bmatrix} 1 & 0 \\ 0 & 0 \end{bmatrix} \begin{bmatrix} e^{-j\frac{\Gamma}{2}}\cos^2\theta + e^{j\frac{\Gamma}{2}}\sin^2\theta & \pm j\sin\frac{\Gamma}{2}\sin 2\theta \\ \pm j\sin\frac{\Gamma}{2}\sin 2\theta & e^{-j\frac{\Gamma}{2}}\cos^2\theta + e^{j\frac{\Gamma}{2}}\sin^2\theta \end{bmatrix} \begin{bmatrix} E_x \\ E_y \end{bmatrix} \tag{38}$$

which reduces to

$$\begin{bmatrix} E'_x \\ E'_y \end{bmatrix} = \begin{bmatrix} \pm j\sin\frac{\Gamma}{2}\sin 2\theta \\ 0 \end{bmatrix} \begin{bmatrix} E_x \\ E_y \end{bmatrix} \tag{39}$$

It follows that the diffraction efficiency of the FLC material is determined by

$$\eta_{FLC} = \sin^2(2\theta)\sin^2\left(\frac{\Gamma}{2}\right) \tag{40}$$

where the optical retardation $\Gamma = \frac{2\pi d\Delta n}{\lambda}$, with d the thickness of the LC layer and Δn its birefringence. It is clear from equations 39 and 40 that in order to maximize the diffraction efficiency then the LC material switching angle must be $2\theta = \pi$ radians; the pixels of a microdisplay employing such a material could then independently impose phase shifts of either 0 or π radians, giving $\varphi_{uv} \sim [0, \pi]$ as required.

It is clear from equation 39 that, given a LC material switching angle of π radians, the pixels of such a device can independently impose a phase shift of either 0 or π radians, giving $\varphi_{uv} \sim [0, \pi]$ as required. Development devices with a switching angle of 88° in the smectic C* phase (SmC*) at operating temperature have previously been demonstrated by Heggarty et al. (2004) and have been deployed as phase modulators in optical switching applications.

A commonly encountered issue with ferroelectric LC devices is the need to DC balance the device by displaying inverse compensating images, during which time the device cannot be illuminated. When used in an imaging architecture, O'Callaghan et al. (2009) has shown that this requirement can effectively halve the maximum achievable optical efficiency. In a phase-modulating system employing the OSPR algorithm, however, since the holograms can be chosen to be automatically DC balanced and because the hologram and its inverse both result in the same image, the device can be illuminated during both the valid and compensating fields resulting in the maximum optical efficiency.

3.4.5 Optical system

Figure 7 shows the simplest optical architecture for a holographic projector. The lens pair of L_1 and L_2 form a telescope, which expands the laser beam to capture the entire hologram pattern so that low-pass filtering of the resultant intensity RPF V_{xy} does not result. The reverse arrangement is used for the lens pair of L_3 and L_4, which acts to demagnify the hologram pixels and consequently increase the diffraction angle Δ as described by Buckley et al. (2006). The demagnification D is set by the ratio of focal lengths f_3 to f_4 and, due to the properties of Fraunhofer diffraction, the images remain in focus at all distances from L_4.

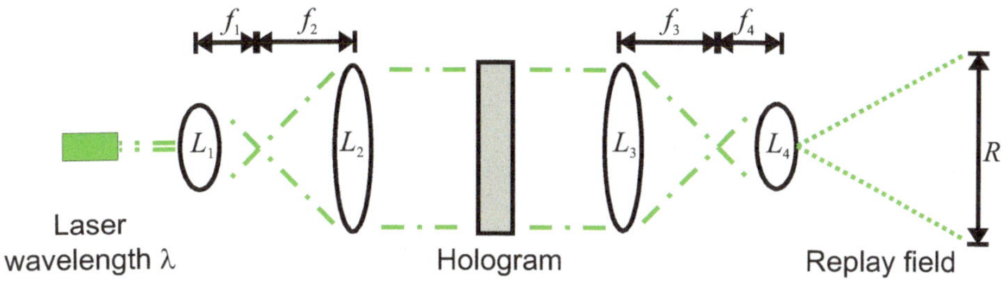

Fig. 7. Optical design for a simple holographic projector. Beam-expansion of the laser diode is performed by lenses L_1 and L_2, and demagnification by lenses L_3 and L_4.

3.4.6 Color architecture

The realization of a color holographic projector is relatively straightforward. A desired image is converted into sets of holograms and displayed on a phase modulating microdisplay illuminated by red, green and blue coherent light. Color images can be formed either by spatially segmenting the microdisplay as per Ito & Okano (2004), designing multi-focal CGHs by the method of Makowski et al. (2008), or by employing the frame-sequential color of Buckley (2008a; 2011a), which has the advantage of maximizing the output resolution. In the latter case, sets of holograms $h_{uv}^{(i)}$, $i = 1, \cdots, N$, are calculated and displayed for each wavelength λ_r, λ_g and λ_b, with the RPF scaled to account for the wavelength-dependent diffraction angle. The subsequent diffraction patterns pass through the simple lens pair L_3 and L_4, which increases the projection angle by demagnifying the microdisplay pixel size Δ. Since the color planes are displayed and illuminated at the subframe rate, the color-sequential approach does not suffer from color breakup.

Figure 8 shows an image obtained from a phase-only holographic projection system, employing the techniques described in this paper, manufactured by Light Blue Optics Ltd. The projector was imaged onto a commercially available rear-projection screen and the resultant image was captured by a digital camera. The nominal resolution at the projection screen was approximately WVGA (850 $\times$ 480 pixels.) It is clear that the image exhibits the highly saturated primaries associated with laser-based display system, but that the speckle artefacts traditionally associated with this method of projection are substantially suppressed.

Fig. 8. Projected image at WVGA resolution resulting from a phase-only holographic projection system, employing the techniques described in this chapter, manufactured by Light Blue Optics Ltd. In this instance, $\lambda_r = 642$ nm, $\lambda_g = 532$ nm and $\lambda_b = 445$ nm.

Several methods can be combined in a holographic projector in order to reduce speckle. In particular, the use of multiple holograms per video frame is beneficial to the speckle contrast; since N phase-independent subframes per video frame are shown within the eye's integration period, then the eye acts to add N independent speckle patterns on an intensity basis, and the contrast of the low-frequency components of the speckle in the field V_{xy} falls as $N^{1/2}$. Due to computational and LC switching speed limitations, N cannot be increased indefinitely so additional methods can be combined to further reduce the speckle contrast. The presence of an intermediate image plane between the lens pair L_3 and L_4 makes it straightforward to employ optical speckle reduction techniques, as previously presented by Buckley (2008c).

4. 2D Fresnel holography

Previous sections have been concerned with far-field (or Fraunhofer) diffraction, in which the RPF F_{xy} and hologram h_{uv} are related by the Fourier transform:

$$F_{xy} = \mathcal{F}[h_{uv}] \tag{41}$$

In the near-field (or Fresnel) propagation regime, RPF and hologram are related by the Fresnel transform which, using the same notation, can be written as

$$F_{xy} = \mathcal{FR}[h_{uv}] \tag{42}$$

and the RPF F_{xy} at a distance z is related to the $P \times Q$-pixel hologram h_{uv} of feature size $\Delta_x \times \Delta_y$ by Fresnel diffraction so that

$$F_{xy} = \mathcal{FR}[h_{uv}] = F_{xy}^{(1)} \cdot \mathcal{F}\left[f_{uv}^{(2)} h_{uv}\right] \tag{43}$$

where

$$F_{xy}^{(1)} = \frac{\Delta_x \Delta_y}{j\lambda z} \exp \frac{j2\pi z}{\lambda} \exp \frac{j\pi}{\lambda z} \left[\left(\frac{x}{P\Delta_x} \right)^2 + \left(\frac{y}{Q\Delta_y} \right)^2 \right] \tag{44}$$

and

$$f_{uv}^{(2)} = \exp \frac{j\pi}{\lambda z} \left(u^2 \Delta_x^2 + v^2 \Delta_y^2 \right). \tag{45}$$

so that the dimensions of the RPF are $\frac{\lambda z}{\Delta_x} \times \frac{\lambda z}{\Delta_y}$, consistent with the size of RPF in the Fraunhofer diffraction regime as per Schnars & Juptner (2002). The Fresnel diffraction geometry is illustrated in Figure 9.

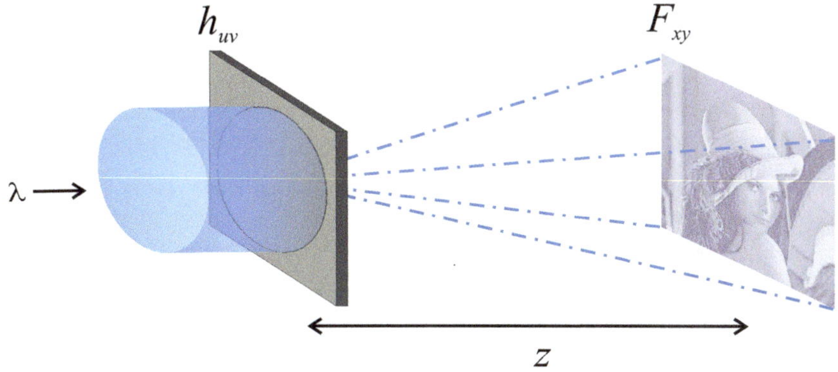

Fig. 9. Fresnel diffraction geometry. When the hologram h_{uv} is illuminated by coherent light, the RPF F_{xy} at a distance z is determined by Fresnel (or near-field) diffraction.

As previously shown by Dorsch et al. (1994); Fetthauer et al. (1995), it is straightforward to generalize hologram generation algorithms to the case of calculating Fresnel holograms. Here, the OSPR algorithm 1 is employed, replacing the conventional Fourier transform step by the discrete Fresnel transform of equation 43. The samples of the discrete Fresnel transform are found to be distributed as

$$\begin{aligned} \Re[F_{xy}(0)] &\sim \mathrm{N}[\mu_r P, \sigma_r^2 P' Q'] \\ \Im[F_{xy}(0)] &\sim \mathrm{N}[\mu_i P, \sigma_i^2 P' Q'] \\ \Re[F_{xy}] &\sim \mathrm{N}[0, (\sigma_r^2 + \sigma_i^2) P' Q' / 2] \\ \Im[F_{xy}] &\sim \mathrm{N}[0, (\sigma_r^2 + \sigma_i^2) P' Q' / 2]. \end{aligned} \tag{46}$$

where $P' = P\Delta_x^2$ and $Q' = Q\Delta_y^2$.

The use of Fresnel holography has in two beneficial effects. Firstly, the diffracted near-field at the propagation distance z does not contain the conjugate image evident in the Fraunhofer region, in which z is necessarily greater than the Goodman (1996) distance. Second, because Fresnel propagation is characterized by a distance z, it is evident that the hologram incorporates lens power determined by the properties of the computed hologram, rather than the optical system. It therefore follows that the lens count in a holographic projection system could be reduced simply by removing L_3 of Figure 7, employing instead a Fresnel hologram which encodes the equivalent lens power $z = f_3$.

4.1 Holographic projector with variable demagnification

In the Fourier projection system of Figure 7 the demagnification D of the hologram pixels, and the concomitant enlargement of the RPF, is determined optically and is given by the ratio $D = f_3/f_4$. The use of a Fresnel hologram displayed on a dynamically addressable microdisplay, however, would allow for a novel variable demagnification effect since the effective focal length of the Fresnel hologram encoding L_3 could be varied simply by recomputing the hologram.

An experimental verification of this variable demagnification principle was performed by removing L_3 of Figure 7 and setting $f_4 = 100$ mm. Three Fresnel holograms were calculated using OSPR with $N = 24$ subframes, each of which were designed to form a target image in the planes $z = 100$ mm, $z = 200$ mm and $z = 400$ mm. A microdisplay with pixel pitch $\Delta_x = \Delta_y = 13.62\mu$m was used to display the holograms, and the resulting RPFs - which were reconstructed at $\lambda = 532$ nm and imaged onto a non-diffusing screen - were captured with a digital camera. The results are shown in Figure 10, and clearly show the RPF scaling caused by the variable demagnification introduced by each of the Fresnel holograms.

(a) $z = 100$ mm (b) $z = 200$ mm (c) $z = 400$ mm

Fig. 10. Experimental results of the variable demagnification principle described. The scale of the RPFs (a) to (c) is determined by the effective focal length z of $N = 24$ sets of Fresnel holograms displayed on a dynamically addressable microdisplay.

4.2 Lens sharing in a holographic projector

In the previous section, it was shown that lens L_3 of the demagnification lens pair could be removed by encoding the equivalent lens power into the hologram. From inspection of Figure 7, it is clear that the same argument could also be applied to L_2 of the beam-expansion lens pair. It follows that, if $f_2 = f_3$, the common lens can be shared between the beam-expansion and demagnification assemblies by encoding it into a Fresnel hologram displayed on a reflective microdisplay. The remaining lens L_4 is typically the smallest in the optical path in order to maximize the demagnification D.

An experimental projector was constructed to demonstrate the lens-sharing concept, and the optical configuration is shown in Figure 11(a). A fiber-coupled laser was used to illuminate the same reflective microdisplay, which displayed $N = 24$ sets of Fresnel holograms each with $z = 100$ mm. Since the light from the fiber end was highly divergent, this removed the need for lens L_1. The output lens L_4 had a focal length of $f_4 = 36$ mm, giving a demagnification D

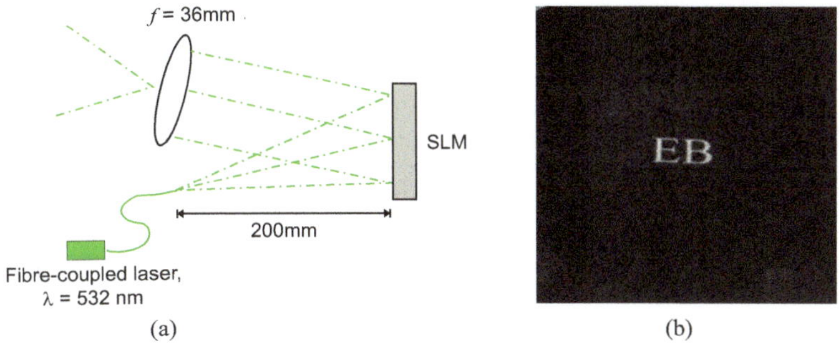

Fig. 11. Optical setup (a) and resultant RPF (b) of a lens-sharing projector design, utilizing a Fresnel hologram with $z = 100$mm displayed on the microdisplay. Polarizers are omitted for clarity. The demagnification caused by the combination of L_4 and the hologram causes optical enlargement of the RPF by a factor of approximately three.

of approximately three. Polarizers were used to remove the large zero order associated with Fresnel diffraction, but have been omitted from Figure 11(a) for clarity. The angle of reflection was also kept small to avoid defocus aberrations.

An example image, projected on a screen and captured in low-light conditions with a digital camera, is shown in Figure 11(b). The RPF has been optically enlarged by factor of approximately three due to the demagnification of the hologram pixels and, as the architecture is functionally equivalent to the simple holographic projector of Figure 7, the image is in focus at all points and, due to the use of Fresnel holography, the conjugate image is absent.

5. 3D holography

A 3D hologram of an object is simply a recording of the complex electromagnetic field (produced by light scattered by the object) at a plane in front of the object. By Huygens' principle as detailed in Hecht (1998), if we know the EM field distribution on a plane P, we can propagate Huygens wavelets through space to evaluate the field at any point in 3D space. As such, the plane hologram encodes all the information necessary to view the object from any position and angle in front of the plane and hence is, in theory, optically indistinguishable from the object. In practice, limitations in the pixel resolution of the recording medium restricts the viewing angle which, as in the 2D case, varies inversely with the pixel size Δ, as given by equation 37.

Consider a plane, perpendicular to the z-axis, intersecting the origin, and one point source emitter of wavelength λ and amplitude A at position (x, y, z) behind it. The field $h(u, v)$ present at the plane $(u, v, z = 0)$ - i.e. the hologram $h(u, v)$ - is given by

$$h(u,v) = \frac{ZA}{j\lambda r^2} \exp\left(\frac{2j\pi}{\lambda} r\right) \text{ with } r = \sqrt{(u-x)^2 + (v-y)^2 + z^2} \tag{47}$$

If we regard a 3D scene as M sources of amplitude A_i at (x_i, y_i, z_i), the linear nature of EM propagation results in the total field hologram $h(u,v)$

$$h(u,v) = \sum_{i=1}^{M} \frac{z_i A_i}{j\lambda r_i^2} \exp\left(\frac{2j\pi}{\lambda} r_i\right) \text{ with } r = \sqrt{(u-x_i)^2 + (v-y_i)^2 + z_i^2} \quad (48)$$

If we wish to sample $h(u,v)$ over the region $u_{min} \leq u \leq u_{max}, v_{min} \leq v \leq v_{max}$ to form a $P \times P$ hologram h_{uv}, then r_i becomes:

$$r_i = \sqrt{\left(u_{min} + u\frac{u_{max} - umin}{P} - x_i\right)^2 + \left(v_{min} + v\frac{v_{max} - vmin}{P} - y_i\right)^2 + z_i^2} \quad (49)$$

In Algorithm 2 we present a version of OSPR that generates N full-parallax 3D holograms $h_{uv}^{(n)}, n = 1 \cdots N$, for a given set of M point sources A_i, $i = 1 \cdots M$, at positions (x_i, y_i, z_i) in the image plane (x, y, z).

inputs: M point sources of amplitude A_i at position (x_i, y_i, z_i), M, N
output: N binary phase holograms $h_{uv}^{(n)}$ of size $P \times P$ pixels

for $n \leftarrow 1$ **to** $N/2$ **do**

Let $h_{uv}^{(n)} = \sum_{i=1}^{M} \frac{Z_i A_i}{j\lambda r_i^2} \exp\left(j\Phi_i^{(n)} + \frac{2j\pi}{\lambda} r_i\right)$ with r_i as equation 49 and where $\Phi_i^{(n)}$ is uniformly distributed in the interval $[0; 2\pi]$

Let $m_{uv}^{(n)} = \Re\left\{g_{uv}^{(n)}\right\}$ where $\Re\{\cdot\}$ represents the real part

Let $m_{uv}^{(n+N/2)} = \Im\left\{g_{uv}^{(n)}\right\}$ where $\Im\{\cdot\}$ represents the imaginary part

Let $h_{uv}^{(n)} = \begin{cases} -1 \text{ if } m_{uv}^{(n)} < 0 \\ 1 \text{ if } m_{uv}^{(n)} \geq 0 \end{cases}$

end

Algorithm 2: The OSPR algorithm modified to calculate N $P \times P$ pixel full-parallax 3D holograms h_{uv} for a given set of M point sources A_i.

To test this algorithm, we consider the calculation of $N = 8$ holograms of resolution 512×512 and size 2 mm $\times$ 2 mm centered at the origin of our plane P, giving a pixel size of $\Delta = 4\,\mu m$ and hence a viewing angle of around 9 degrees under coherent red illumination $\lambda = 632$ nm. The 3D scene used was a set of $M = 944$ point sources that formed a wireframe cuboid of dimensions 12 cm $\times$ 12 cm $\times$ 18 cm, located at a distance of 1.91 m from the plane.
The simulated RPFs produced were calculated by propagating Huygens wavelets from the N holograms $h_{uv}^{(i)}$ in turn through a pinhole aperture K onto a virtual screen (a plane perpendicular to the line from the center of the cube to the pinhole), and recording the intensity distribution on the screen $|F_{xy}^{(i)}|^2$; as before, the time-averaged percept is $V_{xy} = \frac{1}{N}\sum_{i=1}^{N}\left|F_{xy}^{(i)}\right|^2$. Simulated views of the hologram from two positions - $K_1 = (0.20, -0.39, 1.95)$ and $K_2 = (0.39, -0.39, 1.92)$ - are shown in Figures 12(a)-(b) together with experimental results from Cable (2006) in Figures 12(c)-(e).

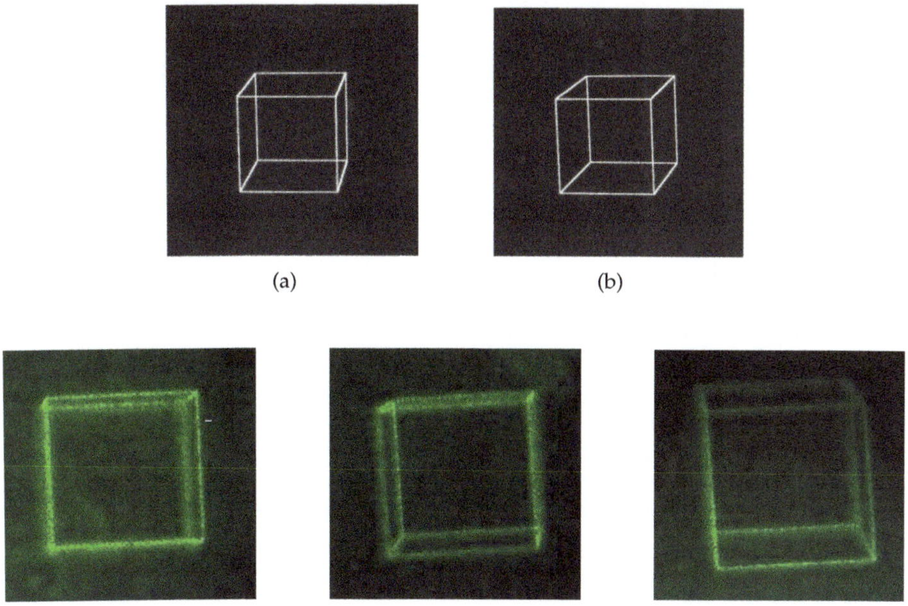

(a) (b)

(c) (d) (e)

Fig. 12. Simulated RPFs produced at pinhole aperture locations K_1 (a), and K_2 (b), and experimental results (c)-(e) from Cable (2006).

6. Conclusion

This chapter has described a number of technical innovations that have enabled the realization of a real-time, phase-only holographic projection technology.

By defining a new psychometrically determined optimization metric that is far more suited to human perception than the conventional MSE measure, a method for the generation of phase-only holograms which results in perceptually pleasing video-style images was demonstrated. This allows the realization of phase-only holographic video projection systems which, for the first time, overcome the twin barriers of the computational complexity of calculating diffraction patterns in real time and the poor quality of the resultant images.

Using these techniques, the chapter has demonstrated algorithms and methods for the generation of 2D and 3D images in the Fraunhofer and Fresnel regimes. As shown in simulation and by preliminary experiment, the RPFs produced by the calculated holograms exhibit a substantial improvement in quality and a reduction in computation time on the scale of orders of magnitude compared to the other techniques demonstrated thus far.

A number of commercially available products, notably from Light Blue Optics Inc. (2010), now employ variants of this technology. This chapter, and the information contained herein, contains a thorough description of state-of-the-art holographic projection technology and provides a complete reference to enable an interested reader to simulate, construct and characterize a 2D or 3D phase-only holographic projector.

7. References

Arrizón, V. & Testorf, M. (1997). Efficiency limit of spatially quantized Fourier array illuminators, *Opt. Lett.* 22(4): 197–199.

Bengtsson, J., Eriksson, N. & Larsson, A. (1996). Small-feature size fan-out kinoform etched in GaAs, *Appl. Opt.* 35(5): 801–806.

Bernhardt, M., Wyrowski, F. & Bryngdahl, O. (1991). Iterative techniques to integrate different optical functions in a diffractive phase element, *Appl. Opt.* 30(32): 4629–4635.

Bhatia, V., Hempstead, M., Grochocinski, J., Sekiguchi, N., Okada, A. & Loeber, D. (2009). Compact and efficient green lasers for mobile projector applications, *J. Soc. Inf. Disp.* 17(1): 47–52.

Bhatia, V., Sekiguchi, N., Hempstead, M., Okada, A. & Grochochinski, J. (2007). High efficiency green lasers for mobile projectors, *Proc. IDW Workshop*, pp. 2349–2352.

Brauer, R. & Bryngdahl, O. (1997). From evanescent waves to specified diffraction orders, *Opt. Lett.* 22(11): 754–756.

Brauer, R., Wojak, U., Wyrowski, F. & Bryngdahl, O. (1991). Digital diffusers for optical holography, *Opt. Lett.* 16(18): 1427–1429.

Buckley, E. (2006). *Computer-generated holograms for real-time image display and sensor applications*, PhD thesis, Cambridge University.

Buckley, E. (2008a). Holographic laser projection technology, *Proc. SID Symposium 2008*, number 70.2, pp. 1074–1078.

Buckley, E. (2008b). Holographic laser projection technology (invited article), *Information Display* 24(12): 22–25.

Buckley, E. (2008c). Speckle-reduced holographic laser projector, *Proc. International Display Workshop*, Niigata, Japan.

Buckley, E. (2011a). Holographic laser projection, *J. Display Technol.* 7(3): 135–140.

Buckley, E. (2011b). Real-time error diffusion processor for increasing contrast ratio in holographic projection, *J. Display Technol.* 7(2): 70–76.

Buckley, E., Cable, A., Wilkinson, T. & Lawrence, N. (2006). Viewing angle enhancement for two- and three-dimensional holographic displays using random super-resolution phase masks, *Appl. Opt.* 45(28): 7334–7341.

Buckley, E., Corbett, A., Surman, P. & Sexton, I. (2008). Multi- viewer autostereoscopic display with dynamically-addressable holographic backlight, *Proc. SID Symposium*, number 25.1, pp. 340–344.

Buckley, E., Isele, R. & Stindt, D. (2009). Novel Human-Machine Interface (HMI) design enabled by holographic laser projection, *Proc. SID Symposium*, number 14.4, pp. 172–177.

Buckley, E. & Wilkinson, T. (2006). Laser-printer computer-generated holograms exhibiting suppressed higher orders in the replay field, *Opt. Lett.* 31(10): 1397–1398.

Buckley, E. & Wilkinson, T. (2007). Low image reconstruction error using continuous real axis modulation, *Opt. Eng.* 46(2): 025802.

Cable, A. (2006). *Real-time high-quality two and three dimensional holographic video projection using the one-step phase retrieval (OSPR) approach*, PhD thesis, Cambridge University.

Cable, A. J., Buckley, E., Mash, P., Lawrence, N. A. & Wilkinson, T. D. (2004). Real-time binary hologram generation for high-quality video projection applications, *Proc. SID Symposium*, number 53.1, pp. 1–3.

Case, S. K., Haugen, P. R. & Lokberg, O. J. (1981). Multifacet holographic optical elements for wave front transformations, *Appl. Opt.* 20(15): 2670–2675.

Dames, M. P., Dowling, R. J., McKee, P. & Wood, D. (1991). Efficient optical elements to generate intensity weighted spot arrays: design and fabrication, *Appl. Opt.* 30(19): 2685, 2688–2691.

Dong, B.-Z., Liu, R., Yang, G.-Z. & Gu, B.-Y. (1998). Design of diffractive phase elements that generate monochromatic or color point and ring patterns, *J. Opt. Soc. Am. A* 15(2): 480–486.

Dong, B.-Z., Zhang, G.-Q., Yang, G.-Z., Gu, B.-Y., Zheng, S.-H., Li, D.-H., Chen, Y.-S., Cui, X.-M., Chen, M.-L. & Liu, H.-D. (1996). Design and fabrication of a diffractive phase element for wavelength demultiplexing and spatial focusing simultaneously, *Appl. Opt.* 35(35): 6859–6864.

Dorsch, R., Lohmann, A. W. & Sinzinger, S. (1994). Fresnel ping-pong algorithm for two-plane computer-generated hologram display, *Appl. Opt.* 33(5): 869–875.

Dresel, T., Beyerlein, M. & Schwider, J. (1996). Design of computer-generated beam-shaping holograms by iterative finite-element mesh adaption, *Appl. Opt.* 35(35): 6865–6874.

Elschner, J. & Schmidt, G. (1998). Analysis and mumerics for the optical design of binary diffractive gratings, *Math. Meth. Appl. Sci* 21: 1297–1342.

Fetthauer, F., Weissbach, S. & Bryngdahl, O. (1995). Computer-generated Fresnel holograms: quantization with the error diffusion algorithm, *Opt. Commun.* 114(3-4): 230–234.

Gallacher, N. C. (1977). Optimum quantization in digital holography, *Appl. Opt.* 17(1): 109–115.

Georgiou, A., Christmas, J., Moore, J., Jeziorska-Chapman, A., Davey, A., Collings, N. & Crossland, W. A. (2008). Liquid crystal over silicon device characteristics for holographic projection of high-definition television images, *Appl. Opt.* 47(26): 4793–4803.

Gerchberg, R. W. & Saxton, W. O. (1972). A practical algorithm for the determination of phase from image and diffraction plane pictures, *Optik* 35(2): 237–246.

Gillet, J.-N. & Sheng, Y. (2003). Multiplexed computer-generated holograms with polygonal-aperture layouts optimized by genetic algorithm, *Appl. Opt.* 42(20): 298–307.

Goodman, J. W. (1996). *Introduction to Fourier Optics*, 2 edn, McGraw-Hill, New York, chapter The Fraunhofer approximation, pp. 73–75.

Goodman, J. W. & Silvestri, A. M. (1970). Some effects of Fourier-domain phase quantization, *IBM J. Research and Dev.* 14(5): 478–484.

Greene, J. & d'Oliveira, M. (1999). *Learning to Use Statistical Tests in Psychology*, Open University Press, Buckingham, UK, chapter Part 1: introduction to experimental design.

Gu, B., Yang, G. & Dong, B. (1986). General theory for performing an optical transform, *Appl. Opt.* 25(18): 3197–3206.

Gupta, M. C. & Peng, S. T. (1991). Diffraction of a light beam by doubly periodic structures, *Opt. Lett.* 16(17): 1301–1303.

Hecht, E. (1998). *Optics*, 3rd edn, Addison-Wesley, San Francisco, California.

Heggarty, K. & Chevalier, R. (1998). Signal window minimum average error algorithm for computer-generated holograms, *J. Opt. Soc. Am. A* 15(3): 625–635.

Heggarty, K., Fracasso, B., de Bougrenet de la Tocaye, J.-L., Birch, M. & Krueerke, D. (2004). A silicon backplane FLC spatial light modulator for uses within an optical telecommunications environment, *Ferroelectrics* 312: 39–55.

Ito, T., Masuda, N., Yoshimura, K., Shiraki, A., Shimobaba, T. & Sugie, T. (2005). Special-purpose computer HORN-5 for a real-time electroholography, *Opt. Express* 13(6): 1923–1932.

Ito, T. & Okano, K. (2004). Color electroholography by three colored reference lights simultaneously incident upon one hologram panel, *Opt. Express* 12(8): 4320–4325.

Ito, T. & Shimobaba, T. (2004). One-unit system for electroholography by use of a special-purpose computational chip with a high-resolution liquid-crystal display toward a three-dimensional television, *Opt. Express* 12(9): 1788–1793.

Jean-Numa Gillet, Y. S. (1999). Irregular spot array generator with trapezoidal apertures of varyng heights, *Opt. Commun.* 166(1-6): 1–7.

Jesacher, A., Furhapter, S., Bernet, S. & Ritsch-Marte, M. (2004). Diffractive optical tweezers in the Fresnel regime, *Opt. Express* 12(10): 2243–2250.

Jones, K. & Harrison, B. (2007). The impact of changing TV technologies and market trends on the energy consumption of TVs and the need for a better TV energy test method, *Technical report*, International Energy Agency.

Keller, P. E. & Gmitro, A. F. (1993). Computer-generated holograms for optical neural networks: on-axis versus off-axis geometry, *Appl. Opt.* 32(8): 1304–1310.

Kelly, D. H. & van Norren, D. (1977). Two-band model of heterochromatic flicker, *J. Opt. Soc. Am.* 67(8): 1081–1090.

Kirk, A., Powell, K. & Hall, T. (1992). A generalisation of the error diffusion method for binary computer generated hologram design, *Opt. Commun.* 92(1-3): 12–18.

Kowarz, M. W. (1995). Homogeneous and evanescent contributions in scalar near-field diffraction, *Appl. Opt.* 34(17): 3055–3063.

Layet, B., Cormack, I. G. & Taghizadeh, M. R. (1999). Stripe color separation with diffractive optics, *Appl. Opt.* 38(35): 7193–7201.

Lee, B. H., Chung, W. J., Jim, T. J., Kim, T. S., Seok, J. G. & Jung, Y. S. (2009). Invited paper: Analysis for reduction of power consumption in PDP, *Proc. SID Symposium* (7.1): 54–57.

Leemis, L. M. & McQueston, J. T. (2008). Univariate distribution relationships, *The American Statistician* 62(1): 45–53.

Lesem, L. B. & Hirsch, P. M. (1969). The Kinoform: A new wavefront reconstruction device, *IBM J. Res. and Dev.* 13: 150–155.

Light Blue Optics Inc. (2010). Light Touch specifications. URL:`http://www.lightblueoptics.com/products/light-touch/specifications/`

Liu, W.-C. & Kowarz, M. W. (1998). Vector diffraction from subwavelength optical disk structures: Two-dimensional near-field profiles, *Opt. Express* 2(5): 191–197.

Liu, W.-C. & Kowarz, M. W. (1999). Vector diffraction from subwavelength optical disk structures: two-dimensional modelling of near-field profiles, far-field intensities, and detector signals from a dvd, *Appl. Opt.* 38(17): 3787–3797.

Lofving, B. (1997). Self-adjusting dynamic binary phase holograms, *Appl. Opt.* 36(11): 2347–2352.

Madrazo, A. & Nieto-Vesperinas, M. (1997). Model near field calcuations for optical data storage readout, *Appl. Phys. Lett.* 70(31): 31–33.

Mait, J. N. (1990). Design of binary-phase and multiphase Fourier gratings for array generation, *J. Opt. Soc. Am. A* 7(8): 1514–1528.

Makowski, M., Sypek, M. & Kolodziejczyk, A. (2008). Colorful reconstructions from a thin multi-plane phase hologram, *Opt. Express* 16(15): 11618–11623.

Meister, M. & Winfield, R. J. (2002). Local improvement of the signal-to-noise ratio for diffractive optical elements designed by unidirectional optimization methods, *Appl. Opt.* 41(35): 7390–7396.

Mok, F., Diep, J., Liu, H.-K. & Psaltis, D. (1986). Real-time computer-generated hologram by means of liquid-crystal television spatial light modulator, *Opt. Lett.* 11(11): 748–750.

O'Brien, D. C., Mears, R. J., Wilkinson, T. D. & Crossland, W. A. (1994). Dynamic holographic interconnects that use ferroelectric liquid-crystal spatial light modulators, *Appl. Opt.* 33(14): 2795–2803.

O'Callaghan, M. J., Ferguson, R., Vohra, R., Thurmes, W., Harant, A. W., Pecinovsky, C. S., Zhang, Y., Yang, S., O'Neill, M. & Handschy, M. A. (2009). Bistable FLCOS devices for doubled-brightness micro-projectors, *Proc. SID Symposium*, number 18.2, pp. 232–235.

Papazoglou, D. G., Loulakis, M., Signakis, G. & Vainos, N. A. (2002). Holographic read-write projector of video images, *Opt. Express* 10(6): 280–285.

Poon, T.-C., Schilling, B. W., Wu, M. H., Shinoda, K. & Suzuki, Y. (1993). Real-time two-dimensional holographic imaging by using an electron-beam-addressed spatial light modulator, *Opt. Lett.* 18(1): 63–65.

Roux, F. S. (1991). Intensity distribution transformation for rotationally symmetric beam shaping, *Opt. Eng.* 30(5): 529–536.

Roux, F. S. (1993). Implementation of general point tranforms with diffractive optics, *Appl. Opt.* 32(26): 4972–4978.

Sando, Y., Itoh, M. & Yatagai, T. (2004). Color computer-generated holograms from projection images, *Opt. Express* 12(11): 2487–2493.

Schmitz, M., Brauer, R. & Bryngdahl, O. (1996). Comment on numerical stability of rigorous differential methods of diffraction, *Opt. Commun.* 124(1-2): 1–8.

Schnars, U. & Juptner, W. P. O. (2002). Digital recording and numerical reconstruction of holograms, *J. Meas. Sci. Technol.* 13(9): 85–101.

Seldowitz, M. A., Allebach, J. P. & Sweeney, D. W. (1987). Synthesis of digital holograms by direct binary search, *Appl. Opt.* 26(14): 2788–2798.

Sinclair, G., Leach, J., Jordan, P., Gibson, G., Yao, E., Laczik, Z. J., Padgett, M. J. & Courtial, J. (2004). Interactive application in holographic optical tweezers of a multi-plane Gerchberg-Saxton algorithm for three-dimensional light shaping, *Opt. Express* 12(8): 1665–1670.

Stobbe, L., Nissen, N. & Middendorf, A. (2008). Evaluation of TV power measurements, *Technical report*, Fraunhofer IZM.

Stuff, M. A. & Cederquist, J. N. (1990). Coordinate transformations realizable with multiple holographic optical elements, *J. Opt. Soc. Am. A* 7(6): 977–981.

Taghizadeh, M. (1998). Best technical diffractive optical element, *Diffractive Beauty Competition, OSA Diffractive Optics and Micro-Optics Conf.*, Hawaii.

Taghizadeh, M. (2000). Best technical diffractive optical element, *Diffractive Beauty Competition, OSA Diffractive Optics and Micro-Optics Conf.*, Quebec City.

Takaki, Y. & Hojo, J. (1999). Computer-generated holograms to produce high-density intensity patterns, *Appl. Opt.* 38(11): 2189–2195.

Tang, Y.-T. & Chang, M.-W. (1992). Testing of aspheric surfaces with computer-generated hologram by using a novel interferometric set-up, *Chinese Journal of Physics* 30(3): 339–348.

Thompson, E. J., Brass, E. D., Samuel, K., Bullock, S. R., Lindesay, J. & Lyons, D. R. (1999). Formation of phase gratings on the end of gradient-index lenses with ultraviolet ablation at 193 nm, *Appl. Opt.* 38(31): 6494–6497.

Weber, L. (2005). Challenge of measuring annual energy consumption of TV sets, *Proc. IDW*, pp. 1429–1432.

Wyrowksi, F., Hauck, R. & Bryngdahl, O. (1986). Computer-generated holograms: defect resistance by object phase manipulation, *J. Opt. Soc. Am. A* 3(11): 1840–1845.

Wyrowski, F. (1991). Upper bound of the diffraction efficiency of diffractive phase elements, *Opt. Lett.* 16(24): 1915–1917.

Wyrowski, F. (1992). Modulation schemes of phase gratings, *Opt. Eng.* 31(02): 251–257.

Wyrowski, F. & Bryngdahl, O. (1988). Iterative Fourier-transform algorithm applied to computer holography, *J. Opt. Soc. Am. A* 5(7): 1058–1065.

Yang, G., Dong, B., Gu, B., Zhuang, J. & Ersoy, O. K. (1994). Gerchberg-Saxton and Yang-Gu algorithms for phase retrieval in a nonunitary transform system: a comparison, *Appl. Opt.* 33(2): 209–218.

Part 4

Seeing

15

A Contribution to Virtual Experimentation in Optics

Javier Gamo
Department of Electronics - University of Alcalá
Spain

1. Introduction

Parallel to the development and popularization of Internet, the emergence of resources for on-line learning is becoming quite common in all scientific disciplines. Optics is a good example of this telelearning. Today there exist several utilities, which allow to perform virtual experiments in Optics, from a computer connected to the Internet. These web-based tools, usually developed in Java (Carnicer, 2010), are suitable to get a first introduction to the optical phenomenon, but do not allow interaction with real, on-going experimentation in the laboratory. On the other hand, websites with videos of real experiments in Optics are also common (Carreño, 2010), but usually they do not allow interaction of the user via simulation. To match both worlds, a set of software tools has been developed at the University of Alcalá, allowing *virtual* and *physical* laboratory testing of different optical phenomena through the same software platform. Developed in MATLAB (Matlab, 2010), these tools cover different topics in Optics. They are aimed to complement "classical" classroom teaching on engineering studies. Current developments include Diffraction, Radiometry and Photometry, Acousto-Optics interaction, Moiré phenomenon, and Computer-Generated Holograms (CGHs)[1]. In this chapter, only diffraction-related phenomena will be described.

2. Module structure

Each software tool is developed under the following structure:

- Theoretical module
- Simulation module
- Laboratory experimentation module

The **theoretical background** module introduces the student to the optical phenomenon. Then, by using the **simulation** module, the user can make simulations of the phenomenon using the power and flexibility of MATLAB. Interaction and/or comparison with real, physical experiments can be achieved through the **laboratory experimentation** module. At any time, the user can get interactive help on each module.

[1] The list of covered optical phenomena is being extended, to include more optical topics such as Theory of Color, Geometrical Optics and Optical Fibers, among others

3. Diffraction tool

The first tool developed explains the phenomenon of optical diffraction. Fig. 1 (a) shows the main window of this module. By clicking on the corresponding icon, the user can choose to study diffraction patterns produced by several classical apertures: square, circular, slit (single, or multiple). Pre-stored and/or user-defined images can be also used as diffractive objects.

3.1 Theoretical module

The basis of diffraction is explained. Fresnel and Fraunhofer diffraction is explained using some typical apertures. For instance, Fig. 1(b) shows diffraction patterns for a single slit, multiple slits, and circular aperture, which in turn produces the well-known Airy disc. The user can navigate back and forth through a multi-page environment using the [Next] and [Previous] buttons. [APPLICATION] button brings the user to the Simulation module described in section 3.2. The [MANUAL] button launches the users's manual of the tool.

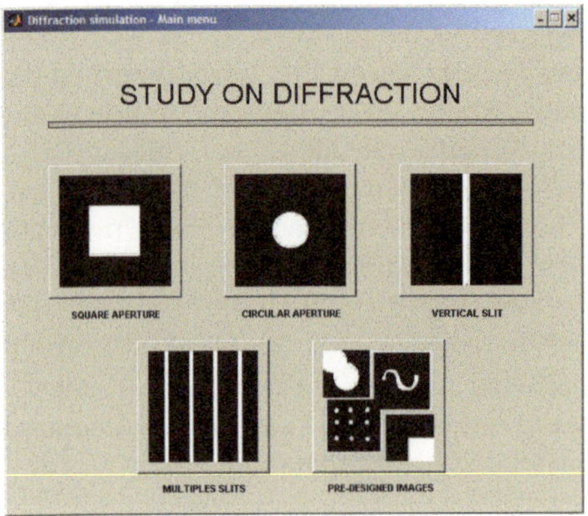

(a) Main menu of the diffraction tool

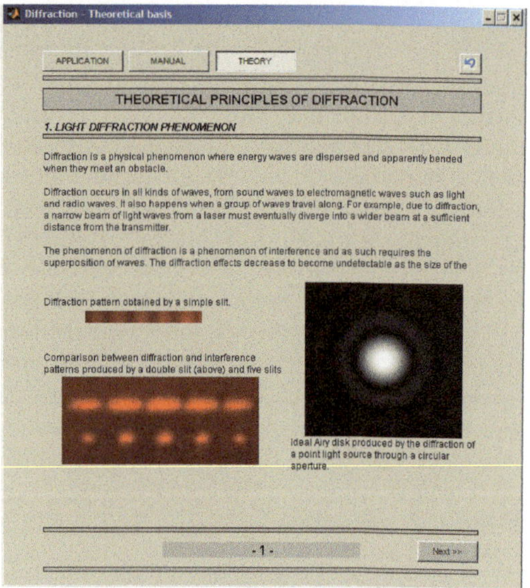

(b) Theoretical explanation of diffraction

Fig. 1. Diffraction tool: main screen and example of the theoretical module

3.2 Simulation module

Once the user has selected the diffracting object, the corresponding computing window is displayed. For instance, Fig. 2 (a) shows the interface for calculating the diffraction pattern from vertical slits. As it can be seen, the main parameters are, in this case: number of slits, distance between slits, slit width, illumination wavelength, and observation distance for near-field (Fresnel) or far-field (Fraunhofer) diffraction computing. Similar parameters are shown for the rest of diffracting objects available.

The intensity of the computed diffraction pattern is shown on the right-side image frame in Fig. 2 (a). Such diffraction pattern can be sent to a printer, or stored as an image file on the PC for further processing. By clicking on `Cross section`, a profile of the intensity diffraction pattern can be plotted, as shown in Fig. 2 (b).

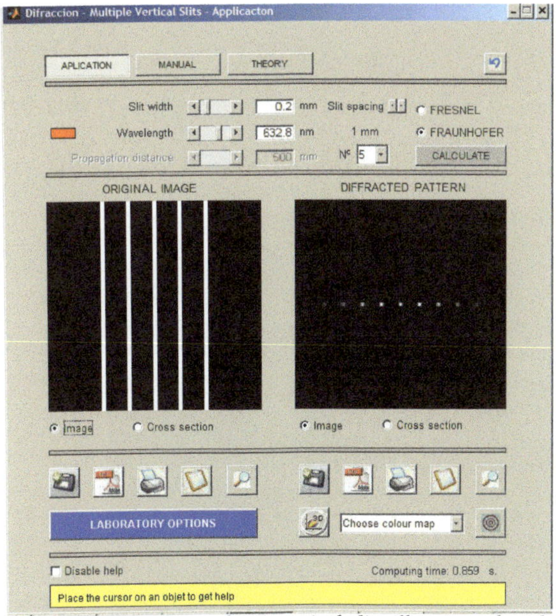

(a) Fraunhofer pattern for 5 vertical slits illuminated with red light

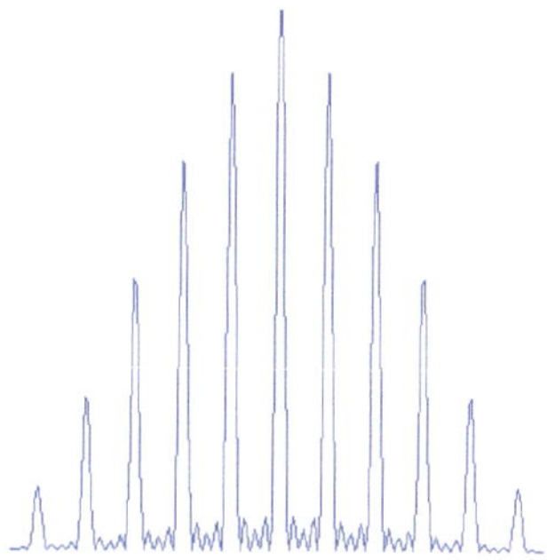

(b) Cross-section of the computed diffraction pattern

Fig. 2. Example of far-field diffraction pattern computed for 5 vertical slits

3.3 Laboratory experimentation module

One of the most interesting features of the tool is the possibility to compare simulations with real experiments. Fig. 3 shows the basic laboratory setup. A collimated He-Ne laser illuminates the object (aperture), and the corresponding diffraction pattern projected on the output screen is stored by a video camera.

Images captured by the video camera can be stored and processed by the laboratory experiment module shown in Fig. 4.

4. Computer generated holograms tool

The idea of using computers to define and generate holograms was proposed by the middle of the 60's (Brown & Lohmann, 1966). Some essential aspects of CGHs are:

- The object does not need to exist.
- The construction of the hologram is done away from the usual diffracting objects: the image is defined, and the diffracting object that produces such an image is then calculated.

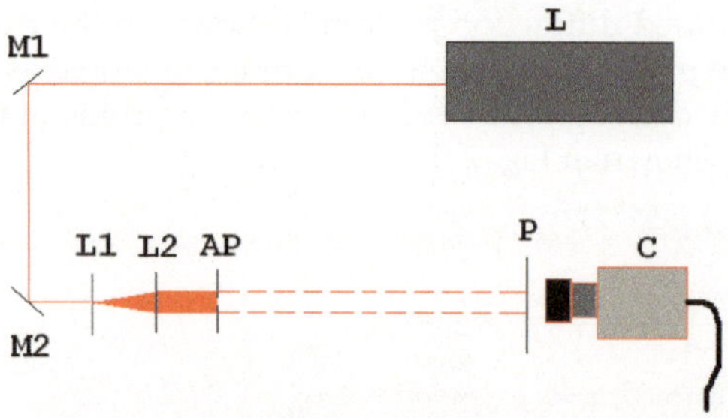

(a) Laboratory setup

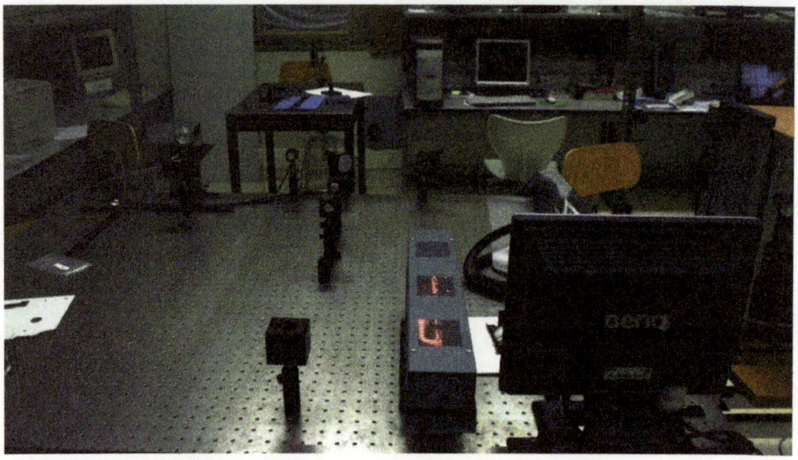

(b) Picture of the actual laboratory assembly

Fig. 3. Laboratory experiments on diffraction

CGHs have long been used in optical processing information, optical interconnects, interferometry and health diagnosis, to name just a few relevant applications (Yaroslavsky & Astola, 2009). Advances in computer power, together with the development of sophisticated manufacturing methods, are causing CGHs to become increasingly efficient and complex.

4.1 Theoretical module

CGHs treated in this module are Fourier holograms, main application is optical interconnects (fan-out). The reproduction setup is shown in Fig. 5 (a).
At its simplest, a CGH is basically a diffraction grating composed by a 2D unit cell, repeated along the hologram surface. Let's consider the diffraction grating shown in Fig. 5 (b).
When this grating is illuminated by a collimated monochromatic beam, a Fraunhofer diffraction pattern is obtained in the focal plane of the lens L (*i.e.* Fourier transform of the complex amplitude transmittance of the diffraction grating). Spacing between diffraction peaks is given by (Hecht, 1988):

$$x = \frac{\lambda f}{d} \tag{1}$$

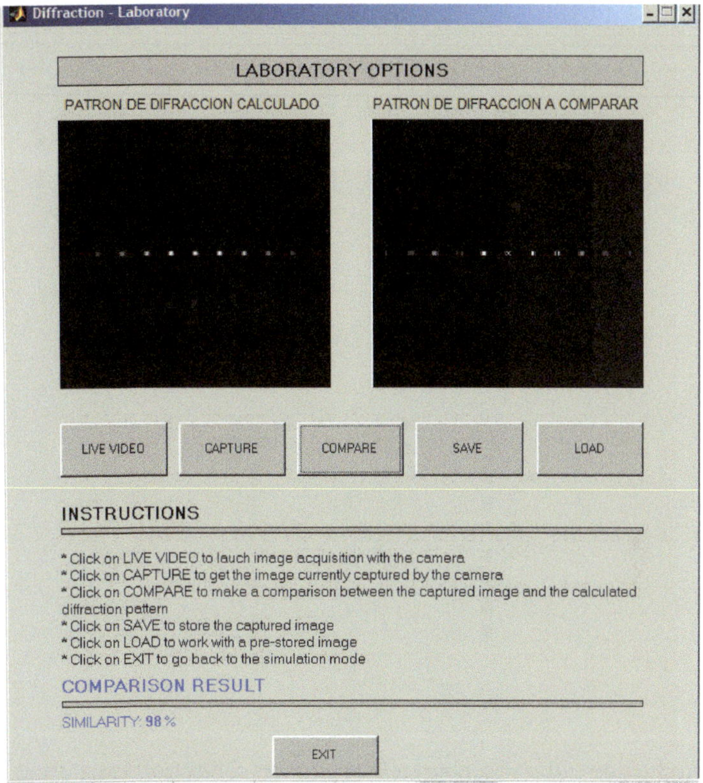

Fig. 4. Main interface for laboratory experiments

where d is the grating spacing, f is the focal length of the Fourier lens, and λ is wavelength of the collimated light beam. Relative intensities of the diffraction orders are proportional to the grating structure a/d. In any CGH, it is important to maximize light on the desired diffraction order (usually +1). Diffraction efficiency DE, defined as:

$$\mathrm{DE} = \frac{\text{Light in the desired diffracted order}}{\text{Total incident light}} \tag{2}$$

measures such light maximization. To this end, when the basic CGH cell is calculated, such 2D cell is repeated to form the CGH structure; spatial invariance of the Fourier transform means that every single cell produces the same diffraction pattern at the output plane, contributing to increased DE. The type of optical material in which the CGH is implemented also contributes to have a good DE.

CGHs calculated in section 4.2 are binary-only (*i.e.* pixels have only two possible values, black or white). Depending on the type of physical object where the CGH is implemented, it can be a Binary Amplitude Hologram (*i.e.* photographic film, where black pixels correspond to dark zones and white pixels are transparent zones in the film) or a Phase-only Hologram (for instance, using a Liquid Crystal Spatial Light Modulator, where the two possible orientation of the liquid crystal molecules correspond to phase states differing by π).

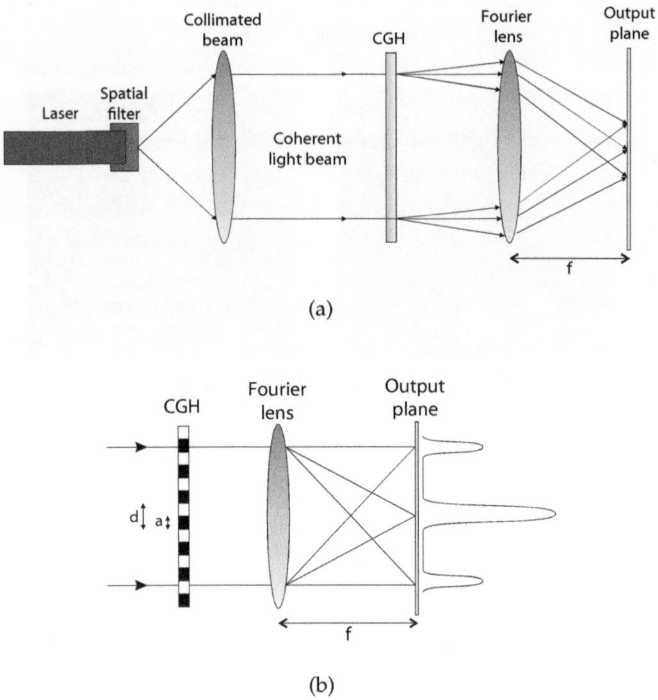

Fig. 5. (a) Reproduction setup of a Fourier CGH (b) Simple diffraction grating

The software developed includes a theoretical module to explain the basis of the CGHs to the user. Fig 6 shows one of the multi-page theoretical screens. Applications, numerical computing algorithms, and setup schemes for physical reproduction of CGHs are fully explained in this module.

4.2 Simulation module

Fig. 7(a) shows the main window for CGH calculation. The left-side image frame displays the object image, while the right-side frame stores the calculated CGH. A blue background indicates which is the active frame, either CGH or object image.
Holograms are calculated from pre-stored object images loaded on the left-side image frame on Fig. 7(a)). The following algorithms can be selected, with their corresponding parameters, as shown in Fig. 7(b):

- Fourier transform (Wyrowsky & Bryngdahl, 1988)
- Detour phase (Levy et al., 1998)
- Simulated annealing (Kirkpatrick, 1983)

The resolution of the image clearly influences the hologram computing time. The CGH calculated is represented on the right-side image frame on Fig. 7(a), and can be stored on the computer using different format files (e.g. .BMP, .JPG, ...).

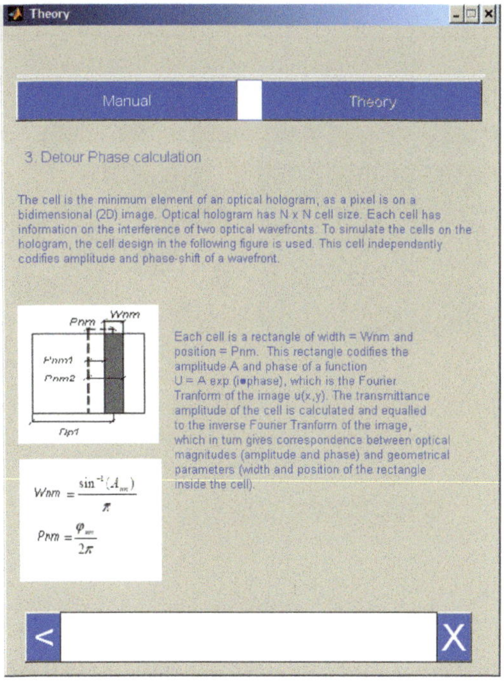

Fig. 6. Main interface for laboratory experiments

Once the CGH has been calculated, the performance of such CGH can be also checked from this window. To this end, a valid CGH image (either freshly calculated or pre-loaded from the PC) must be displayed on the right-side image frame in Fig. 7(a). By pressing on "DEPECA" icon (or, alternatively, selecting `Tools, Image calculation` on the Menu, shortcut *Ctrl + I*), the inverse Fourier transform is performed, and the object image is then reconstructed.

Fig. 8 shows two CGH masks built from the same unit cell, calculated using the simulated annealing algorithm in Fig. 7. Each mask has 2×2 and 4×4 times the unit cell, respectively. Target output for such unit cell is a 4×4 fan-out array of light spots at the Fourier plane in Fig. 5.

When the target output is symmetrical (as in Fig. 8), due to the symmetry properties of the Fourier Transform binary CGHs, only half of the output target is needed to be defined (specular image appears automatically at the output plane) (Morrison, 1992). This reduces significantly the computing time for the CGH (Samus, 1995).

Fig. 8 (a.2) and (b.2) show simulations of the optical power spectrum detected by physical sensors. This power spectrum is proportional to the square of the 2-D complex Fourier transform amplitude of the CGH (Hecht, 1988). Low spatial frequencies are near the center of the spectrum (zero spatial frequency), and high frequencies are farther away from the centre, as usual in optical power spectrum representations (Poon & Banerjee, 2001). Simulations show that spot spacing is inversely proportional to the corresponding size cell, as shown in Equation 1, where d is the size of the unit cell.

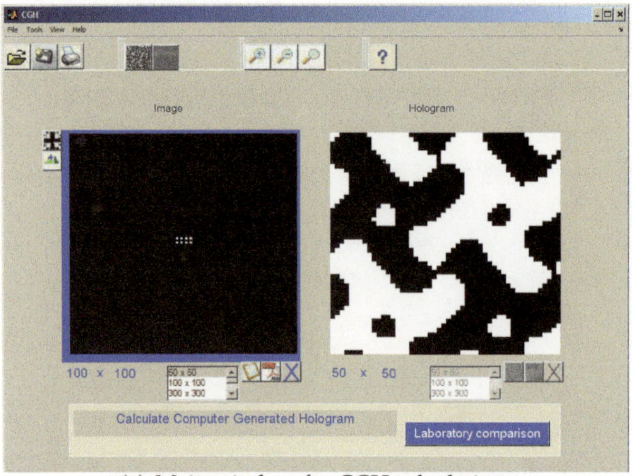

(a) Main window for CGH calculation

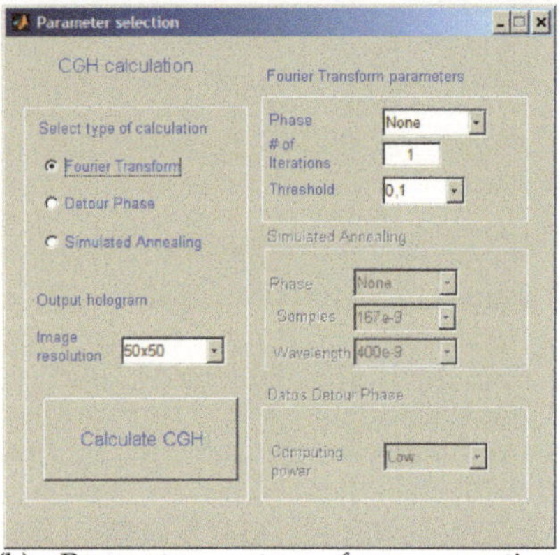

(b) Parameter setup for computing algorithms

Fig. 7. CGH simulation module

4.3 Laboratory experimentation module

Fig. 9 displays the actual implementation of the optical setup from Fig. 5.

A webcam is placed at the output (Fourier) plane to collect the diffraction pattern produced by a CGH illuminated with a coherent, collimated optical beam from a He-Ne laser.

To show the performance of the simulation module, some fan-out CGHs have been implemented on a photographic film, as shown in Fig. 10. The presence of non-developed grains in the emulsion increases the undesired central DC spot in the diffraction pattern, as shown in Fig. 11.

Image acquisition and processing is managed from the interface shown in Fig. 12. The user can also make a comparison between theoretical CGH output, simulated output using the computing tool in Section 4.2, and real CGH output captured by the webcam placed at the Fourier plane in Fig. 9. Comparison is made in two ways: similarity (i.e. digital substraction of images), and correlation between images.

(a.1) 2 × 2

(b.1) 4 × 4

(a.2)

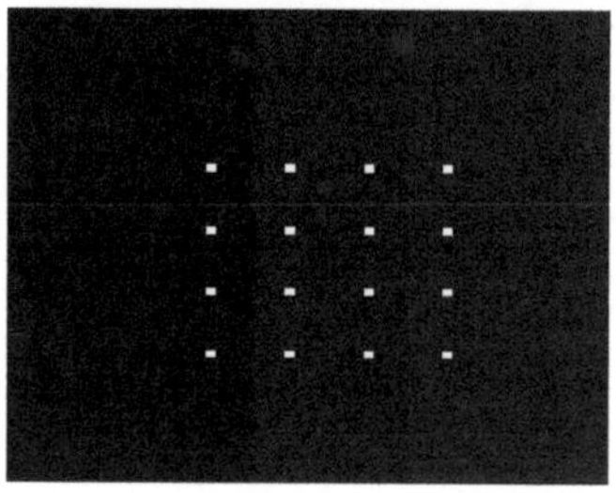

(b.2)

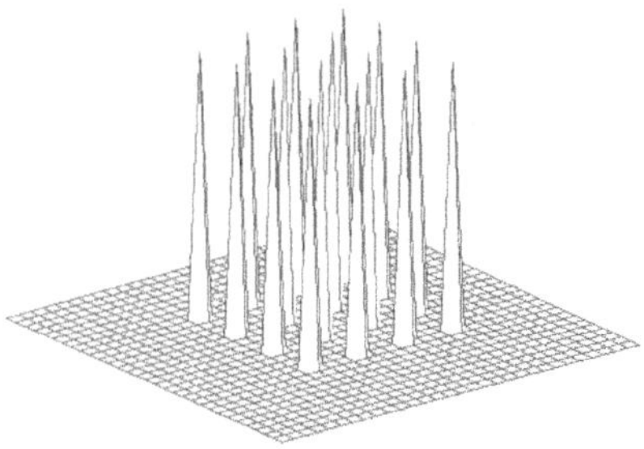

(a.3)

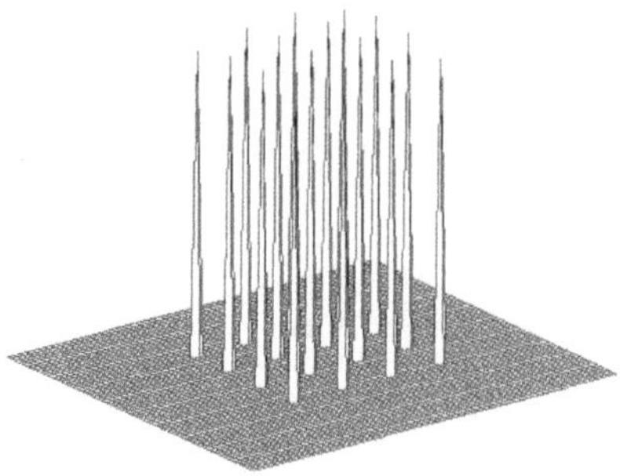

(b.3)

Fig. 8. CGH cells replicated (a.1) 2 × 2 and (b.1) 4 × 4 times; (a.2) and (b.2) simulated optical output (a.3) and (b.3) 3-D intensity profiles

5. Acousto-optical effect

Acousto-optics describes the interaction of sound with light (Saleh & Teich, 1991). A sound wave unleashed on an optical medium creates a disturbance in the refractive index of the medium. The sound can then control a light beam hitting the medium. This fact, known as acousto-optical (AO) effect is used in various devices, such as beam deflectors, optical modulators, filters, isolators and spectrum analyzers, among others.

Fig. 9. CGH experimental setup

(a) CGH on photographic film

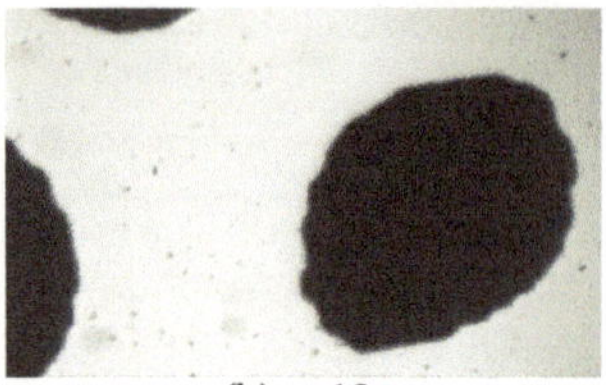

(b) ×10

(c) ×40

Fig. 10. (a) CGH implemented on photographic film; (c), (d) Microscopic details showing development imperfections

5.1 Theoretical module

This module explains the basis of AO phenomena: Raman-Nath and Bragg configurations, AO materials, figures of merit, etc. The user can navigate interactively through different tutorials, as shown in Fig. 13. Documentation and a user's manual are also available, including datasheets of the components used in section 5.3.

5.2 Simulation module

Acousto-optical virtual experiments are performed within this module. Fig. 14 shows the main interface of the simulation module. The user can choose one of the two interaction regimes (Bragg or Raman-Nath), explained in the previous theoretical module. The application allows to select the light source wavelength (λ_O), the acousto-optical material

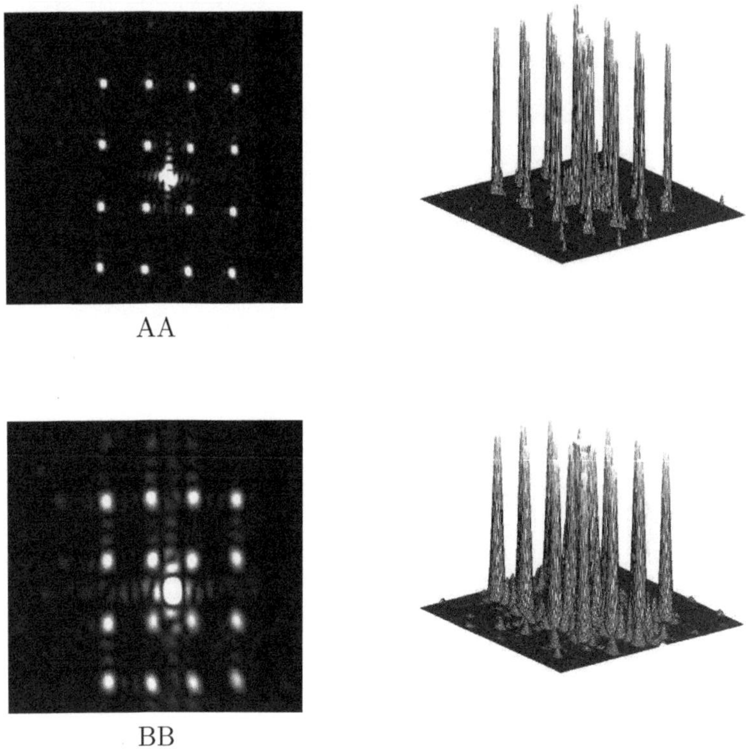

Fig. 11. (Left) Output at Fourier plane of 4 × 4 dots fan-out CGHs consisting of AA = 4 × 4 and BB = 2 × 2 unit cell. (Right) Corresponding 3D intensity profiles

(characterized by the refractive index, n, and the acoustic velocity in the material, V_S), and the RF waveform (characterized by its amplitude A and frequency f_S). By these conditions, the first-order deflection angle under the Bragg regime is given by:

$$\theta_D = 2\theta_B = 2\sin^{-1}\frac{\lambda_O}{2n\Lambda_S} \approx \frac{\lambda_O f_S}{nV_S} \tag{3}$$

When pressing the CALCULATE DEVIATION button, the deflection angle as a function of input parameters is calculated. The simulation module can also create a video sequence, which shows graphically the phenomenon of acousto-optical interaction from a 2D or 3D perspective, as shown in Figure 14 (b). The video is also saved as AVI file for later viewing.

5.3 Laboratory experimentation module

Figure 15 shows the block diagram of the experimental setup. A radio frequency (RF) electronics module generates a 24 MHz wave which impinges the acousto-optical device (AOD), thus producing a diffraction grating along the material. The acousto-optical interaction is observed when a laser beam coming out of a laser pointer crosses the AOD

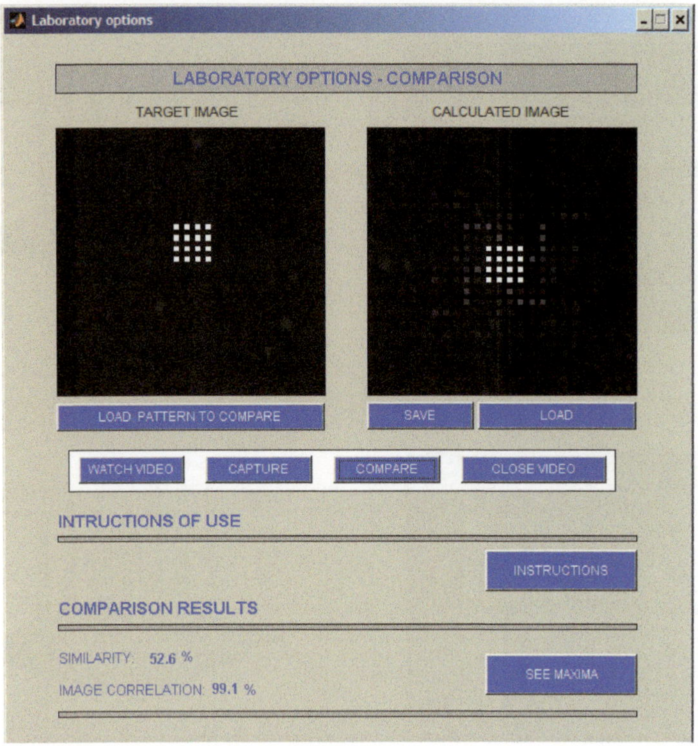

Fig. 12. CGH laboratory module: main window

Parameter	**Value**
Interaction Material	Fused Silica
Wavelength	1047 Ű 1064nm
RF Frequency	24 MHz
Clear Aperture	8×10 mm
Acoustic Mode	Compressional or Shear
Input impedance	$5\,\Omega$

Table 1. Main specifications of the AOD employed

while the RF wave is on; under such circumstances, the light beam is deviated, according to Equation 3.
Laser ignition and RF waveform generation are controlled by the software module through the RS232 port. Finally, the diffracted beam is projected on a screen for direct viewing and/or detection by a suitable sensor (*i.e.* photodetector, optical power meter, etc.).
The acousto-optical device (AOD) shown in Fig 16 has been employed. This AOD was acting as Q-switch on a dismantled Nd:YAG laser. Table 1 shows main parameters of this AOD.
For this application, a red laser pointer with wavelength $\lambda = 670$ nm has been used. According to Equation 3, the theoretical deflection angle for this red light source is:

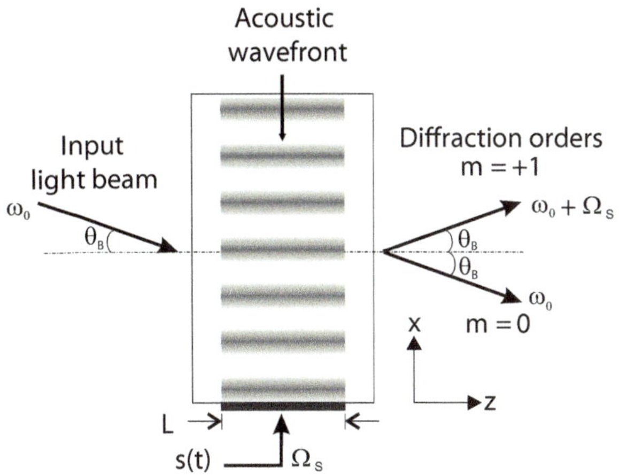

(a) AO interaction under Bragg regime

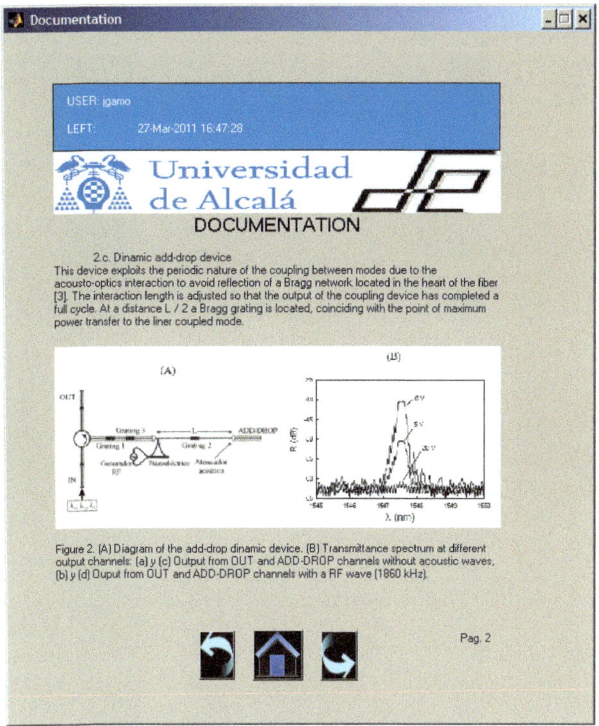

(b) Main window of the AO theory module

Fig. 13. AO theoretical module

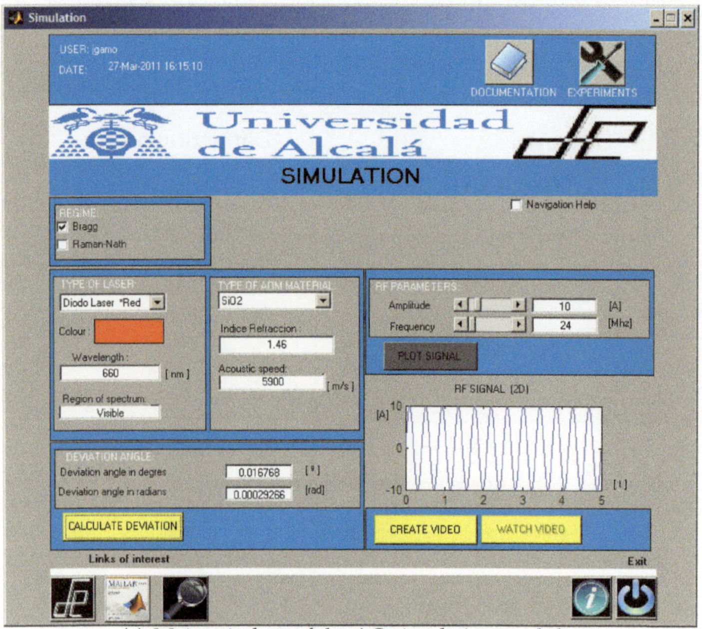

(a) Main window of the AO simulation module

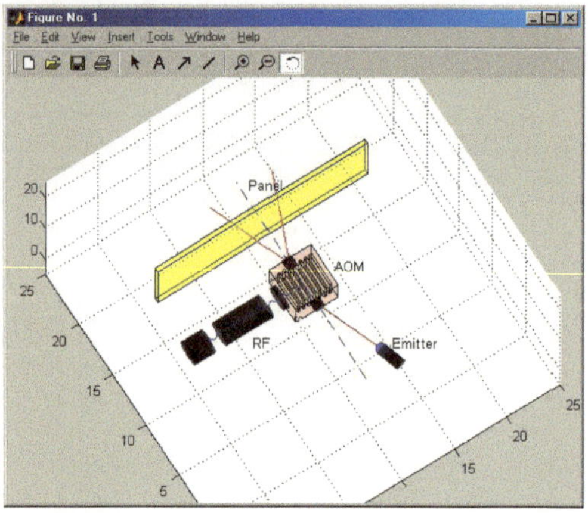

(b) 3D video animation illustrating AO interaction

Fig. 14. AO simulation module

$$\theta_D = 0.00091943 \text{ rad} \tag{4}$$

Since this angle is small enough, the following assumption can be made:

$$\sin\theta_D \approx \theta_D = \frac{\Delta x}{L} \tag{5}$$

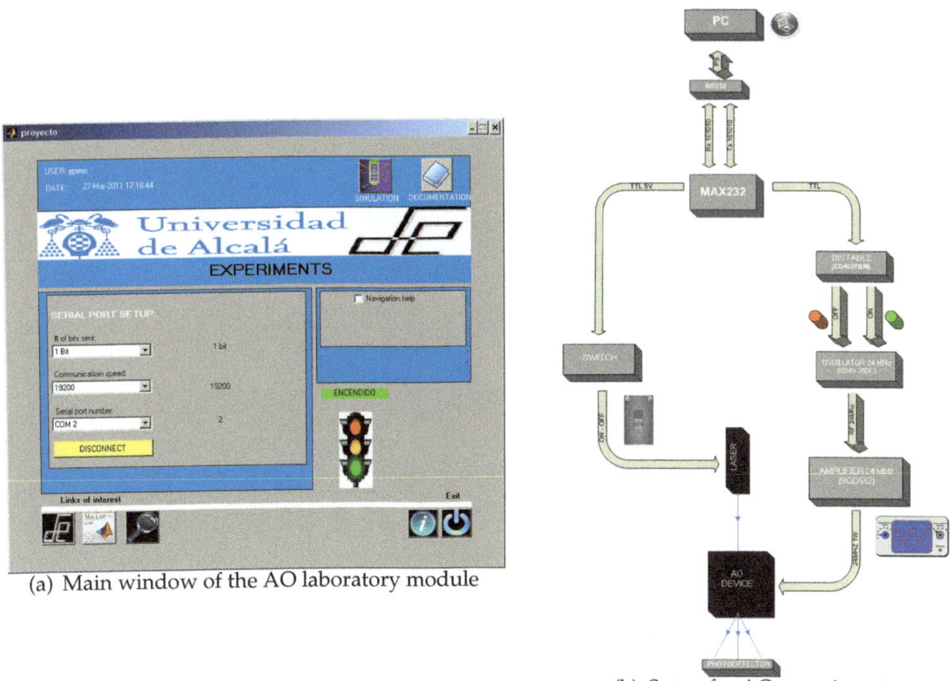

(a) Main window of the AO laboratory module

(b) Setup for AO experiments

Fig. 15. Acousto-optical laboratory experimentation module

A separation $\Delta x = 1$ cm between the diffracted and non-diffracted beams is achieved at a distance $L = 10$ m away from the AOD output.

The incoming light beam crossing the AOD is deviated as long as a proper RF signal is injected through the AOD transducer. To this end, a 24 MHz signal is generated using a IQXO-350-C, which gives an RF output power of:

$$P = V \cdot I = 1{,}8\,\mathrm{V} \cdot 26{,}2\,\mathrm{mA} = 47{,}16\,\mathrm{mW} \tag{6}$$

This RF power output is too low to properly observe the acousto-optical phenomenon. A good figure for this RF power output is 1 W, which gives a gain G_{dB} of:

$$G_{dB} = 10 \log \frac{P_{out}}{P_{in}} = 10 \log \frac{1000}{47{,}16} = 13{,}26\, dB \tag{7}$$

Therefore, a BGD502 amplifier with G_{dB} = 18.25 dB is employed to amplify the RF output signal.

RF signal generation and laser diode injection are managed via the RS232 port on the control PC. Tx and Rx signals RS232 send bit patterns with voltage levels of +15 V or -15 V. Since the RF generator circuit needs TTL levels (0 V - 5 V), a RS232-TTL level adaptation is implemented through a MAX232 integrated circuit. The flip-flop CD4027BM transforms the last bit of the TTL data frame provided by the MAX232 to a fixed voltage level (0V or 5V) which enables

Fig. 16. AOD used for experimental verification

(5V = logic 1) or disables (0V = logical 0) the generation of RF waveform and turning on / off the laser diode transmitter.

Fig. 17 shows the output light beam after passing through the AOD when the RF signal is activated. As predicted by theory, the diffraction efficiency of the diffracted beam depends on the geometry of interaction between the beam and the AO material. In the case of normal incidence (Raman-Nath regime, Fig. 17 (a)), two diffraction orders occur, one on each side of the central DC spot corresponding to the non-diffracted light beam. Rotating the AOD to get

(a) Raman-Nath regime

(b) Bragg regime

Fig. 17. Output beam diffracted by the AOD under different interaction regimes

Regime	I_i (μ W)	I_i (n W)	DE (%)
Raman-Nath	4.55	68	1,5
Bragg	4.55	98	2,2

Table 2. DE measured on Fig. 17 experiments

the Bragg angle, the Bragg regime shown in Fig. 17 (b) is obtained, and the single diffracted beam reaches its greatest intensity, as shown in Table 2.

6. Conclusions and further work

The work presented in this chapter aims to explain some optical phenomena from an educational perspective. The tools implemented combine virtual and real experimentation, and they can be used as laboratory material for technical studies in science and engineering. All the modules described in this chapter share in common the diffracting behavior of light. Nevertheless, there are other modules available (*i.e.* Moiré effect, Radiometry and Photometry) or under development (*i.e.* Theory of color, Geometrical Optics).
To date, these software tools need to be installed on the PC. Nevertheless, there is an on-going project to run them directly from a web browser. Other further improvements include:

Computer Generated Holograms tool:

- Implementation of new computing algorithms.
- Interaction with a Spatial Light Modulator to dynamically display the calculated CGHs.

Acousto-optics tool:

- Integration of a photodetector within the Laboratory module for measuring and plotting the diffracted light beams.
- Optical design to reduce the total optical path and improve the detection at the output plane.
- Characterization of the AOD (bandwidth, rise-time, etc.) by modulating the 24MHz RF waveform.

7. References

Brown, B. R. & Lohmann, A. W. (1966). Complex spatial filtering with binary masks, *Applied Optics*, Vol. 5, Issue 6, 967 – 969.

Carnicer, A. (2010). *The JOptics Course*, `www.ub.es/javaoptics`, University of Barcelona.

Carreño, F. (2010). *Group of Teaching Optics*, `www.ucm.es/info/opticaf`, University Complutense of Madrid.

Hecht, E. (1998). Optics, Addison-Wesley.

Kirkpatrick, S. (1983). Optimisation by simulated annealing filtering with binary masks, *Science*, Vol. 220, 671 – 680.

Levy, U; Marom, E. & Mendlovic, D. (1998). Modifications of detour phase computer-generated holograms, *Applied Optics*, Vol. 37, Issue 14, 3044 – 3052.

MATLAB - The Language Of Technical Computing, `www.mathworks.com/products/matlab`.

Morrison, R. L. (1992). Symmetries that simplify the design of spot array phase gratings, *Journal of the Optical Society of America - A*, Vol. 9, Issue 3, 464 – 471.

Poon, T.C. & Banerjee, P.P (2001). *Contemporary Optical Image Processing with Matlab*, Elsevier Science.
Saleh, B. E. A. & Teich, M. (1991). *Fundamental of Photonics*, John Wiley & Sons.
Samus, S. (1995). *Computer Design and Optimisation of Holographic Phase Elements*, Ph.D. Thesis, The University of Edinburgh.
Wyrowsky, F. & Bryngdahl, O. (1988). Iterative Fourier-transform algorithm applied to computer holography, *J. Opt. Soc. Am. A*, Vol. 5, 1058 – 1065.
Yaroslavsky, L. & Astola, J. (2009). *Introduction to Digital Holography*, Bentham Science.

16

The Visual Language of Holograms

Paula Dawson
University of New South Wales
Australia

1. Introduction

Any medium which represents three-dimensional space via a two-dimensional plane employs a representational system. Perspective, a system in which lines, which recede from the foreground to the horizon, create the illusion of depth by converging to the vanishing point has become the dominant means of representing space in western civilisation. Just like other representational systems, display holograms represent space in an abstract way through a series of conventions, which are defined by recording procedures and replay methods. The holographic representational system is so radically different from perspective, enabling the subject to appear in front of, straddling, or behind the hologram plane, that at times it seems invisible to viewers who believe they see the subject by a kind of magic. Even astute theorists, Umberto Eco (Eco,1987) who saw holograms as *hyperreal*, and Jean Baudrillard (Baudrillard,1997) who thought of them as *simulacra*, were implying that the hologram did not have a representational system, that it was a replica of reality. Though holograms can have a high level of spatial verisimilitude, this is variable, as is the resolution, spectral range, blur and the shape and location of the viewing zone.

A primary aim of this chapter is to introduce the properties of the representational systems of several hologram types, including HOE holographic optical elements, laser transmission, white light reflection, 'rainbow' white light transmission, white-light transmission holographic stereograms and printed synthetic holograms. Secondly the chapter will trace how the expressive potential of pictorial qualities such as light and darkness, transparency, reflections, colour and impossible forms operate within these representational systems. Case studies by artists such as, Margaret Benyon, Rudie Berkhout, Brigitte Burgmer, Salvador Dali, Paula Dawson, Jacques Desbiens, Mary Harman, Dieter Jung, Eduardo Kac, Martina Mrongovius, Seth Riskin, Andy Pepper, Martin Richardson, James Turrell, Doris Vila and Sally Weber will be examined.

Occasional comparison will be made between the effect of using the same pictorial qualities - light, darkness, transparency, reflections, colour and impossible forms in holograms and in other traditional media. Thirdly, the chapter looks at some factors, which are likely to impact on viewer reception and interpretation of the visual language of holograms: the bodily experience of interaction with the holographic image, the physiological and psychological aspects of depth perception and other influences such as historic genres and the associations of related media.

2. Representational systems

By representing space one makes programmatic statements about the kind of relationship between subject and space that is at stake when one makes an image (Alphen, 2005).
The holographic image is capable of completely renouncing illusion... moreover holography confronts the principle of illusion perception with a totally new aesthetic effect and a new model of perception, whose special characteristic is the correlation of physiological perception and psychological perception (Zec, 1989).
Holography is not just a medium to explore 3D space. We now have a medium in which it is possible to explore our own perceptions, the subtleties of human awareness (Law, 2008).
The Broken Window (Fig. 1) by Canadian artist Jacques Desbiens evokes the spaces of Chinese scroll painting, perspective and the holographic representational system. (Desbiens, 2009) The viewer is confronted with a visual paradox by the holographic plate containing the scroll, which contains the window, which is traversed by tree-branch and moving leaves. The scroll is positioned at the plane of the holographic plate and rolls open and closes rather like a pair of theatre curtains. Though not continuously unraveling space by one side unwinding while the other winds up, the all-important function of the scroll in revealing and hiding sections of the image is clearly indicated. The windowpanes, also located at the picture plane of the hologram, are framed as an aperture to a scene.
The window on the right, framing the scene behind, evokes a perspectival notion of the image as a window onto a world. Whereas the window pane on the left is broken and through it comes a more beautiful and intense light, a protruding branch which extends from behind the picture plane into the viewer's space and a floating leaf. These elements, the broken picture plane, the strong presence of light and the spatial and temporal distribution of the image, are key points to the holographic representational system.

Fig. 1. Jacques Desbiens, *The Broken Window*, 2006, Printed Synthetic hologram, computer graphic (not the actual hologram) of the 1100th point of view (total of 1280 points of view) 140 x 38cm.

However, the functioning holographic representational system of this synthetic reflection hologram, which enables the viewer to see the subject matter of the scroll, the window the branch and falling leaf, is not directly visible. To conceive of this representational system, like any other, it is necessary to strip away the subject matter to reveal its underpinning structure.

Beaudrillard's and Eco's discourses, which set the tone for the appreciation of display holography, do not describe the specifics of holographic representational space. Rather they describe the effect as a *simulacra* or a part of a kitsch trend towards the *hyperreal*:

...as you shift your gaze you can see those parts of the object that you were prevented from glimpsing by the laws of perspective... Holography could only prosper in America, a country obsessed with realism, where, if a reconstitution is to be credible, it must be absolutely iconic, a perfect likeness, a "real' copy of the reality being represented (Eco, 1987).

The hologram simply does not have the intelligence of trompe l'oeil, which is one of seduction, of always proceeding, according to the rules of appearances, through allusion to and ellipsis of presence (Baudrillard, 1997).

Underpinning the arguments of simulacra and hyperreal is the assumption that a hologram replicates or simulates reality, leaving no room for cultural interventions such as expression and abstraction, or a representational system, and by extension no potential for conceptual content.

By contrast, from the above example, *The Broken Window*, it is clear that the hologram is not necessarily a replication or simulation of a physically real thing, that its referents can be highly conceptual and in fact engaged with the very nature of its representational capabilities. The potential offered by the synthetic hologram's representational system clearly extends the scope of the functionality of the hologram well beyond *simulacra* and *hyperreal*. The representational systems of laser transmission holograms, holographic optical elements, one and two step reflection holograms, rainbow holograms and white-light transmission stereograms each operate in a different way and therefore make available other types of compositions. But before examining the specifics of holographic representational systems it is useful to first establish the basics of representational systems in general. These are firstly their formal abstract properties, secondly the way in which these interface with the beholder and finally the meanings which are ascribed to them.

Fundamentally, representational systems are abstract codified means by which spatial and temporal information can be represented and interpreted. They underpin the majority of the images we see. Representational systems can be employed in hand-made works such as drawings and paintings, or through optical/digital photography, film, computer generated environments, video and diffractive holograms. Depending on the representational system used, the same subject matter can have quite a different appearance. For example, visually, the image of a house using the representational system of perspective would differ from the same house rendered in oblique projection, because in perspective the rules of abstraction dictate that the lines running away from the viewer will converge to a vanishing point and in oblique projection these lines remain parallel. Both achieve the objective of abstracting the three-dimensional space in such a way that the viewer can deduce that the two-dimensional image signifies a house.

In each case the representational system shapes the way in which information about the subject will be formed into an image. In some cases the properties of one representational system can be used across many media. Perspective for example can be applied to drawings, painting, computer graphic environments and holograms. Traits of some representational systems can also be quoted or appropriated in others.

We have seen that there is a range of examples of holograms, which engage in a didactic or overt way with other representational systems, such as *The Broken Window*. Early optical holograms such as Bridget Baumer's work *Leonardo's Baby* a "Denisyuk" reflection hologram employs a physical model based on a line drawing by Leonardo da Vinci from the Codex Atlanticus. This work is engaged with the representation of anamorphic distortion, one of the artifacts of perspective within the holographic diffractive environment.

...the hologram is distorted and the plastic anamorphosis grows enormously in the pseudoscopic, so that it becomes more anthropomorphic than the model (Burgmer, 1987).

Also concerned with re-stating perspective space, but within one step laser transmission holograms, is the work by Salvador Dali, *Card Players,* which refers to the spatial composition of a painting by Diego Velasquez by employing physically built sets and live actors. Using laser transmission holography, Margaret Benyon in 1969 made a work entitled *Picasso* in which the elements of *Les demoiselles d'Avignon,* considered the essential lexicon of cubism, were represented by cardboard cutouts of the subjects of the painting placed at angles to a physical cube. As Coyle and Hayward remark:

Cubism provided one particular starting point in this area and [Benyon] responded to Cubist attempts to render three-dimensional materiality without recourse to traditional painting techniques of perspective or colouristic logic (Coyle, 1995).

Andrew Pepper used the oblique projection system to make a projected drawing, *Drawing 1 (from a series of 5)* which is recorded as an optical reflection hologram. David Pizzanelli transposed the animation process of Muybridge's original gravure photographic prints by alignment of side by side photographic views in a mirror-backed holographic stereogram on glass, *Bruno Walking to the Left,* 20 x 13cm, 1989. (Pizzanelli, 1989) When the beholder walks left to right Bruno walks to the left. When the viewer walks right to left Bruno walks backwards.

Muybridge was working several years before the invention of the cine camera, so in order to expose his plates in quick succession a number of individual cameras were triggered by the subject's passing laterally in front of each camera in turn, recording not just the motion but different perspective views of the subject...synthesized into a single three-dimensional animated scene, showing all the depth and solidity of the original event (Pepper, 1995).

The visibility of the sprockets and frame numbers in Patrick Boyd's work *Bartus takes the Downtown train* (Fig. 2) alludes to the illusion of time being generated though the series of still frames of the super 8 film. As the viewer moves across the train moves either in or out along the z axis.

The purposes of representation systems vary enormously. Some generate illusions. Others transpose salient factors of the subject to an image, which clearly can be recognized as referencing the subject, but in an abstract way. The representational systems of Cubism or a plan view do not entail what we commonly call illusions, yet they refer effectively to the subject in a symbolic way. All representational systems in some way address and reciprocate perceptual and conceptual capabilities of the beholder, for it is through these that the image is readable.

In humans the information for perception of spatial environments comes from approximately nine sources - occlusion, relative size, relative density, height in the visual field, aerial perspective, motion parallax, binocular disparities, convergence and accommodation. The relative importance of these has been shown to vary according to the size of the space the beholder occupies (Cutting, 1995). In examining holographic representational systems of different types, which operate within the zones of personal space or action space, it will be seen that some cater more to certain source types of three-dimensional information than others.

Fig. 2. Patrick Boyd , *Bartus Takes a Downtown Train*, 1990, two colour reflection hologram in a box, 10" x 8"x 3".

The diverse motivations behind the early development of the various range of holographic representational systems has been thoroughly documented by Sean Johnston (Johnson, 2006). Though varied, the underlying commonality of "display" holographic representational systems is that they generate new types of pictorial space in which representation can take place. The technical basis of these holographic representational systems is extensively set out by Benton, Bove et al in *Holographic Imaging* (Benton, 2007). These techniques enable unique arrangements of the viewing zone in respect to the support (holographic plate) and the image of the subject.

Ernst Van Alphen in the chapter The Representation of Space and the Space of Representation has argued that :

The space of representation is certainly not a fixed entity. The history of art can even be seen as a sequence of changing conceptions of the space of representation. There have been periods in which representational space was defined as illusionistic space; in polemical reaction to this many twentieth century artists have devoted their careers to fighting illusionism and have instead explored the flatness of the painted image (Alphen, 2005).

The three kinds of representational spaces are identified by Van Alphen as operational in two dimensional and shallow relief works, namely: the illusionistic space behind the picture surface; secondly the two-dimensional space of the picture surface and thirdly the lived space of the beholder between the image and the viewer. To describe holographic representational systems additional elements are needed - the replay light, the image space in front and through the support and also the viewing zone. The holographic viewing zone, (the volume of space in which the viewer must be located in order to see the image) is rarely the same as the volume of physical space in front of the holographic picture plane that the beholder can occupy. The beholder, as is customary, looks towards the support (holographic picture plane) in order to see the subject. The holographic image may appear to extend behind and or out into the viewer's space but it can only be seen when the beholder is inside the prescribed viewing zone.

Van Alphen has argued that though perspective's requirement of stability of the viewing position, which causes the viewer to become stationary and fixed to the ground, limits the viewer's response, compositions which bring attention to the surface exclude the viewer from pictorial space. The limited nature of the *ideal* viewing zone, or point, for perspective imagery does not prevent the beholder from seeing the image from a complete range of angles because the image and its support occupy the same two-dimensional space. However due to the physical properties of the hologram there is a limit to the range of angles through which light can be diffracted to produce an observable image and so if the beholder of a hologram walks outside this viewing zone, though still looking at the holographic picture plane, they will not see the subject at all. The beholder becomes conscious of the shape of the viewing zone and its boundaries when its edges occlude the subject, inside the angle subtended by the viewer's eyes and the perimeter of the plate.

The following examples demonstrate the way in which aspects of holographic representational systems such as the viewing zone, the hologram plane and the position of the subject are used in conjunction with other pictorial agents to convey ideas. These are works which must be experienced, as they mobilise the beholder in a complex interplay of embodied engagement over time, with spatially distributed visual information.

3. Holographic Optical Element (HOE)

The holographic optical element (HOE) diffracts light as though it had passed through some type of optical device such as a lens, focusing, fragmenting, expanding or collimating the light. The diversity of forms available with HOEs is so extensive that it is not possible to generalize however the two following examples give a sense of the scope of this representational system.

Seth Riskin *Figure With Crowns* (Fig. 3 and 4) employs the fundamental relationship of the replay light of the hologram to the support (the holographic plate) in determining the nature of the reconstructed image, to generate a dynamic performance.

Holograms are different from other image types in the way in which the image is formed at the moment of viewer observation. The image is frequently referred to as a "reconstruction" shorthand for the term wave front reconstruction, the event in which a wavefront of light is changed usually from being quite simple to complex through the process of diffraction. It is significant that all holograms are seen by the diffraction of light from a replay source from an interference pattern. The distance, orientation and wavelength of the replay source effects the way in which the image appears. The same diffraction pattern could produce quite different qualities in an image if the lighting used in replaying were to be altered. These changes in the image can range from a complete change in colour, to slight shifts in spectral range towards reds or blues, changes in scale of the subject from large to small or vice versa, the introduction of distortions and appearance of the image as virtual or pseudoscopic (inside out). Such changes to images which are painted, photographed or drawn cannot take place as these images and the surfaces to which they are recorded (the support) are synonymous.

Fig. 3. Seth Riskin, *Figure with Crowns*, 2002, HOE, installation, Sky Art Conference, 24" x 20", Ikaria, Greece.

"Display" holographic representational systems are seen though the conjunction of a light and some sort of screen or plate, which holds the interference pattern. These two elements, one intangible and the other imperceptable, the light and the interference pattern, are employed sending the diffracted light away from the screen or plate to other locations to form the image. In addition, there is residual light passing through or being reflected from the surface of the plate, depending upon on whether the hologram is played back with light passing from behind (transmission) or in front of the plate (reflection). Because the replay light is redirected by the interference pattern distributed within the volume of the photosensitive material attached to the substrate,which in the case of painting or drawing would be called the support, it is necessary for the viewer to look in the direction of this plate or film to see the image. The subject of the holographic image may appear to extend out into the viewers space but the subject can only appear in locations, which fall within the angle subtended by the viewer's eyes and the perimeter of the plate. For this reason the hologram viewing space constantly changes shape according to the position of the viewer. In most instances the replay beam is stationary and the beholder moves, whereas in this case the reference beam source, the laser diode, is attached to the moving performer who is viewed simultaneously with the changing holographic image he is influencing.

Fig. 4. Seth Riskin, *Figure with Crowns* , 2002, HOE, 24" X 20", Installation, Sky Art Conference, Ikaria, Greece

In Riskin's *Figure With Crowns* performance the transparent hologram support is held on a tripod, which is placed at a vantage point overlooking the shoreline. The performer is up to 30′ away from the hologram support on the shoreline. The performance and viewing of the HOE takes place at sunset to enable a mixing of the orange and golden hues of the natural environment with the blue monochromatic subject of the HOE. As described by Riskin:

The performer casts an enlarged beam of blue light onto the hologram as viewers look through it onto the figure, the shore and the sea. In the eyes of the viewer, two rings of blue light hover, flanking the body, "attached" to it as the figure moves in the landscape (Riskin, 2011).

The spatial locations of these discs are in a plane with the light source which is attached to the performers body and appear as extensions of the performers body. The subject of the HOE is therefore not a spatial object but rather it embodies the changing spatial relationship between the beholder and the performer through the translucent subject.

Alignment, by Sally Weber, is a HOE that projects blue, green and red vertical lines of light 6 to 8 feet in front from a 7-foot high curved acrylic structure. In the following discussion the z axis is a line perpendicular to the plane of the holographic plate. The beholder can mix colours and alter their intensity by moving towards and away from the acrylic support within the long z-axis viewing zone. The image of the coloured bars is available at every height on the y, (height) axis. An important characteristic of this viewing zone is the change in the subject, which is perceived as the viewer's distance from the holographic picture plane changes along the z-axis. But most importantly the beholder can stand *in* these bands of light with the sensation of the light being at the eye. When this physical shift occurs a phenomenological transition takes place in the perception of viewing from exterior to interior. The light is not shone on the beholder, rather it is perceived as being inside the body of the beholder. The product of the large freedom of movement of the body in relation to the coloured bands is not so much the experience of entering a rainbow as of having a rainbow enter the body.

When the piece is viewed from a distance, the focal lines drift and merge as the viewer moves from side to side in front of the sculpture until he or she is aligned with the three lines. At this point the colours blend towards an achromatic balance. This is a calm place, a balance point, where harmony is symbolized by a shift towards whiteness. Moving closer, the viewer sees the blending and shifting off the hues become more vibrant. When the viewer is aligned with the lines at the focus of the holographic image, he or she is surrounded by whirling, brilliant fans of colours that flow across the curve of the sculpture...Standing in this axis of light, the viewer experiences a physiological response as the light seems to rush upward through the spinal chord (St. Cyr, 1989).

Perhaps the only other artwork to achieve the sensation of a light image existing inside the body is the projected poetic words by Centre for Advanced Visual Studies MIT, Fellow, Elizabeth Goldring, which employ a scanning laser opthalmoloscope to draw imagery directly on the retina. The beholder is tethered to the apparatus and therefore the light can only exist as an image inside the eye, unlike *Alignment* in which the beholder roams between experiencing the holographic image as an exterior and as an interior phenomenon.

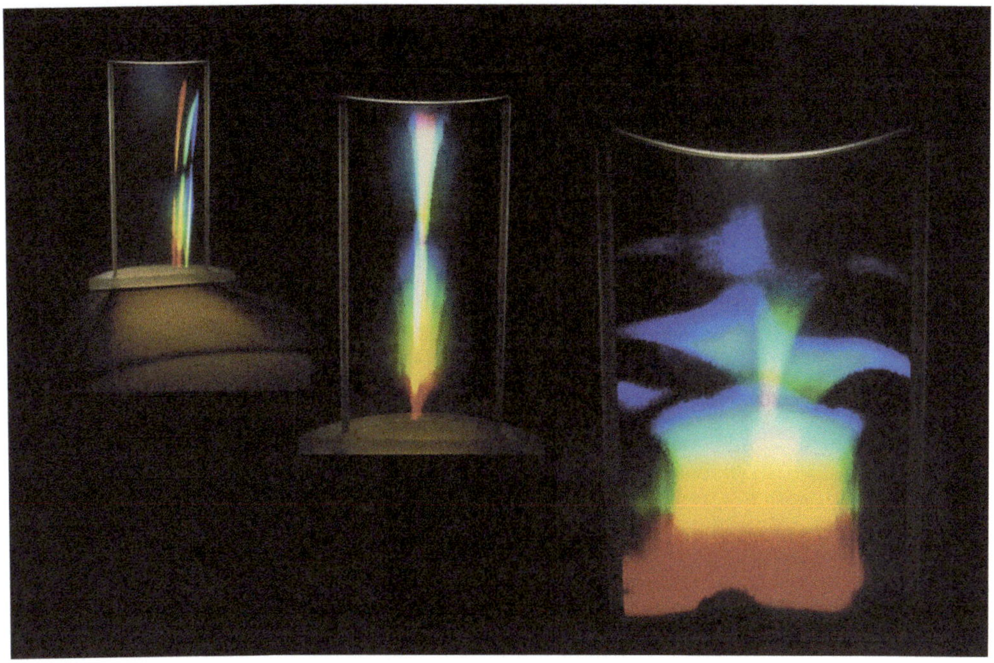

Fig. 5. Sally Weber, *Alignment*, 1987, HOE, 7′x 3′

Peter Zec in his paper for the International Congress on Art and Holography 1991 commented:
As light is not only a generative principal but also a subject and the basic substance of the holographic reconstruction as well, the self -referencing of light represents an essential form for the articulation of the holographic message (Zec, 1990).

4. Laser transmission hologram

The space represented in the laser transmission image extends from the entire surface of the holographic plate back to the extremities of the subject, exactly in the same way that a windowpane acts as a boundary to a view. This is also true for one step optical reflection holograms.The free viewing zone is the same as viewing an object from any vantage point in through a window, the viewing zone is 180 degrees horizontally and vertically from the plate. The image can be seen just as easily from near or far. The space is unfocused and undistorted - it is an exact spatial replica of the subject and elicits perceptual cues of the physiological type which are precisely in accord with the perception of real objects.
The one- to- one scale of the representation of space occurs at a resolution higher than is perceptible by the human visual system. Other representational systems use scaling, whereby the original image is larger, distorted or smaller than the original subject. All the sources for information of depth come to the beholder as they would for a real world object separated by a piece of glass, in particular the physiological cues of accommodation and fusion which are critical at close range. It stands to reason that a key effect of the one step optical transmission holograms representational system is a strong phenomenological link to its real-world referent.
The dominant textural property of laser transmission holograms is the grainy quality of laser light known as laser speckle. This minute dotting in three-dimensional volume lends a somewhat worn, aged, yet constantly mobile micro surface to all image elements. A correspondence between physiological operations of the eye in the perception of laser speckle and the need for alterations in aperture of the camera in documenting laser transmission holograms can be deduced from Graham Saxby's explanation:
Laser viewable holograms are particularly difficult to photograph successfully owing to the presence of obtrusive speckle and to the very high contrast they often possess… the speckle size is large and more obtrusive for small lens apertures and in order to minimize the effect it is important to use the largest possible lens aperture (Saxby, 1988).
To the beholder it may seem that the real-time movement of the laser speckle directly reflects the constant shifting of their eyes. This gives the effect of a deep silence, which Karl Frederick Reuterswald refers to in his finger language - the silent movement of fish in an aquarium.
Holographic space has its own time and depth; yet it also has a profound silence, as in an aquarium. To me the silence of the aquarium is more impressive than the colour or the movement of the fish (Reuterswald, 1989).
Laser transmission holograms are generally monochromatic and replayed in the same wavelength in which they were recorded though they can be full colour by using a combination of lasers to record and replay.

4.1 Boundary of the holographic picture plane

Because laser light is less bright than the typical level of gallery lighting this draws the viewer to the holographic plate. In the case of large format, large-scale holograms in which the reconstructed image is confined to the space behind the plate, viewers physically lean onto the plate in order to look at the image more closely. The result of the image being very spatially realistic and also behind the picture plane is to necessarily heighten the feeling of exclusion from the scene depicted. The holographic picture plane is not always interpreted as a physical boundary, except in one -step optical, and laser transmission and reflection holograms.

On a physical level the boundary of the plate can cause the viewer to want to occupy the space of the image or a conceptual interpretation could be that the image is a metaphor for an aspect of memory- imagery being present yet physical accessibility being blocked. The interaction of the spectator with the subject with some holograms, which have the image in front of the plate may relate to the way Alex Potts has characterized the work of the sculptor Richard Serra, in contrast to the way that this type of viewing zone behind the plate invites a response analogous to the work of other contemporary sculptors:

Serra was well aware that physically involving the viewer ...in an interplay between exteriority and interiority involved not just formal or perceptual effects, but carried a certain psychological charge...but this psychological dimension was a muted one as he conceived it. It is striking how his work which allows the viewer to enter produces a much less psychologically charged sense of interiority than works such as Hesse's or even Judd's where the viewer feels drawn towards an interior space from which he or she is barred, and feels it to be strangely alluring and inviting yet separate (Potts, 2000).

Additionally the distance over which the eyes are able to travel in the x, y and particularly the z-axis has a bearing on beholder reception. Laser transmission holograms can depict scenes up to approximately ten meters in the z axis such as *To Absent Friends* by Paula Dawson (Popper, 1993). The experience of looking towards a distant space can be similar to daydreaming. The size of the support (holographic plate) also determines the distance at which the beholder will decide to stand. As has been discussed, resolution also plays a role in selecting a suitable viewing distance. Therefore z axis movement of the beholder before the hologram picture plane, similar to the viewing of two-dimensional imagery, is primarily dependent on the size of the support.

4.1.1 Darkness

Darkness, which has always has been a pictorial device of great importance in traditional media has also frequently been employed by artists to create ambiguities in optical holograms. Michael Baxandall in *Shadows and Enlightenment* pointed to an example of a drawing of a Roman soldier by Tiepolo in which the overall lighting pattern has nested within it smaller zones in which the lighting behaves in a completely different way (Baxandall 1995). This enables the viewer to make two possible interpretations as to where the leg of the soldier is located. This ambiguity as to the position of the leg engages the viewer in perceptual switching. In Margaret Benyon's, one- step laser transmission work *Hot Air*, 1970 laser transmission hologram 20 X 25cm, the beholder sees within one composition still - life objects, which display a full range of tonal values, and also a hand-shaped black void area. Similarly to the Tiepolo example, there is one overarching lighting arrangement of the objects in the scene, however, within this arrangement is the totally black hand, which is inconsistent with the overall pattern. The primacy of the artist's presence being registered through a negative rendering of their hand is immediately brought to mind, but in reverse, for the precise volumetric rendition of the hand instead creates a poignant sense of absence through the three-dimensionality of the trace.

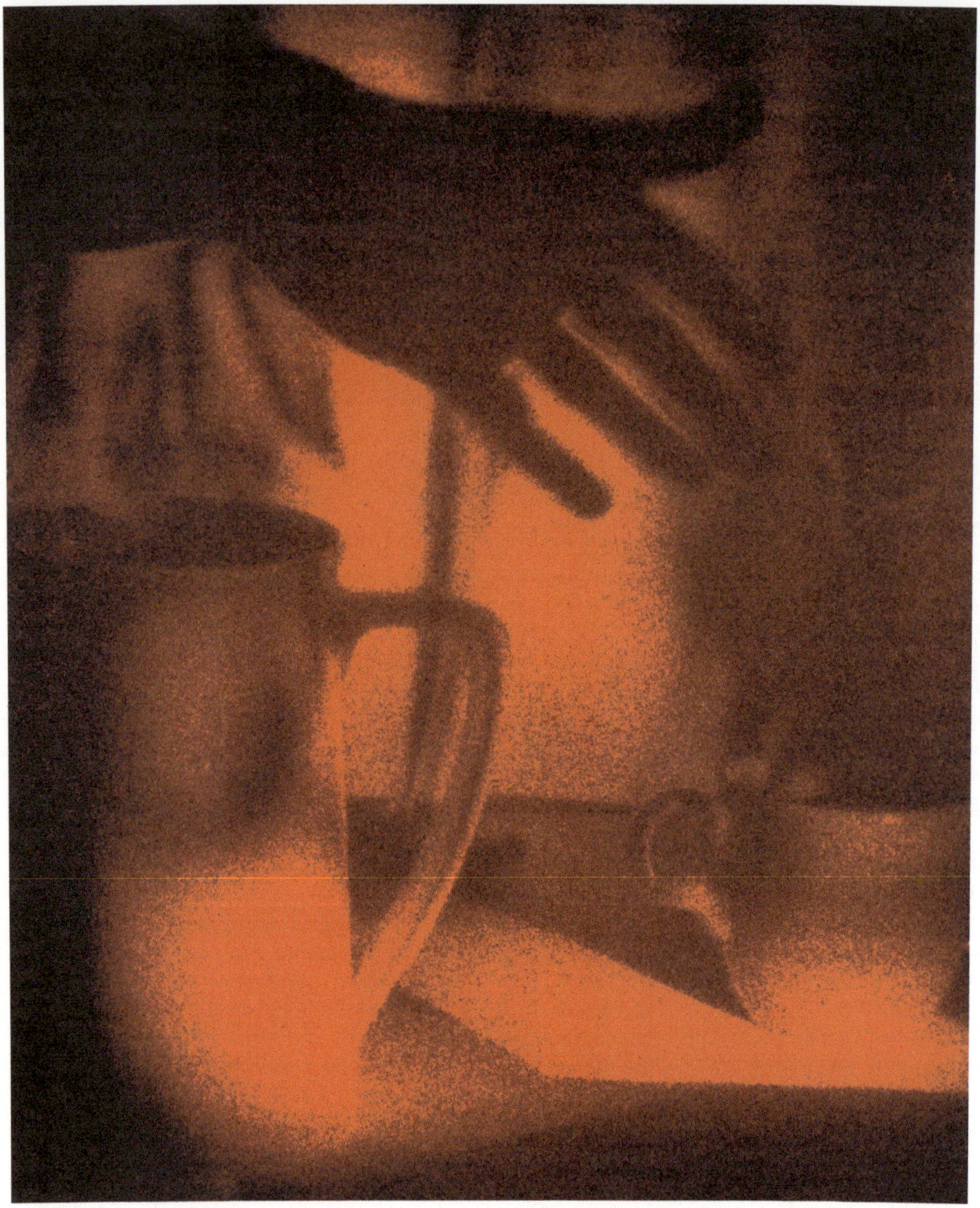

Fig. 6. Margaret Benyon, *Hot Air*, 1970, laser transmission hologram, 20 X 25cm

Although the hand-shaped black hole in this hologram has little visual detail, it subjectively captures the essence of the hand, which was once there and stimulates questions about the owner of the invisible flesh…something beyond the 'object', something which is intangible and untouchable - the roots of an idea or conceptual improbability (Pepper, 2008).

The six panels of *Working Model One*, laser transmission holograms by Paula Dawson, each show a plaster frame straddled by a curved section of plaster cornice. The lighting of the overall scene of the plaster cornice frames seems consistent, however, as the viewer moves, a darkness, which cannot be ascribed to shadowing, also moves over the three-dimensional space of the curved piece of plaster. This unusual soft edged zone of darkness traveling in unpredictable ways through the image volume lends a sense of mystery. Dawson achieved this effect by applying the holo-diagram, a graphic invention by Nils Abramson for plotting physical zones of the repetition of laser coherence and incoherence (lightness and darkness) (Abramson, 1981). Abramson's application of the holo-diagram enabled him to avoid the dark zones and thereby record a large piece of machinery all in light with a short coherence laser whereas Dawson intentionally located the zones of darkness in a different way.

In order for the interference pattern to be recorded as a hologram, all elements involved in the recording process including the optics and the subject must remain absolutely static throughout the duration of the exposure time. Unless a pulsed laser is used, a living hand cannot be static to the required extent. Benyon's effect was achieved by intensionally using the long exposute of a continuous wave laser to allow the stable inanimate objects to be successfully recorded as bright light forms and the living hand to not be recorded and therefore become dark. Darkness of a holographic subject is generally regarded as a mistake. Saxby's fault- finding chart proposes that the only way to get around the problem is to "use a more rigid subject" (Saxby, 1991). As Benyon points out in discovering means for manipulating this representational system:

The physicist tries to get an approximation close to reality, to eliminate properties of the medium that distort, whereas it could be exactly those properties that make it unique to the artist (Benyon, 1994).

5. Reflection hologram – one and two step

Andrew Pepper in his article *Holographic Space: A Generalised Definition*, conceptualizes the spaces available for the subject of a hologram as being box like, behind the picture plane "Box 1", in front of the holographic picture plane "Box 2"and then a third space which spans these two. As Pepper notes:

When objects in the hologram are in a position between "Box 1" and "Box 2" they straddle the flat holographic plate. Because we know that a solid object cannot pass through a solid holographic plate, we visually ignore the plate and are no longer confined by it.

The previous examples of HOE's were either "Box 1" Seth Riskin or " Box 2" Sally Weber. Similarly, one step laser transmission imagery of Margaret Benyon and Paula Dawson occupied "Box 1" (Pepper , 1989).

Andrew Pepper's one-step optical reflection hologram protruding drawings are a perfect example of the opening of the third type of subject space, one that traverses the support (holographic picture plane).

Pepper explains the process:

The drawing series was produced using a simple single beam Denisyuk system. The 'drawings' (isometric projections of a cube), were placed into the single reference beam, causing its shadow to fall onto the holographic plate, pass through it, and 'fall' onto the featureless white background situated behind the holographic plate. Once processed, the hologram was rotated around its horizontal axis so the shadow of the cube (originally visible behind the holographic plate) becomes visible as a pseudoscopic image, protruding out of the hologram and manifest in the space between the observer and the plate. The drawing exists in two places simultaneously: one on the surface of the holographic plate; another slightly in front of the plate. The two' images' naturally interfere with each other visually and spatially (Pepper, 2011).

Fig. 7. Andrew Pepper, Drawing 1 (from a series of 5), 1987, "Denisyuk" one step reflection, 20.3 x 25.4cm.

Drawing A, 20.3 x 25.4 cm, sets up a dialogue of paradox with the beholder by using a two-dimensional drawing which employs the representational system of oblique projection to represent three-dimensional space in two dimensions as its subject. Though the image is lit from the front, a shadow hovers above the plate. It is the shadow of the drawing seen in the hologram. The shadow, which can be seen from almost any possible viewpoint, occludes the drawing giving the existence of the shadow a priority over the image. It is in fact the three-dimensional location of the shadow, which is so confounding. The beholders sense of illusionism is challenged, partially as illusions of space are usually linked to the representation of three-dimensional objects. Here the subject is completely flat, a drawing, so in a sense there is no spatial illusion.

The historical antecedent of this work is Jan Dibbets', *Perspective Correction (My Studio1, 1-Square on a Wall)* 1969 black and white photograph on photographic canvas 110 X 110cm. Dibbets draws a distorted shape on a wall of his studio which when photographed becomes a square. The drawn square appears to have separated from the wall within the foreshortened photographic space and rests on the picture plane of the image.

5.1 H1, H2

One method for producing the effect of the subject passing though the box 1 and box 2 spaces is to make a hologram (known as H2) of the pseudoscopic image reconstructed from another hologram (known as H1 or Master hologram), the two steps of the process being first the making of the H1 and then the recording of the H2 from it. When the H2 is produced, the subject position is determined by the placement of the H1. When the beholder observes the subject in the H2 they do so by looking through the image of the H1 view port. The size of the H1 support becomes the aperture size of the viewing zone of the H2. The subject can appear on both sides of the holographic picture plane.

Mary Harman's *Figure with Metal Base* 1992 is a two step reflection hologram of a hand-made miniature reclining figure which is projected through the holographic picture plane to match up spatially with a highly reflective metal cube. The cuboid viewing zone restricts the viewers point of view, encouraging them to focus on the highly reflective cube placed behind the holographic picture plane. As the hologram is monochromatic the subject seems as a figure made from light but with a convincing spatial presence though it passes through the boundary of the hologram plate. The awkwardness of the figure pose and its lack of physicality are at odds with the materiality and formality of the cube, and this creates a tension. The close space of the viewing zone almost grotto like, draws the beholder into the miniature world. As the beholder moves, the translucent figure is not occluded by the physical metal cube instead it appears to pass through it.

...the aesthetic effect of holographic space is based on the loss of materialization rather then on the addition of a third dimension (Zec, 1990).

In the *Cosmetic Series: Margot* Margaret Benyon employs the small rectilinear viewing zone through which the three- dimensional portrait can be seen in two ways. The limited viewing zone acts as an extended spatial frame, which contains the movement of the viewer, encouraging a face- to- face orientation to the subject who returns this frontal gaze. The effect of moving sideways out of the spatial viewing zone is that the image moves in the z axis from the three-dimensional space behind and in front of the holographic picture plane to its two dimensional flat surface. This is accomplished by having a gouache painting of the subject directly behind the hologram plane. The beholder can see the painting of the subject from any orientation but it is only when the beholder is directly facing the subject that the subject projects forward towards the beholder. A constant dialogue between the reflected and diffracted colour of the two and three-dimensional surfaces are encouraged by the use of painting, both on the surface of the subject's skin and what appears to be on the holographic picture plane.

The holographic images are frontal and central, partly to give a classical, hieratic quality...a spatial image had to be merged with a flat image on the image plane, and a full frontal view is flatter than any other orientation of the head (Benyon, 1989).

Benyon's work employs a synthesis between the monochromatic hologram and its underlay of a brightly coloured gauge painting. Full colour, either highly realistic or abstract through the use of multiple exposures with different colored lasers or by swelling the emulsion to different thicknesses between exposures is also a feature of reflection holograms.

The latter technique,'pseudo colour' was employed to great effect in *Healing of Broken Hearts* (1985) by Melissa Crenshaw. Rather than hanging on the wall the work is mounted in a display case so it can be viewed from above. This evokes in the beholder the method of looking at precious items in a display case. The play of light over the intensely coloured surfaces is reminiscent of light playing on insect wings, or other Lepidoptera, a favorite subject of this type of hologram to demonstrate verisimilitude. The colours of Melissa Crenshaw's work are highly saturated, green, blue, violet, red and black and the composition remains static as the viewer moves. The formal arrangement of the colour along diagonals which intersect with a broken circle also encourages the beholder to circle the plate and interpret the composition from multiple view points.

Crenshaw introduces us to a plastic, lineal and geometric holographic universe inspired on the first third of the twentieth century's constructivist and neo-plasticist avant gardes, a field restituted here by light. Colour acquires a structural and emotive function in an obvious way (Carreton, 1996).

6. Rainbow – white light transmission hologram

In this representational system the viewing zone is much smaller than for optical one- step holograms or HOEs. The viewing zone is of appreciable angular spread in one direction only in one direction, usually horizontal; in all other directions is is very narrow, only a few degrees.

Also there is a specific distance along the z-axis between the plate and the viewer at which the image seems best. As the viewer walks towards the holographic plane they tend to stop at this point, well away from the support. The basic characteristic of the image is that the subject appears in 3 D only in the horizontal plane, known as HOP horizontal parallax only. The image presents spatial information about the subject which accords with motion parallax, stereopsis, but not with the physiological cues of accommodation and fusion. The image is astigmatic, and, except in the case of the achromatic rainbow hologram, the colours change when the viewer moves along the y-axis.

Once the viewer walks out of the viewing zone the image appears to wipe off the plate. The viewer usually turns back and walks towards the other extremity of the viewing zone. Similarly, if the viewer moves above or below the narrow horizontal viewing aperture the image vanishes. The shape of this invisible aperture and viewing zone has been explained as a letterbox, a narrow wide slit aperture onto a volumetric scene.

It is not a straightforward matter for the beholder to locate themselves within this type of viewing zone. Sometimes a prescriptive approach has been taken. As Mark Diamond comments of early exhibitions at the Museum of Holography in New York.

...we did a lot of placing of little stickers on the ground... like those games where they show you how to tango, and where to put your feet; it was instructions on how to view the holograms, because people were unfamiliar with how to view the space (De Freitas, 2003).

These holograms are usually played back with a "white light" point source. Because of the dispersion of wavelengths over distance, when a white light source is used for replay, the image will be most focused at the hologram plane and become less focused in both + and - z.. This effect is more noticeable with large depth images. The decrease in resolution the closer the image protrudes into the + z space in front of the holographic plate is contrary to the experience of aerial perspective where the resolution of the image becomes less as it becomes more distant. Therefore although the cue of aerial perspective is not consistent with real-world experience, in practice it has little significance, because this cue is only significant over long distances and typically rainbow holograms are of a size that fits within personal or action space.

Rudi Berkhout's three colour rainbow holograms such as *Future Memories, Event Horizon* and *Toba* which include HOEs are types of landscapes in which elements undergo spatial transformations as the beholder moves from side to side. There is a use of a combination of types of H1, some of which portray abstract forms with a constant position such as fields of bubbles and others, which warp space in unpredictable ways.

In *Event Horizon* I was able to generate an image element totally by optical means. This was a breakthrough for me, for the first time I was able to draw with light. Starting with a point of laser light, I stretched multiplied and curved coil-like structures that I placed over a sphere (from a second master) in the final transfer. The third master used in this piece was of a forced -perspective field made of receding lines (Berkhout, 1989).

The spatial stability of randomly dispersed elements provides an alternate framework to the grid for navigating the space. Though on supports of modest size, like landscape paintings, the beholder's experience is that of traversing an immense and complex space.

For the most part these artificial light realities oppose the onlooker's perspective and visual habits. One reason for this is Berkhout's particular partiality for apparently impossible spatial arrangements of geometric bodies and forms... here Berkhout speaks of the creation of hyperperspective spaces (Lipp, 1985).

The complexity of these compositions has the effect of prolonged beholder engagement.

...for me part of the uniqueness of the holography is that the image is simultaneously there and not there, depending on the viewer's position...being able to see the work from the sides and from behind allows the viewer to examine and consider all angles of the image (Berkhout, 1989).

6.1 Colour

The most striking feature of the rainbow hologram is that the same subject is seen in different colours according to the height in the viewing zone. The relationships of the different colours of separate subject elements are preserved, yet the colours themselves change. Colour behavior, because of its decoupling from stable relationships with surfaces of forms, probably cannot be processed by the brain in the usual way. As Zeki has shown the brain processes colour information in a very specific way:

The brain is principally interested in acquiring knowledge about the constant and invariant characteristics of a surface, namely reflectance. This it does by comparing the wavelength composition of the light reflected from it with the wavelength composition of the light reflected from its surrounds. By doing so, the brain is able to discard all the variation in the wavelength energy composition of the light reflected from a surface and assign a constant colour to it (Zeki, 2009).

If the colour of the form does not match the colour in memory then information is processed in another area of the brain. Zeki's results from functional MRI imagery of brain activity while viewing Fauvist paintings may serve as a model for understanding viewer response to rainbow holograms:

… I would like to draw another conclusion from the …survey of the Fauvist brain, which activates a distinct part of the monitoring system in the frontal lobes. I do not imply that this part is devoted to seeing Fauvist art, rather, it is an area that monitors the incoming information for any conflict with previous experiences (Zeki, 1999).

From these data it can be assumed that the way in which colour changes according to viewer position in rainbow holograms already activates the part of the brain which is concerned with abstract problem solving and resolving ambiguities.

In Dieter Jung's *Present Space* 1984 (Fig. 8.) the beholder sees an array of coloured horizontal bars of varying height and width- more widely spaced at the top and bottom (Fehr, 1991). The colour appears to exist in space without any material surface and is translucent. As the beholder moves from side to side the planes of colour overlap each other to generate new colour mixtures. There is no observable modulation of form of the colours in the z-axis. Jung makes sense of the wiping off of the image at the side edges of the viewing zone by locating two large black rectangles, one at the top and the other at the bottom of the composition, which are also empty.

The horizontality of the composition encourages a side-to-side motion in the viewer. The physical shifting of the beholder and the corresponding dynamic immaterial colour mixing of the composition is intended as an analogy to the "oscillation structure of the activity of our consciousness" (Fehr , 1991).

Doris Vila's *Chart: Space-Time-Sex-Money-Continium* monotype 1985 employs a one step rainbow hologram technique. The beholder of this very long 17 ½ x40 " image becomes immersed in layers of simultaneous narrative. The layers appear floating in the space between the viewer and the support, each of a different colour. The visual field holds in transparent layers overlapping stencils and found objects in multiple exposures; the translucency of these negative forms suspends them from concrete interpretation. The narrative metaphor is extended by the replacement of the usual white light replay source with a film loop of money falling into the image.

As Vila comments:

Most importantly the narrative line of the imagery was synthesized in the process of shooting, with many of the objects existing only for the moment of the shot; therefore the only record of their existence is the shadow they made at the moment of exposure. Some of the objects used were stencils of simply cut shapes that allude to natural phenomena, transparent forms of glass or plastic, fabrics and found objects…viewing the hologram is like adopting a memory, where each person's keys to the experience have a unique sequence (Vila, 1989).

Fig. 8. Dieter Jung, *Present Space*, 1986, Benton hologram, 42 X32cm.

7. White light transmission holographic stereograms

The representational system of the holographic stereogram is characterized by the spatial arrangement of an array of two-dimensional views, which correspond, to the position of the beholder relative to the subject across a horizontal field of view. Typical source image types for holographic stereograms are film footage, photographs and computer graphics all of which use perspective. A batch of any type of two dimensional perspective images is exposed firstly to the H1, which is then transferred as an H2 in order to locate the array of 2D images away from the holographic picture plane. Typically this process introduces many optical aberrations and the output stereograms have distortions. The sequences of 2D images can be of sequential side-to-side viewpoints of a static three-dimensional space or sequential side-to-side viewpoints of a dynamic space over time.

The first type of representation, of a consistent, static space can be seen in James Turrell Pace Wildenstein 2009 exhibition of "transmission light works". In these works the position of a single shape--a triangle, rectangle, ring or a conjunction of two forms, remains static in the centre of the composition. The stereogram is replayed with monochromatic light lending the form a specific hue almost reminiscent of Bauhaus (Kandinsky) and Russian constructivist (Malevich and Kliun) correspondences between hues and forms. The conditions for stereopsis and motion parallax are satisfied. The lack of occlusion of the subject gives it a less material feeling while holding a stable spatial position (Gallery, 2009).

The second type of imagery, that of a dynamic field of information necessarily deprives the viewer both of motion parallax and stereopsis as side-by-side views now carry information about objects in displaced positions. In order to be able to track the image elements the amount of motion has to be very restricted. Too much movement of the subject in the initial 2D images will result in "time smear" where the parts of the subject do not match up in side-by-side views and consequently cause retinal rivalry. Because the animation of the holographic stereogram is activated by the walking of the viewer side to side, the motion of the subject will be read in forward and reverse.

The fluid motion of floating text in Eduardo Kac's, *Adhuk* , 1991 encourages the viewer to walk right and left to locate the point at which the letters coalesce into a word. At this point the viewer pauses. Instead of the beholder observing the three-dimensional representation of a space, the imagery instead designates that the beholder will be in a specific space in relation to the subject and the hologram plane.

Language plays a fundamental role in the constitution of our experiential world. To question the structure of language is to investigate how realities are constructed. My holograms define a linguistic experience that takes place outside syntax and conceptualise instability as a key signifying agent (Pepper, 1995).

Similar to the rainbow hologram the optimum viewing position within the viewing zone for the beholder of the rainbow stereogram lies along a line, which is at the same height as the centre of the holographic plate. The imagery can appear behind, in front of, or straddling the hologram plane. The viewing zone generally extends to the right or left of the plate or film. The locations of the horizontal boundaries of the viewing zone are variable depending on how the hologram is made.

The confines of this type of viewing zone have been used to great effect in works by Martina Mongrovious. "*Pater Noster*", 300 X 400 mm stereogram reflection print, was first exhibited at the centre for the Holographic Arts in New York and then in the Stairwell Gallery Melbourne.

Fig. 9. Martina Mrongovius, *Jumping Jellies*, 2007, rainbow stereogram, 30 X 40 and 60 x cm

In both these installations I wanted the viewer to mimic ascending /descending in the image. The subject's (my) holding of the camera links the viewer's proprioception to the subject, establishing an emotional and imaginative connection of shared experience…(Mrongovius, 2009).

In *Figure 8* the beholder is propelled at an increase speed in an office wheelie chair along a horizontal line parallel to the plate at the optimum viewing distance and in *Jelly*, the beholder jumps on a trampoline to fly past the small aperture located on the vertical y-axis. Here there is a deliberate reference to the inability to see the stereogram except from a specific height.

Falling, running, jumping and skipping are descriptions of movements but also come with strong emotive connotations. In the *Jumping Jellies* installations it is the physical mimesis between your own sense of blubber and the jellyfish movement that establishes the image. The intention of this work is both playful and to illustrate how proprioception can create a reading through experience (Mrongovius, 2009).

8. Printed synthetic hologram

Printed synthetic reflection holograms are composed from thousands of perspective views of a static or moving 3D scene. The viewing zone for these images is deep, the images being readable from both near and far with no particular point along the z-axis from which they seems best. The subject can appear in full colour and holograms are played back with a white light source. The side- to- side movement available to the viewer within the viewing zone is wider than the holographic plate, approximately 25 degrees, which encourages the viewer to move in a random way.

The recording process of this hologram type involves the sequential exposure at the film plane of tiny holographic elements, or holographic pixels known as 'hogels', which are recorded in such a way, that corresponding yet different views of the same scene are distributed to each of the beholders eyes. The parallax and disparity effects enable the beholder to perceive the scene as three-dimensional. The smaller the size of the 'hogel', the higher the resolution of the image.

In *Holographic Imaging,* Michael Klug and Mark Holzbach, the authors of Chapter 20 Holographic Stereograms and Printing refer to the approach:

Holographic printing represents an amalgamation of 3-D hardcopy concepts proposed, developed and refined over the last one hundred years...In addition, image processing algorithms, derived from models of optical recording system and applied to perspective imagery made it possible to anticipate distortions and pre-distort component imagery to preclude their effects (Benton, 2007).

A three channel, synthetic, optically formed, fringe digital hologram *Shadowy Figures* by Paula Dawson simulates three scenarios of light and darkness identified by Michael Baxandall as important case studies in the analysis of shadow(Baxandall, 1995). If the beholder is standing to the left hand third of the image, at every position, high and low and side to side, the light relative to the figure seems to originate from the position of the viewers eyes. This effect was achieved in the CG environment by moving the light source to the position of the capture camera for each of the holographic pixels making up the image. If the viewer stands in the centre of the field of view, a static light from the back left shines towards the front right and illuminates the side of the figure,while on the right a source of darkness is projected from in front of the hologram plane very low out of the field of view. Thus by moving side to side the role of the beholder is radically transformed. On the right the viewer witnesses by movement of their body the interaction of the darkness source obliterating the light. In the centre they are a passive witness to a scene in which the lighting and subject are locked together whereas to the left, the effect of the light source being decoupled from the scene and instead attached to the viewer position has the effect of implying that the beholder is a light source. Conceptually, luminosity of the beholder is indicated by the changing composition (Gage, 2003).

Martin Richardson's *Over the Rainbow* incorporates within an overall spatial scene, a montage of animations, which are replayed out at the speed that the viewer moves in front of the hologram film plane. The embedded animations activate and disrupt the pictorial space. The spatial field is in the first instance composed of disparate montage elements. These retain their spatial fidelity while introducing an uneven flow of space-time over the image surface. A woman is holding crumpled surfaces in front of her eyes, while behind her is a landscape – a still of the yellow brick road from the *Wizard of Oz.*

Returning to *Broken Window,* the key elements : the broken window, the light and the tree-branch signifying the holographic picture plane, the replay light, and the subject traversing the space of the picture plane have been discussed. The motion of the scroll is of critical importance as it is possible to track through this visual element the moving boundary of the image zone.

In each of the three above-mentioned examples there is very little z-axis depth and only shallow information on the side edges, which could potentially be cropped by the edges of the viewing zone. Richardson introduces a large black border in the background making the wipe of the image zone less noticeable, Dawson uses 2D imagery for a large surrounding of the figure, which is in a niche, and Debiens employs the rolling scroll. The tapering of the z-axis space near to the edge of the holographic plate where it will be cropped has some similarly to the tolerance of the framing of film scenes. As Peter Greenaway comments:

Fig. 10. Paula Dawson, *Shadowy Figures*, 2003, printed synthetic optically formed fringe digital hologram, 120 X 120cm.

Fig. 11. Martin Richardson, *Over the Rainbow*, 2009, iLumogram image taped to an invisible plane. The solidity of space is built strongly on occlusion. The taped landscape of the back plane being occluded by the torso, which is occluded by the hands holding up the crumpled surfaces. As the viewer moves side- to- side the surfaces reveal movie animations from the *Wizard of Oz* of the Tin Man kissing Dorothy and the great Oz appearing from the flames. The work, which is a subtle comment on the current place of the rainbow hologram and acknowledgement of *Black Rainbow* by Benyon and Benton, is aligned with what Timothy Murray has called the Digital Baroque as "enfolding the user in the energetic present, as articulated in relation to the analogue past while bearing on the digital future" (Murray, 2008).

The painter can pick his own frame, the filmmaker is not so lucky. Having made a selection form the smaller number of film ratios possible, careful composition of a picture into the corners of the frame and right to the very edges is still not dependable when projected. Exact symmetries are not to be relied upon. Tolerances must be permitted. Hence the concept of the floating edge (Greenaway, 1990).

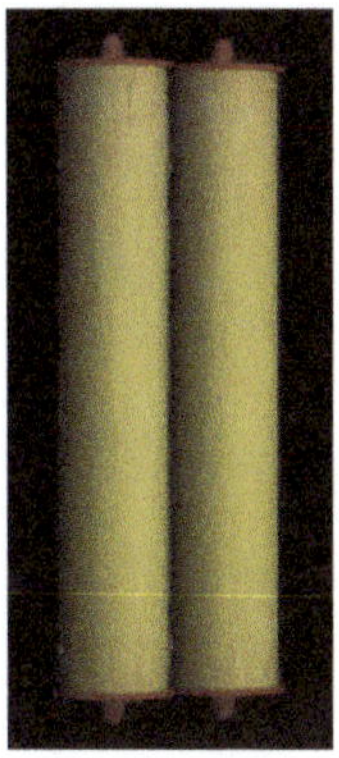

Fig. 12. Jacques Desbiens, *The Broken Window*, 2006, Printed Synthetic hologram, CG render from the left, the first point of view, 140 X 38cm.

The space is deepest at the centre of the holographic picture plane which has the widest angle of view. This central placement of a protruding image element to avoid cropping as the viewer moves side-to-side is common to many holograms. What is quite unique about this representational system is the ability of altering in very subtle ways, hogel by hogel, the structure of the subject as the beholder moves.

Synthetic holography offers a solution to the age old distortion problem of accidental linear perspective, the addition of a wide field of view and dynamic observation ...movement of the eyes obviously, but also movement of the body, free wandering of observation, nomadism of points of view (Desbiens, 2009).

9. Conclusion

Holographic representational systems types such as the HOE, laser transmission hologram, reflection, rainbow, stereogram and printed synthetic stereogram have expanded the space of representation. They each accomplish this by using the support (holographic plane) the viewing zone, the replay light, the location of the image (behind, in front of or intersected by the support) and the properties of the subject to engage with the beholder in different spatial and temporal ways.

Holographic representational systems are chameleon-like; they can operate directly with real-world subject matter or draw from other pictorial systems rendered in a wide diversity of media ranging from film, photographs or arrays of computer graphic rendered frames composited within the holographic diffractive optic environment. The types of aperturing of the viewing zone between the beholder and the holographic plate introduce a non-visible formal element of great importance to the holographic representational system. In some cases the shapes of the viewing zones are echoed in the compositional structure of the work in abstract and literal ways.

Like perspective, the range of positions available for viewing the subject is directly tied to the representation of the subject. These are inseparable elements, which form the image space/ viewing space continuum. However unlike perspective, where the ideal space for viewing is confined to a specific point along the x-axis, holographic representational systems range from full X, Y, Z mobility of the beholder to narrow long corridors, thin wide strips and trapezoids.

The fundamental property which all holographic representational systems share is that they precipitate and reciprocate the movement of the beholder with further visual information distributed in space and time. From the above mentioned examples it is clear that the additional information gained by the beholder from moving though the viewing zone, rather than confirming the *simulacra* and *hyperreality* of the subject, can equally be used to introduce ambiguities of tone, colour, space and time which challenge initial assumptions.

As Zeki argues:

...the capacity to give multiple interpretations is not a separate faculty invented or used by the artist. It is instead tied to a general capacity of the brain to give several interpretations, to instill meaning by applying several concepts, a capacity that is important for art in its role of acquiring knowledge. It is on this physiological basis that the prized quality of ambiguity in art is built (Zeki, 2009).

10. Acknowledgements

Firstly, thanks to Sylvia Ross, Head of the School of Art, College of Fine Arts, The University of New South Wales, Sydney for generously approving research time to undertake the writing of this chapter. Next, thanks to several artists for willingly and speedily providing information and imagery of their works, Margaret Benyon, Patrick Boyd, Melissa Cranshaw, Jacques Desbiens, Dieter Jung, Martina Mrongovius, Seth Riskin, Andy Pepper, Martin Richardson and Sally Weber. Also thanks to Jonathan Ross for providing information on Patrick Boyd and David Pizzanelli. Profound thanks go to the inventors and developers of the representational systems discussed and to the many artists who developed the visual language of holograms. Finally, thanks to John Gage for his constant support and for reading and commenting on several drafts.

11. References

Abramson, N. (1981) *The Making and Evaluation of Holograms,* Academic Press, ISBN120428202 London, NY.

Alphen, E. v. (2005) *Art in Mind How Contemporary Images Shape Thought,*University of Chicago Press, ISBN0226015289, Chicago and London.

Baudrillard, J. (1997) *Simulacra and Simulation,*The University of Michigan Press, ISBN0472065211, Michigan.

Baxandall, M. (1995) *Shadows and Enlightenment,*Yale University Press, New Haven & London, ISBN0-3000-05979-5,

Benton, S. A., Bove, V.M (2007) *Holographic imaging,*Wiley,ISBN0-470-22412-6, New York.

Benyon, M. (1989) 'Cosmetic Series 1986-1987', *Leonardo,* 22(3)

Benyon, M. (1994) *How is Holography Art?*, unpublished thesis (Thesis in Holography), Royal College of Art.

Berkhout, R. (1989) 'Exploring a New Art Realm- Shaping Empty Space with Light', *Leonardo,* 22(3,4), 313-316

Burgmer, B. (1987) *Holographic Art Perception Evolution Future,*Daniel Weiss, ISBN8440416180, La Coruna.

Carreton, V. (1996) 'Melissa Crenshaw', *Melissa Crenshaw* [online], available: http://www.holonet.khm.de/Holographers/Carreton_Vincente/text/MELISSA%20CRENSHAW.html [accessed

Coyle , R. H., P (1995) *Apparition,*Power Publications, ISBN0 909952272, Sydney.

Cutting, J. V., P. (1995) 'Perceiving layout and knowing distances: The integration,relative potency,and contextual use of different information about depth' in Epstein, W., Rogers,S., ed. *Handbook of perception and cognition,Vol 5 ;Perception of space and motion(pp69-117),* San Diego,CA: Academic Press, 69-117.

De Freitas, F. (2003) 'Holotalk internet radio interview with Mark Diamond', *Holotalk internet radio interview with Mark Diamond* [online], available: www.holoworld.com [accessed

Desbiens, J. (2009) 'The Perspectives of Synthetic Holography', *International Symposium on Display Holography,* ISBN, Shenzhen, China, 2009

Eco, U. (1987) *Travels in Hyperreality,*Richard Clay Ltd, ISBN0330296671, Suffolk.

Fehr, M. G., S. Horn,Gabriele. Jund,D. Piene,O. Eberhard,R. (1991) 'Dieter Jung Bilder Zeichnungen Hologramme',

Gage, J. (2003) 'Light from Shadow', *Light from Shadow* [online], available: http://shadowyfigures.com/index2.html [accessed

Gallery, P. W. (2009) 'James Turrell, "Large Holograms"', *James Turrell, "Large Holograms"* [online], available: http://artnews.org/gallery.php?i=166&exi=16370&Pace_Wildenstein&James_Turrell [accessed

Greenaway, P. (1990) *Papers = papiers,*Dis Voir,ISBN2906571202, Paris.

Johnson, S. F. (2006) *Holographic Visions,*Oxford University Press, ISBN139780198571223, New York.

Law, L. (2008) 'DIMENSIONAL ART', *Computer Graphics World,* 31(7), 26 -31,ISSN02714159.

Lipp, R. A. Z., P (1985) *Mehr Licht (More Light),*Ernst Kabel Verlag GmbH, ISBN3822500151, Hamburg.

Mrongovius, M. (2009) 'Folding Spaces ,Unfolding Action', *International Symposium on Display Holography,* ISBN, Shenhzen,China, 2009

Murray, T. (2008) *Digital Baroque : New media Art and Cinematic Folds,*University of Minnesota Press,ISBN9780816634019, Minneapolis, London.

Pepper, A. (1989) 'Holographic Space: A Generalised Graphic Definition', *Leonardo,* 22(3, 4), 295-298,ISSN0024094X.

Pepper, A. (1995) *The Creative Hologaphy Index,*Monand Press, ISSN0942-735X, Bergisch Gladbach.

Pepper, A. (2008) 'The Perception of Reality - Looking at Looking', *Holography in the Modern Museum,* ISBN, De Montford University, UK, 2008

(2011) *Drawing 1* Saturday, 26 February 2011 2:23 AM, available: [accessed].

Pizzanelli, D. (1989) 'Bruno,Walking to the Left', *Bruno,Walking to the Left* [online], available: http://www.jrholocollection.com/forsale/pizzavideo/bruno.html [accessed

Popper, F. (1993) *Art of the Electronic Age,*Thames and Hudson,ISBN050023650X, London.

Potts, A. (2000) *The Sculptural Imagination:Figurative,Modernist,Minimalist,*Yale University Press,ISBN0-300-08801-9, New Haven and London.

Reuterswald, C. F. (1989) 'Rubies and Rubbish:An Artist's Notes on Lasers and Holography', *Leonardo,* 22(3,4), 343-344,ISSN0024094X.

(2011) *Figure with Crowns* Saturday, 26 February 2011 9:47 AM available: [accessed].

Saxby, G. (1988) *Practical Holography,*Prentice Hall ISBN0136937977 New York.

Saxby, G. (1991) *Manual of Practical Holography,*Butterworth - Heinemann Ltd,ISBN0240513053, Oxford.

St.Cyr, S. a. W., S (1989) 'Treading on the Tail of the Tiger:A Collaborative Effort in Large-Format Holography', *Leonardo,* 22(3,4), 357-364,ISSN0024094X.

Vila, D. (1989) 'Chasing Rainbows: One Holographer's Approach', *Leonardo,* 22(3.4), 345 - 348,ISSN0024094X.

Zec, P. (1989) 'The Aesthetic message of Holography', *Leonardo,* 22(3,4), 425-430

Zec, P. (1990) *For a Theory of Holography,* International Congress on Art and Hologrpahy, Indiana, USA: 1990.

Zeki, S. (1999) *Inner Vision,*Oxford University Press,ISBN0 19 850519 1,

Zeki, S. (2009) *Splendors and Miseries of the Brain : Love, Creativity, and the Quest for Human Happiness* Wiley-Blackwell ISBN1405185589, Chichester.

Permissions

The contributors of this book come from diverse backgrounds, making this book a truly international effort. This book will bring forth new frontiers with its revolutionizing research information and detailed analysis of the nascent developments around the world.

We would like to thank Dr. Izabela Naydenova, for lending her expertise to make the book truly unique. She has played a crucial role in the development of this book. Without her invaluable contribution this book wouldn't have been possible. She has made vital efforts to compile up to date information on the varied aspects of this subject to make this book a valuable addition to the collection of many professionals and students.

This book was conceptualized with the vision of imparting up-to-date information and advanced data in this field. To ensure the same, a matchless editorial board was set up. Every individual on the board went through rigorous rounds of assessment to prove their worth. After which they invested a large part of their time researching and compiling the most relevant data for our readers. Conferences and sessions were held from time to time between the editorial board and the contributing authors to present the data in the most comprehensible form. The editorial team has worked tirelessly to provide valuable and valid information to help people across the globe.

Every chapter published in this book has been scrutinized by our experts. Their significance has been extensively debated. The topics covered herein carry significant findings which will fuel the growth of the discipline. They may even be implemented as practical applications or may be referred to as a beginning point for another development. Chapters in this book were first published by InTech; hereby published with permission under the Creative Commons Attribution License or equivalent.

The editorial board has been involved in producing this book since its inception. They have spent rigorous hours researching and exploring the diverse topics which have resulted in the successful publishing of this book. They have passed on their knowledge of decades through this book. To expedite this challenging task, the publisher supported the team at every step. A small team of assistant editors was also appointed to further simplify the editing procedure and attain best results for the readers.

Our editorial team has been hand-picked from every corner of the world. Their multi-ethnicity adds dynamic inputs to the discussions which result in innovative outcomes. These outcomes are then further discussed with the researchers and contributors who give their valuable feedback and opinion regarding the same. The feedback is then collaborated with the researches and they are edited in a comprehensive manner to aid the understanding of the subject.

Apart from the editorial board, the designing team has also invested a significant amount of their time in understanding the subject and creating the most relevant covers. They scrutinized every image to scout for the most suitable representation of the subject and create an appropriate cover for the book.

The publishing team has been involved in this book since its early stages. They were actively engaged in every process, be it collecting the data, connecting with the contributors or procuring relevant information. The team has been an ardent support to the editorial, designing and production team. Their endless efforts to recruit the best for this project, has resulted in the accomplishment of this book. They are a veteran in the field of academics and their pool of knowledge is as vast as their experience in printing. Their expertise and guidance has proved useful at every step. Their uncompromising quality standards have made this book an exceptional effort. Their encouragement from time to time has been an inspiration for everyone.

The publisher and the editorial board hope that this book will prove to be a valuable piece of knowledge for researchers, students, practitioners and scholars across the globe.

List of Contributors

Yoshiharu Morimoto and Motoharu Fujigaki
Moire Institute Inc. & Wakayama University, Japan

Jean-Michel Desse
Office National d'Etudes et Recherches Aérospatiales (ONERA) Lille, France

Hamdy Wahba and Mamdouh Shams El-Din
Physics Department, Faculty of Science, 34517 Damietta, University of Mansoura, Egypt

Atsushi Wada
National Defense Academy, Japan

Yasuhiro Harada, AizuddinWan and Hiroyasu Sone
Kitami Institute of Technology, Japan

Roman Romashko and Yuri Kulchin
Institute of Automation and Control Processes FEB RAS, Russia

Sergio De Nicola, Andrea Geltrude, Massimiliano Locatelli, Kais Al-Naimee, Riccardo Meucci and F.Tito Arecchi
Istituto Nazionale di Ottica - CNR Firenze, Italy

Kaige Wang
Department of Physics, Applied Optics Beijing Area Major Laboratory, Beijing Normal University, Beijing, China

Francisco Palacios
University of Oriente, Cuba

Angel Lizana, Laura Lobato, Juan Campos and María J. Yzuel
Universitat Autònoma de Barcelona, Spain

Andrés Márquez
Universidad de Alicante, Spain

Claudio Iemmi
Universidad de Buenos Aires, Argentina

Ignacio Moreno
Universidad Miguel Hernández, Spain

Jakub Svoboda, Marek Škereň and Pavel Fiala
Czech Technical University in Prague, Czech Republic

José J. Lunazzi
Campinas State University, Brazil

P. W. Wachulak
Military University of Technology, Institute of Optoelectronics, Poland

M. C. Marconi
NSF ERC for Extreme Ultraviolet Science & Technology and Department of Electrical and Computer Engineering, Colorado State University, USA

Edward Buckley
Pixtronix Inc., USA

Javier Gamo
Department of Electronics - University of Alcalá, Spain

Paula Dawson
University of New South Wales, Australia